PLANT GROWTH SUBSTANCES IN AGRICULTURE

PLANT GROWTH SUBSTANCES IN AGRICULTURE

Robert J. Weaver

UNIVERSITY OF CALIFORNIA, DAVIS

W. H. FREEMAN AND COMPANY
San Francisco

International Standard Book Number: 0–7167–0824–8
Library of Congress Catalog Card Number: 71–166964

9 8 7 6 5 4 3 2 1

To Lucinda Jean Weaver

CONTENTS

PREFACE

The minute amounts of naturally occurring growth substances that are present in plants control their growth and development. Such processes as root initiation, onset and termination of dormancy and rest, flowering, fruit set and development, abscission, senescence, and rate of growth are under hormonal control. In many agricultural plants, these processes can often be altered to man's benefit by proper application of plant growth substances, and it is quite possible that, in time, all physiological processes in plants will be controlled by application of growth substances.

Since the first plant hormones, the auxins, were identified approximately fifty years ago, rapid advances have been made, in both fundamental and applied research, in the field of plant growth substances. Three new general classes of hormones, the gibberellins, cytokinins, and inhibitors, including abscisic acid, have been recognized, and ethylene has also been recognized as a plant hormone.

This book is written largely from an agricultural point of view and stresses both present and possible future important commercial uses of growth substances in agriculture. However, coverage of the field of growth substances is made more complete by inclusion of chapters on their nomenclature and history, chemical and biological determination, occurrence and chemical nature, and biological effects and mechanism of action. Some of the theory explaining the effects of exogenous regulators on plants is interwoven throughout the book. A chapter on growth substances in weed control is included since this is still the greatest commercial use for growth substances.

Spray recommendations are given for some crops, but I cannot assume responsibility for success or failure of a growth regulator treatment. Growers should always consult their farm advisor or extension agent before using growth regulators, since recommendations are constantly changing and vary according to local climate and soil.

Discussion of uses for plant regulators for agricultural crops does not necessarily imply that the United States Food and Drug Administration has approved all usages or that the compounds have been registered with the appropriate state or federal agencies. At the present time, with so much emphasis on improvement of the environment and prevention of pollution, registrations are in a state of flux— some of the old compounds are being eliminated and new ones are being added. Care should be taken that compounds are applied according to directions on the manufacturer's label as registered under the Federal Insecticide, Fungicide, and Rodenticide Act. The recommendations on the label should be carefully followed and the local county agent or farm advisor should be consulted if a question or problem arises.

Trade names are sometimes used in the text in combination with chemical names or abbreviations. This practice does not imply endorsement of any single product; trade names are given only as a convenience to the reader.

This textbook is designed principally for classroom instruction and should be appropriate for a one-semester college course in plant growth substances. From my teaching experience I would suggest that Chapter 2 ("Biological and Chemical Determination") can serve as the basis for a laboratory course in growth substances. The figures in this chapter that depict the procedures used in some of the bioassays should also prove helpful for such a course.

An extensive documentation of the literature has been included to provide the reader with sufficient references for an in-depth study of any subject of particular interest. With the present trend toward increased independent study, an extensive literature review is essential.

I am grateful for help afforded by many of my colleagues. Some of the chapters, in draft form, were reviewed by the following scientists: F. T. Addicott, University of California, Davis; B. Baldev, Indian Agricultural Research Institute, New Delhi; H. M. Cathey, United

States Department of Agriculture, Agricultural Research Service, Beltsville, Maryland; C. W. Coggins, Jr., University of California, Riverside; A. S. Crafts, University of California, Davis; F. G. Dennis, Michigan State University, East Lansing; L. J. Edgerton, Cornell University, Ithaca, New York; H. T. Hartmann, University of California, Davis; A. H. Halevy, Hebrew University, Rehovoth, Israel; R. L. Jones, University of California, Berkeley; L. C. Luckwill, Long Ashton Research Station, England; J. MacMillan, University of Bristol, England; G. C. Martin, University of California, Davis; J. W. Mitchell, United States Department of Agriculture, Agricultural Research Service, Beltsville, Maryland; S. K. Mukherjee, Botanical Survey of India, Calcutta; Y. Murakami, National Institute of Agricultural Sciences, Tokyo; L. G. Paleg, Waite Agricultural Research Institute, Glen Osmond, South Australia; J. D. Quinlan, East Malling Research Station, Kent, England; G. S. Randhawa, Institute of Horticultural Research, Hessaraghatta, Bangalore, India; R. M. Sachs, University of California, Davis; and K. V. Thimann, University of California, Santa Cruz.

May, 1972   ROBERT J. WEAVER

ABBREVIATIONS
AND TRADE NAMES

The first-listed common name or abbreviation is the one most commonly used in the text. The synonyms in parentheses are also found in the literature. Trade names (initially capitalized) are sometimes also used as common names.

ABA 3-methyl-5-(1'-hydroxy-4'-oxo-2',6,6-trimethyl-2'-cyclo-hexen-1' yl)-*cis,trans*-2,4-pentadienoic acid
(*syn.* abscisic acid; abscisin II; dormin)
abscisic acid *see* ABA
abscisin II *see* ABA
ACPC *see* Amo-1618
Actidione *see* cycloheximide
Alanap *see* NPA
Alar *see* SADH
Amchem 66–329 *see* ethephon
amiben 3-amino-2,5-dichlorobenzoic acid
(*syn.* chloramben)
amitrole 3-amino-*s*-triazole
(*syn.* 3-amino-1*H*-1,2,4-triazole; amino triazole; ATA)
amino triazole *see* amitrole

Amo-1618 Ammonium (5-hydroxycarvacryl)trimethyl chloride piperidine carboxylate
(*syn.* 2-isopropyl-4-dimethyl-amino-5-methylphenyl 1-piperidine-carboxylate-methochloride, ACPC)
ATA *see* amitrole
BA 6-benzylamino purine
(*syn.* benzyladenine; BAP; Verdan)
Banvel D *see* dicamba
BAP *see* BA
benzyladenine *see* BA
B-Nine, B-9, B-995 *see* SADH
BNOA β-naphthoxyacetic acid
BOA benzothiazole-2-oxyacetic acid
(*syn.* BTOA)
BOH β-hydroxyethylhydrazine

BTOA *see* BOA
BTP *see* PBA
Captan *N*-trichloromethylmercapto-4-cyclohexene-1,2-dicarboximide
carbaryl *see* Sevin
Cardavan 3-isopropyl-4-dimethylamino-6-methyl-phenyl 1-piperidine carboxylate methyl chloride
CBBP *see* Phosfon-D
CCC (2-chloroethyl)trimethylammonium chloride
(*syn.* chlorocholine chloride; chlormequat; Cycocel)
CEPA *see* ethephon
chloramben *see* amiben
chlormequat *see* CCC
chlorocholine chloride *see* CCC
chloroflurenol (a morphactin) 2-chloro-9-hydroxyfluorene-9-carboxylic acid
chloropropham *see* CIPC
CIPC isopropyl *m*-chlorocarbanilate
(*syn.* isopropyl *N*-(3-chlorophenyl)carbamate; chloropropham)
CO11 *N*-dimethylamino maleamic acid
3-CP 3-chlorophenoxy-*a*-propionic acid
3-CPA 3-chlorophenoxy-*a*-propionamide
4-CPA 4-chlorophenoxyacetic acid
(*syn.* *p*-chlorophenoxyacetic acid; PCPA)
cycloheximide 3-[2-(3,5-dimethyl-2-oxocyclohexyl)-2-hydroxyethyl]-glutarimide
(*syn.* isocycloheximide; Actidione)
Cycocel *see* CCC
2,4-D 2,4-dichlorophenoxyacetic acid
dalapon 2,2-dichloropropionic acid
2,4-DB 4-(2,4-dichlorophenoxy)butyric acid

2,4-DEB 2-(2,4-dichlorophenoxy)ethyl benzoate
DEF tributylphosphorotrithioate
2,4-DEP tris[2-(2,4-dichlorophenoxy)ethyl] phosphite
(*syn.* Falone; Falodin)
2,4-DES *see* sesone
2,6-D 2,6-dichlorophenoxyacetic acid
dicamba 3,6-dichloro-*o*-anisic acid
(*syn.* Banvel D; mediben; 2-methoxy-3,6-dichloro benzoic acid)
dichlorprop *see* 2,4-DP
diuron 3-(3,4-dichlorophenyl)1,1-dimethylurea
DMSO dimethylsulfoxide
DNA desoxyribonucleic acid
DNOC 4,6-dinitro-*o*-cresol or sodium 4,6-dinitro-*o*-cresylate
dormin *see* ABA
2,4-DP 2-(2,4-dichlorophenoxy)propionic acid
(*syn.* dichloroprop)
Duraset *N*-*meta*-tolyl phthalamic acid
(*syn.* 7R5)
EDNA *N*,*N*′-dinitroethylenediamine
EDTA ethylenediaminetetraacetic acid
(*syn.* (ethylenedinitrilo)tetraacetic acid)
EHPP ethylhydrogen 1-propylphosphonate
Elgetol sodium 4,6-dinitro-*o*-cresylate; *see* DNOC
endothall 7-oxabicyclo(2.2.1)-heptane-2,3-dicarboxylic acid
(*syn.* 3,6-endoxohexahydrophthalic acid)
ethephon (2-chloroethyl)phosphonic acid
(*syn.* Ethrel; CEPA; Amchem 66-329)
Ethrel *see* ethephon
Falodin *see* 2,4-DEP
Falone *see* 2,4-DEP

FAP *see* kinetin
FeEDDA Fe-ethylenediamine-di-(*o*-hydroxyphenol) acetic acid
fenoprop *see* 2,4,5-TP
Ferbam ferric dimethyl dithiocarbamate
FW-450 sodium salt of α,β-dichloroisobutyric acid
(*syn.* Mendok)
GA₃ gibberellic acid (2β,4a,7-trihydroxy-1-methyl-8-methylene-4aα,4bβ,-gibb-3-ene-1α,10β-dicarboxylic acid, 1,4a-lactone) (Subscripts indicate specific analogues, such as GA₁, GA₂.)
Gibberellin(s) One or more of the known gibberellins (see GA₃). General references to exogenous gibberellins may be assumed to refer to GA₃ or KGA₃, commercially available compounds that have equal activity when used on an acid equivalent basis.
HAN a petroleum fraction of high aromatic content (approximately 87 percent by weight)
heteroauxin *see* IAA
IAA indoleacetic acid
(*syn.* indole-3-acetic acid; 3-indoleacetic acid; indolylacetic acid; heteroauxin)
IAAld indoleacetaldehyde
(*syn.* indole-3-acetaldehyde)
IAEt ethylindoleacetate
(*syn.* ethyl-3-indoleacetate; indole-3-ethylacetate)
IAMe methylindoleacetate
(*syn.* methyl-3-indoleacetate; indole-3-methylacetate)
IAN indoleacetonitrile
(*syn.* indole-3-acetonitrile)
IBA indolebutyric acid
(*syn.* indole-3-butyric acid)
indoleacetic acid *see* IAA
2iP 6-(γ,γ-dimethylallylamino)-purine

IPA indolepropionic acid
IPC isopropyl carbanilate
(*syn.* isopropyl *N*-phenylcarbamate, propham)
IPyA indolepyruvic acid
(*syn.* indole-3-pyruvic acid)
isocycloheximide *see* cylcoheximide
IT 3233 (a morphactin) *N*-butyl-9-hydroxyfluorene-9-carboxylate
IT 3456 (a morphactin) methyl-2-chloro-9-hydroxyfluorene-9-carboxylate
KGA₃ potassium gibberellate (potassium salt of GA₃)
kinetin 6-furfurylamino purine
(*syn.* *N*-furfuryladenine, FAP)
maleic hydrazide *see* MH
MCPA [(4-chloro-*o*-tolyl)oxy]-acetic acid
(*syn.* 4-chloro-2-methylphenoxyacetic acid, Methoxone)
MCPB 4-[(4-chloro-*o*-tolyl)oxy]-butyric acid
(*syn.* 4-chloro-2-methylphenoxybutyric acid)
MCPES 2-[(4-chloro-*o*-tolyl)-oxy]ethyl sodium sulfate
MCPP 2-[(4-chloro-*o*-tolyl)oxy]-propionic acid
(*syn.* 2-(2-methyl-4-chlorophenoxy)propionic acid, mecoprop)
mecoprop *see* MCPP
mediben *see* dicamba
MENA methyl ester of naphthaleneacetic acid
Mendok *see* FW-450
Methoxone *see* MCPA
MH 1,2-dihydro-3,6-pyridazinedione
(*syn.* 6-hydroxy-3-($2H$)-pyridazinone, maleic hydrazide)
NAA naphthaleneacetic acid
(*syn.* α-naphthaleneacetic acid)
NAAm naphthaleneacetamide
(*syn.* NAD, NAAmide)
NAAmide *see* NAAm
Nacconol NR an alkylarylsulfonate

NAD   *see* NAAm
naptalam   *see* NPA
NPA   N-1-naphthylphthalamic
acid
   (*syn.* naptalam, Alanap)
paraquat   1,1'-dimethyl-4,4'-bi-
pyridinium ion
PBA   6-(benzylamino)-9-(2-tet-
rahydropyranyl)-9H-purine
   (*syn.* SD8339, BTP)
PCPA   *see* 4-CPA
Phosfon   *see* Phosfon-D
Phosfon-D   2,4-dichlorobenzyl-
tributylphosphonium chloride
   (*syn.* Phosfon, CBBP)
Phosfon-S   ammonium   analogue
of Phosfon-D
Phygon XL   2,3-dichloro-1,4-
naphthoquinone
picloram   4-amino-3,5,6-trichlo-
ropicolinic acid
   (*syn.* Tordon)
POA   phenoxyacetic acid
propham   *see* IPC
QC   8-hydroxyquinoline citrate
7R5   *see* Duraset
RNA   ribonucleic acid
SADH   succinic acid-2,2-dimethyl-
hydrazide
   (*syn.* N,N-dimethylaminosuccin-
amic acid, Alar, B-Nine, B-9,
B-995)
SAPL   N-pyrrolidino-succinamic

acid
SD8339   *see* PBA
sesone   2-(2,4-dichlorophenoxy)-
ethyl sodium sulfate
   (*syn.* 2,4-DES)
Sevin   1-naphthyl N-methyl carbamate
   (*syn.* carbaryl)
silvex   *see* 2,4,5-TP
simazine   2-chloro-4,6-bis(ethyl-
amino)-s-triazine
2,4,5-T   2,4,5-trichlorophenoxya-
cetic acid
2,4,5-TB   2,4,5-trichlorophenoxy-
butyric acid
2,3,5,6-TBA   2,3,5,6-tetrachloro-
benzoic acid
2,3,6-TBA   2,3,6-trichlorobenzoic
acid
2,4,5-TES   sodium 2-(2,4,5-tri-
chlorophenoxy)ethyl sulfate
TIBA   2,3,5-triiodobenzoic acid
Tordon   *see* picloram
2,4,5-TP   2-(2,4,5-trichlorophe-
noxy)propionic acid
   (*syn.* fenoprop, silvex)
UDMH   unsymmetrical dimethyl-
hydrazine
Verdan   *see* BA
zeatin   6-(4-hydroxy-3-methyl-2-
butenylamino)purine

PLANT GROWTH SUBSTANCES
IN AGRICULTURE

Chapter 1

NOMENCLATURE AND HISTORICAL ASPECTS

Plant growth substances play a major role in plant growth and development. This fact was stated by Went many years ago in his now-famous pronouncement, *Ohne Wuchstoff, kein Wachstum* ("Without growth substance, no growth"). Went found that to grow in length, tissues must receive growth substances. Although the naturally occurring (endogenous) growth substances normally control plant growth, modifications of growth can be produced by applications of exogenous growth substances, some of which may produce effects beneficial to man.

Research on naturally occurring growth substances is gradually revealing the hormonal control mechanisms of plant growth and development. Both these experimental studies and basic research have led to the use of synthetic growth substances in agriculture, where they have assumed an importance equal to that of pesticides and fungicides. Plant regulators are now widely used in weed control, control of fruit development, defoliation, propagation, and size control.

NOMENCLATURE OF PLANT GROWTH SUBSTANCES

General Terms

A definition of plant growth substances is indeed a prerequisite to any study of their historical development and is of utmost importance for those desiring to communicate intelligently on the subject. At present, four general classes of plant hormones are recognized: auxins, gibberellins, cytokinins, and inhibitors. Inhibitors will be considered to be a fourth class of plant regulators here, although they do not comprise such a clearly delineated group as the other three. The hormonal properties of ethylene have also been recognized.

The word "hormone" was first coined by animal physiologists (Bayliss and Starling, 1904). Historically there has been considerable confusion about growth substance terminology. In 1951 K. V. Thimann, President of the American Society of Plant Physiologists, resolved to clarify this terminology and suggested the appointment of a committee to propose a uniform nomenclature for growth substances (van Overbeek *et al.*, 1954). The definitions recommended by that committee will be used in this text. For nomenclature introduced subsequent to the work of this committee, the most commonly accepted definitions will be used.

Plant regulators are defined as organic compounds—other than nutrients—which, in small amounts, promote, inhibit, or otherwise modify any plant physiological process. *Nutrients* are defined as materials that supply the plants with energy or essential mineral elements. *Plant hormones* (synonym: *phytohormones*) are regulators produced by plants, which, in low concentrations, regulate plant physiological processes. Hormones usually move within the plant from a site of production to a site of action.

The term "hormone," correctly used, is restricted to naturally occurring plant products. The term "regulator," however, is not necessarily restricted to synthetic compounds but can also include hormones. The term "regulator" has very wide boundaries. It can apply to any material that modifies a plant physiological process. The term should be used instead of "hormone" when referring to agricultural chemicals used for crop control. The word "regulator" may be further defined by adding to it the name of the process that it influences. For example, *growth regulators* (synonym: *growth substances*), regulators that affect growth, include the auxins. *Growth hormones,* hormones that regulate growth, include, among others, the B-complex vitamins, which are required for root growth and are produced primarily in the shoot. *Flowering regulators*, if they exist, are regulators that induce flowering; *flowering hormones*, if they exist, are hormones that initiate the formation of floral primordia or promote their development.

Auxins

Auxin is a generic term for a group of compounds characterized by their capacity to induce elongation in shoot cells. Some auxins are naturally occurring, others are produced synthetically. Auxins resemble IAA in the physiological action they induce within plant cells, the most critical of which is elongation. These compounds are generally either acids with an unsaturated cyclic nucleus or derivatives of such acids. *Auxin precursors* are compounds that can be converted to auxins in the plant. *Antiauxins* are compounds that inhibit the action of auxins, presumably by competing with them for the same points of attachment on a receptor substance or substances. By increasing the concentration of auxin, the inhibitory effect of some antiauxins can be completely overcome. It should be emphasized that none of these terms are mutually exclusive; for example, under certain circumstances some compounds can be auxins as well as antiauxins.

There are other inhibitors of auxin action that cannot be classified as antiauxins and are not believed to retard auxin action by direct competition with auxin for the same points of attachment. Increasing the concentration of auxin cannot completely overcome their inhibitory effect.

Gibberellins

In the years since the committee on terminology met in the early 1950's, there has been rapid development of three other main types of plant regulators: gibberellins, cytokinins, and inhibitors. Ethylene has also been recognized as a plant hormone. A *gibberellin* may be defined as a compound that has a gibbane skeleton (Fig. 3–3) and that stimulates cell division or cell elongation or both (Paleg, 1965). Gibberellin can cause a striking increase in shoot elongation in many species that is especially marked when certain dwarf mutants are treated. For example, when some dwarf mutants of maize and peas are treated with gibberellin, they grow very rapidly and become as tall as untreated normal plants. Apparently, the failure of these dwarf mutants to produce enough gibberellin for normal growth can be remedied by application of gibberellin to the plants. It is essential that any tests be performed on mutants in which type of growth is dependent upon a single gene and in which the growth response is specific for the gibberellins (Phinney and West, 1960). Another nearly specific test for gibberellins is the stimulation of synthesis of certain enzymes in seeds.

Antigibberellins are compounds that competitively inhibit the action of gibberellins, and *gibberellin precursors* are compounds in the plant that can be converted into gibberellins. *Gibberellin-like* substances are those that have unknown chemical configurations but that produce the requisite biological activity in appropriate dwarf mutant tests.

Cytokinins

Cytokinins are plant growth substances that cause cell division in plants. (The synonym *phytokinin* is less generally used.) Many exogenous and all endogenous cytokinins are probably derivatives of adenine, a purine nitrogenous base. The first cytokinin was discovered in a sample of aged DNA by Professor Folke Skoog's group at the University of Wisconsin in the 1950's. In 1962 a group of polypeptides that possess hormone activity in animal tissue were discovered; these polypeptides were termed *kinins* (Collier, 1962). To avoid the possibility of confusion with animal mechanisms, the term "cytokinin" was introduced, to be applied specifically to plants, although the word "kinin" is sometimes also used for plant cytokinins.

Kinetin belongs to the general group of cytokinins. There are also anticytokinins, which competitively inhibit the action of cytokinins, and *cytokinin precursor* compounds, which can be converted into cytokinins.

Inhibitors

The *inhibitors* are a rather diverse group of plant growth substances that inhibit or retard a physiological or biochemical process in plants. Some endogenous inhibitors appear to be plant hormones on the basis of their physiological properties. Various naturally occurring inhibitors can have different actions; for example, they may be growth inhibitors, auxin inhibitors, gibberellin inhibitors, or germination inhibitors.

In recent years new types of organic chemicals, the *plant growth retardants*, have been discovered that retard cell division and cell elongation in shoot tissues and thus regulate plant height physiologically without causing malformation of leaves and stems. These compounds also intensify the green color of leaves and indirectly affect flowering. The growth of plants treated with growth retardants is not completely suppressed. Thus far most of these compounds have been produced synthetically. Other exogenous inhibitors can retard the growth of plants, but they nevertheless do not fit the definition of growth retardants. For example, MH is a growth inhibitor, but this compound suppresses growth most effectively by killing shoot tips. Most auxins, when applied in a supraoptimal concentration, inhibit plant growth but are likely to cause a visible malformation of the leaves, stems, and flowers of treated plants.

HISTORICAL ASPECTS OF AUXINS

Early Work

Approximately two hundred years ago it was believed that sap movement in the plant produces correlation (the effect of one part of a plant on growth or development of another part) between different plant parts—in other words, that sap produced in one part of the plant is moved to another part to control the growth in some manner. Duhamel du Monceau (1758) concluded from his experiments that there was one sap that moved downward and one that moved upward. The downward-moving sap was said to originate in the leaves and then to move down and control the nutrition of the roots. If the downward-moving sap was interrupted by ringing, callus and root formation occurred above the ring.

Over a century later Sachs (1880, 1882) revised Duhamel du Monceau's theory and assumed that there were root-forming and other substances that moved in different directions in the plant. Sachs claimed that these organ-forming substances were polarly distributed (that is, moved in only one direction in a plant organ) and that they could control plant growth. Furthermore, he believed that their distribution could be modified by such environmental factors as light and gravity. Sachs postulated that root-forming substances originated in the leaves, then moved to the base of the stems; he demonstrated that a cut made in the twig would stop the downward movement of the root-forming substances, and the roots would form above the cut. However, many years elapsed before the organ-forming and root-forming substances were shown to be hormones.

The first article written about the organ-forming substances was one by Beyerinck (1888) on the subject of galls. Beyerinck studied the willow gall that was caused by the leaf wasp *Nematus capreae*. The

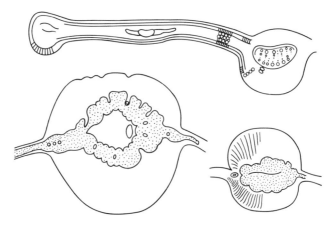

FIGURE 1-1

Gall of *Nematus capreae* on leaf of *Salix*. *Top,* egg is deposited with some mucilage in mesophyll of young leaf; *bottom left,* mature gall before larvae are hatched; *bottom right,* gall in which, by accident, no egg was deposited. Note that mucilage alone can produce a large gall but that a larger gall forms when eggs are also deposited. (After M. W. Beyerinck, 1888.)

wasp deposits eggs along with some mucilage in the mesophyll of the leaves, and a large gall develops (Fig. 1–1). Beyerinck suggested that "growth enzymes" move out from the egg and cause the gall to form. Not until approximately forty years later were several plant hormones, including IAA, identified.

For almost a half century following the findings of Sachs, researchers assumed that nutritional factors were more important than hormonal ones in producing correlations in plants. Probably the best example of the nutritional approach is the classical work of Kraus and Kraybill (1918), who studied the effects of the carbon-to-nitrogen ratio on the growth of tomato plants. They demonstrated that normal fruiting and vegetative growth occurs in plants growing under conditions that lead to a high level of carbohydrates and a relatively low supply of nitrogen. A high level of nitrogen favors succulent vegetative growth but decreases fruitfulness. When levels of both carbohydrate and nitrogen are low, vegetative and reproductive growth is greatly limited. Their experiments stimulated a vast amount of work on the effect of the carbon-to-nitrogen ratio on flowering and other types of plant growth. Some investigators misinterpreted the work of Kraus and Kraybill and believed the ratio to be a direct cause of flowering. However, subsequent work showed, confirming the work done on tomatoes, that although a particular carbon-to-nitrogen ratio may be associated with a certain type of growth, this ratio is in no way a direct cause of flowering.

Correlative Work with Coleoptiles

DARWIN. Many scientists consider that modern plant hormone research began with the ingenious experiments conducted by Darwin (1880), in which he studied the effect of light on the coleoptile (tubular first leaf) of seedlings of *Phalaris canariensis* and *Avena sativa*. Darwin showed that when a coleoptile of *Phalaris canariensis* was unilaterally illuminated, a strong phototropic positive curvature was produced (Fig. 1–2). When the tip of the coleoptile was covered with a tinfoil cap, no such curvature was produced; moreover, when the tip was removed, the coleoptile did not react phototropically. From these and other experiments Darwin concluded that when coleoptiles are exposed to unilateral light, some influence is transmitted from the tip to the lower parts of the coleoptile, causing the lower part to bend. In demonstrating localized sensitivity to light in the hypocotyls of *Beta vulgaris* and *Brassica oleracea*, Darwin concluded that the force of gravity is exerted specifically on the root tips, which transmit a stimulus that causes the root to bend downward.

Darwin's experiments set in motion a chain of events that led, approximately fifty years later, to the discovery of the plant hormones that are now called auxins. During this period there was a gradual and steady advance in knowledge of the hormone, even though work on the nutritive aspects of plant growth was more popular. A brief discussion of some of the key experiments follows.

ROTHERT AND FITTING. Rothert (1894) confirmed and extended Darwin's experiments and also demonstrated that the phototropic stimulus is conducted in the parenchyma tissue of the coleoptile. Fitting (1907) showed that the growth rate of *Avena* coleoptiles is not affected by the making of unilateral incisions (Fig. 1–2). He carried out his work in a room saturated with moisture so that the cut surfaces did not dry before or after they were firmly pressed together. If the cut surfaces had been allowed to dry, he would have obtained different results because of the barrier to translocation thus formed. Fitting also showed that the phototropic response was not affected by lateral incisions, regardless of their positions with regard to the light. He concluded that the stimulus is transmitted through the living material and goes around the incisions. Fitting explained the positive phototropic response by suggesting that the light sets up a polarity in the cells of the tip and that the stimulus is transmitted from the cells of the unilaterally illuminated tips to the cells in the darkened basal portion.

BOYSEN-JENSEN.   Boysen-Jensen (1913) demonstrated that the phototropic stimulus can be carried through nonliving material or across a wound gap (Fig. 1–2). He cut off the tips of the *Avena* coleoptiles and then inserted a block of gelatin between the tips and the severed stump. When he illuminated only the tip, curvature was obtained in the basal portion below the gelatin. Boysen-Jensen also obtained a curvature when he made an incision on the back side of the coleoptile, but no curvature was produced when he placed a piece of mica high up in the side of the coleoptile away from the light. He concluded that the stimulus is conducted down the shaded side of the coleoptile and can be transmitted across an incision.

PAÁL AND STARK.   Paál (1918) confirmed Boysen-Jensen's findings and further showed that some diffusible substance produced in the tip controls the growth of the coleoptile. He found that if the tip of a coleop-

FIGURE 1-2

Historical outline of some early discoveries concerning plant growth hormones.
   (*a*) Unilateral light causes positive phototropic curvature of *Phalaris* grass coleoptile (Darwin).
   (*b*) Removal of coleoptile tip prevents phototropic response (Darwin).
   (*c*) Lateral incisions do not prevent phototropic response (Fitting).
   (*d*) Insertion of gelatin between coleoptile tip and stump does not prevent positive phototropic curvature (Boysen-Jensen).
   (*e*) *Left,* insertion of mica plates on shaded side prevents phototropic curvatures. *Right,* insertion of plates on illuminated side does not prevent curvature (Boysen-Jensen).
   (*f*) Negative curvature results when an excised tip is placed on one side of the cut surface (Paál).
   (*g*) Decapitation slows growth of *Avena* coleoptile. Almost-normal growth resumes when tip is replaced (Söding).
   (*h*) *Center,* agar blocks containing expressed sap from various tissues cause positive curvature (Stark). *Right,* agar blocks containing diastase, saliva, or malt extract cause negative curvature (Seubert).
   (*i*) *Left,* decapitation causes cessation of growth of *Avena* coleoptile. *Center,* addition of a plain agar block has no effect. *Right,* addition of a block containing juice from excised tip renews growth (Went).
   (*j*) *Left,* growth hormone diffuses from coleoptile. Curvature results when a block of agar is placed unilaterally on a decapitated *Avena* coleoptile. *Right,* curvature results that, within limits, is proportional to the concentration of growth hormone present (Went).

(Adapted from P. Boysen-Jensen, *Growth Hormones in Plants,* trans. by G. S. Avery, Jr., and P. R. Burkholder. Copyright 1936. McGraw-Hill Book Co., Inc. Used by permission.)

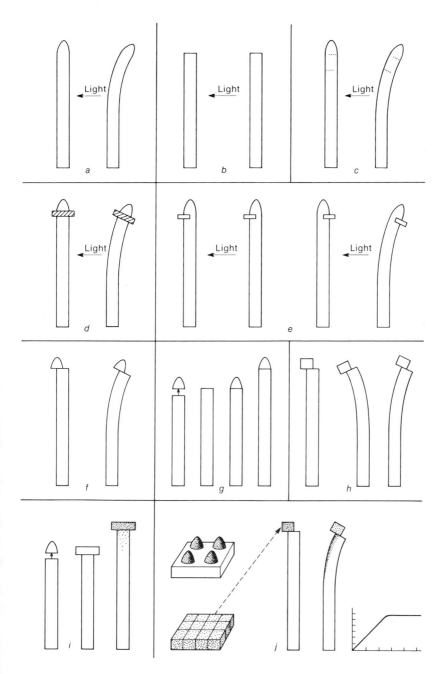

9

tile is removed and then placed on one side of the cut surface, negative curvatures (that is, away from the agar block) are induced in the base (Fig. 1–2). This experiment introduced the idea of a growth hormone. Söding (1925) extended Paál's work; using straight-growth tests, he made accurate growth measurements and showed that if he replaced the cut tip of the coleoptile, the coleoptile resumed almost normal growth.

Stark (1921) extended Paál's work and placed small blocks of agar that contained various tissue extracts on one side of the cut surface of a decapitated coleoptile. Positive curvatures (that is, toward the agar block) almost inevitably resulted. (In the positive curvature a substance is assumed to pass downward from the agar block and to inhibit cell elongation on the side of coleoptile under the block, resulting in an inhibition of coleoptile elongation under the block.) However, in other experiments in which Stark used expressed sap from *Avena* coleoptiles, he obtained no curvatures. Seubert (1925) used Stark's technique to show that agar blocks containing diastase, saliva, or malt extract cause a negative curvature (that is, a growth promotion) and thus demonstrated for the first time that a growth substance exists outside of plants.

WENT. F. W. Went, working in Utrecht, Holland, in 1926, obtained the active chemical substance from the coleoptile tip by placing coleoptile tips on blocks of agar, cutting the agar into smaller blocks, and then placing each of them on one side of the cut surface of the stump of a decapitated coleoptile (Fig. 1–2). Went's (1928) biggest contribution to the technique was the development of a quantitative hormone test, the *Avena* curvature test. This test indicated that the curvatures were proportional, within limits, to the amount of the active hormonal substance (Fig. 1–2).

Went's invention of the *Avena* test greatly stimulated hormone research. It was now possible to ascertain how much hormone, if any, a plant part contained. Much of the current information regarding hormones is based on results obtained by means of the *Avena* test. Within a few years after this discovery, Kögl and his colleagues Haagen-Smit and Erxleben (1934a, 1934b), also working in Utrecht, isolated in pure form three highly active substances, auxins a and b and IAA. (The discovery of IAA was certainly a milestone in hormone work, but auxins a and b have subsequently been shown not to be natural plant products; see Vliegenthart and Vliegenthart, 1966.) Concurrently, IAA was isolated from human urine (Kögl *et. al.*, 1934b), from yeast plasmolyzate (Kögl and Kostermans, 1934), and from cultures of *Rhizopus suinus* (Thimann, 1935). IAA was then termed "heteroauxin" because the compound was believed to be a product of micro-organisms and not a hormone of higher plants, but this term was discontinued when IAA was realized to be one of the most widespread auxins in higher plants. The compound was soon isolated in pure form from alkali-hydrolyzed corn meal (Haagen-Smit *et al.*, 1942) and from the endosperm of immature corn grains (Haagen-Smit *et al.*, 1946).

Other Advances before 1942

While work with coleoptiles was progressing, Fitting (1910) demonstrated that orchid pollen produces a swelling of the orchid ovary. Fitting is credited with introducing the word "hormone" into plant physiology. Although he realized that hormones are stimulating substances, he did not stress the fact that these compounds are capable of being transported from one plant part to another.

Between 1910 and 1942 the occurrence and distribution of auxins in plants were studied and the polarity of transport of auxins in the coleoptile was revealed. The downward distribution of auxins under gravity was shown to result in a geotropic curvature (curvature of a plant organ in response to gravity), which, in roots, was explained by the fact that auxins inhibit root growth and usually do not stimulate it. Studies were made on the mechanism of action of the growth substance, and many correlations in plants were shown to be caused by growth substances. The demonstration of the inhibitory effects of terminal buds on the development of lateral buds by Snow (1925) led to the work by Thimann and Skoog (1933) that demonstrated the role of auxins in apical dominance. Bouillenne and Went (1933) showed that growth substances from leaves and buds control the rooting of cuttings. Other important correlative phenomena shown to be caused by auxins included fruit parthenocarpy (Yasuda, 1934; Gustafson, 1936), cambial growth (Snow, 1935), callus formation, and retardation of leaf abscission (La Rue, 1936). The existence of a wound hormone was also postulated (Bonner and English, 1938). After these developments in the auxin field, Thimann (1963) stated that it was clear that "there is no aspect of plant growth and development in which auxin does not play an important role."

Within a few months after IAA was identified as an auxin, it was synthesized and made readily available, thus stimulating rapid development in the auxin field. IAA was found not to be a good material for use in agriculture because it is a rather unstable compound. A search was therefore made for compounds whose activities were similar to those of IAA but that were more stable and more active. The many new compounds that were discovered are discussed in Chapter 3.

Research Conducted from 1942 to 1946

UNITED STATES.   The early work with auxin had shown that in a low concentration it stimulates growth but that in a high concentration it inhibits growth. Furthermore, there is a close relationship between the quantity or concentration of the acid used and the amount and type of plant response. For example, if too high a concentration of auxin is added to a nutrient solution, short, stunted, stubby roots result. This was regarded as an interesting physiological response at the time, but no one thought that it might be of practical importance. Zimmerman (1942) was one of the first investigators to note the growth abnormalities in plants caused by the herbicide 2,4-D. With the advent of World War II the idea had developed that auxins might be used in high concentrations to kill crops or limit crop yield. Another suggestion, that auxin might be used for herbicidal purposes (Kraus and Mitchell, 1947), led to the initiation of tests to study inhibitory potency, some of which were conducted at the Special Projects Division, Chemical Warfare Service, Fort Detrick, Maryland, in 1944 and 1945, under the leadership of A. G. Norman (1946). Many new compounds were tested under this program; some of them were synthesized by M. S. Newman at Ohio State University.

Other laboratories participated in this work on a contract basis. Early in 1944 Mitchell and Kraus performed experiments at the University of Chicago that provided proof of the high herbicidal activity of a few synthetic auxins. Publication of this material was incompatible with wartime security policies, but the June 1946 issue of the *Botanical Gazette* contained eighteen papers originating from Norman's group at Fort Detrick.

ENGLAND.   During the war similar work was being conducted in England, where much agricultural research was being carried out in an attempt to increase food production. At the Jealott's Hill Research Station of the Imperial Chemical Industry, Ltd., Slade *et al.* (1945) were performing experiments on the effect of auxin on the organic matter in soils. In one experiment they discovered that NAA at the rate of twenty-five pounds per acre killed seedlings of the weed yellow charlock (*Brassica arvensis*) but that oats were not injured. They subsequently proved that cereals were generally insensitive to auxins but that the broad-leaved plants, the dicotyledons, were very sensitive. In their search for selective herbicides of greater potency they discovered that MCPA is extremely toxic. Again, because of wartime security, the publication of results was postponed until 1945.

Also during the war period studies on nodule formation in clover roots conducted by Nutman *et al.* (1945) at Rothamsted, England, showed that red clover and sugar beets were very sensitive to auxin but that wheat was quite insensitive. This finding led to further experimental use of hormones as toxicants on plants. The work of these two British teams was later given to another group headed up by Blackman (1945), who carried on an intensive program of field trials with 2,4-D and MCPA.

PUERTO RICO. A study of the effects of 2,4-D on weeds grown in Puerto Rico's tropical climate was initiated in 1945 and produced highly successful results (van Overbeek, 1947).

Advances after 1945

Following the publication of the early papers on the use of herbicidal regulators on plants, a vast amount of research was initiated on the application of various synthetic auxins. Their use in weed control has now assumed an important role in modern agriculture. Research on the use of regulators in such other phases of agriculture as fruit set, fruit growth, propagation, and growth inhibition was also greatly expanded. These subjects will be discussed in Chapters 5 through 11.

Work in the fundamental field of hormone physiology has also been expanded. New scientific methods, including chromatography, mass spectroscopy, and nuclear magnetic resonance have provided tools by which the various auxin compounds can be separated and identified. The tremendous advances being made in the work on plant hormones are reflected in the number of papers on that subject being published yearly. In their book *Phytohormones*, published in 1937, Went and Thimann refer to seventy-seven publications issued in 1936 that describe such work. At present that many are published every two weeks. After the early fundamental work had pointed out possibilities for practical use, it was not long before much greater effort was expended on practical than on fundamental work. However, work of both a fundamental and a practical nature is now in progress at nearly all biological institutes and experiment stations. It is in the fundamental area that the best possibilities lie for future progress in this field.

After 1945 gibberellins and cytokinins emerged as other main classes of growth hormones. In addition, the significance of the inhibitors, including ABA, was realized and they can now be considered a fourth main class of plant regulators. As previously mentioned, ethylene also is now considered to be a plant hormone.

HISTORICAL ASPECTS OF GIBBERELLINS

Research Conducted before 1946

It is interesting that modern auxin research arose from the observations of the bending of coleoptiles by Darwin and that gibberellins were discovered by the Japanese as a consequence of their observations and interest in the "bakanae" disease of rice (*Oryza sativa*). Kurosawa (1926), a plant pathologist who worked on rice diseases in Formosa, is given credit for the discovery of gibberellins. The bakanae disease had been observed for over 150 years in Japan. In the early stages of the disease, the affected plants were often 50 percent or more taller than adjacent healthy plants but formed less seed (Fig. 1–3). Thus, the name bakanae ("foolish seedling") was coined for the disease, which results from an ascomycetous fungus. (The sexual form is known as *Gibberella fujikuroi* and the asexual stage is named *Fusarium moniliforme*.) In 1926 Kurosawa found that the medium in which the fungus had been grown would stimulate growth of rice and maize seedlings, even though these seedlings were not infected by the fungus. He showed that a heat-stable substance was the cause, and he and his colleagues delineated the gross chemical properties of the active material. Their work failed to arouse interest in the West even though auxins had been detected in fungi and were attracting considerable attention.

The isolation of the bakanae-producing substance was hindered by the presence of a growth-inhibiting material, fusaric acid. However, in 1935 Yabuta obtained an active preparation and named it gibberellin after the fungus from which it was isolated. Subsequent work was hampered because of the difficulty of culturing large quantities of the fungus at that time.

Research Conducted after 1946

Details of the early Japanese work were published in Japanese in obscure Japanese agricultural periodicals, and therefore the information failed to reach the West for many years. During World War II meth-

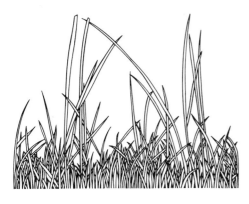

FIGURE 1-3

Section of rice paddy showing growth response of seedlings infected with *Gibberella fuji-kuroi*. (After T. Tanimoto and L. G. Nickell. In 1967 *Reports, Hawaiian Sugar Technologists.* Jour. Series Exptl. Sta. HSPA Paper no. 199, p. 137. Drawing used by permission.)

ods were developed in the West for mass culture of fungi for antibiotic production. At this time it was also realized that hormones other than auxin affected plant growth and development. About 1950, when Western scientists learned of the Japanese experiments, substantial research on gibberellins was immediately undertaken.

In the United States the first work on gibberellins was performed at the Biological Warfare Center at Fort Detrick, Maryland (Mitchell and Angel, 1950) and soon thereafter Stodola and his colleagues at the Northern Regional Research Laboratory of the United States Department of Agriculture began a large-scale isolation of the materials. Stodola (1958) also prepared a source book on gibberellin containing abstracts of papers on that topic. Simultaneously, at Imperial Chemical Industries in England, Borrow *et al.* (1955) were carrying on similar research and developmental work. Gibberellins were soon made available for experimental and commercial purposes since many pharmaceutical firms were able to adapt the equipment used to develop antibiotic cultures to the production of gibberellins.

It was noted that gibberellin application markedly increases stem length in plants. Brian and Hemming (1955) showed that dwarf peas attain a normal growth rate as a result of application of GA_3; soon after this finding Phinney (1956) demonstrated that certain single-gene dwarf mutants of maize grow to normal height following gibberellin application. Some dwarf plants have been found useful for gibberellin bioassay, and this discovery has accelerated progress in the gibberellin field.

Many researchers have been seeking plant responses to gibberellins that would be of commercial interest. Within five years after gibberellins were first tested on grapes in 1957, almost all 'Thompson Seedless' grapes grown in California and destined for table use were commercially sprayed with gibberellin in order to produce larger, elongated berries. Gibberellins are also used in the malting industry to produce a more rapid formation of diastase. Other commercial possibilities and uses will be discussed in later chapters.

Rapid progress has been made in studies of the effect of gibberellins on the physiology of plant growth. It is necessary not only to study the gibberellin responses but also to interrelate them with those of the other types of hormones. Lang (1956) showed that GA_3 induces flowering in *Hyoscyamus,* a plant requiring long nights to flower, even when it is grown during a noninductive photoperiod. Several investigators also showed that gibberellin can replace cold treatment (vernalization) in some biennial plants. The cold or light or darkness treatment required for seed germination of some species can also be replaced by gibberellin treatment. Gibberellin affects fruit set and fruit growth in many species and stimulates cell division, although some auxin must also be present. These subjects and other physiological features of gibberellin will be discussed in Chapters 5 through 11.

Studies have been made of the biosynthesis and identification of various gibberellins; already at least thirty-seven different gibberellins have been identified from the *Gibberella* fungus or from higher plants. There are also many other gibberellin-like compounds of unknown chemical nature that behave like gibberellins.

HISTORICAL ASPECTS OF CYTOKININS

The cytokinins play a part in cell-division processes. The early work of Haberlandt (1913) revealed the existence of types of hormones other than auxins that stimulate cell division. He showed that cell division in potato tuber pieces occurs only in the presence of a vein containing phloem tissue and that the influence inducing cell division originating in the phloem penetrates a thin film of agar. Haberlandt (1921) also showed that cell division can be induced on cut surfaces of certain succulent plants by spreading on these surfaces crushed tissues of other leaves. However, he found that rinsing the cut surface elutes the stimulus. In 1938 Bonner and English isolated the fatty acid traumatin ($C_{12}H_{20}O_4$) from bean fruits. Application of this compound to bean pods was found to cause localized cell division and enlargement. Since traumatin is present on cut surfaces or in crushed cells, it has been termed a "wound" hormone.

In the early 1940's van Overbeek *et al.* (1942) showed that the growth of embryos in tissue cultures is greatly stimulated by coconut milk.

The discovery of the first cytokinin, kinetin, was made during experiments that were an outgrowth of tissue culture investigations in the laboratories of Skoog and Strong at the University of Wisconsin. Segments of tobacco (*Nicotiana tabacum* 'Wisconsin 38') stems were grown on a synthetic medium. A callus tissue (a mass of soft, undifferentiated tissue that often forms at the cut surface of living tissues) grew quickly but soon slowed and stopped. Addition of IAA failed to cause growth to resume, but the addition of coconut milk or yeast extract in combination with IAA stimulated new cell divisions. After much testing, a substance capable of inducing cell division was extracted from yeast DNA, which is composed of the organic nitrogenous compounds, purines. This led to the conclusion that the active compound might be a purine. The active compound, kinetin, was later identified as 6-furfurylamino purine (Miller *et al.*, 1955, 1956). Purines are important because they are part of most of the synthetic processes that occur as a cell or plant grows.

Although cytokinins are especially associated with the stimulation of cell division, they have also been shown to have other effects on growth. Skoog and Miller (1957) showed that kinetin was effective in forming buds on tobacco pith tissue cultures. Skoog and Tsui (1948) had previously shown that the purine adenine was also slightly effective. Leaf size was shown to be increased by kinetin as a result of cell enlargement (Scott and Liverman, 1956; Kuraishi and Okhumura, 1956). Wickson and Thimann (1958) showed that apical dominance is overcome as a result of applications of kinetin to *Pisum sativum* 'Alaska.' The breaking of dormancy in seeds of lettuce, tobacco, white clover, and other plants and the stimulation of growth of intact plants of duckweed as a result of applying a cytokinin have also been reported. Investigation by Mothes *et al.* (1959) has shown that the mobilization of nutrients is enhanced by application of cytokinin.

It was not long before new and more active cytokinins were synthesized. One of the first of these was 6-benzylamino purine. This compound, also known as Verdan, was shown to be effective in prolonging the shelf life of leafy vegetables by slowing the breakdown of protein and other plant constituents (van Overbeek and Loeffler, 1962) and was also shown to cause fruit enlargement (Weaver *et al.*, 1966). More effective compounds have subsequently been introduced.

Letham (1964) found a cytokinin material in plum fruitlets but lacked sufficient information to be able to characterize it fully. In the same year he extracted the material from maize endosperm, isolated it in crystalline form, characterized it, and called it zeatin. This compound was the first naturally occurring cytokinin to be isolated from a higher plant. Subsequently other naturally occurring cytokinins were discovered (see Helgeson, 1968).

HISTORICAL ASPECTS OF INHIBITORS

The view that endogenous inhibitors function as growth substances is quite old, but until recently there was considerable resistance to their acceptance as a functional group of plant hormones like the auxins and cytokinins (see Wareing, 1966). Inhibitors are very diverse, and it is often difficult to distinguish between inhibitory substances that have a physiological function in plants and those that do not.

The plant hormone ABA, 3-methyl-5-(1'-hydroxy-4'-oxo-2',6',6'-tri-methyl-2'-cyclohexen-1'-yl) $cis,trans$-2,-4-pentadienoic acid), is one probably of the most important and widespread inhibitors found in plants (Addicott and Lyon, 1969). Addicott and his colleagues (1964) isolated this compound from cotton fruit and named it abscisin II. It was found to be identical to dormin, discovered in England at about the same time (Wareing et $al.$, 1964). To avoid confusion of nomenclature plant scientists active in ABA research agreed on the name "abscisic acid" and reported their recommendation to a Special Session of the Sixth International Conference on Plant Growth Substances meeting in Ottawa in 1967; the name was subsequently approved by that body (Addicott et $al.$, 1968). The uses of ABA in agriculture have not been established because the compound is available only in small amounts, and relatively little information is available on its effects on crop plants.

The phenols and the phenolic flavonoids are also inhibitors that are found in a wide range of plants and are of physiological importance.

The early researchers who used bioassays to determine the presence of auxin in plant tissues often discovered far less auxin than they had anticipated. Other substances were then found to be present that in some way masked the action of the auxin. Purified plant extracts often evidenced much more activity in the bioassay than did the impure crude extracts from which they were prepared. The growth-inhibiting substances probably play a role in the control of growth and development of plants in conjunction with the other hormones. Inhibitory substances help to keep the plant from becoming too large and affect dormancy, growth, reproduction, and other plant activities.

Inhibitory compounds have been isolated from plants for many years. Köckemann (1934) discovered inhibitors in seeds and Audus (1947) found that coumarin, as well as other lactones that limit growth rates and germination, exist in roots. Hemberg found large amounts of inhibitors in dormant buds of potatoes (1949a) and ash trees (1949b) and showed that inhibitor content decreases toward the termination of rest. Hemberg's work stimulated research in this field that led to the concept that inhibitors are a part of the spectrum of growth-controlling chemicals. Inhibitors can control a wide range of growth processes. Many plants that cease growth when photoperiods are shortened show a simultaneous increase in their level of growth inhibitors (Nitsch, 1957). Many growth-inhibiting substances have been found in both fruits and vegetative organs (leaves and stems).

It has been shown that roots of certain plants can excrete inhibitors and that decomposition products of plants release inhibitors. In 1928 Davis showed that the chemical juglone was the toxic product produced in the roots and branches of walnut trees that inhibits the growth of plants growing near these trees. Bonner and Galston (1944) showed that guayule plants produce cinnamic acid, which inhibits the growth of adjacent plants. Similar inhibitors associated with other plants have also been found (Bonner, 1944).

Plant Growth Retardants

New synthetic organic chemicals that retard stem elongation, increase green color of leaves, and indirectly affect flowering without causing plant malformation have been introduced since 1949 and have proved to be valuable tools to control plant size. These compounds slow cell division and cell elongation in shoots and control height of the plant without causing stem bending or leaf malformation. Growth of treated plants is not completely suppressed, and neither their rate of organ development nor their vigor is affected. This advantage separates growth retardants from such synthetic growth inhibitors as MH, which suppresses cell division in the apical meristem. High concentrations of auxin also inhibit plant growth, but malformations occur in the leaves, stems, and flowers. Auxins also disrupt flowering. Herbicides cannot be considered to be growth retardants because, at low concentrations, they inhibit the expansion of all plant parts, thus producing a dwarf plant; they usually also cause some foliar damage.

The first plant growth retardants discovered were the nicotiniums (Mitchell et al., 1949), which were found to reduce stem elongation of bean plants. A year later it was shown that a series of quaternary ammonium carbamates retards the growth of snap beans (Wirwille and Mitchell, 1950). The compound designated as Amo-1618 was found to be the most active of the group. This and related chemicals were discovered in screening tests begun by the National Academy of Sciences in cooperation with the United States Department of Agriculture (Anonymous, 1955).

In 1960 Tolbert reported the existence of another series of quaternary ammonium compounds. The most active of these was designated CCC and was found to retard the growth of more species than any of the previously reported compounds. In 1962 Riddell and his colleagues showed that maleamic and succinamic acid act as growth retardants on legumes, potatoes, vines, and ornamental plants and that the compounds designated as CO11 and SADH retard the growth of many species.

The existence of another group of growth retardants, the phosphoniums, was first reported in 1955 (Anonymous). The most active compound of the group, Phosfon-D, affects the growth of a wider range of species than does Amo-1618 (Preston and Link, 1958).

As soon as growth retardants became available for experimental use, many studies were initiated to ascertain their physiological action and to find agricultural applications for them. The numerous applications of growth retardants as well as of other types of inhibitors will be described in suceeding chapters (see also Cathey, 1964).

The hydrazines constitute another group of growth retardants. BOH was the first of several compounds of this group to be found capable of inducing pineapple to flower (Gowing and Leeper, 1955).

HISTORICAL ASPECTS OF ETHYLENE

More than a century ago it was noted that illuminating gas and various smokes cause severe injury to and have other harmful effects on plants. For example, in Germany trees growing in the vicinity of broken gas mains were injured, and in the United States leaky gas pipes caused injury to greenhouse carnations and other plants (see Burg, 1962). In 1901 Neljubow showed that ethylene is the active portion of illuminating gas that evokes a biological response in plants. Formerly the green color had been removed from lemons and oranges by burning oil or kerosene in the same room with them. Packinghouse managers at first attributed the enhanced coloration to the heat emanating from the burners and to the high humidity. However, when more modern, steam-heated "sweat rooms" were utilized, the growers and packinghouse managers were disappointed because no ripening of fruit occurred. The research of many workers has shown that the agent responsible for the coloring is ethylene, which is produced when oil or kerosene is burned.

Numerous experiments performed on various harvested fruits have indicated that ethylene is a ripening agent (Pratt and Goeschl, 1969). Maxie and Crane (1968) showed that ethylene hastens the ripening of fig fruits while they are still attached to the tree.

Ethylene was not accepted as a hormone until the 1960's, in spite of the tremendous amount of data that was revealed showing that minute quantities of the gas have marked physiological effects on plants and in spite of doubts as to its capability of being translocated, which is one of the attributes of a plant hormone. Exogenous auxins have been found to stimulate plant tissues to produce ethylene, and it may be that other growth regulators exert their effects on plants with ethylene as an intermediate.

It is now believed that the ethylene produced at the apex of a shoot can diffuse down the shoot. It may well be that ethylene is the fruit-ripening hormone in plants and that it has other plant functions as well (Pratt and Goeschl, 1969). Perhaps the ethylene produced in the center portion moves outward, in the same direction as the ripening of many fruits (from center outward), and stimulates the unripe tissues to ripen.

Much agricultural interest has recently been stimulated by ethephon, a material that is available commercially and that provides a convenient method for making field treatments with ethylene. Ethephon breaks down in plant tissue and releases ethylene close to the site of action. Its effects are similar to those of ethylene on fruit ripening, abscission, and other growth phenomena (see Pratt and Goeschl, 1969).

FUTURE OUTLOOK

In the years since Darwin made his observations that led to the discovery of auxin, the hormone picture has become more and more complex. There are now four main classes of plant regulators, but there may well be many more in the future. Mitchell and his colleagues (1970) isolated a family of new plant hormones, which they called brassins, from the pollen of several species of plants. Brassins appear to have a glyceride structure. Several other hormones have been postulated but their existence remains to be established (flower-forming hormones, root hormones that stimulate growth of roots, and senescence-controlling hormones, for example).

Advances in molecular genetics have made it possible to study growth and development in relation to the effect of hormones on gene activity. Thimann (1963) stated that "Sometime soon we shall be visualizing any one of the organs of a plant as a veritable Times Square of intersecting streams of traffic, with specific hormones crossing and recrossing on predictable paths, some entering a cell together, there to activate specific biochemical processes, others accumulating or decaying, and every external influence playing its part in changing their fate."

Research on modifying plant growth by the use of exogenous growth regulators is also rapidly expanding, and consequently the number of commercial and practical uses for these regulators is constantly increasing. It is definitely within the realm of possibility that at some time in the future all plant growth processes will be controlled by plant regulators.

Supplementary readings on the general topic of nomenclature and historical background of plant growth substances are listed in the Bibliography: Addicott, F. T., and Lyon, J. L., 1969; Audus, L. J., 1959 and 1968; Avery, G. S., and Johnson, E. B., 1947; Boysen-Jensen, P., 1935; Cathey, H. M., 1964; Crocker, W., 1948; Darwin, C., 1880; Galston, A. W., and Davies, P. J., 1970; Leopold, A. C., 1955 and 1964; Letham, D. S., 1967; Miller, C. O., 1961; Mitchell, J. W., 1966; Mitchell, J. W., and Marth, P. C., 1947; Phillips, I. D. J., 1971; Pilet, P. E., 1961; Pincus, G., and Thimann, K. V., 1948; Popoff, K. J., ed., 1969; Pratt, H. K., and Goeschl, J. D., 1969; Rappaport, L., 1970; Ruhland, W., ed., 1961; Skoog, F., and Armstrong, D. J., 1970; Stodola, F. H., 1958; Tukey, H. B., ed., 1954; Wareing, P. F., Eagles, C. F., and Robinson, P. M., 1964; Wareing, P. F., and Phillips, I. D. J., 1970; Kent, F. W., and Thimann, K. V., 1937; and Wilkins, M. B., ed., 1969.

Chapter 2

BIOLOGICAL AND CHEMICAL DETERMINATION

Auxin was not discovered until the 1920's. It may seem curious that a substance that is so widespread in the plant kingdom and that is of such great importance could have been so long overlooked. One reason is that auxin occurs in such minute amounts that special methods were necessary to demonstrate its presence. Furthermore, early plant physiologists were more interested in other phases of plant growth. Attempts were made to explain plant growth in a simpler manner then, and many scientists were very skeptical of the existence of plant hormones. After Darwin demonstrated that young grass shoots bend toward a unilateral light source, there was a step-like progression of experiments that ultimately led to the proof that a hormone, auxin, affected plant tropisms. Approximately fifty years after Darwin's observations, Went developed the *Avena* test, a quantitative bioassay that could be used for further studies of the distribution and identification of auxin. Kögl and his colleagues were then able to isolate auxin and to identify it chemically.

Several methods have been developed to identify plant hormones and regulators, including chemical and physical means and bioassay. The book by Klein and Klein (1970) contains excellent descriptions of the separation and analysis of plant components, and that by

Mitchell and Livingston (1968) is an excellent source book that delineates the procedures used in bioassays.

AUXINS

Extraction Methods

Auxins and their precursors are present in several different forms in plants. Since the chemical nature of these forms is not fully understood, there is no one best method of extraction. The extraction technique that will obtain the maximum amount of auxin must be determined for each type of tissue or plant. Readily available or "free" auxins can be obtained by the diffusion technique. Extraction of tissues with cold solvents for a short period (two hours or less) probably yields about the same amount of auxins. Additional auxin, termed "bound" auxin, is obtained by prolonged extraction. However, in general, hydrolysis or enzymic action must destroy the bond attaching the auxin before the free unbound auxin is removed (see Thimann, 1969).

DIFFUSION.   Diffusion can be utilized to obtain auxin from intact tissues or from plant parts (Avery et al., 1937; Kramer and Went, 1949). The physiological bases of the plant parts are first dipped in a 10 percent solution of gelatin at 30°C and are then placed in contact with agar, usually at a concentration of 1 to 2 percent. The material is allowed to stand for from several minutes to several hours in an atmosphere saturated with water vapor. The agar block can then be cut into small blocks and used for assay in the *Avena* curvature test. The diffusion method is valuable for showing the relative amounts of auxin present in the different parts of the plant. However, this technique measures only the net auxin content; often the inhibitor content will mask most of the auxin activity. Diffusion is a poor choice if the auxin is to be analyzed chemically; auxins and inhibitors are obtained in such small amounts in the diffusion process that their chemical separation is difficult. Some auxin inactivation can also occur at the cut surface of plant parts, although there are methods for preventing this.

SOLVENT EXTRACTION.   This method is the one most commonly used to obtain auxin from plant materials. There is always a possibility that enzymatic changes may occur during preparation of the material for extraction or that chemical changes may occur either during the extraction process or during manipulation of the extract. Such changes may lead to the formation of artifacts, active compounds that do not naturally occur. Since the chemical nature of the substances being examined is not completely understood, artifacts may be difficult to recognize.

There are several ways to prepare plant tissue for extraction. Enzyme action can be prevented by freeze-drying the tissues (lyophilization), but the frozen tissues must not thaw before they are completely dry because thawing results in very rapid destruction of auxins and other hormones and the formation of artifacts. A packet of dehydrating agent, such as silica gel, should be included in the bottle or other airtight container used for storing the dried material. A further precaution is to store the material in a nitrogen atmosphere, always at below-freezing temperature. Material that is frozen fresh should be similarly stored, but since material will slowly lose its hormone activity even under optimal storage conditions, frozen fresh samples should not be kept longer than six months.

Another method of killing plant material is to subject the tissues to heat or boiling water. After tissues are boiled and extracted, only free auxin is obtained, since the auxin-producing enzymes are destroyed (Thimann and Skoog, 1940). Vlitos and Meudt (1953) found that inhibitors are not released when tissues are boiled to prevent enzymatic release or destruction of IAA.

Some researchers have recommended the addition of chemicals to inhibit the enzymes that destroy IAA. Terpstra (1953) recommended 0.01 percent, N-diethyl-dithiocarbamate, and cyanide was recommended by Steeves et al. (1953). However, the latter substance interferes with biological testing because it can seldom be removed completely with $FeSO_4$. Absolute alcohol inactivates enzymes and was recommended by Bennet-Clark and Kefford (1953). The pH is an important factor during extraction. A pH of less than 2.5 can destroy IAA (Larsen, 1955a). Furthermore, because auxin complexes can be hydrolyzed at an acid pH, the plant scientist may be working with the hydrolysis products of the auxins instead of the naturally occurring auxins.

Organic solvents, such as chloroform, ethanol, methanol, and diethyl ether, have been widely used. Most of the indole auxins are soluble in these solvents. Nitsch (1956) found that methanol is an excellent solvent for preventing the activity of polyphenol oxidase enzyme systems in Jerusalem artichoke tuber extracts.

The best way to obtain the free auxins is to extract them at a low temperature (0°C) for a short period (two hours or less). During a longer period the bound auxin complexes or precursors may be changed to free auxins. Although two hours or less are often considered necessary to extract the free auxins, the length of time needed to extract significant amounts of bound auxin varies with the tissue and the temperature. At higher temperatures bound auxin is converted to free auxin much more rapidly; the rate is often doubled by a ten-degree rise in temperature. For example, auxin can be extracted from *Nicotiana* ovaries at 0°C with ether for periods of up to sixteen hours without excessive formation of new auxin, but at higher temperatures auxin is produced enzymatically (Wildman and Muir, 1949). However, van Overbeek *et al.* (1945) found that after being extracted from sugar cane for a half-hour at −5°C, auxin was produced enzymatically. When Gustafson (1941) extracted auxin from dried tomato leaves or Iris ovaries at 90°C, yields were reduced. He believed that this reduction was caused by the destruction of an enzyme system that was capable of either activating or liberating auxin.

The separate roles of free and bound auxins in plant growth are still not clear, although it now appears that bound rather than free auxin may be the major factor controlling growth.

There are water-soluble (ether-insoluble) auxins that can break down to form the ether-soluble free auxins. These materials may be in a bound form and may require water for hydrolysis before extraction of auxin is possible. Water has seldom been used as the extracting solvent because many other materials, such as sugars, are soluble in it. Pohl (1951a) extracted auxin with water and then electrodialyzed the extracts.

Purification of Extracts

Once the extract is obtained the next step is to purify it, since the crude extract is a mixture of auxins and inhibitor compounds. A bioassay of a crude plant extract may show no activity because the auxin action may be completely masked by that of inhibitors. Separation of these individual constituents is therefore necessary.

SOLVENT FRACTIONATION.   After the extract is obtained, the first step usually is to use solvent fractionation as a preliminary purification method. The materials can be partitioned between ether and water at different pH values. If an ether solution is shaken with a solution of sodium bicarbonate at a pH of 9.0, the neutral auxins and many of the pigments are retained in the ether. The acidic auxins move into the aqueous phase and can then be recovered by acidifying the bicarbonate solution and re-extracting the auxins with ether. Hydrochloric acid is very satisfactory for acidification, and IAA (an acidic auxin) is recoverable from an aqueous solution by ether in a pH range of 2.5 to 5.0 (Larsen, 1955b).

PAPER CHROMATOGRAPHY.   This is an excellent technique for obtaining individual auxins. The advent of paper chromatography is one of the most significant advances that has been made in the field of auxin research. The method was first introduced in 1944 but was not used for auxins until the early 1950's. Before paper chromatography the extracts obtained from plant tissues were thought to contain just one auxin—IAA. Paper chromatographic studies have shown, however, that what is commonly referred to as "auxin" is a rather complex mixture of growth-promoting substances. The extract may contain not merely one chemical substance but as many as six or more. Numerous compounds are also usually present that may be readily converted into auxins.

Ordinary filter paper, usually Whatman No. 1, is used in paper chromatography. The paper is often square or cut in long narrow strips (Fig. 2–1) and is enclosed in a tightly closed chamber to provide a solvent-saturated atmosphere. A glass plate or a cork is often used to cover the top of the chamber to prevent evaporation of the solvent. In *ascending chromatography*, test substances are applied along a line drawn across the paper one to two inches from the bottom edge. The spot (or streak) of extract can be applied with a pipette, hypodermic, or capillary, but an automatic applicator is convenient, rapid, and accurate. Approximately one-half inch of solvent is placed in the container and the paper is lowered into the solvent so that the treated area is one-half to one inch above the level of the solvent. The chamber is immediately closed and the chromatogram is allowed to develop until the solvent front is near the top edge of the paper. In *descending chromatography*, the material is spotted or streaked near the upper end of the paper, which is suspended into a tray of development solvent. The solvent front moves downward until it is near the bottom of the paper.

Many different solvent systems can be used equally well. A commonly used solvent is a mixture of isopropanol-water-28 percent ammonia (80:10:10). The ammonia tends to keep the spot small, although it may destroy some auxins and thus produce artifacts.

During development the test substances migrate from the point of application. If the extract contains a mixture of materials, it will be separated wholly or partially into its components. With a given solvent

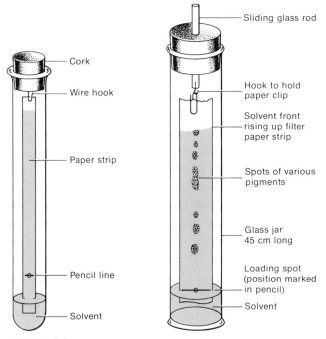

FIGURE 2-1

Test tube and cylinder chambers used for ascending paper chromatography. (After R. M. and D. T. Klein, *Research methods in plant science.* Garden City, N.Y.: Natural History Press, Fig. 7–7 on p. 414. © 1970 by Richard M. Klein. Reprinted by permission of Doubleday & Company, Inc.)

system, a particular chemical will migrate to a certain position on the paper. The ratio of the distance from the starting point to the auxin to the distance from the starting point to the solvent front is known as the R_f. (For example, IAA in isopropanol-water-ammonia always has an R_f of 0.35 to 0.40). This is one way of tentatively identifying chemical compounds. A known material may be chromatographed and its R_f may be compared with that of the unknown material. A compound cannot be identified solely by chromatography, however, because dissimilar compounds may show identical R_f's in several solvent systems.

There are two types of paper chromatography: one-dimensional and two-dimensional. The one-dimensional types were described in the preceding paragraphs. In the two-dimensional type, the material is placed near one corner of the paper and is developed in one direction in a solvent. The paper is then dried and developed in a perpendicular direction in another solvent. This procedure is an effective means of separating certain components that run close together or at the same R_f when developed in one solvent system only.

A one-dimensional type of double chromatography was developed by Nitsch (1960). In this system a chromatograph strip is developed in an ascending manner to 20 cm in a mixture of isopropyl alcohol, 28 percent ammonium hydroxide, and water (80.0, 0.1, and 19.9 parts by volume, respectively) and is then removed and dried. A second development is then made to 25 cm using a mixture of hexane, chloroform, and water (75, 15, and 10 parts by volume, respectively). This method effectively separates neutral auxins from acidic types.

After the chromatogram is developed, the location of the hormones on the paper must be determined. If they are present in sufficient quantities, they may be sprayed with chemicals that will react with the indole auxins present to form colored compounds. However, often the auxins are present in too low a concentration to be detected by spray reagents. Therefore, the chromatograms are usually cut into segments and the auxins are eluted from the paper and bioassayed. Certain areas may also be cut from the chromatograms, eluted, and bioassayed. A straight-growth test might be used, in which wheat coleoptile cylinders are prepared and placed in test tubes containing water, and a segment of the chromatogram is added. After twenty hours the cylinders are measured and the percentages of growth in comparison with that of control cylinders are determined. Enough chromatograms of the unknown extract should be run so that several color tests and bioassays can be made.

Since many of the auxins are unstable in light and air, treatment should be as rapid as possible and the paper, if not sprayed, dipped, or bioassayed immediately after running, should be stored in the dark in a nitrogen atmosphere at a low temperature. Kefford (1955a) showed that about 60 percent of the IAA is lost on chromatograms that are run in air and then air-dried overnight before spraying. In his experiments, J. P. Nitsch found that extracts chromatographed in a nitrogen atmosphere produce superior results.

For further reports on results of chromatographic analysis of plant extracts, see Gordon (1954), Larsen (1955b), and Bentley (1958). Surveys of the most suitable solvents have been made by Stowe and Thimann (1954) and by Sen and Leopold (1954).

THIN-LAYER CHROMATOGRAPHY.   This method was introduced in the late 1930's but was not used extensively in hormone work until the early 1960's. Crude extracts of many plant species have been found to possess auxin-type biological properties. Thin-layer chromatography provides a convenient and sensitive method for separating and identifying the known auxins and their derivatives. In this technique an applicator is used to spread a uniformly thin layer of adsorbent, such as kieselguhr or Silica Gel G on a glass plate. After the adsorbent has dried and has sometimes been activated by heating, the plate is stored in a desiccator to protect it from absorbing too much moisture. All other operations in thin-layer chromatography are the same as those in paper chromatography: Samples are spotted or streaked at one end of the plate, the plate is placed end down in the solvent, and the solvent migrates over the plate, causing the compounds to separate. However, thin-layer chromatography requires much less time for development (often less than one hour) and achieves a better resolution than does paper chromatography. Since very small plates (even microscope slides) can be used, much less solvent material and space are required. Another advantage is that smaller amounts of auxin can be determined by this method. Furthermore, the thin layers can be sprayed with chemicals to identify the auxins at concentrations of spray that would destroy paper. After the chromatogram has dried, spots or bands can be scraped off and bioassayed for activity. A disadvantage of thin-layer chromatography is that many of the materials it uses are toxic to bioassays.

COLUMN CHROMATOGRAPHY.   This technique is usually required when a chemical identification is desired because it allows large quantities of auxin to be separated and purified. Columns of aluminum oxide or silica gel are often used. The auxin extracts are placed at the top of the column and as the solvent flows through the column, the auxins move down in bands to different regions of the column (Powell, 1960). These bands can often be seen under ultraviolet light. If the solvent is passed through the column, the auxins can be eluted and collected for bioassay.

A combination of column chromatography and paper chromatography yields a better separation of auxins than does either technique used separately (Linser et al., 1954). Paper or thin-layer electrophoresis, a technique by which compounds are separated by their characteristic ionic charge, is valuable when used as a supplement to chromatographic methods. It is especially valuable in determining auxin complexes and precursors since these are readily broken down under conditions of alkaline chromatography (Denffer et al., 1952).

CHARCOAL ABSORPTION AND DISPLACEMENT.   In this method aromatic compounds are adsorbed on deactivated charcoal and then eluted with phenol (Dalgliesh, 1955). Active water-soluble auxins can be obtained that are difficult to extract directly by using organic solvents. Since this procedure can be carried out at a neutral pH under relatively mild conditions, there is much less danger of auxin destruction.

SPECTROPHOTOFLUOROMETRY AND GAS CHROMATOGRAPHY.   Most auxins are probably indolic in nature. Spectrophotofluorometry is an extremely sensitive means for measuring indoles, but plant extracts contain numerous fluorescent impurities that cause much interference (Stowe and Schilke, 1964). Therefore, for practical use this procedure must be combined with a highly efficient purification system. Gas chromatography can achieve excellent single-step purification if the compounds are sufficiently volatile and show suitable selective retention values when partitioned between the liquid substrate and the gas phase. However, despite the utility of gas chromatography, preliminary purification of the plant extract is usually required before this method can produce sufficiently clean preparations (Powell, 1964). Powell performed a preliminary purification by first fractionating the indoles into acidic, basic, neutral, and water-soluble compounds and then partially purifying the indoles in each group by silica gel column chromatography. The neutral and acidic indoles were further purified by gas chromatography, after which the fluorescence intensity was measured.

A method of volatilizing acidic indoles is to esterify them by diazomethylation. Grunwald *et al.* (1968) esterified eleven indole compounds by this method and then successfully separated the esters by gas chromatography. They also suggested that trimethylsilycation shows promise as an alternate method of synthesizing volatile indole derivatives for gas chromatography.

Biological Determination of Auxins

Although chemical methods are very useful in the determination of auxins in plants, they have the disadvantage that relatively large amounts of compounds are required. Very often only minute amounts of the hormone occur in plant tissues and chemical techniques are not sensitive enough to be of value. Bioassays, which are generally more sensitive than chemical methods, measure the response of a plant or plant part to a known amount of auxin or to tissue extracts. Some auxin tests have been devised to determine the amount of the hormone obtained from natural sources, whereas others have been adopted mainly for the determination of synthetic compounds. The latter tests can sometimes be used for endogenous auxins. Both types of tests will be discussed here.

Auxins have the capacity to increase the rate of cell elongation in coleoptiles and stems. They also influence other physiological processes, such as fruit development and root formation. A low concentration of auxin stimulates cell extension, but an extremely high concentration may cause inhibition. Generally, however, the amount of auxin obtained in plant extracts is not high enough to cause inhibition.

Gibberellins are also capable of promoting cell elongation, particularly in intact plants, and this ability may pose a difficulty when auxin determinations are made. (For example, gibberellin is active in the first-internode test and the coleoptile section test.) Furthermore, in certain tests gibberellins can enhance auxin activity or modify its response. Therefore, when crude extracts are tested in some bioassays, the total growth stimulation may be caused by two or more hormones rather than by the activity of any particular auxin or hormone and a combination of various biological, chemical, and physical tests must be used to determine the chemical nature of the active material.

Auxin has a pronounced characteristic of polar transport within the plant. When early researchers applied auxin to one side of a decapitated coleoptile, the auxin moved downward and produced cell elongation and thus a bending of the coleoptile. This property of polar transport led to the development by Went (1928) of the *Avena* coleoptile curvature test for auxin determination. Many of the known facts and theoretical conclusions about auxins are based on results of this test.

AVENA CURVATURE TEST.   A genetically uniform variety of *Avena sativum* should be utilized in this test. An understanding of the morphology and anatomy of the *Avena* seedling reveals why it is such a good test object. When an *Avena* seed is placed in conditions favorable for growth, the primary root and the shoot begin to elongate at about the same time. The shoot consists of a very short stem with two partially developed leaves, which are surrounded by a sheath called the coleoptile. In daylight, the coleoptile remains short and is soon pierced by the primary leaf. However, in darkness, the coleoptile may reach a length of 6 cm or more before it ceases growth and is pierced by the primary leaf. The coleoptile is considered a leaf structure, and the stem portion is referred to as the "first internode" by most auxin researchers.

Since cell multiplication ceases early in the growth process, the coleoptile is an excellent material for the study of cell elongation. In darkness, after the coleoptile is approximately 10 mm long, elongation is caused solely by cell extension. After this stage coleoptile growth is not complicated by further cell division, and the results obtained can be attributed only to cell elongation. In transverse section the coleoptile is elliptical and has two small vascular bundles running along either side. The cells at the tip of the coleoptile are not elongated but are almost isodiametric. The growth of the primary leaf closely follows that of the coleoptile. Approximately seventy hours after planting there is a maximum growth rate of 1 mm per hour. After the coleoptile is 20 mm long, the 2 mm apical portion scarcely elongates at all, and the zone of maximum cell extension is 10 mm below the tip. Coleoptiles ranging in length from 25 to 30 mm and growing at a rapid rate are often used for bioassay.

The *Avena* curvature test must be carried out in a temperature-controlled darkroom so that the plant material will not undergo phototropic curvatures during handling. Phototropic induction can be obviated by utilizing light filtered through an orange-red filter that cuts off wave lengths shorter than 550 mμ. Light in the region of from 550 to 700 mμ is transmitted and suppresses elongation of the first internode. More uniform results can be obtained by using a green safelight that emits light between 500 and 550 mμ (Withrow and Price, 1957). A similar type of green safelight is recommended by Nitsch and Nitsch (1956), but a green light will not suppress the elongation of the first internode. A constant relative humidity of 85 to 89 percent is necessary because if the humidity is too high, water exudes from the cut surfaces of the decapitated coleoptiles and may wash off the agar block or spread the hormone diffusing out of the agar onto the sides of the coleoptile. If the humidity is too low the agar dries out, resulting in a loss of contact between the block and the cut surfaces. A suitable humidity can be maintained in small chambers by lining them with wet filter paper. In a bioassay room, flooding the floor with water may result in a satisfactory humidity. A temperature of 25°C is used by most researchers.

Test plants are sensitive to smog. Since the amount of smog in the air affect the results obtained from test plants, filtering systems must be utilized to eliminate smog from the hormone room (Went, 1957).

Several varieties of oats are available but a hulless variety is preferable because it does not require the tedious work of freeing the seeds from their husks. Went's (1928) method is to soak the seeds in water for two or three hours, lay them with the grooves downward on wet filter paper in petri dishes, and keep them in darkness for thirty hours. They are then exposed to red light, which suppresses the growth of the first internode. The seedlings are fitted onto sockets and fixed in rows

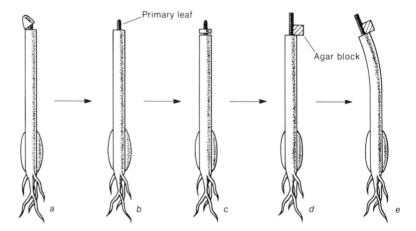

FIGURE 2-2

Steps in *Avena* curvature test: *a*, Tip containing auxin-producing center is removed; *b*, plant is allowed to remain decapitated for 3 hours to deplete it of auxin; *c*, in a second decapitation, short piece of stump is removed; *d*, agar block containing auxin is placed on cut surface adjacent to leaf; *e*, after 90 minutes, curvature results from asymmetric growth of 2 sides. (After J. Bonner and A. W. Galston, 1952.)

by means of brass clips that fit into grooves in a wooden rack. Both the clips and the holder can be adjusted so that the plant is in an upright position and the roots are in a trough containing water. Forty-eight hours after planting, the coleoptiles are 20 to 30 mm long and ready to use. The first step is to cut off each tip with a razor (Fig. 2–2). Three hours later the apical 4 mm of the stump is removed by making a cut on one side of the coleoptile without cutting the primary leaf. This second decapitation is necessary because after ninety minutes the severed tip of the coleoptile again begins to produce auxin. (For maximum sensitivity it is best to have little or no auxin in the coleoptile.) The primary leaf is then pulled up so that it is broken at the base. The block of agar to be tested is placed on one side of the cut surface resting against the leaf, and after ninety minutes the curvatures are recorded by placing a piece of bromide paper behind the plants and illuminating them from the side that reveals the amount of curvature. The curvature on the resulting shadowgraph (that is, the angle between the tangent to the curved tip and straight base) can then be measured by a protractor.

Skoog (1937) showed that he could effectively reduce the level of endogenous auxin in the coleoptile by "deseeding" the plants. In this method, twenty-four hours before the test, when the coleoptile is 1.5 cm long, the endosperm is removed from the seedling. When the tip is cut off there is no regeneration of the auxin-producing capacity of the coleoptile. Since the auxin or auxin precursors in the seed are removed, only one decapitation of the coleoptile is necessary.

Went (1928) found that when decapitated coleoptile tips are placed on 3 percent agar plates for two hours, the growth substance diffuses into the agar. Agar plates containing extract are cut into blocks each measuring 2 mm × 2 mm × 1 mm. Freshly severed portions of leaves, buds, and other plant parts can also be used. When the agar block is placed on one side of the coleoptile tip, auxin moves downward and stimulates extension growth on the same side. The resulting curvatures are proportional to the concentration of growth substance in the agar, in the range of 1° to 20°. A series of tests using IAA must be run in conjunction with the unknown substance, the amount of which can be expressed as the IAA-μg equivalents. Went's original techniques are still used today, with only minor modifications.

A method commonly used in Japan is effective and requires much less equipment than Went's technique. This simplified procedure is based on work by Thimann (1952) and Shibaoka and Yamaki (1959). Oat seeds are placed embryo up in petri dishes containing 5 ml of distilled water and incubated at 25° to 27°C for twenty-four hours. Vials are filled with sand and enough water is added merely to wet the sand; excess water should be avoided. Seeds with coleoptiles 2 to 3 mm long are selected and planted at a 45° angle in the sand, embryo downward, so that the coleoptiles are exactly vertical. The seeds should be pushed down so that the endosperm end of each seed is level with the surface of the sand. The vials are incubated in the dark at 27°C for twenty hours. The seedlings are exposed to red light for three hours and are then returned to the dark for another twenty hours.

The seedlings should be 10 to 25 mm long; shorter seedlings are more sensitive to IAA but are more difficult to manipulate and therefore are generally not used. Any seedlings that are not straight should be discarded. Subsequent decapitations and other operations are the same as those previously described for Went's method.

The Japanese method for making agar blocks and infusing them with hormone is as follows. One percent agar is prepared by dissolving 1 g of agar in 100 ml of water. The water is heated with stirring almost to boiling, using care not to burn the agar. The temperature of the melted agar must be held above 50°C or the agar will solidify. The agar is then poured over a clean microscope slide to a depth of 2 mm and allowed to solidify. The slide is placed on a piece of millimeter graph paper, and the agar is cut into 2 mm × 2 mm squares with a razor blade, using the paper as a measurement guide.

Twelve of the agar blocks are placed in a petri dish containing 1 to 2 ml of a test solution (as little as 0.1 ml may be used). Paper chromatogram segments may be added to the water and blocks in the dish for direct elution. The blocks should equilibrate with the test solution for 1.5 to 2.0 hours at room temperature.

AVENA COLEOPTILE STRAIGHT-GROWTH TEST. The physiological basis of this test is the stimulation of cell extension by floating sections of *Avena* or *Triticum* coleoptiles in solutions containing auxin. A good method is to soak seeds of a "hulless" oat variety, such as 'Brighton,' for several hours and then to plant them on moist filter paper (Nitsch and Nitsch, 1956). At an early stage of growth the seedlings are subjected to some red light to suppress the elongation of the first internode and to stimulate the growth of the coleoptile. After approximately seventy-two hours, many coleoptiles are 20 to 30 mm long and ready for the test. A good tool for cutting the sections is a guillotine, as shown in Figure 2–3. Knives set perpendicularly on a wooden block may also be used.

Most researchers leave the leaf inside the coleoptile because it makes sectioning easier. First the apical 3 mm of the coleoptile is removed, then a 4 mm section is cut off. The coleoptile sections are floated in distilled or deionized water for two hours to lower the endogenous auxin content and are then placed in a test solution (containing 2 percent sucrose and buffered to pH 5 with a citric:phosphate buffer) in a beaker or test tube that is positioned on a test tube roller to prevent geotropic curvature. The sections are incubated for twenty hours at a temperature of 25°C. They then can be either measured immediately under a dissecting microscope or photographed. (Enlargements or projected images can be measured when convenient.) In this test, controlled humidity is not required because the sections are floated in water.

Coleoptile straight-growth tests are sensitive to inhibitors that retard growth as well as to auxins that stimulate elongation.

AVENA FIRST-INTERODE TEST. This test was developed by Nitsch and Nitsch in 1956 in their search for an auxin test that was sufficiently simple and sensitive to be used in routine assays of paper chromatograms. The seeds of 'Brighton' oats are soaked in water in darkness for two hours; all subsequent operations should be performed under a yel-

FIGURE 2-3

Guillotine used for cutting coleoptile or first-internode sections to determine presence of auxin.

low-green light so that the elongation of the first internode is not suppressed by light. The seeds are germinated on moist sawdust at 25°C, and after three days the first internodes are 25 mm long and ready for testing; the coleoptiles are only 5 mm long (Fig. 2–4). Each coleoptile is cut off and a 2 mm section is removed from the top of the first internode; an adjacent 4 mm section is utilized for bioassay, and the remainder of the internode is discarded. The internode sections are kept in glass-distilled water for one hour but are placed on cheesecloth that is stretched on plastic rings so that only the surface of the water is broken by them. Ten internode sections are placed in a pyrex test tube containing 0.5 ml of either standard or test solution. [The test solution contains 2 percent sucrose and is buffered to pH 5.0 with K_2HPO_4 (1.794 g/liter) and citric acid (1.019 g/liter); paper chromatogram squares of 1 cm^2 can be eluted directly in this solution. A suitable range of concentrations for the standard solution is 0.00017 to 1.0 μg per test tube.] The test tubes are then placed in a test tube roller that rotates at a speed of 1 rpm and prevents geotropic curvature of the sections. After the test tubes have been rotated for twenty hours in darkness at 25°C, the length of the sections is measured.

The first-internode test is a more sensitive indicator of auxin than the *Avena* curvature test and has the added advantage of indicating the presence of inhibitors.

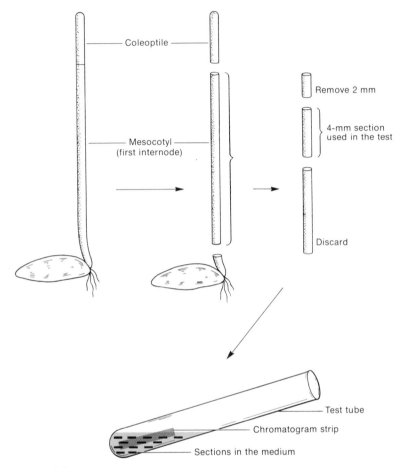

FIGURE 2-4
Steps in the *Avena* first-internode test.

SLIT-PEA TEST. The physiological basis for this test is that auxin stimulates cell elongation in the epidermal cells more than it does in the inner cortical cells, thus causing an inward curvature. Wounding may also reduce the response of cells near the cut surface of the tissue. In the slit-pea test stems are slit longitudinally in the growing zone and placed in water. The two halves curve outward in water and inward in auxin solution. The outward curvature is a result of a release of the tissue tensions that are present in the stem at the point where the pith is compressed. Went (1934b) studied the reaction of slit pea stems to auxin (see also Kent and Gortner, 1951; Thimann and Schneider,

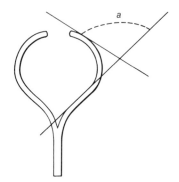

FIGURE 2-5

Shadowgraph of slit segment of pea stem showing curvature that developed after stem was exposed to growth-regulating chemical and method of measuring angle of curvature (*a*). (After J. W. Mitchell and G. A. Livingston, 1968; diagram suggested by F. W. Went.)

1938; and Went and Thimann, 1937). The slit-pea test has been used widely in the study of synthetic compounds, but it is not generally used to study naturally occurring compounds because it requires relatively large amounts of auxin.

In the test, peas 'Alaska' are soaked in water for six hours, planted in moist vermiculite, and kept in darkness for seven days. The sensitivity of the test can be increased by exposing plants to red light for four hours approximately thirty-two hours before the start of the test. By the time of testing, the plants should be 10 to 12 cm high and have two nodes, each bearing a scale, and one node at the apex bearing a leaf. The plants should also have an internode of less than 5 mm between the apical leaf and the terminal bud. The top of each plant is cut off 5 mm below the apical node and the cut portion is discarded. A 3 cm slit is then made downward from the cut surface of each section and the stem is cut off a few millimeters below the slit section. After the sections have been rinsed in glass-distilled water for one hour, five or ten of them are placed in 10 to 20 ml of a test solution that has been buffered to a pH of 4.0. (A higher pH may interfere with auxin-induced curvatures by causing opposite, that is, outward, curvatures.) The samples may be prepared in diffuse white light without affecting curvatures. Plant materials are kept in the dark for six to twenty-four hours. The response can be measured directly on the sections with a protractor or on photographs or shadow prints. The angle often measured is shown in Figure 2–5. Within limits, a straight line is obtained when the curvature is plotted against the logarithm of the concentration. Suitable standard concentrations of IAA for this test are 0, .0175, 0.175, and 1.75 ppm. A drawback to this bioassay is the difficulty of slitting the section in the middle so that symmetrical curvatures are obtained in both halves. The slit-pea test has been used mainly in studies of synthetic growth regulators.

PEA-STEM SECTION TEST.   Stem sections of dicotyledonous plants can be used for auxin tests in the same way that sections from grass seedling internodes or coleoptile sections are used. Thimann and Schneider (1939) used the epicotyls of *Pisum sativum*. Seedlings that are grown in darkness for seven to ten days are ready for use when the third internode is 15 to 40 mm long. Sections 3 to 5 mm long are removed from a region beginning 5 to 6 mm down the second internode. (If the seedlings have been subjected to red light during growth, the first internode is greatly shortened.) The plants should be sectioned when the third internode is 50 to 70 mm long or when the fourth internode is 6 to 12 mm long. Sections are cut from the subapical portion of the third internode and are then placed in either unknown or known (standard) solutions containing 2 percent sucrose. The increase in length is measured after a suitable interval. It is best to stand the epicotyls with their bases in a buffered 2 percent sucrose solution for thirty minutes before putting them in the test solutions to improve their response. Galston and Baker (1953) found that sections of stems from plants grown in complete darkness are more sensitive to auxin than those that receive red light during the seedling stage.

The pea-stem section test is a slightly more sensitive indicator of auxin than the *Avena* coleoptile test and is also a sensitive indicator of gibberellin, as are most of the straight-growth tests. Stowe and Yamaki (1959) demonstrated that auxin-induced elongation of pea-stem sections is stimulated by certain fatty acid esters.

AGERATUM PETIOLE CURVATURE TEST.   Petioles of *Ageratum houstonianum* curve when placed in solutions of IAA (Bottelier, 1956). The amount of curvature is related to the auxin concentration. (Approximately 10 μg of IAA per liter is the lowest active concentration.) Petioles are removed from the third, fourth, and fifth leaves of the apex of normal green plants and rinsed in distilled water in darkness for one hour. Ten uniform petioles are placed in 50 ml of test solution contained in 150 ml beakers. The degree of curvature is measured twenty to twenty-four hours later either with a protractor or by means of a shadowgraph. A light of uniform quantity must be utilized for the shadowgraphs as the response is highly dependent on illumination; Bottelier used 4,500 foot-candles of light from an incandescent bulb. Curvatures of 30° to 60°, depending on the age of the petioles, are produced. Curvatures of less than 30° developing in auxin-free solutions are probably a result of endogenous auxin. This test is used mainly to measure the activity of exogenous or synthetic auxins.

ROOT-GROWTH TEST.   Young roots are far more sensitive to auxin than are stems or coleoptiles. Many techniques have been developed to take advantage of root sensitivity. Unfortunately, root-growth tests often produce erratic results. The usual result of an application of exogenous auxin is a retardation of extension, but auxin sometimes produces a slight increase in rate of growth. Research has also been conducted on excised root sections and decapitated roots. The level of endogenous auxin could be expected to be lower in excised than in intact roots. For further information on root-growth tests, the student should refer to the review of Larsen (1961).

ROOT-FORMATION TEST.   There are many bioassays whose physiological basis is the formation of roots in plants (see review by Libbert, 1957). The advantage of root-formation tests is that they are not sensitive to inhibitors. Went (1934a) described a quantitative method using *Pisum* seedlings. These seedlings are grown in the dark and shoots 10 to 12 cm long are cut off just above the first scale node. The bases are washed in water and potassium permanganate is used as a disinfectant. After the terminal buds are removed, the apical ends of the stems are slit downward for 1 to 2 cm and are immersed in 1 ml of the test solution for fifteen hours, during which the auxin moves polarly from apex toward the base of the shoots. The cuttings are then placed base down in a 2 percent sucrose solution for seven days, after which the sucrose solution is replaced with distilled water. One week later the number of roots formed per cutting is counted. The greater the number of roots initiated, the greater is the concentration of auxin in the test solution.

A more recent test is based on the initiation of roots on cuttings taken from the seedlings of the mung bean (Hess, 1964). An adaptation of this test is to sterilize seeds for three minutes in a solution of one part Clorox to sixteen parts water and then to rinse them under running tap water for twenty-four to twenty-six hours. The seeds are planted in moist vermiculite at 26°C at a relative humidity of 60 percent. The light source is 700–750 foot-candles at plant level. After ten to twelve days, the cuttings are prepared by removing the root system 3 cm below the cotyledonary node (Fig. 2–6). The cotyledons are also removed if they have not yet abscissed. The cutting then consists of 3 cm of hypocotyl, the epicotyl, the primary leaves, and the trifoliate bud. Five to ten cuttings are placed in vials measuring 20 mm × 68 mm that contain chromatogram sections or eluate plus 10 ml distilled water. Distilled water is added every twenty-four hours to retain the original volume. Six days after the cuttings are treated, the number of roots per

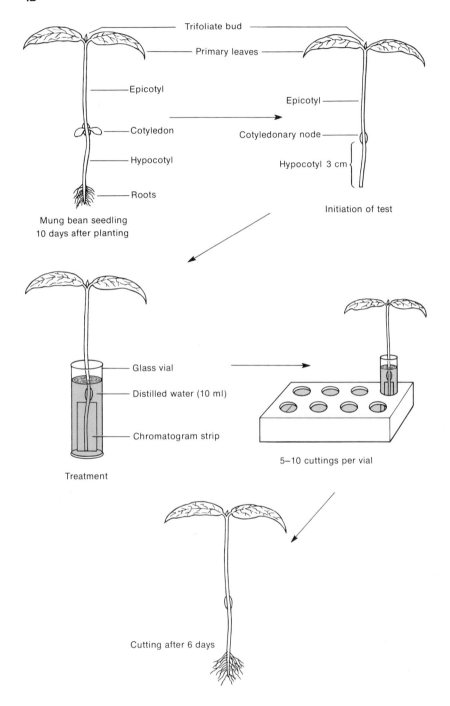

Trifoliate bud

Primary leaves

Epicotyl

Cotyledon

Hypocotyl

Roots

Mung bean seedling
10 days after planting

Epicotyl

Cotyledonary node

Hypocotyl 3 cm

Initiation of test

Glass vial

Distilled water (10 ml)

Chromatogram strip

Treatment

5–10 cuttings per vial

Cutting after 6 days

cutting is counted and compared with the number of roots of controls grown in water. The number of roots formed is proportional to the amount of auxin in the solutions.

The radish root-formation test developed in Japan is more a rapid method (Mitsuhashi and Shibaoka, 1965). Seeds of radish (*Raphanus sativus* var. *acanthiformis* 'Riso Daikon') are planted and allowed to grow to a height of 5 cm in darkness at 27°C. Hypocotyl cuttings are made, consisting of the whole hypocotyl together with the cotyledons and a small bud. The basal parts of twenty cuttings are soaked in 10 ml of the test solution. After a forty-five-hour incubation in the dark at 27°C, the central cylinders of the hypocotyl are pulled out and the number of rootlets protruding from them are counted.

ROOT-INHIBITION TEST.   Inhibition of root growth is the physiological basis for this assay method, in which the roots of corn, cucumber, and watercress are most often used. The growth of most roots is inhibited even by very low auxin concentrations. A test using corn seeds was developed by Swanson (1946). Seeds of a corn hybrid, 'Silver King,' are sterilized in hypochlorite solution for three minutes and are then thoroughly rinsed in tap water. The seeds are placed on moist filter paper, embryo down, in petri dishes and are germinated in the dark. When the roots are 15 to 25 mm long, they are measured and transferred to new petri dishes (twenty-five per dish) lined with moist filter paper containing 15 ml of the solution containing auxin. After forty-eight hours in the dark, root lengths are again measured and the percentage of elongation or retardation, compared with that of the controls, is determined. This method is used mainly for determining the activity of synthetic compounds. Unknown concentrations of such compounds, ranging from 0.01 to 2.00 ppm, can be accurately determined by this method.

TOMATO OVARY TEST.   For a study of the natural auxins that are found in fruits and their relationship to fruit setting and fruit growth, none of the previously mentioned tests is quite adequate. The fact that an auxin can cause bending of an *Avena* coleoptile or curvature of pea-stem sections does not necessarily mean that such auxin will induce fruit set or stimulate fruit growth. IAA, for example, vigorously stimulates cell elongation but does relatively little to stimulate fruit set.

FIGURE 2-6
Steps in the mung bean root-formation test.

Luckwill (1948a) developed a bioassay based on the actual setting and growth of tomato ovaries. He used mainly the cultivar 'Blaby,' but other cultivars are also suitable. Seeds are sown in boxes and are later transplanted to pots. One or two days before flowering, two flower buds on the first truss are emasculated, and other flowers and buds on the truss are cut off. Twenty-four hours later, aqueous solutions of growth substances or of water only are applied to twenty of the ovaries with a small pipette. (A 1 ml hypodermic syringe fitted with a No. 17 needle is also satisfactory.) No wetting agent is used so that the liquid will not run off the ovary. Each ovary is treated with three drops (22.5 μl) of solution containing the required amount of growth substance. Aqueous solutions of BNOA are applied in amounts ranging from 0 to 0.90 μg per ovary. The fruitlets are cut off six days after treatment and their diameters are measured with a binocular dissecting microscope. The growth is proportional to the logarithm of the dose of BNOA applied, over a limited range of dosage. A disadvantage of this test is that a large supply of tomato plants is required, necessitating the availability of considerable greenhouse space.

COTTON EXPLANT ABSCISSION TEST.   Many plant materials can be used as sources for abscission zone explants, but cotton, beans, and coleus (Luckwill, 1956) have been most widely used. The cotton abscission test was developed by Addicott and his coworkers (1964) during the isolation of ABA from cotton plants. The test may be used to detect abscission-accelerating as well as abscission-delaying agents.

Cotton seedlings are grown in vermiculite under fluorescent light (2,000–2,400 foot-candles) at 32°C for fourteen days. (The conditions under which the seedlings are grown are most important.) Uniform cotton seedlings are selected and the explants are prepared (Fig. 2–7). Each explant consists of 3 mm stumps of the cotyledonary petioles, the stem, and the upper 10 mm of the hypocotyl. Each explant contains two cotyledonary abscission zones. To test abscission, chemicals and extracts are applied to the cotyledonary petiole stumps in the form of a 5 μl drop of 1.1 percent agar. The explants are inserted in special holders that are placed in petri dishes containing 2 percent agar. Concentrations of 0.1 mg/ml of IAA and 0.025 mg/ml of ABA are suitable for inhibiting and stimulating abscission, respectively. The petri dishes are placed in the dark at 30°C and are tested for abscission once every twenty-four hours by the application of a 10 g force to each petiole stump (Mitchell and Livingston, 1968). A balance set at 10 g is used to calibrate force to use, and the percent of abscission that occurs every twenty-four hours is recorded.

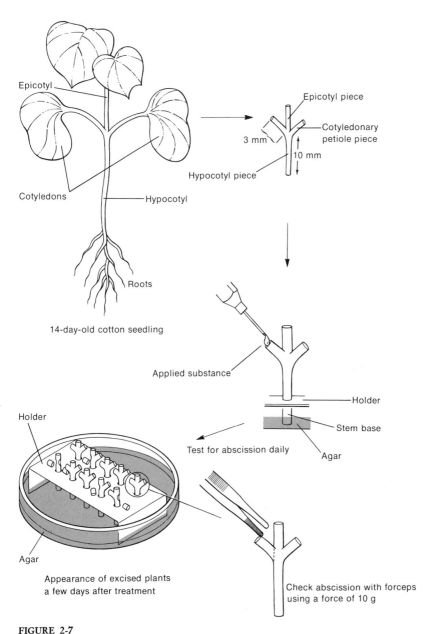

FIGURE 2-7

Technique used in cotton explant abscission test. (After F. T. Addicott, H. R. Carns, J. L. Lyon, O. E. Smith, and J. L. McMeans, 1964.)

Before the explants are cut, tools should be washed with 95 percent ethanol. Explants should be handled with forceps. When readings are taken, tools should be washed with ethyl alcohol after each treatment to avoid cross-contamination of hormones.

Abscission tests are rather poor indicators of auxin because many naturally occurring inhibitors also cause abscission. In addition, gibberellins can cause abscission in the cotton explant test. Therefore, these tests are valid only under limited circumstances.

Chemical Determination of Auxins

Biological tests for auxins are often relatively complex and frequently there is considerable variability in results. Therefore, several investigators have developed colorimetric methods for the quantitative determination of auxins. The presence of indole compounds (IAA and its derivatives) can be determined by chemical methods. Several chemical reactions, in which characteristic colors indicate the presence of indole compounds, can take place in solutions of auxin or on paper or thin-layer chromatograms.

In general, the techniques used for chemical determination of indole auxins are suitable only for indole auxins that previously have been partially purified to remove interfering substances. The primary use of the indole reagents is usually to locate auxin on chromatograms, although some researchers have used the indole reagents to make quantitative auxin determinations of plant extracts. Perley and Stowe (1966) have pointed out the difficulties of using indole reagents on crude plant extracts. Two disadvantages of the chemical methods are that they are much less sensitive than the bioassays and they are not specific for active compounds.

SALKOWSKI REACTION. This method is the most common and is based on a reaction originally observed by Salkowski in 1885. Even though the technique has been widely used and studied, the chemistry of the reaction is poorly understood. Several modifications have been proposed, but most of them require the use of high concentrations of perchlorate or persulphate, two highly corrosive chemicals that rapidly cause the paper to disintegrate. In the less dangerous method of Sen and Leopold (1954), 2 ml of 0.05 M $FeCl_3$ is added to 100 ml of 5 percent $HClO_4$. The paper is sprayed or dipped in the solution and dried for one hour. The presence of the various auxins is revealed by the color developed by the indole present.

EHRLICH REACTION.   The Ehrlich reaction can detect as little as 1 μg of IAA. It is less dangerous than the Salkowski reagent, but the HCl in the spray causes the paper to deteriorate. In addition, this method does not produce the wide range of colors with different indoles that the Salkowski reagent does. Chromatograms are sprayed or dipped in a solution of 2 percent p-dimethylaminobenzyaldehyde dissolved in 2 N HCl in 80 percent alcohol. The Ehrlich reagent causes indole auxins to turn red, purple, or blue (except for some hydroxyindoleacetic acids). Anthony and Street (1970) found that the specificity of the test can be increased by extracting the color complex into chloroform; their technique of using trichloroacetic acid to obtain maximum sensitivity has proved satisfactory for estimating the recovery of indoles from paper and thin-layer chromatograms and from silica gel columns.

QUANTITATIVE METHODS.   When nonpurified plant extracts are measured by color tests, the results can only be qualitative because of the possible interference of other compounds. The Salkowski reaction is affected by the reduction of such substances as phenols (Platt and Thimann, 1956). However, purification of the extract by chromatography enables the researcher to make fairly accurate quantitative determinations.

A fluorimetric method to determine the presence of IAA in solutions that is very accurate and sensitive is based on the occurrence of fluorescence when H_2SO_4 and $CuSO_4$ are added to the solution. The method of Ebert (1955) is to add 2 ml of .05 M $CuSO_4$ to 10 ml of an aqueous solution containing IAA at concentrations ranging from 0.2 to 12.5 ppm. Then 5 ml of concentrated H_2SO_4 is added and mixed well. The mixture is heated for five minutes in boiling water, cooled under running tap water for five minutes, and then placed in the dark until it reaches room temperature. The amount of fluorescence obtained under ultraviolet light is measured photoelectrically with a fluorimeter. The fluorescence is green at low and yellow at high concentrations.

When chromatograms are sprayed with a ferric-perchloric acid reagent, there is a linear relationship between spot area and logarithm of quantity of IAA of 0.25 to 8.0 μg. For IBA and indolepropionic acid, the relationship holds from 2.0 to 20.0 μg.

Absorption spectra are useful in making quantitative estimations of auxins. Measurements can be made on chromatograms by spraying the paper with paraffin oil to make it translucent (Nitsch, 1956). Using a spectrophotometer, Nitsch obtained an absorption curve in the ultraviolet region that corresponded closely with the curve of IAA in solution. Another method is to react indolic auxins with a chemical, thus producing a color reaction. The absorption spectrum in visible light can then be determined either on the paper or after elution. A good method of determining the spectrum is to elute the spot three times with 2 ml portions of methanol and then to add 1 ml of 0.5 N $FeCl_3$ in 50 ml of 35 percent $HClO_4$. Maximum color intensity occurs approximately sixty minutes later. More intense and stable color results when ether is used to dissolve the IAA (Gordon and Paleg, 1957).

For quantification of spots developed by Ehrlich's reagent, a densitometer has been found to be a satisfactory means of measuring the optical density of the spots (Vlitos and Meudt, 1953). There is a straight-line relationship between the logarithm of concentration and the percent transmission of light. This method is an improvement over the area-spot method.

Another technique has been developed by which a transmission densitometer is adapted to measure fluorescent substances on chromatographic paper (Mavrodineanu et al., 1955). The instrument consists of an ultraviolet light source, an ultraviolet-transmitting quartz filter, a filter that transmits light only in the visible spectrum, and a photomultiplier tube connected to the measuring unit. This test is sensitive to 0.5 μg of IAA.

GIBBERELLINS

Extraction and Purification Methods

The two solvents most commonly used to extract gibberellins are acetone and methyl alcohol. Several British workers, as well as West and Phinney (1959) working in America, have used acetone. Methyl alcohol has been used by Sembdner and Schreiber (1965), Hayashi et al. (1962), Weaver and Pool (1965b), and others. MacMillan et al. (1961) have also used 70 percent aqueous ethyl alcohol.

The following method was used by West and Reilly (1961) to extract and purify gibberellin-like substances from immature seeds of the wild cucumber (Echinocystis macrocarpa). Immature seeds were extracted with equal parts of acetone and water. After evaporation, two liters of the viscous endosperm extract was adjusted to pH 3 with sulphuric acid. The suspension was then extracted several times with ethyl acetate. The extracts that were combined and concentrated contained most of the biologically active material, as determined by bioassay with dwarf mutants of maize. The combined extracts were extracted with 5 percent aqueous sodium carbonate solution to remove acidic substances. The biologically active material was removed to the aqueous phase after acidifying this phase to pH 3, and the gibberellin-like substances were extracted with ethyl acetate. The residue from the ethyl acetate layer (2.5 g) was chromatographed on a charcoal-Celite (1:2) column developed with increasing concentrations of acetone and water. Paper chromatograms of 1 to 100 μg portions of the biologically active solids obtained from the charcoal column were developed with n-butyl alcohol-1.5 N ammonium hydroxide (3:1) (upper phase) as the developing solvent.

MacMillan et al. (1961) extracted seed of the scarlet runner bean (Phaseolus multiflorus) with 70 percent aqueous ethyl alcohol. The ethyl alcohol was then removed at a low temperature and the residue was separated by buffer into an acid and a nonacid fraction, according to the method of West and Reilly described in the preceding paragraph. Next, the crude acid fraction was fractionated by column chromatography according to the following method used by West and Phinney (1959). The crude acidic extract is first chromatographed on a column of charcoal-Celite that is eluted by gradient elution so that the concentration of acetone and water increases almost linearly from 0 to 95 percent. Gibberellins A_8, A_1, A_6, and A_5 were eluted with the following percentages of acetone and water: 27 to 38, 38 to 41, 43 to 49, and 51 to 56, respectively. Although the procedure separates the gibberellins from each other, it does not separate them from other acids, so further chromatography was performed on a column of silicic acid and Celite using increasing amounts of ethyl acetate in chloroform, and pure crystals were obtained.

Sembdner and Schreiber (1965) extracted flower buds and shoot tips of tobacco (Nicotiana tabacum 'Virgin Gold A') in methanol. Fifteen liters of extract was obtained from 2.9 kg of fresh material and was then concentrated to the water phase (approximately 3 liters) under vacuum at 40°C. The water was acidified to pH 2.6 to 2.8 and was then extracted, first with ethyl acetate and then with n-butanol. Both fractions were tested, utilizing chromatography, for gibberellin-like activity.

Other investigators have used more elaborate solvent fraction procedures (Hayashi et al., 1962; Weaver and Pool, 1965b).

DIFFUSION TECHNIQUE.   Although the technique of diffusing hormones from plant parts has been used widely in auxin research, it was not until 1964 that Jones and Phillips demonstrated that gibberellins readily diffuse from apical buds and that this technique can be used to quantitate the gibberellin content of plants. Diffusion into agar enables the researcher to estimate hormone production over a specific period, whereas extraction indicates only the gibberellin content of the extracted plant part at a single point in time. Jones (1968) found no quantitative differences in the levels of diffusible and extractable gibberellin present in peas.

Identification Methods

After the gibberellins have been removed from most of the other chemical components of the plant tissues, they must be separated and identified. In this section some of the important chemical and physical techniques used to accomplish this goal will be discussed.

THIN-LAYER CHROMATOGRAPHY.   Thin-layer chromatography has become a valuable tool for identifying gibberellins. Paper chromatography is also suitable, but it results in only limited separation of the gibberellins. At least seventeen of the known gibberellins have been separated by thin-layer chromatography and observed by the use of differential heating treatment of the thin-layer plates and different sprays (Cavell *et al.*, 1967). After spraying the thin-layer plates with the H_2SO_4 reagent, GA_1 and GA_7 fluoresce under ultraviolet light. The other gibberellins produce characteristic colors under ultraviolet light after the plate is heated at 120°C for ten minutes.

Before the advent of thin-layer chromatography, gibberellins could be separated only by countercurrent distribution or column chromatographic techniques. These methods are quite time-consuming, require relatively large samples, and necessitate the use of highly trained personnel. Thin-layer chromatography can sometimes avoid all these objections. In addition it allows the use of caustic spray detection reagents that drastically lower the threshold of detection for gibberellins. For example, the minimum level of GA_3 previously detectable in residue studies was 0.1 ppm. By separating the gibberellins with thin-layer chromatography and spraying the thin-layer plates with concentrated H_2SO_4, as little as 0.003 ppm of GA_3 may be detected.

A simple procedure is to prepare a slurry of water and silica gel. A uniform layer 0.25 mm thick is made on the glass plates; when the gel has set, the plates are activated by heating at 110°C for thirty minutes. After the development of the chromatogram, the usual procedure is to spray the plate with one of the reagents described by MacMillan and Suter (1963) as, for example, concentrated H_2SO_4:95 percent EtOH (5:95). The sprayed plate is then heated for ten to thirty minutes; the fluorescence produced under ultraviolet light can then be observed. Two-dimensional chromatography is another useful technique. Schneider *et al.* (1965b) have reported that thin-layer electrophoresis of gibberellins on Silica Gel G is also feasible.

Mitchell *et al.* (1969) noted that a mixture of self-condensation products of acetone can be formed during the extraction or isolation of gibberellins. These products can also be formed by combining acetone with Silica Gel H or aluminum oxide G powder and stirring for a brief period. Two of the condensation products tested, phorone and triacetonediol (three molecules of acetone), were very active on the bean second-internode assay, and 5 μg of either compound produced responses similar to those induced by equal amounts of GA_3. The investigators state that some condensation products of acetone represent a new family of growth substances. These facts must be kept in mind when acetone is used in extraction and purification processes.

GAS CHROMATOGRAPHY AND MASS SPECTROMETRY.   The use of thin-layer chromatographic methods for the identification of gibberellins has two disadvantages: The results can only be considered to be circumstantial, and access to pure samples of known gibberellins is required.

In 1963 Ikekawa *et al.* demonstrated the feasibility of using gas chromatographic techniques to separate gibberellins. Cavell and his coworkers (1967) separated the fifteen then-known gibberellins by performing chromatography utilizing their methyl esters or, in certain instances, by using the trimethylsilyl ethers of the methyl esters.

More recently, gas-liquid chromatography has been used, followed by mass spectrometry (MacMillan *et al.*, 1968). In this method diazomethane is used to form volatile methyl esters of the gibberellins. Individual gibberellins separated by gas-liquid chromatography are then fragmented and analyzed by a mass spectrometer.

OTHER IDENTIFICATION METHODS.   In order to separate large amounts of gibberellins, column chromatography must be utilized. The method of West and Reilly (1961) has been previously described. Khalifah *et al.* (1965) fractionated a mixture of seven known gibberellins using two different gradient elution systems in one continuous operation. Another method of separating large amounts of gibberellin is the countercurrent distribution method of Sumiki and Kawarada (1961).

Theriault *et al.* (1961) developed a fluorimetric assay for GA_3 in the presence of GA_1. At 0°C, GA_3 fluoresces strongly when exposed to ultraviolet light after treatment with cold concentrated sulphuric acid; GA_1 exhibits practically no fluorescence. At this temperature, the assay is reproducible and specific for GA_3. It agrees well with other quantitative methods for GA_3 currently in use and is simple and applicable to processing samples of tissue.

Infrared spectra techniques have been developed using methyl esters of the gibberellins (Sumiki and Kawarada, 1961).

Biological Determination of Gibberellins

There are many tests for gibberellins; a few of the important ones will be discussed here. It is best to use a bioassay that is specific for gibberellin—one that is not influenced by other growth regulators. For example, the *Avena* first-internode test is sensitive to gibberellins, but it also responds to auxins and inhibitors. Sometimes an efficient method is first to check an unknown extract by the *Avena* first-internode or coleoptile test to detect presence of activity. The active spots can then be checked by means of a bioassay specific for gibberellins.

DWARF CORN TEST.   One of the main plant responses to gibberellins is shoot elongation. Single-gene dwarf mutants that react specifically to gibberellins are very useful for bioassays. Phinney and his coworkers (Neely, 1959) developed a quantitative bioassay for gibberellins using extension of the leaf sheaths of seedlings of dwarf *Zea mays*. The dwarf corn or maize test has the advantages of specificity, sensitivity, and simplicity. It is unaffected by auxins, cytokinins, leucoanthocyanins, or other compounds.

The following procedures are adapted from the method of Phinney. Dwarf corn seed, D-1 and D-5, is soaked for twenty-four hours in tap water. The dwarf corn has a single recessive dwarfing gene and the seedlings segregate three tall to one dwarf, so six to eight times as many seeds as seedlings desired must be planted for the test. A sterilized soil mixture is used, consisting of two parts sand, one part loam, and one part vermiculite. Approximately 300 to 350 seeds should be planted in a flat 1.0 foot × 1.5 feet, as this method produces from 50 to 65 seedlings suitable for testing. The seeds are planted 0.5 inch deep in rows 2.0 inches apart in single file but touching. Seedlings can be grown in a greenhouse or in a controlled-temperature chamber.

One day before treatment the tall plants, misshapen dwarf plants, and any dwarfs within 0.75 inch of each other, are removed. The remaining plants are carefully watered and water is then withheld until a day after treatment to facilitate absorption of the gibberellin. Plants are usually treated approximately six days after planting when the dwarfs are 0.75 to 1.25 inches tall and before the first and second leaves have expanded. Treatment should be made during the afternoon to reduce interference from guttation water. One-hundred μl of solution is applied to each plant with a micropipette into the cup formed by the leaves at the top of the shoots. If the leaves are sufficiently expanded, solution is placed at the base of the first and second leaves. Tween 20 at 0.05 percent v/v is added to all solutions to increase test sensitivity. GA$_3$ at concentrations of 0, 0.002, 0.02, and 0.2 ppm is used for the standards. After seven or eight days the plants are uprooted and the length of the base of the first and second leaf is measured to the nearest millimeter. This is the distance from the ligule (collar) to the nearest prop root. The average length for the 0 standard (the control) is subtracted from that of the treated plants, giving a measure of elongation over the control. To this value 10 mm is added and the logarithm of this value is plotted against the log of the concentration of the standard. With this method, the range of concentrations can be plotted in a straight-line relation.

A more recent adaptation of this method by Phinney (unpublished) utilizes water culture to grow the seedlings. In this method seeds are germinated in vermiculite and as soon as the dwarfs can be distinguished from the normal plants, the dwarfs are transplanted to water culture under red light using apparatus similar to that described for the dwarf pea test.

PISUM SATIVUM ASSAY.   This assay was developed in 1955 by Brian and Hemming. The physiological basis of this test is that the single-gene dwarf mutant of *Pisum sativum* responds to a wide range of gibberellin concentrations. Dwarf cultivars of unknown genotypes can also be utilized. The following method is an adaptation by McComb and Carr (1958). Seeds of the cultivar 'Meteor' are soaked for five or six hours and are then planted in Perlite and covered with coarse gravel. Plants are grown at a day length of twelve hours with a day temperature of 22°C and a night temperature of 18°C. By the tenth day after planting the fourth internode is rapidly elongating and the seedlings are ready for bioassay. Plants that have uniform internode lengths between the third and fifth nodes should be selected; fifteen plants are used per treatment. Solution in the amount of 4 μl for each concentration is placed in the axils of the leaf subtending the third node. Six days later the distance between the third and sixth nodes is measured and the logarithm of the percent increment over that of the controls is plotted against the logarithm of the concentration of the standards. There is a good supply of dwarf cultivar *Pisum* seeds available, although only one mutant-dwarfing gene, *le,* is known to affect the lengths of internodes.

Another test has been described by Cohen *et al.* (1966) in which the elongation response of dark-grown pea seedlings to the amount of GA_3 applied in 10 μl agar disks is measured. When plants of dwarf pea 'W. F. Massey' have developed so that the third internode is 10 to 20 cm long, two marks 6 mm apart are made on the stem, one of them at the point of formation of the hook. The plants are decapitated by cutting through the upper mark; agar blocks are then placed on the cut surface. The marked section is measured after the section has been incubated for twenty-four to forty-eight hours at 25° to 30°C. The plants should not be exposed to light of an intensity greater than 10 foot-candles before or during the treatment period. This test allows the measurement of diffusible gibberellins and thus removes the necessity of extracting tissues, which may cause changes in the endogenous gibberellins.

LETTUCE HYPOCOTYL TEST. The requirements for a good bioassay system are sensitivity and specificity of response. In addition the bioassay should be as rapid as possible and only a minimum amount of special equipment or skills should be required. Many of the classic bioassay systems were selected for their extreme sensitivity or specificity but have the disadvantage of requiring trained personnel and much tedious labor. To overcome these shortcomings many plant scientists have tried to develop rapid systems for screening potentially active materials. One test that resulted from this search is the lettuce hypocotyl test (Frankland and Wareing, 1960). This test is not extremely sensitive, but it is quite specific for gibberellins (except for inhibitors), is rapid, and does not require highly purified extracts. It also can be used to assay chromatograms without prior elution.

Thompson's method (1962) is as follows. Strip chromatograms are developed by the usual techniques. The chromatograms are air-dried and cut into equal sections, and each section is placed in a petri dish 2.75 inches in diameter. Then 0.5 ml of buffer solution (or 0.5 ml of GA_3 standard in buffer solution) is pipetted into each dish. Pieces of untreated chromatography paper cut to the size of the chromatogram pieces are added to the petri dishes, which are then covered and left overnight to reach equilibrium. Twelve seeds of the lettuce cultivar 'Grand Rapids' are then sown on each strip. The dishes are placed in an incubator at 16°C and four or five days later the length of the hypocotyl of the ten most advanced seedlings is measured. The minimum sensitivity is 0.005 μg GA$_3$ and the test has a linear response from 0.0125 to 0.25 μg GA$_3$. Gibberellins A$_3$ and A$_7$ are the most active, A$_4$, A$_1$, and A$_9$ show less activity, and A$_5$ shows very little activity. Gibberellin A$_8$ is practically inactive.

CUCUMBER HYPOCOTYL TEST. This test was developed by Brian and Hemming (1961) of the Imperial Chemical Industries group in England. Although all the known gibberellins are active in this test, it is used primarily to distinguish the gibberellins that are characterized by the presence of a hydroxyl group at the seven position (GA$_4$, GA$_7$, GA$_9$, and GA$_{11}$) from the other gibberellins. Gibberellins A$_4$ and A$_7$ give approximately twice the response of A$_3$ at the same dosage, and A$_9$ is even more active. In addition to its advantage of differential sensitivity, the test also is used because it requires no facilities other than those found in any ordinary laboratory. (No greenhouse or growth chambers are required.) Its chief disadvantage is that it is relatively insensitive; the minimum dosage for response (0.2 μg) compares unfavorably with that of most other bioassay systems.

Seeds of the cucumber cultivars 'Long Green Ridge' or 'Perfection Ridge' (twenty seeds per treatment) are germinated on moist filter paper in the dark at 25°C. After forty-eight hours the seed coat is removed and the seeds are transplanted to 1 percent agar in water. The test solution is applied as 2 μl drops to the cotyledon. After two or three days of growth in diffuse white light, the hypocotyl length is measured to the nearest millimeter. A standard curve is made by plotting the logarithm of the gibberellin dosage against the logarithm of the hypocotyl length.

RED-LIGHT INHIBITION TEST ON DWARF PEA. The dwarf pea bioassay adapted from the method of Köhler and Lang (1963) is based on the observation of Lockhart (1956) that gibberellins reverse the red-light inhibition of elongation of dwarf pea. The peas (Pisum sativum 'Morse's Progress No. 9') are soaked for eight hours in running water and are then planted in vermiculite in the dark at 27°C. Four days later seedlings ranging in length from 2.5 to 2.7 cm are transferred to water culture under red light (Fig. 2–8). A convenient method is to insert the pea roots through the holes in a perforated plastic sheet that fits into the pan containing the water. A layer of rubber material with perforations corresponding to the holes in the plastic plate is placed over the plastic plate to support the peas in an upright position. The radiation source, which is approximately 70 cm above the plants, consists of four red fluorescent tubes placed above a sheet of red cello-

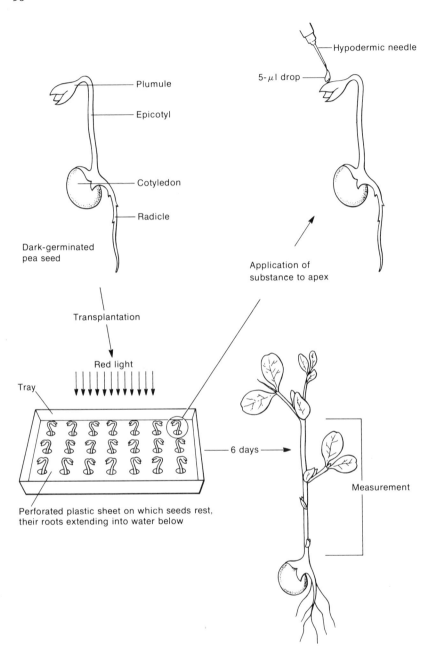

FIGURE 2-8

Steps in the red-light inhibition test to determine presence of gibberellin using dwarf pea.

phane. Twenty-four hours after transfer the plants are treated with 5 μl of extract containing an unknown gibberellin or with known gibberellin standards. For the control series, 0, 0.001, 0.01, or 0.1 μg of GA_3 is used. Six days after treatment the height of the plants from the seed to the second node is measured. If ten replicate plants are used per treatment, activity greater than that produced by 0.001 μg of GA_3 per plant is usually significant at the 5 percent level. It must be emphasized that because the effect of different gibberellins on dwarf pea varies, only relative levels of gibberellin-like activity can be determined. (Results would also vary with the variety of dwarf pea.)

BARLEY ENDOSPERM TEST. The physiological basis of this bioassay is the production of reducing sugar by embryoless pieces of barley endosperm in response to gibberellin. Gibberellin triggers a hormonal mechanism in the aleurone layer, the outermost layer of the endosperm, causing the release of amylase into the starchy endosperm. Starch is hydrolyzed to reducing sugar in amounts proportional to the concentrations of applied gibberellin.

Either the enzyme induced or the sugars produced as a result of the enzyme action may be determined and correlated with gibberellin concentration. Many researchers utilize a simple procedure to measure the sugar produced. However, because extracts often contain endogenous reducing substances and because the residues of some organic solvents are capable of inducing the production of reducing substances in barley embryos, the measurement of sugar production can sometimes give misleading results. The measurement of the enzyme is less subject to these sources of error.

An adaptation of the method used by Paleg (1960) is to soak seeds in distilled water for forty-eight hours and sterilize them in 1 percent hypochlorite for twenty minutes (Fig. 2–9). The seeds are then rinsed several times with distilled water using sterile technique. Each seed is cut 4 mm from the distal end, and the embryoless pieces are stored in moistened filter paper in a petri dish. Three endosperm pieces are added to test tubes containing 1 ml of solution consisting of standards or unknowns and 0.05 percent Tween 80. Chromatograms may be eluted directly in the tubes. The tubes are then rotated horizontally in a test-tube roller for forty-eight hours at 1 rpm. At the end of this incubation period the tubes may be frozen for later analysis.

According to Monselise et al. (1967), reducing substances can be analyzed by pipetting from 0.1 to 1.0 ml of the liquid into a test tube and adding water to make a total of 1.0 ml. One ml of Sumner solution is added and the tube is placed in a boiling water bath for five min-

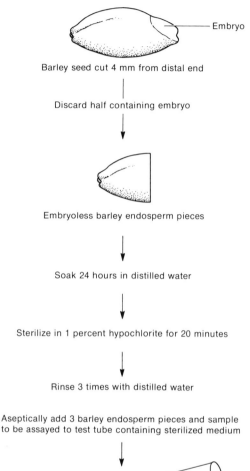

Barley seed cut 4 mm from distal end

Discard half containing embryo

Embryoless barley endosperm pieces

Soak 24 hours in distilled water

Sterilize in 1 percent hypochlorite for 20 minutes

Rinse 3 times with distilled water

Aseptically add 3 barley endosperm pieces and sample
to be assayed to test tube containing sterilized medium

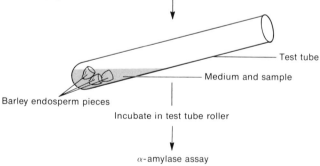

Incubate in test tube roller

α-amylase assay

FIGURE 2-9

Steps in determining presence of gibberellin using barley endosperm.

utes. The tube is then cooled and the solution is diluted to 20 ml and read at 550 mμ in a colorimeter.

To measure α-amylase, starch labeled covalently with Remazol-brilliant Blue R, a dye, can be used as a substrate (Rinderknecht *et al.*, 1967). The substrate is 15 ml of 2 percent RBB-starch suspended in phosphate buffer pH 7.0. As the starch is solubilized by the enzyme, the dye is released, thus coloring the solution. The reaction is started by adding incubate to the water to make a total of 0.5 ml. The mixture is shaken at 37°C for fifteen minutes; 2.0 ml dilute acetic acid is then added to stop the reaction (pH < 4). The solution is filtered and read at 595 mμ against a reagent blank. The absorbance is proportional to enzyme concentration in a range of 0 to 0.6 OD. An alternate method to measure the enzyme is that of Jones and Varner (1967), based on the reduction in the blue color formed by the addition of iodine as the starch is degraded by the enzyme.

This bioassay, which is sensitive to concentrations of GA$_3$ of from 10^{-5} M to 3×10^{-11} M, is a hundredfold more sensitive than any of the commonly employed tests. Another advantage is that only twenty-four hours are required to complete the test. The method has been used successfully for both crude and chromatographed extracts.

RICE SEEDLING TEST. The physiological basis for this bioassay, widely used by Japanese scientists, is stimulation of the growth of rice plants. An adaptation of the method used by Murakami (1959) is to soak rice seeds, of a variety sensitive to gibberellin, for two to three days in water at 30°C, after which the coleoptile is 0.5 mm long and the seeds are ready for use (Fig. 2–10). Sterile cotton 2 to 3 mm thick is placed in the bottom of petri dishes 3 to 5 cm in diameter and 3 ml of gibberellin or extract is added to the cotton. (For the standards, 0.01, 0.1, 1, or 10 ppm of GA$_3$ is utilized.) After the cotton has air-dried, twelve germinated rice seeds are placed in each petri dish and 3 to 5 ml of water or half-strength Hoagland solution is added. The petri dishes are covered and placed in the light for two days. Then 2 to 3 ml of water is added and the petri dishes are placed, uncovered, in a moist chamber under light. After five to seven days the length of the shoot from the mesocotyl to the second leaf sheath is measured. The longest and shortest plants are discarded and the lengths of the other ten plants are averaged.

In Japan, Ogawa (1963) found the dwarf mutant 'Tawanishiki' to be the most suitable variety for gibberellin bioassay. It responded to neither kinetin nor IAA in concentrations of 0.01 to 10 mg/liter.

In 1968 Murakami described a new rice seedling test for gibberellin

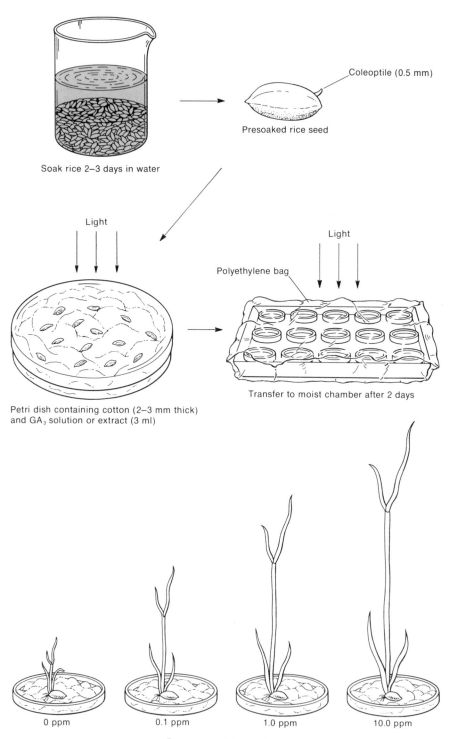

Coleoptile (0.5 mm)

Presoaked rice seed

Soak rice 2–3 days in water

Light

Light

Polyethylene bag

Transfer to moist chamber after 2 days

Petri dish containing cotton (2–3 mm thick)
and GA_3 solution or extract (3 ml)

0 ppm   0.1 ppm   1.0 ppm   10.0 ppm

Response to GA_3 after 5 days

based on the elongation response of the second leaf sheath. Coleoptiles of rice seedlings growing at the one-leaf stage are treated with a 1.0 μl drop of solution containing known or unknown gibberellin. (Agar blocks with gibberellin can be used instead of the drops.) After several days the length of the second leaf sheath is measured. The test measures the sensitivity of the dwarf cultivar 'Tan-ginbozu' to GA_3 in amounts of 0.05 to 100 μg. The varieties of rice dwarfs used in Murakami's test differ in their specificity of response to gibberellins A_1 to A_5. This rice test using the microdrop application method has proved less sensitive to such inhibiting substances as ABA and fusarinic acid and to IAA than a rice test that applies the test compound to the growing medium.

RETARDATION OF LEAF SENESCENCE IN RUMEX.   The physiological basis of this bioassay is that gibberellin application causes a delay in leaf senescence in the broad-leaved dock plant (*Rumex obtusifolius*) (Whyte and Luckwill, 1966). The oldest leaves, retaining a uniform color, of dock plants growing in pots in a greenhouse are detached and put, with their petioles in water, in the dark for twenty-four hours at room temperature. Disks 7 mm in diameter are cut from the main leaves, avoiding the large veins, and are placed, abaxial surface downward, on 2.5 cm disks of filter paper moistened with 0.3 ml of test solution. Four leaf disks from one leaf are placed on each filter paper disk, and each treatment is replicated four times. The disks are incubated in darkness at 25°C under conditions of high humidity. After four or five days, when the color has almost completely disappeared from control disks, the four disks from each replicate are placed overnight in 6 ml of methanol or acetone to extract the chlorophyll. The optical density of the solution at 665 mμ is then measured with a spectrophotometer and the chlorophyll content is expressed as a percentage of the control. The *Rumex* test is very sensitive. (The lower limit is 0.00001 μg GA_3 per ml; the upper limit is 0.1 μg per ml.) Kinetin and IAA at levels of 6 μg per ml produced no response in this test.

FIGURE 2-10
Steps in determining presence of gibberellin using rice seedlings.

CYTOKININS

Extraction and Purification Methods

Miller (1961) extracted tobacco callus with 95 percent ethanol. The active principle was precipitated by silver salt under acidic conditions. A small fraction of the material precipitated by silver was extractable into diethyl ether. Chromatography of the ether extract resulted in an active fraction with an absorption peak in ethyl alcohol at 268 mμ. The presence or absence of the active fraction was always correlated with activity. Letham (1964) also extracted a cytokinin, zeatin, from plum fruitlets with ethanol.

GAS CHROMATOGRAPHY. This technique has been successful for separating some of the naturally occurring cytokinins. Upper et al. (1970) found that ethyl acetate extraction of yeast transfer RNA hydrolysates and of culture filtrates of *Agrobacterium tumefaciens* resulted in sufficient concentration of the naturally occurring cytokinins for immediate gas-liquid chromatography.

Biological Determination of Cytokinins

Because no wholly adequate quantitative determinations for cytokinins are available, research in this field has proceeded relatively slowly. Some of the most effective available tests are discussed in the following paragraphs.

RADISH OR XANTHIUM TEST. The fact that cytokinins stimulate protein synthesis and delay senescence of leaves is useful for bioassays. A method used by Osborne and McCalla (1961) is to plant *Xanthium* (cocklebur) or radish seeds in clay pots. The leaves are ready to use when fully expanded. They are then excised and stored, with their petioles in water, in covered boxes in dim light for three days. After storage, leaves with a uniform light green color are selected. An aqueous solution of 0.5 ml of kinetin at concentrations of 0.063 to 16.0 mg/liter is applied to 4.25 mm disks of Whatman No. 1 filter paper placed in the bottom of petri dishes. Water (0.5 ml) is used for the control. Twelve 1 mm disks of leaf tissue are cut from the intervenal region of the leaf blade, and four of them are placed on each piece of filter paper so that the lower side of the blade is in contact with the paper. The dishes are stacked in large pans lined with filter paper to maintain a saturated atmosphere. The pans are covered with aluminum foil and stored at 25°C for forty-eight hours. Each group of disks is then gently boiled in a water bath containing 5 to 6 ml of 80 percent ethyl alcohol until the chlorophyll is extracted. The tubes are cooled and their contents diluted to 10 ml with 80 percent ethyl alcohol, and the absorbance of chlorophyll is read at 665 mμ. Absorbance is plotted against the logarithm of the concentration of kinetin; the curve is linear from 0.1 to 10 mg/liter of kinetin.

Loeffler and van Overbeek (1964) modified this test at follows. Radish seeds are planted in soil in a greenhouse and grown until the leaves are mature and slightly yellowed. Leaves are cut from the plant and are immediately placed on wet blotting paper. A standard drop (10 to 50 μl) or a drop containing cytokinin is placed on the blade of each leaf, midway between the margin and the midvein. The leaves are covered with black plastic until they have turned completely yellow. The amount of cytokinin can be measured by comparing it with that obtained in the standard treatments or by extracting the chlorophyll with 80 percent ethyl alcohol and measuring the absorbance. The mobility of the cytokinin is indicated by width of the spot; growth promotion is indicated by a puckering of the spot.

BEAN TEST FOR CYTOKININ MOBILITY.   The method of van Overbeek (unpublished) is to grow bush-type beans to the stage at which each plant has one trifoliate leaf. A section of stem containing one leaf is then cut from the plant. The base of the section is placed in 20 ml of a standard or an unknown solution. A mobile cytokinin such as BA is utilized for the standard. After five days the length of the bud or axillary shoot is measured. To suppress gibberellin activity, IBA at a concentration of 1 ppm is added to the solution. If two series are tested (one with and one without IBA), the activity of both gibberellin and cytokinin can be measured. Senescence may be measured after two weeks either by visually comparing the greenness of the leaves treated with an unknown with leaves treated with standard solutions or by extracting the chlorophyll and measuring the amount that has been absorbed.

RADISH COTYLEDON EXPANSION TEST.   This test was devised by Letham (1968) and is based on the growth response of radish leaves to cytokinins. It is fairly rapid, taking two to three days, and is very suitable for testing large numbers of samples. Letham has used it extensively in the isolation of naturally occurring cytokinins. This test is slightly sensitive to gibberellin, but not so sensitive that it interferes with the response to cytokinin.

Seeds of 'Long Scarlet' (other varieties are also satisfactory) are sieved to a uniform size and the larger seeds are reserved. Filter paper is placed on the bottom of a plastic tray and wetted with distilled water. (There should be a slight excess of water.) An amount of seeds equal to thirty times the amount of cotyledons required is scattered on the paper. The tray is covered with a plastic bag, sealed, and incubated in the dark at 25°C for thirty-six to forty-eight hours. Seedlings are then carefully selected for uniformity of cotyledon size. Each cotyledon is broken off and placed in a tray of distilled water. When a large excess of cotyledons has been harvested, the medium is stirred and uniform cotyledons are again selected and placed in petri dishes 9 cm in diameter containing filter paper that has been wetted with the cytokinin solution to be tested. Twelve leaves are placed in each dish and the dishes are incubated at 25°C under fluorescent light for three days, after which either the fresh weight of the leaves can be determined or the leaf area measured. The test will detect as little as 10 μg/liter of kinetin and will indicate a response of up to 25 mg/liter.

TOBACCO TISSUE CULTURE ASSAY.   The tobacco callus test has played the same role in the development of the knowledge of cytokinin physiology as has the *Avena* curvature test for auxin and the dwarf corn test for gibberellin. It is the "classic" test and is based on the fact that growth of callus (a mass of undifferentiated meristematic tissue) from tobacco pith is limited by the cytokinin concentration after all other growth requirements of the callus are satisfied. The test was developed by Professor Skoog and his coworkers at the University of Wisconsin (Skoog and Tsui, 1948).

To prepare the medium Linsmaier and Skoog (1965) make four stock solutions. Stock solution A consists of NH_4NO_3 (33 g), KNO_3 (38 g), $CaCl_2 \cdot 2H_2O$ (8.8 g), $MgSO_4 \cdot 7H_2O$ (7.4 g), and 1.0 liter double-distilled water. Stock solution B consists of KH_2PO_4 (4.25 g) and 250 ml double-distilled water. Stock solution C consists of $FeSO_4$ $\cdot 7H_2O$ (1.39 g), Na_2EDTA (1.86 g), and 250 ml double-distilled water. Stock solution D consists of H_3BO_3 (620 mg), $MnSO_4 \cdot 4H_2O$ (2.23 g), $ZnSO_4 \cdot 4H_2O$ (860 mg), KCl (83 mg), $Na_2MoO_4 \cdot 2H_2O$ (25 mg), $CuSO_4 \cdot 5H_2O$ (2.5 mg), $CoCl_2 \cdot 6H_2O$ (2.5 mg), and 1.0 liter double-distilled water. To 900 ml double-distilled water, 50 ml Stock A, 10 ml Stock B, 5 ml Stock C, and 10 ml Stock D are added. The pH is adjusted to 5.6 with 1 N NaOH, and the organic constituents sucrose (30 g), agar (10 g), IAA (2 mg), thiamine hydrochloride (1 mg), and myoinositol (100 mg) are added.

First, the stock solutions are mixed. The agar is dissolved by heating with stirring, and then the other organic constituents, which have been previously prepared, are added. The medium should not be allowed to cool or the agar will solidify. Fifty milliliters of medium is used per 125 ml erlenmeyer flask. Each flask is plugged with cotton, covered loosely with aluminum foil, and autoclaved for fifteen minutes at fifteen pounds of pressure.

To prepare the pith tissues, several portions of the internodes of tobacco ('Wisconsin No. 38') are selected from plants that are four months old and two to three feet high. After the internodes are washed, surface sterilizing is accomplished by swabbing them with 70 percent ethanol and then immersing them for twenty minutes in 0.1 percent $HgCl_2$. All subsequent operations must be done using sterile technique. The epidermis is peeled and the internodes are cut into 7 mm segments (the end segments are discarded), each of which is then cut into four wedge-shaped pieces. Three pieces are transferred to each erlenmeyer flask. The flasks are covered with white paper to reduce light intensity and then placed in the growth chamber. When adequate growth of callus has occurred, the callus pieces are weighed.

SOYBEAN CALLUS TEST.   This test is another tissue culture bioassay. The method of Miller (1963) is to sterilize seeds of soybean (*Glycine max* var. *acme*) by soaking them in 0.1 percent mercuric chloride for fifteen minutes, and then to rinse them four times with sterile distilled water. Three seeds are planted in each 125 ml erlenmeyer flask, which contains 50 ml of the medium consisting of $Ca(NO_3)_2$ at 347 mg/liter, KNO_3 at 1,000, NH_4NO_3 at 1,000, KH_2PO_4 at 300, $MgSO_4$ at 35, KCl at 65, sodium ferric ethylenediaminetetraacetate at 32, $MnSO_4$ at 4.4, $ZnSO_4$ at 1.5, H_3BO_3 at 1.6, KI at 0.8, glycine at 2, nicotinic acid at 0.5, thiamine hydrochloride at 0.1, pyridoxine hydrochloride at 0.1, sucrose at 30,000, and agar at 10,000 mg/liter. The *p*H is adjusted to 5.8.

After the seeds have germinated, the cotyledons are cut off and sliced into pieces measuring 4 mm × 4 mm × 2 mm. These slices are then added to the previously described medium to which have also been added IAA at 5 mg/liter and kinetin at 0.5 mg/liter. The wound callus that develops in approximately three weeks is subcultured to more of the same medium. This rapidly growing callus is used as a stock culture for testing. When kinetin is omitted growth quickly ceases.

Small pieces of callus measuring 2 mm × 2 mm × 2 mm are planted in 125 ml erlenmeyer flasks, each of which contains 50 ml of medium. The flasks are placed beneath 25 foot-candles of fluorescent light at 27° C. After at least two weeks the fresh weight of the pieces is determined. The growth response is directly proportional to the logarithm of the kinetin concentration from 0.004 to 10 mg/liter.

The soybean callus test can detect cytokinins over a wide range of concentrations and at very low concentrations. There is little, if any, of the cell enlargement response to cytokinins that sometimes occurs in the tobacco tissue culture assay (Miller, 1963).

INHIBITORS

Extraction and Purification of Inhibitors

Since most plant tissues contain stimulators as well as inhibitors, the biological determination of the latter may be difficult. Furthermore, the concentration of an inhibitor frequently is masked by growth-enhancing compounds. Chromatographic techniques are the most rapid means of separating inhibitors and other growth regulators. Since an inhibitor and other growth regulators can have the same R_f, two-dimensional chromatography is often necessary to produce a complete separation of the materials. In general, inhibitors can be separated and purified by many of the techniques used for auxins and other growth regulators as described here.

ABSCISIC ACID. Many organic solvents can be used to extract ABA from plant tissues; the best solvent must be determined for each plant tissue. After extraction most investigators have made an acid-base fractionation to yield organic acids (Addicott and Lyon, 1969), followed by a gradient elution of carbon and/or silicic acid column chromatograms. The final step is to use paper and/or thin-layer chromatography followed by crystallization of the eluate.

Purified extracts can be chemically assayed for ABA by spectropolarometric (Cornforth *et al.*, 1966) or gas chromatographic methods (Davis *et al.*, 1968). Seelby and Powell (1970) demonstrated that an electron capture detector can be used with gas chromatography to detect small amounts of ABA.

QUATERNARY AMMONIUM COMPOUNDS. Bayzer (1964) has found that aliphatic, quaternary ammonium compounds including CCC can be separated by thin-layer chromatography on cellulose layers. The best solvent system is chloroform-methanol-water (75:22:3). Thin-layer electrophoresis has also been used successfully to separate these compounds. Bayzer (1966) used silica gel layers and pyridine acetate buffer pH 3.6 or cellulose layers and pyridine acetate buffer pH 6.5 and found that two-dimensional separation of aliphatic, quaternary compounds can be accomplished by a combination of thin-layer electrophoresis and chromatography.

Biological Determination of Inhibitors

Various auxin tests can be used to determine the presence of inhibitors; inhibition, rather than stimulation, is indicated. For example, the *Avena* curvature test can be used. An agar block containing extract from the plant organ is applied in the usual manner to one side of the cut surface of a decapitated coleoptile. The inhibitory substances diffuse downward and inhibit cell extension on the treated side of the coleoptile. A curvature is thus obtained toward the side on which the agar block has been placed, or just opposite the curvature that results from auxin. Inhibition of elongation of sections of various plant parts, such as *Avena* coleoptiles, wheat coleoptiles, or the mescocotyl sections (first internode) of *Avena*, also indicates the presence of inhibitors. Cylinders of wheat elongate more than do cylinders of *Avena*, so the former are generally preferable because the effects of the inhibitors are more visible. The pea straight-growth test is also useful (Thimann and Schneider, 1939).

The retardation of root growth in such species as *Lepidium* (Moewus, 1949), rye (Vogt, 1951), wheat (Lexander, 1953), and pea (Kefford, 1955b) indicates the presence of inhibitors. Other tests have studied the inhibition of the growth of pollen tubes (Pohl, 1951b) and the inhibition of seed germination (Libbert, 1955).

CRESS SEED GERMINATION TEST.   This test was first devised by Libbert in 1954. Masuda (1962) performed the test as follows. Filter paper or chromatogram segments are placed in petri dishes 7 cm in diameter. Approximately 120 seeds are scattered uniformly on the filter paper after it has been moistened with 4 ml of distilled water. The dishes are placed in the growth chamber, and when germination of the controls reaches 85 percent (usually in approximately sixteen hours), the percentage of germination in each dish is determined.

TESTS FOR ABSCISIC ACID.   There is no specific bioassay for ABA. A bioassay should be used that fits the biological process that the hormone seems to be influencing. Since ABA can be an abscission-inducing agent, an inhibitor of growth or development, or a senescing agent, several bioassays may be useful, and it is important that several bioassays be used. For example, the cotton explant abscission test, the *Avena* curvature test showing negative curvature, and the straight-growth test showing inhibition of elongation, as well as the tests showing inhibition of seed germination and inhibition of growth of rice seedlings, can be used.

Relation of Inhibitors to Action of Growth-Stimulating Hormones

Most inhibitors have been studied for their inhibiting effects on auxin-stimulated growth. However, it is now known that gibberellin-stimulated growth can also be inhibited by certain naturally extracted materials (Corcoran *et al.*, 1961). An inhibitor that inhibited gibberellin-induced growth was separated from lima beans by Köhler and Lang (1963). It is well known that certain synthetic plant growth retardants antagonize the action of gibberellin. However, work in this field is merely beginning.

Little work has been done on the effect of inhibitors on cytokinins. Van Overbeek (unpublished) has shown that ABA counteracts the stimulating effect that cytokinins have on the leaf growth of *Lemna*.

Supplementary readings on the general topic of biological and chemical determination of plant growth substances are listed in the Bibliography: Bentley, J. A., 1961b and 1961c; Klein, R. M., and Klein, D. T., 1970; Larsen, P., 1961; Linser, H., and Kiermayer, O., 1957; Mitchell, J. W., and Livingston, G. A., 1968; and Phinney, B. O., and West, C. A., 1961.

Chapter 3

OCCURRENCE AND CHEMICAL NATURE OF PLANT GROWTH SUBSTANCES

This chapter will be concerned with the occurrence and chemical nature of the hormones in various plants and plant parts as well as the chemical nature of various exogenous growth regulators.

AUXINS

Early Work

The existence of naturally occurring compounds with auxin activity was first demonstrated by Seubert (1925), who found such activity to be present in various commercial enzyme preparations. Activity was also demonstrated in extracts of cultures of the fungus *Rhizopus suinus* (Nielsen, 1928). Auxins might be expected to have been isolated first from *Avena* coleoptiles because of the early work performed on them, but the quantities of auxin present in these coleoptiles are far too small. The development of the *Avena* curvature test by Went in 1928 made quantitative work with auxins possible and stimulated work on their isolation and identification.

Early researchers identified hormones by their rate of diffusion through agar and their sensitivity to destruction by hot acids and alkalis. Since large molecules diffuse more slowly than do smaller molecules, the size of a hormone molecule can be estimated by its diffusion rate. Using this method Went (1928) assigned a molecular weight of 376 to the auxin obtained from *Avena* coleoptiles. This figure does not agree with the value of IAA, which has a molecular weight of 175. Repetition of this work by Wildman and Bonner (1948), however, showed that if the agar blocks containing the auxin diffusate were extracted with ether and the extract applied to fresh agar blocks, a diffusion constant was obtained that was close to the molecular weight of IAA. The higher value found in the earlier work was probably caused by presence of inhibitors in the diffusates, which would result in lower values in the *Avena* curvature test.

With the advent of paper partition chromatography and other techniques in the 1950's, IAA was detected in many plant tissues. Minute traces of other biologically active substances in addition to IAA were also soon discovered.

Naturally Occurring Auxins

The compounds possessing auxin activity are organic compounds; all of them have hydrogen and oxygen in different proportions and arrangements and some of them also contain nitrogen and chlorine. Some are simple in structure but most are complex. The formulas of several of these compounds are given in Figure 3–1.

IAA (Fig. 3–1, I), one of the main auxins occurring in higher plants, has been detected in a wide range of plant tissues. The level of IAA in plant tissue usually varies with the stage of plant development. For example, the hormone has been found in the colorless, inner leaves of Brussels sprouts but not in the outer green leaves (Linser *et al.*, 1954). IAAld (Fig. 3–1, II) is a neutral substance found in plant tissues that can be rapidly converted to IAA in soil or by aldehyde dehydrogenase preparations (Larsen, 1944). The small auxin effects produced by IAAld are probably due to its conversion to IAA in plant tissues.

IAN (Fig. 3–1, III) was the first growth hormone to be extracted from the leaves and stems of rapidly growing higher plants. Members of the family Cruciferae, such as turnip, radish, and Brussels sprouts, are especially rich in IAN. IAN is present in mature cabbage at a concentration of approximately 0.0002 percent (Jones *et al.*, 1952). Some research has indicated that IAN appears as a breakdown product of glucobrassicin (Kutacek and Prochazka, 1964). Whether IAN should be called a hormone is debatable, since its activity results from its conversion to IAA.

Both IAMe and IAEt are biologically active compounds. The presence of IAEt (Fig. 3–1, IV) has been reported in immature maize kernels, grape seeds, tobacco tubers, and weeping willow. It is very likely, however, that neither of these compounds is a native auxin but that both are artifacts formed during the isolation procedures (Bentley, 1961).

The presence of IPyA (Fig. 3–1, V) has been demonstrated in maize endosperm and in *Salix* and *Nicotiana* tuber tissues. Since this compound is readily broken down into IAA, the importance of its role is difficult to assess.

Several other indole compounds have been identified in plant tissues by means of chromatography. However, many of these have eluded identification and may not be of as great importance.

In addition to substances that can be identified more or less certainly with known indole compounds, there are other types of unidentified auxins (Bentley, 1961), including unknown water-soluble auxins. Most work has concentrated on ether and organic extracts, and therefore much less is known about the water-soluble auxins. They are probably related to the indole substances in some manner currently unknown to science.

Auxins do not exist in plant tissues solely as free molecules of IAA or other auxins. Auxins are found as complexes or in bound forms. For example, IAA may be conjugated with an amino acid or a sugar (see Shantz, 1966). There may be neutral precursors that are not immediately available for growth but that can be converted to IAA. Precursor complexes have been found mainly in storage tissues, such as seeds and tubers. The existence of an ascorbic acid complex that gives rise to IAA as a result of alkaline hydrolysis has also been demonstrated. Indoleacetyl peptides have been formed by adding IAA to many plant tissues and may be a form of bound auxin. Large amounts of glucobrassicin (Fig. 3–1, VI) and related compounds are found in various species of *Brassica* and can undergo a series of both enzymatic and chemical reactions to yield a number of indole compounds. One such compound is ascorbigen (Fig. 3–1, VII), a bound form of ascorbic acid; another is IAN. Glucobrassicin (Fig. 3–1) may serve as a pool of bound auxin or auxin precursors (see Shantz, 1966).

Synthetic Auxins

Soon after IAA was shown to be the most frequently occurring auxin in higher plants, a search was made for synthetic compounds of similar chemical constitution and growth-promoting activity. In 1936 Zimmerman and his colleagues at the Boyce Thompson Institute in Yonkers,

72

FIGURE 3-1
Structural formulas, names, and abbreviations of some auxins and auxin precursors.

I

—CH₂COOH

Indoleacetic acid (IAA)

II

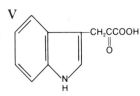

Indoleacetaldehyde (IAAld)

III

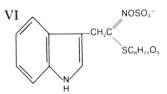

Indoleacetonitrile (IAN)

IV

—CH₂COC₂H₅

Ethylindoleacetate (IAEt)

V

—CH₂CCOOH

Indolepyruvic acid (IPyA)

VI

—CH₂C≡NOSO₃⁻ / SC₆H₁₁O₅

Glucobrassicin

VII

Ascorbigen

VIII

—(CH₂)₃COOH

Indolebutyric acid (IBA)

IX

—CH₂COOH

α-Naphthaleneacetic acid (NAA)

X

—CH₂COOH

β-Naphthaleneacetic acid

XI

—CH₂COOH

Phenylacetic acid

XII

—CH₂COOH

Anthraceneacetic acid

XIII

—OCH₂COOH

β-Naphthoxyacetic acid (BNOA)

XIV

—OCH₂COOH

Phenoxyacetic acid (POA)

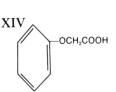

FIGURE 3-1 (*Continued*)

XV

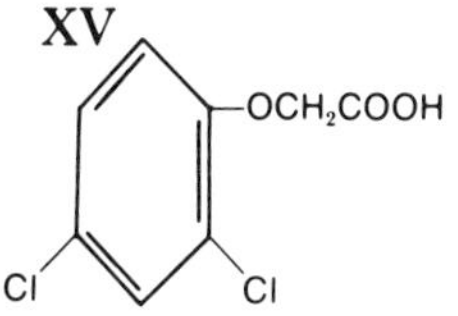

-OCH₂COOH

2,4-Dichlorophenoxyacetic acid (2,4-D)

XVI

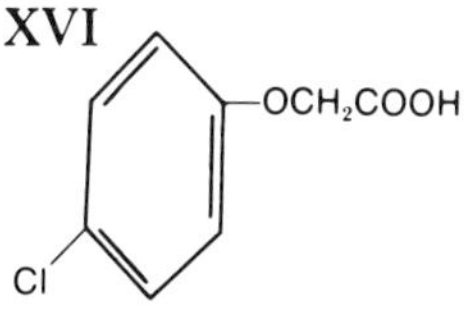

-OCH₂COOH

4-Chlorophenoxyacetic acid (4-CPA)

XVII

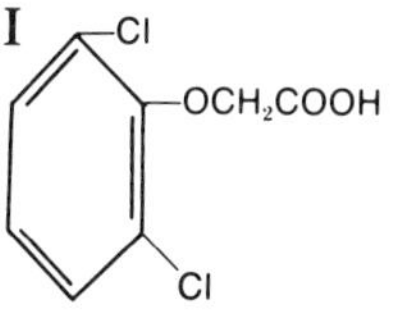

-OCH₂COOH

2,6-Dichlorophenoxyacetic acid (2,6-D)

XVIII

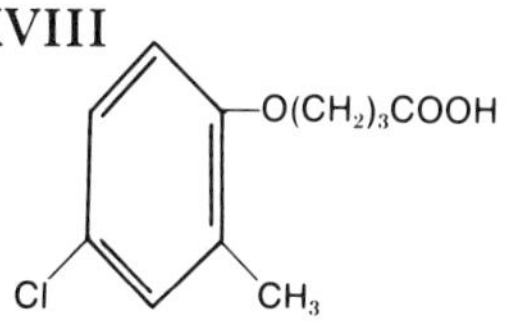

-O(CH₂)₃COOH

4-[(4-Chloro-o-tolyl)oxy]butyric acid (MCPB)

XIX

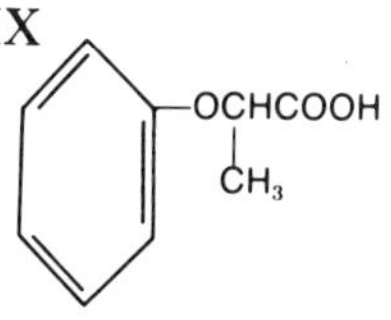

-OCHCOOH
CH₃

2-Phenoxypropionic acid

XX

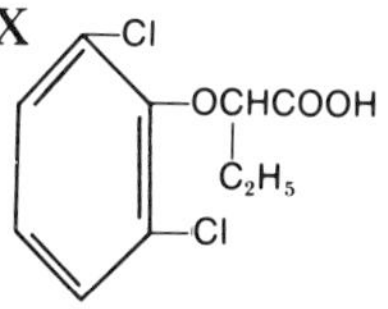

-OCHCOOH
C₂H₅

2-(2,6-Dichlorophenoxy)butyric acid

XXI

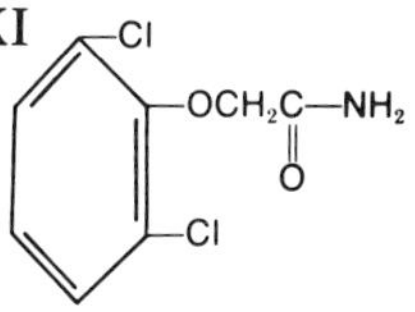

-OCH₂C—NH₂
O

2,6-Dichlorophenoxyacetamide

XXII

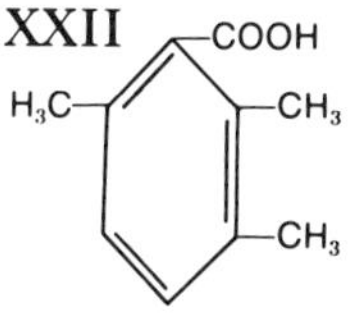

-COOH
H₃C— —CH₃
—CH₃

2,3,6-Trimethylbenzoic acid

XXIII

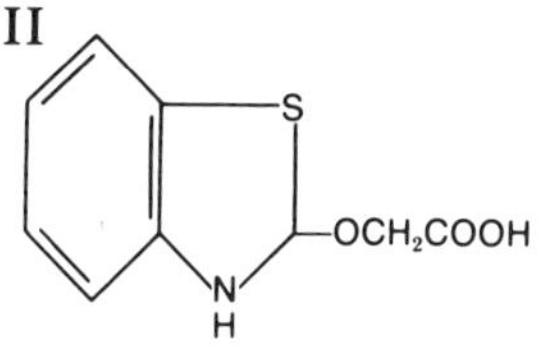

-OCH₂COOH

Benzothiazole-2-oxyacetic acid (BOA)

New York, investigated several new compounds including IBA (Fig. 3–1, VIII), IPA, NAA, and β-naphthaleneacetic acid (Fig. 3–1, IX and X), phenylacetic acid (Fig. 3–1, XI), and anthraceneacetic acid (Fig. 3–1, XII). In 1935 Haagen-Smit and Went demonstrated the biological activity of IPyA, and in 1938 Irvine studied BNOA (Fig. 3–1, XIII). In 1942 Zimmerman and Hitchcock investigated the POA (Fig. 3–1, XIV) series, of which 2,4-D (Fig. 3–1, XV) is a member. Some of the phenoxy compounds tested were 4-CPA (Fig. 3–1, XVI), 2,6-D (Fig. 3–1, XVII), MCPB (Fig. 3–1, XVIII), 2-phenoxypropionic acid (Fig.

3–1, XIX), 2-(2, 6-dichlorophenoxy) butyric acid (Fig. 3–1, XX), and 2,6-dichlorophenoxyacetamide (Fig. 3–1, XXI).

Another important group are the benzoic acid derivatives, such as 2,3,6-trimethylbenzoic acid (Fig. 3–1, XXII). The hormone compounds, including phenoxy and benzoic types, that are of importance in weed control are listed in Table 12–1. BOA (Fig. 3–1, XXIII) is another type of compound that produces interesting auxin effects. Beginning in the 1940's, a vast number of new auxins were synthesized in addition to the ones already mentioned here.

GIBBERELLINS

Early Work

Soon after Kurosawa, working in Japan in 1926, demonstrated the presence of a growth stimulator in culture filtrates of *Fusarium moniliforme,* studies were initiated to establish the chemical nature of the stimulus. In 1939 this research resulted in the isolation of a crystalline material that stimulates growth when applied to the roots of seedlings (Yabuta and Hayashi, 1939). This substance was termed "gibberellin."

Gibberellin is produced by the growth of the fungus *Gibberella fujikuroi* on or in liquid media. The earlier Japanese work was done with surface cultures, but techniques later were developed that utilized the stirred and aerated deep fermentation that is used for antibiotic production. The Japanese workers had obtained gibberellin A, a mixture of GA_1, GA_2, and GA_3 (Stodola, 1958). In 1951 Stodola and his group at the Northern Regional Research Laboratories of the United States Department of Agriculture began an investigation of the gibberellins produced by the fungus and obtained gibberellin A and gibberellin X. Their reinvestigation of the Japanese gibberellin A showed that it is a mixture of GA_1, GA_2, and GA_3. At approximately the same time, researchers at the laboratories of the Imperial Chemical Industries in England obtained much greater yields of GA_3 than did the Japanese or American workers (Cross, 1954). Comparison of the gibberellin X of Stodola and the GA_3 of the British group showed that the two compounds were identical. The name gibberellic acid was agreed on by all three groups. The yield increase obtained by the British group was probably a result of the use of a different strain of fungus and different conditions for fermentation. The British group used a higher sugar content in the fermentation medium (4 percent glucose or fructose) and a longer fermentation period (approximately eighteen days).

Gibberellins in Higher Plants

OCCURRENCE.   Soon after the work done by the Japanese, American, and British groups, gibberellins were found to occur in higher plants and to be one of the important classes of plant hormones that control plant growth. Immature seeds are usually the best sources of naturally occurring gibberellins (Skene, 1970a).

At present there are at least thirty-seven known gibberellins (Fig. 3–2) and the list grows almost every year (MacMillan and Takahashi, 1968; Lang, 1970; MacMillan, private communication). Some gibberellins are found only in the *Gibberella fujikuroi* fungus, some are found only in higher plants, and some are found in both. At least sixteen of the known gibberellins have been isolated from the fungus (GA_1 to GA_4, GA_7, GA_9 to GA_{16}, GA_{24}, GA_{25}, GA_{36}), and twenty-seven have been isolated from higher plants (GA_1 to GA_9, GA_{13}, GA_{17} to GA_{23}, GA_{26} to GA_{35}) (Lang, 1970; Wittwer, 1970; MacMillan, private communication). At least seven gibberellins (GA_1 to GA_4, GA_7, GA_9, GA_{13}) occur both in the fungus and in some higher plants.

Gibberellins have been isolated from a wide variety of plants, and many plants have been shown to contain numerous gibberellins. For example, thirteen gibberellins (GA_1 to GA_5, GA_7 to GA_9, GA_{19}, GA_{20}, GA_{26}, GA_{27}, GA_{29}) have been found in *Pharbitis nil,* and five (GA_1, GA_5, GA_8, GA_9, GA_{13}) have been found in tulip bulbs (Wittwer, 1970). Some other plants or plant parts shown to contain gibberellins are grape (GA_3, GA_4, GA_7), bamboo shoots (GA_{18}, GA_{19}, GA_{20}), hazel seed (GA_1, GA_3), immature apple seeds (A_3, A_4, A_7), sword-bean (GA_{21}, GA_{22}), *Lupinus luteus* (GA_{18}, GA_{23}, GA_{28}), citrus water sprouts and cucumber seeds (GA_1), sugar cane (GA_5), banana (GA_7), peanuts, corn, rhubarb, barley, and wheat (GA_1), *Phaseolus coccineus* (GA_1, GA_3 to GA_6, GA_8, GA_{13}, GA_{17}, GA_{20}), *Rudbeckia bicolor* (GA_1, GA_4, GA_7 to GA_9), and seeds of *Calonyction aculeatum* (GA_{30}, GA_{31}, GA_{33}, GA_{34}). The numbers do not necessarily indicate the order of discovery; for example, GA_3 was the first-characterized gibberellin and GA_{17} was discovered after GA_{18} to GA_{22}.

CHEMICAL NATURE.   Gibberellins have been defined as compounds that contain a gibbane skeleton and have appropriate biological properties (Fig. 3–3). However, a new proposal before the International Union of Pure and Applied Chemistry Committee on Organic Nomenclature recommends that all gibberellins contain the skeleton of the enantiomer of gibberellane (*ent*-gibberellane) (Rowe, 1968). This skeleton has the advantage of using a numbering system cor-

FIGURE 3-2

Structural formulas of some known gibberellins. Heavy lines and wedges indicate bonds lying above the plane of the ring; broken lines indicate bonds lying below this plane. (After A. Lang, 1970.)

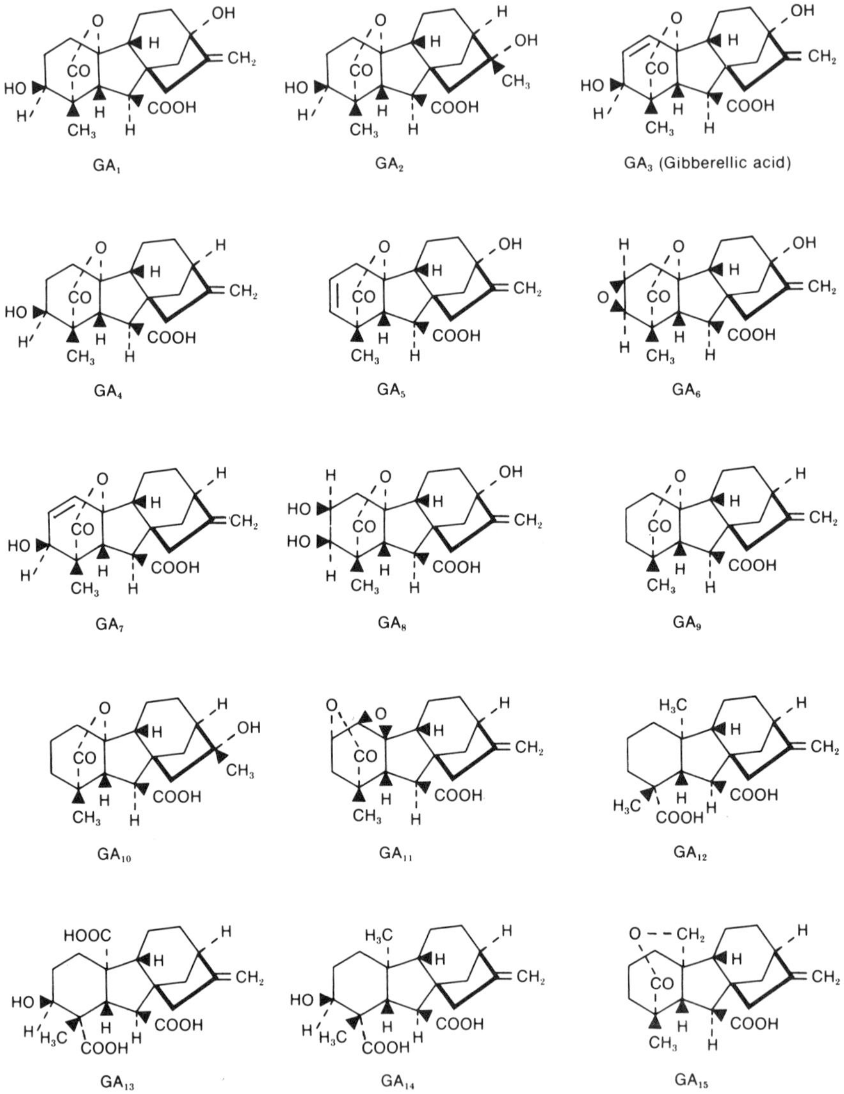

FIGURE 3-2 (*Continued*)

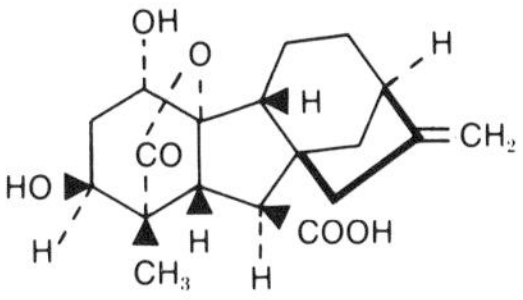

GA$_{16}$

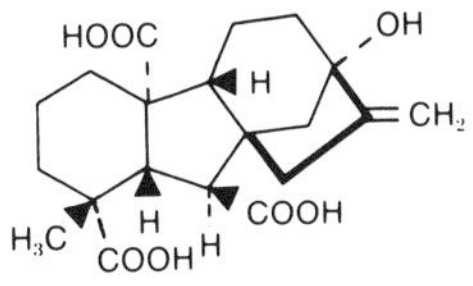

GA$_{17}$

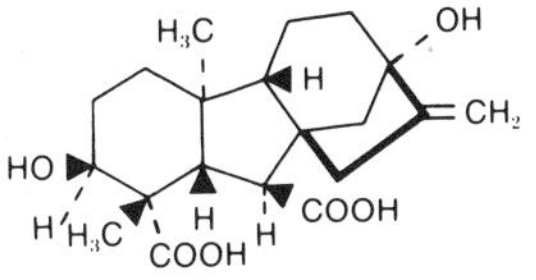

GA$_{18}$ (*Lupinus* gibberellin I)

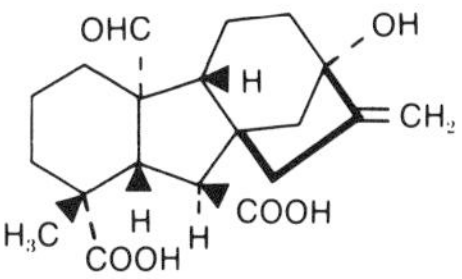

GA$_{19}$ (bamboo gibberellin)

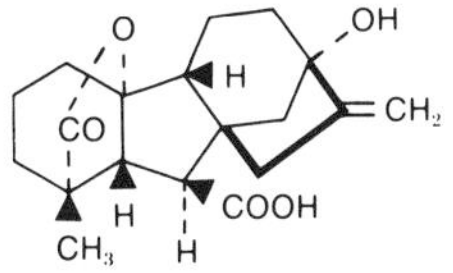

GA$_{20}$ (*Pharbitis* gibberellin I)

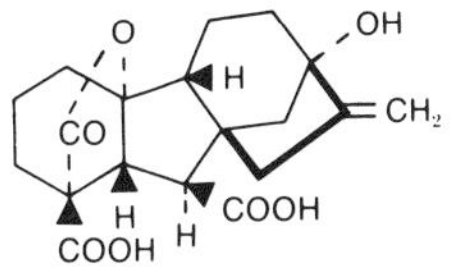

GA$_{21}$ (*Canavalia* gibberellin I)

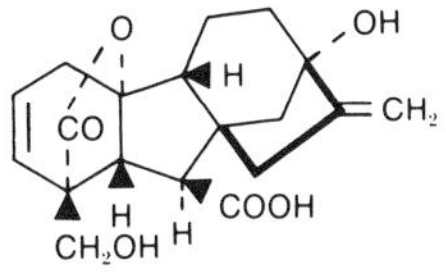

GA$_{22}$ (*Canavalia* gibberellin II)

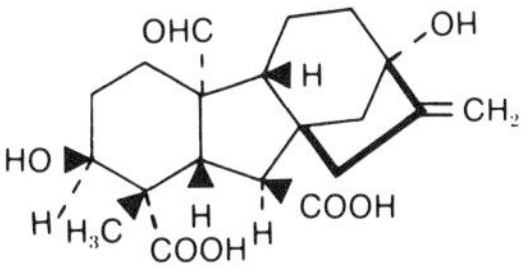

GA$_{23}$ (*Lupinus* gibberellin II)

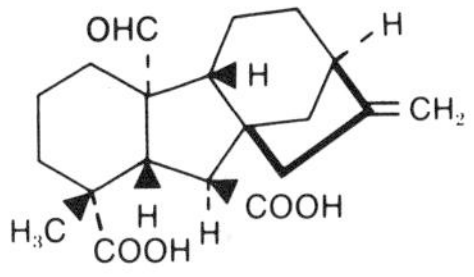

GA$_{24}$

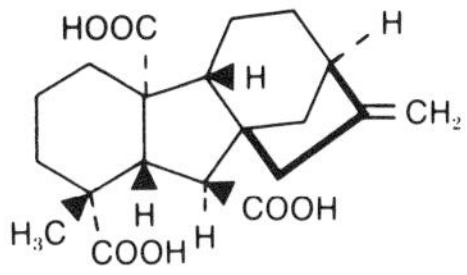

GA$_{25}$

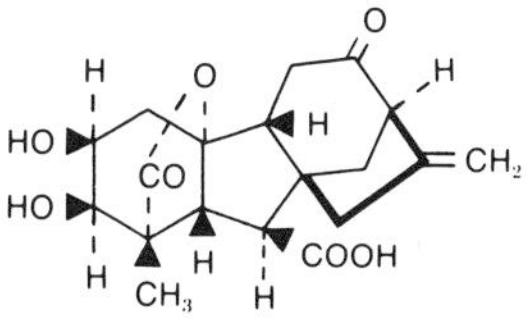

GA$_{26}$

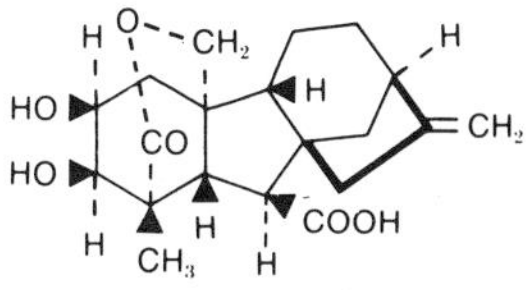

GA$_{27}$

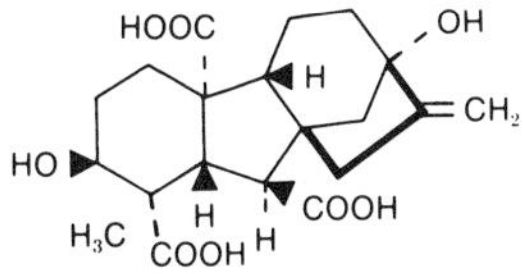

GA$_{28}$

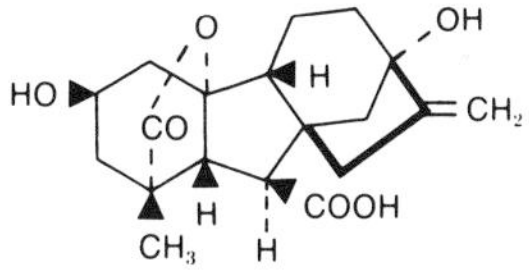

GA$_{29}$

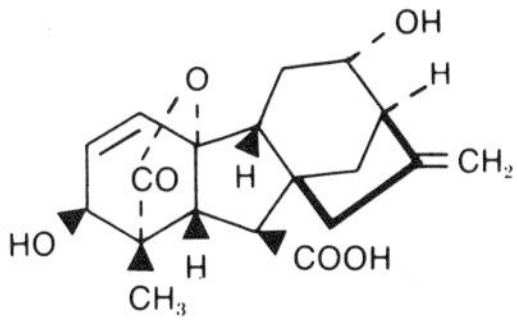

GA$_{30}$

FIGURE 3-2 (*Continued*)

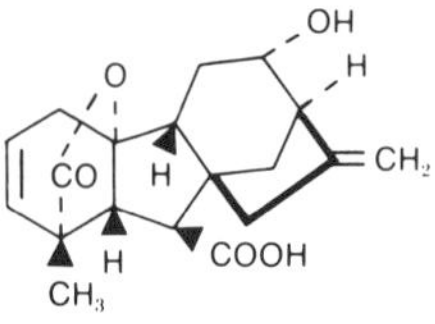

GA$_{31}$ 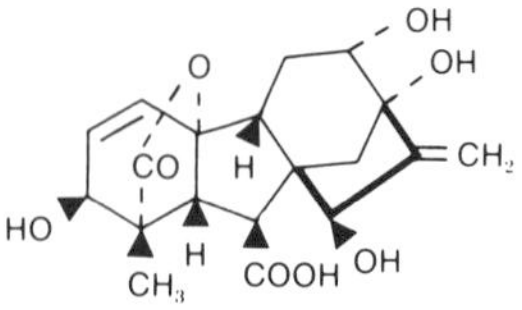   GA$_{32}$ 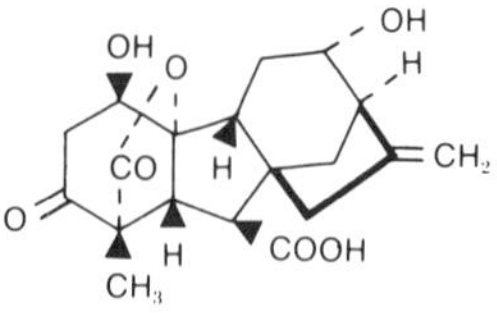   GA$_{33}$

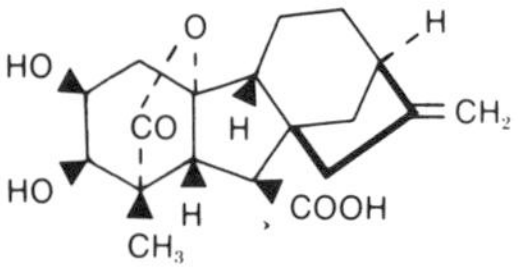

GA$_{34}$ 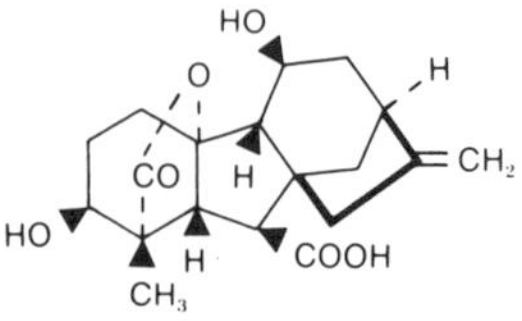   GA$_{35}$ 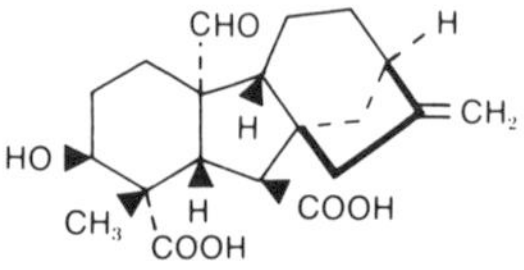   GA$_{36}$

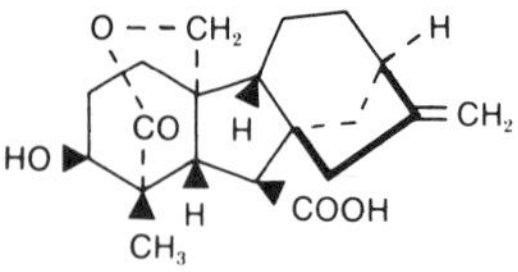

GA$_{37}$

responding to that of other cyclic diterpenes, a class to which all gibberellins belong.

The major differences in the known gibberellins are that (1) some gibberellins have nineteen carbon atoms and others have twenty and (2) hydroxyl groups may be present or absent in positions 3 and 13 (*ent*-gibberellane numbering system). All the gibberellins with nineteen carbon atoms are monocarboxylic acids, have the COOH group in position 7, and have a lactone ring. The structures of many of the gibberellins are almost identical (see Fig. 3–2); for example, GA$_1$ is the same as GA$_3$ except that the double bond between carbon atoms 1 and 2 is absent in GA$_1$ (*ent*-gibberellane numbering system). It is highly unlikely that gibberellins will be synthesized commercially in the near future because GA$_3$, for example, has 8 asymmetric carbon atoms, which means that there are 256 possible isomers. Plant scientists will probably have to depend on the fungus to make their gibberellin for a long time to come.

Some compounds have been found that are precursors of gibberellin and that have shown weak gibberellin activity in several

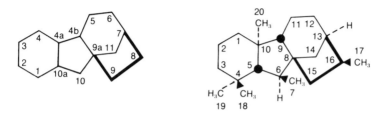

FIGURE 3-3

Structural formulas of the gibbane (*left*) and *ent*-gibberellane (*right*). Heavy lines and wedges indicate bonds lying above the plane of the ring system; broken lines indicate bonds lying below this plane. (After A. Lang, 1970.)

gibberellin bioassays. These compounds are kaurene (Fig. 3–4) or its oxidation products, kaurenol, kaurenal, and kaurenoic acid; their activity probably depends on their conversion to a gibberellin within the test plant as, for example, in wild cucumber, *Echinocystis macrocarpa* (West *et al.*, 1968).

Some compounds that have been extracted from plants show appropriate biological activity but do not possess the gibbane or gibberellane structure and therefore are not classified as gibberellins. Tamura and his colleagues in Japan (Kato *et al.*, 1966) have tested the metabolic products of many fungi. They found that the fungus *Helminthosporium sativum* growing on wheat produces a substance, helminthosporol (Fig. 3–4), that is active on sheath elongation of rice seedlings and on barley endosperm but inactive on dwarf maize and pea.

Steviol is another naturally occurring compound that does not have the gibbane skeleton (Fig. 3–4) but does have some of the biological properties of gibberellins (Ruddat *et al.*, 1963) and may serve as a precursor of gibberellins. A host of unknown compounds in plant tissues also have appropriate gibberellin activity, but until these compounds are definitely chemically identified they are best referred to as "gibberellin-like" substances. Like the auxins, gibberellins are also found in plants in combined form; the β-glucoside of GA_3 has been found in several plants (Murakami, 1961).

CYTOKININS

Cytokinins are naturally occurring or synthetic substances that induce cell division in certain excised plant tissues in the presence of auxin. In their activity they resemble kinetin, the first cytokinin discovered.

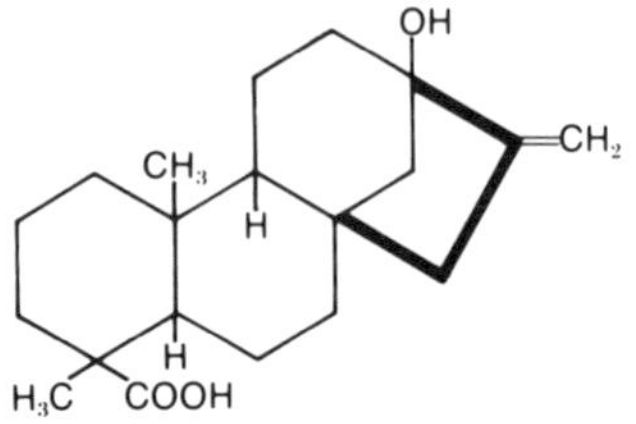

Steviol

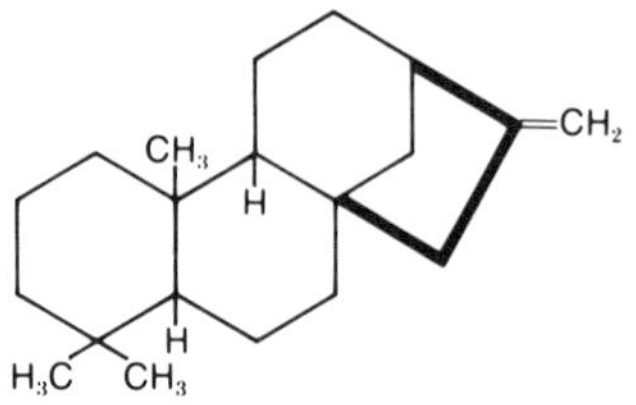

Kaurene

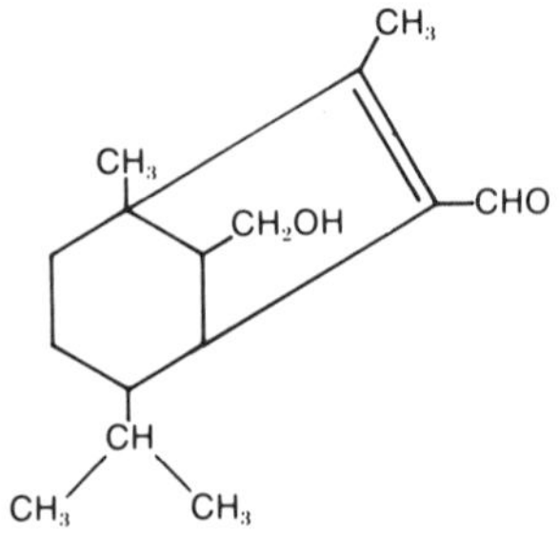

Helminthosporol

FIGURE 3-4

Structural formulas of some compounds that lack a gibbane skeleton but evidence gibberellin activity.

Naturally Occurring Cytokinins

OCCURRENCE. Extracts containing compounds evidencing cytokinin activity have been obtained from more than forty plant species (Letham, 1967). Relatively high levels of the compounds have been found mainly in tissues in which there is active cell division, such as germinating seeds and young fruitlets. For this reason cytokinins are regarded as regulators of cell division. Cytokinin-like substances have been found in yeast extracts and in micro-organisms. Cytokinins have been found in germinating seeds of barley, lettuce, and pea. Developing fruits of quince, apple, plum, peach, pear, and tomato have yielded cytokinins (Letham, 1967). Cytokinins have also been found in seeds of lupine (Koshimizu et al., 1967), watermelon, and pumpkin (Gupta and Maheshwari, 1970). Cytokinin activity usually is correlated with the location of regions of active cell division and the periods of active cell division.

Cytokinins have been found in the bleeding sap of several species, including tobacco, grape, and sunflower. Chromatography has revealed that sap usually contains more than one cytokinin; five different compounds were found in grape sap (Loeffler and van Overbeek, 1964) and at least two were found in sunflower (Kende, 1964). Cytokinins probably are synthesized in the root tips and move through the xylem to the leaves, where they are important in metabolism and senescence.

The nicotinamide derivatives, which have been found in *Vinca rosea* and in dividing cells of tobacco and cactus, may comprise a new class of naturally occurring cell-division-promoting substances, according to Wood and his colleagues (1969). Wood *et al.* suggest that nicotinamide derivatives, rather than purine cytokinins, may promote cell division in higher plants by inducing the synthesis of substances that are directly involved.

CHEMICAL NATURE.   The first cystalline cytokinin was extracted from *Zea mays* seeds and was termed "zeatin" (Letham, 1967). Zeatin, which has been synthesized (Fig. 3–5), is an extremely active compound. (It is approximately ten times more active than kinetin.) A compound whose chemical structure is similar to that of zeatin, 2iP (Fig. 3–5), has been found in cultures of *Cornynebacterium fascians,* a bacterium that causes an "overgrowth" type of development in certain higher plants. This compound has also been isolated as the ribonucleoside from the tyrosine and serine transfer RNA's of yeast and also from soluble RNA hydrolysates of peas, spinach, yeast, and various types of liver (see Helgeson, 1968).

Letham (1968) has reported on the isolation of several new cytokinins in sweet corn extracts. Miller (1968) found that zeatin, ribozeatin, and a nucleotide of zeatin are produced by the fungus *Rhizopogon roseolus.* Several other naturally occurring cytokinins have been discovered, and these findings are reviewed by Skoog and Armstrong (1970).

Synthetic Cytokinins

After Miller and colleagues (1956), working in Professor Folke Skoog's laboratory, isolated kinetin from an aged DNA preparation, it was identified chemically as 6-furfurylaminopurine (Fig. 3–5). Research was then undertaken to discover more-potent compounds. A host of synthetic 6-substituted purines have been studied and many have been found to be more active than kinetin (for example, the compound BA, synthesized by the Shell Development Company; see

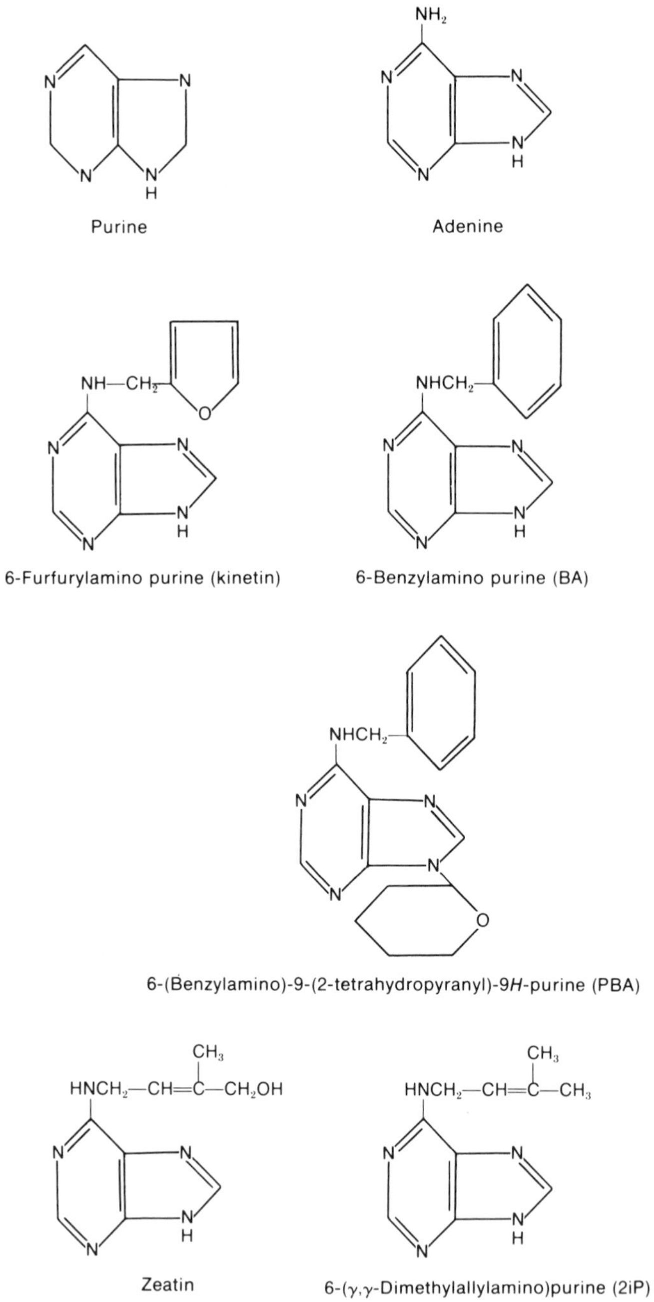

FIGURE 3-5

Structural formulas of purine, adenine, and some cytokinins. Kinetin, BA, and PBA are synthetic; zeatin and 2iP are naturally occurring.

Fig. 3–5). A wide range of substituents can replace the furan ring, thus giving the compounds a wide range of fat solubility. Adenine alone shows some cytokinin activity, although kinetin is approximately 30,000 times more potent. Van Overbeek has referred to kinetin as "adenine with a handle."

Another compound produced by the Shell Development Company, PBA, was found to be even more active in higher plants than BA. The structure of PBA is similar to that of BA except that the hydrogen on nitrogen 9 is replaced by a nonpolar ring structure of tetrahydropyran (Fig. 3–5). The increased activity is probably a result of the compound's greater solubility and penetrability into the plant tissue and to its greater mobility within the plant.

A wide range of chemical types, including several types of aromatic and nonaromatic compounds, are reported to evidence cytokinin activity. Skoog et al. (1967) tested sixty-nine compounds, mostly purine derivatives, for growth promotion and organ formation in the tobacco pith bioassay. Forty-three of these compounds were synthesized; thirteen of the forty-three were being reported for the first time. A total of fifty-one of the compounds, which included a wide range substitutents, showed cytokinin activity.

INHIBITORS

Naturally Occurring Inhibitors

Plants usually contain many inhibitory substances. Such processes as seed germination, suppression of shoot growth, and bud dormancy are controlled at least in part by inhibitors. The importance of inhibitors was revealed by Hemberg in his studies on dormant buds of potatoes (1949a) and ash (1949b), which showed that dormant buds contain high levels of inhibitors but that at termination of rest the concentration rapidly decreases.

Since 1949 inhibitors have generally been classified by scientists as one of the major groups of plant regulators. Inhibitors have been found in almost all plant parts. Inhibitors in stylar tissue can retard pollen tube growth. Such inhibitors as juglone (Fig. 3–6) can be excreted by the roots or leaves of some plants or can appear as a result of the decomposition of plant parts.

Naturally occurring growth inhibitors comprise a very diverse group of compounds although the most common are aromatic organic substances (Fig. 3–6). Many of them are phenyl compounds, including phenols, benzoic acids, and other compounds with longer chains, such as cinnamic and caffeic acid. Gallic acid and shikimic acid are derivatives of benzoic acid. Gallic acid is commonly found in ripening fruits. Ferulic acid and *p*-coumaric acid are cofactors for IAA oxidase. Other compounds include depsids between two acids, such as chlorogenic acid. The plant phenols (especially the flavonoid inhibitors, such as naringenin) usually occur naturally as glycoside esters with sugars or as depsids. Lactone inhibitors include coumarin, scopoletin, and aesculin.

In extracts of many plant tissues an inhibitory area (from R_f 0.7– 0.9, called inhibitor β) appears on paper chromatograms developed with a mixture of isopropanol, ammonia, and water (Bennet-Clark and Kefford, 1953). This inhibitory area consists of a mixture of substances whose complete composition remains to be revealed, but ABA is probably an important component (Millborrow, 1967). The inhibitor β complex also includes phenolic substances.

ABA was isolated from young cotton fruit (*Gossypium hirsutum*) by a group led first by Carns and later by Addicott (Addicott *et al.*, 1968); its structure was determined in 1965 by Ohkuma *et al.* (Fig. 3–7). Other investigators worked with dormancy-inducing substances in deciduous trees. A group led by Wareing and Cornforth isolated ABA from *Acer* (Cornforth *et al.*, 1965). Addicott and Lyon (1969) state that during the same period a program was initiated by Van Steveninck (1959) and continued by Rothwell and Wain (1964) that led to the isolation of ABA in lupines.

The naturally occurring enantiomer of ABA is S-abscisic acid, and the synthetic racemic substance is RS-abscisic acid. Both compounds are highly and about equally active. ABA has been isolated from the leaves, stems, rhizomes, tubers, buds, pollen, fruits, embryos, endosperm, or seed coats of more than thirty plant species, including such diverse plants as potato, bean, apple, avocado, fern, willow, linden, rose, peach, coconut palm, and various grasses. The compound is usually present in mature and senescing tissues, but it has also been found in young fruits and leaves.

Several other naturally occurring substances are related to ABA but evidence less activity (Addicott and Lyon, 1969). These substances include 2-*trans*-abscisic acid, phaseic acid, 2-*trans*-phaseic acid, (+)-abscisyl-β-D-glucopyranoside, and theaspirone (Fig. 3–7). Phaseic acid occurs in seeds of *Phaseolus multiflorus,* the abscisyl glucoside occurs in fruit of *Lupinus luteus,* and theaspirone is an important flavor constituent in tea leaves (Addicott and Lyon, 1969).

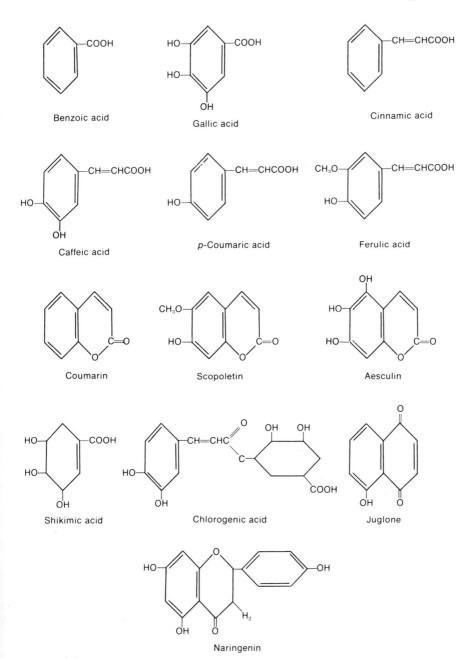

FIGURE 3-6
Structural formulas of some naturally occurring inhibitors.

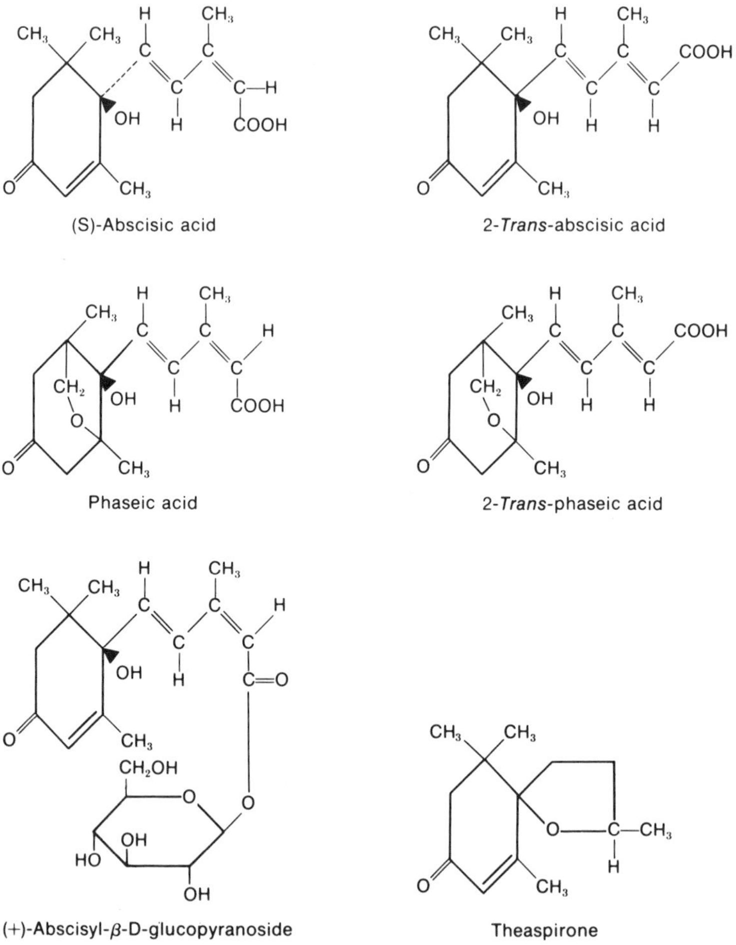

FIGURE 3-7
Structural formulas of ABA and some naturally occurring related substances. (After F. T. Addicott and J. L. Lyon, 1969.)

The 2-*trans* isomer of ABA has been detected in cotton and strawberries, and 2-*trans*-phaseic acid has been discovered in cotton. However, data on the latter compound are very meager.

Synthetic Inhibitors

There are several synthetic inhibitors that are of importance in agriculture (Fig. 3–8). Some are utilized to eliminate problems of overgrowth and others are used to stimulate flower initiation, retard senescence, and control other plant processes.

MALEIC HYDRAZIDE. One of the first growth inhibitors to reach the market was MH (Schoene and Hoffman, 1949; Naylor and Davis, 1905). This compound is widely used to prevent sprouting of potato tubers and onions and suckering of tobacco and is also used to some extent to control excessive growth in grasses. MH is available commercially as the diethanol amine salt.

PLANT GROWTH RETARDANTS. Growth retardants comprise an important group of inhibitors that were discovered in the 1950's and 1960's. Subapical meristematic activity, which is responsible for stem elongation, is retarded by growth retardants, usually without similarly affecting the apical meristem.

The compound Amo-1618, a quaternary ammonium compound based on thymol, was found to be the most active chemical of many quaternary compounds including picolinium, morpholinium, piperidium, pyridinium, quinolinium, and quinaldinium (Wirville and Mitchell, 1950). The structure of Amo-1618 can be divided into the following parts for discussion: the carbamate nitrogen, the terpene ring, the quaternary nitrogen, and the halide salt (Fig. 3–8). The reduction of any portion of this basic molecule removes activity (Cathey, 1964). Because Amo-1618 is very expensive to synthesize, its availability has been restricted.

Phosfon-D is the most active structure of a group of quaternary compounds that contain a phosphonium cation. The tributyl quaternary phosphonium cation is required for activity; for optimal activity the benzene ring should have a substituent in the 4 position that is small, nucleophilic, and nonionizable. Phosfon-S refers to the ammonium analogue of Phosfon-D.

The compound CCC has a quaternary structure and is an analogue of choline. The bromide and the chloride salts are active, and the trimethyl quaternary ammonium cation is necessary for activity.

SADH belongs to the group of succinic acids. Unlike the other plant growth retardants discussed here, it contains no benzene ring, quaternary ammonium, or phosphonium cation, nor any substituents that are small, nucleophilic, and nonionizable. SADH is a free, ionizable acid containing a C-C-N-N system.

There are also several other groups of lesser importance, including the nicotiniums and hydrazines (Cathey, 1964). One of the most

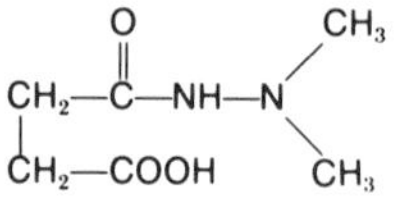

Succinic acid-2,2-dimethylhydrazide
(SADH; B-995; B-9; Alar)

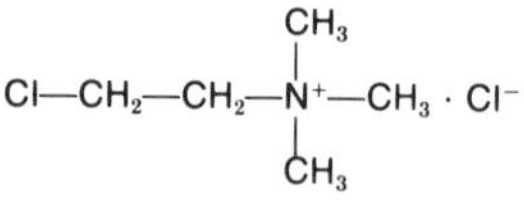

(2-Chloroethyl)trimethyl ammonium
chloride (CCC; Cycocel)

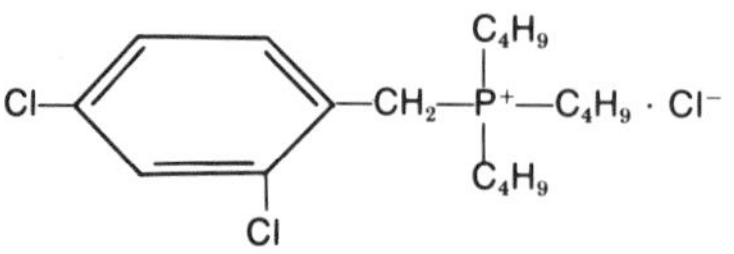

2,4-Dichlorobenzyltributylphosphonium
chloride (Phosfon-D)

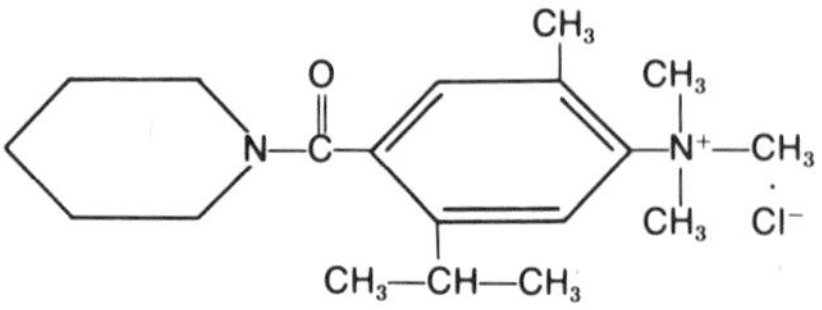

Ammonium (5-hydroxycarvacryl)trimethyl chloride
piperidine carboxylate (Amo-1618)

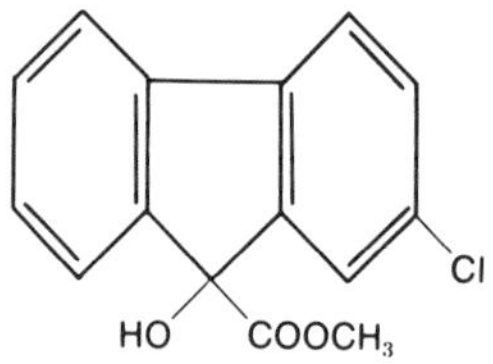

Methyl-2-chloro-9-hydroxyfluorene-9-
carboxylate (IT 3456, a morphactin)

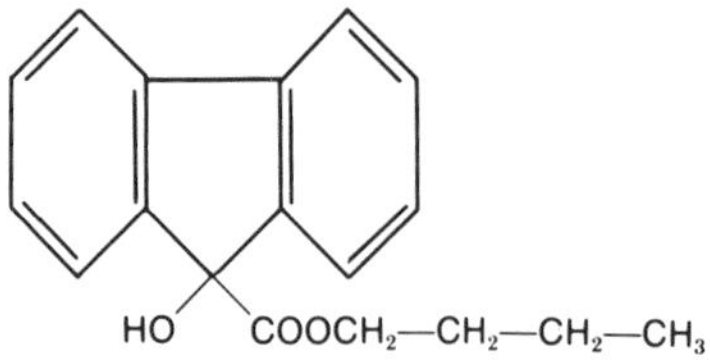

N-Butyl-9-hydroxyfluorene-9-carboxylate
(IT 3233, a morphactin)

HOCH₂CH₂NHNH₂

β-Hydroxyethylhydrazine (BOH)

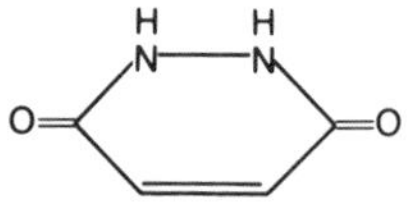

1,2-Dihydro-3,6-pyridazinedione (MH; maleic hydrazide)

FIGURE 3-8

Structural formulas of some synthetic inhibitors. SADH, CCC, Phosfon-D, and Amo-1618 are important plant growth retardants.

important hydrazines is BOH (Fig. 3–8), which also contains the C-C-N-N system found in SADH.

MORPHACTINS. Some of the newer inhibitors are derivatives of fluorene-9-carboxylic acid. The biological activity of fluorene-9-carboxylic acid was first described by Wain in 1958, and the activities of several derivatives of this compound were reported by Schneider in 1964. Since these chemicals produce morphological changes and a striking suppression of growth in many plant species, they were termed "morphactins" (Schneider, 1970). The two most readily available and generally used morphactins are methyl-2-chloro-9-hydroxy-fluorene-9-carboxylate (IT 3456) and *n*-butyl-9-hydroxyfluorene-9-carboxylate (IT 3233)—see Figure 3–8.

ETHYLENE

It is probable that ethylene is produced in all living tissues. Ripening fruits are particularly rich sources of ethylene. *Penicillium digitatum* is a prolific producer of ethylene. Lemons in storage that are infected with this fungus cause degreening of adjacent uninfected fruit.

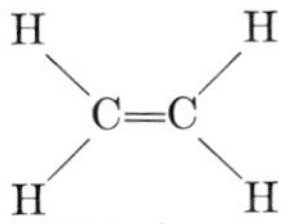

$$H_2C=CH_2$$

Ethylene

Ethephon has been introduced as an agent that apparently will break down in plant tissue, thus releasing ethylene close to a site of action.

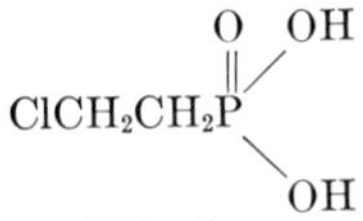

$$ClCH_2CH_2P(=O)(OH)_2$$

Ethephon

Much agricultural interest has been stimulated by ethephon because it can be applied by ordinary agricultural techniques and because its effects often are similar to those of ethylene.

Supplementary readings on the general topic of occurrence and chemical nature of plant growth substances are listed in the Bibliography: Audus, L. J., 1959; Cathey, H. M., 1964; Leopold, A. C., 1964a; and Wilkins, M. B., ed., 1969.

Chapter 4

BIOLOGICAL EFFECTS
AND MECHANISM
OF ACTION

This chapter will discuss some of the general biological effects on plants produced by each class of growth substances, as well as their mechanism of action. It must be stressed that the responses of a plant or plant part to a growth substance may vary with the specie and variety. Even a given variety may respond differently under different environmental conditions. The literature frequently describes the positive results obtained by one investigator and the negative results obtained by another, using the same plant material. Both investigators may have performed accurate work; the varying results must be explained in part by variations in each plant's physiological or developmental stages, differences in the environmental conditions under which each plant is growing, and differences in nutritional status and in absorption and translocation of growth substance. Variations in results may also doubtlessly be caused by fluctuating hormonal contents of plants at different physiological stages, as well as by the manner in which naturally occurring growth substances interact with applied growth substances. When a general rule is made regarding the effect of a growth regulator on plants, exceptions are almost always found. One of the few laws of hormonology that still stands after nearly a half century is Went's statement, *Ohne Wuchstoff, kein Wachstum* ("Without growth substance, no growth").

The biological effects of the different hormone classes often overlap; for example, auxins, gibberellins, and cytokinins can cause cell division and cell enlargement, as well as parthenocarpic fruiting. Besides having common biological effects, each group of growth substances also causes some distinctive responses on which bioassays, specific for one class of hormone, are based.

Research on the mechanism behind auxin action has been pursued since the 1930's. Numerous theories have been proposed, but the exact mechanisms still remain obscure. It is not surprising, then, that the mechanism behind the growth substances that were discovered at about the same time as or subsequent to the auxins also remain unexplained. However, advances in this field have been rapid, especially in recent years.

AUXINS

Biological Effects

Auxins play an important role in the cell enlargement of stems and coleoptiles. The discovery of auxin evolved from studies of the tropistic curvatures of coleoptiles, and the first auxin bioassay, the *Avena* curvature test, utilized decapitated *Avena* coleoptiles. When an agar block containing an auxin is placed on one side of a decapitated coleoptile, the auxin moves downward and stimulates cell elongation, thus causing a (negative) curvature away from the block. For a compound to be classified as an auxin it should cause a negative *Avena* curvature in the manner of IAA. Segments of coleoptile or stem also elongate when immersed in auxin solution. Other auxin bioassays are based on this effect.

Epinasty, or the downward bending of leaves caused by excessive growth of the morphologically upper side, results two hours or less after auxins in sufficient concentrations are applied to plants. Twisting of stems and development of formative effects (stunting, cupping, and development of abnormal venation) on newly developing leaves are other symptoms of auxin treatment (Fig. 4–1). Growers of commercial crops that are sensitive to 2,4-D, such as cotton and grape, often become alarmed at the discovery that even a few leaves show such formative effects. However, if only a few such leaves are present, usually no reduction in crop yield occurs.

FIGURE 4-1
Severe epinastic effects (bending of stem, thickening of hypocotyl, and abnormal leaf venation) observable in red kidney bean plant 6 weeks after spraying with 2,4-D.

Many plants are exceedingly sensitive to 2,4-D. Kasimatis *et al.* (1968) found that a drop containing 0.0001 μg of 2,4-D placed on a young leaf of 'Tokay' grape will later cause formative effects (Fig. 4–2). Thus 0.8 mg of 2,4-D, properly placed, could cause a leaf malformation on every vine in the approximately 20,000 acres of 'Tokay' grapes grown in California.

Auxins also stimulate cell division; for example, auxins often stimulate callus development, from which root-like outgrowths can occur.

Auxins are very effective for initiating root formation in many plant species. This response was the basis for the first practical application of growth substances in agriculture.

Auxins can initiate flowering (for example, in pineapple) and induce fruit set and development in some species. Fruit set, especially in species with many-seeded fruits, such as pepper and cucurbits, is often increased by auxins. The application of auxin to young developing fruits increases their size. The maturation of some fruits, such as fig, is also hastened.

Apical dominance is often enhanced by auxin application. Auxin also stimulates cambial activity and can affect abscission phenomena. It can effect the differentiation of abscission layers but often retards their development.

FIGURE 4-2

Crumpling and abnormal venation of leaves of 'Tokay' grape that were treated with a drop of water containing 0.0001 μg of 2,4-D when young. Such symptoms, usually confined to a limited area of the leaf, commonly occur in vineyards throughout the world. (After A. N. Kasimatis, R. J. Weaver, and R. M. Pool, 1968.)

Mechanism of Action

AUXINS AND CELL ELONGATION. Over the years numerous theories have been advanced to explain the primary mechanism behind the action of auxin in promoting cell extension, but to date none is entirely satisfactory. One of the earliest theories, that auxin increases the plasticity of the cell wall (Heyn, 1931), is still the most satisfactory, but more work is required to reveal the exact mechanisms involved. When extensibility of the wall is increased, the wall pressure around the cell decreases and the turgor pressure caused by osmotic forces in the vacuolar sap causes water to enter the cell, resulting in cell enlargement.

Plasticity is a nonreversible wall deformation that probably is caused by the breaking of cross-links between the cellulose microfibrils of the cell wall (Galston and Davies, 1969). The increase in cell size occurs in two stages. First there is a loosening of the cell wall (a process that requires the presence of auxin and oxygen), followed by an uptake of water and an expansion of the wall.

It is becoming increasingly clear that auxins, as well as other hormones, may work by controlling the type of enzyme produced in the cell. Thimann (1969) suggested that auxin may function by activating a messenger-type of RNA that induces the synthesis of specific enzymes. These enzymes would cause insertion of new materials into the cell wall, resulting in its extension.

A brief explanation of the mechanism of protein synthesis at the molecular level is that DNA (which contains the genetic information) serves as a template for the synthesis of messenger RNA, which moves into the cytoplasm and to the ribosomes, where it imparts its information and controls protein synthesis. Transfer RNA molecules collect the amino acids and carry them to the ribosomes, where they are assembled into protein according to the directions of the messenger RNA. It has been known for several years that the first response to auxin is not growth but some process that later affects the growth process (see Galston and Davies, 1969).

In many plants and plant parts auxin induces and promotes the synthesis of RNA and proteins (Sacher and Salminen, 1969; Soleimani et al., 1970). Such synthesis may be a prerequisite for auxin-induced growth. However, the occurrence of a one-hour lag between application of auxin to pea-stem sections and increase in RNA would seem to disprove this theory. (On the other hand, after one hour the rate of cell elongation is already maximal.) Although the effect of auxin on the synthesis of RNA seems to be quantitative and not qualitative, a messenger RNA that is specific for auxin-induced growth may be present in cells: Upon application of auxin, the code of the messenger RNA is translated into protein (see Key, 1969). Perhaps two RNA's are present in the cell—one that is active in the cell before auxin application and another that is associated with the actual growth response (Galston and Davies, 1969).

Enzymes may be required for auxin-induced wall loosening and for auxin-induced cell elongation. The nature of these enzymes is unknown, but auxin application does result in changes in protein patterns.

Some evidence obtained by the German scientists Nissl and Zenk (1969) casts doubt on the theory that auxins induce or promote protein synthesis. These scientists studied cell elongation of *Avena* coleoptiles using a flow chamber. By increasing the temperature and applying high concentrations of IAA, they found that the lag in the growth response resulting from auxin application could be gradually shortened to zero.

FIGURE 4-3

Red kidney bean plants 6 weeks after being sprayed with growth regulators at the primary leaf stage. *Left to right:* unsprayed control; plant sprayed with GA₃ at a concentration of 100 ppm; plant sprayed with SADH at a concentration of 2,000 ppm; and plant sprayed with 2,4-D at a concentration of 10 ppm. GA₃ greatly stimulated growth and SADH retarded it. Plant sprayed with 2,4-D exhibits phytotoxic effects.

GIBBERELLINS

Biological Effects

The most striking effect of spraying plants with gibberellin is the stimulation of growth (Fig. 4-3). Stems of sprayed plants usually become much longer than normal (Stowe and Yamaki, 1959). Growth is stimulated in the younger internodes and tissues, and frequently the length of the individual internode is increased while the number of internodes remains unchanged. A temporary lightening of the

FIGURE 4-4

Effect of gibberellin on flower formation in an
early variety of carrot. *Left,* control (received
neither cold nor gibberellin); *center,* received no
cold, but was treated with gibberellin (10 µg
daily for approximately 4 weeks); *right,* received
6 weeks of cold but no gibberellin. (After A.
Lang, 1957.)

leaves of many treated plants often is associated with the increase in
leaf area, but normal green color returns within ten days.

Gibberellin can induce flowering in many species requiring cold
temperatures, such as carrot (Fig. 4-4), endive, cabbage, and turnip.

The application of gibberellin to stems produces a pronounced
increase in cell division in the subapical meristem (Sachs *et al.,*
1960) (Fig. 4-5) and causes bolting in rosette plants. The rapid
growth that occurs is a result of both the greater number of cells
formed and an increased elongation of the individual cells.

One of the most remarkable effects of gibberellin is that occurring
in dwarf plants. When some dwarf varieties, such as dwarf pea

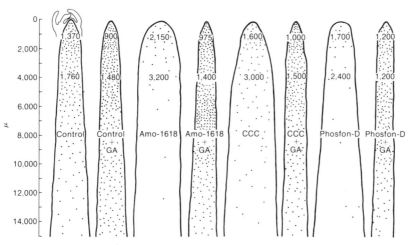

FIGURE 4-5

Density and distribution of mitotic activity in the pith tissue of chrysanthemum stems treated with Amo-1618, CCC, or Phosfon-D in the presence or absence of added gibberellin. The heavy lines are the vascular boundaries; the thin lines surrounding the control plant are outlines of the apex and youngest leaf primordia. The vertical scale is in microns (1,000 microns = 1 mm). Each dot represents one mitotic figure. The numbers at 1 mm and 4 mm below the apex are the transverse diameters of the pith tissue in microns at these levels. Mitotic activity in the retardant-treated stems is greatly reduced and transverse growth is increased; in the gibberellin-treated plants mitotic activity is greatly increased (generally 6 mm and more below the apex) and transverse growth is reduced. (After R. M. Sachs and A. M. Kofranek, 1963.)

(Brian and Hemming, 1955) and certain dwarf maize (Phinney, 1956), are treated with gibberellin they grow to normal height (Fig. 4-6). Since as little as 0.001 μg suffices to increase stem elongation, some dwarf varieties are of value for gibberellin bioassays.

Gibberellin can terminate rest in the seeds of many species. In early work, the seeds of some species were not affected by exogenous gibberellin; subsequent research indicated that the cause frequently was the failure of the substance to penetrate the seed coat.

In many plants apical dominance is enhanced by treatment with gibberellin. Some bushy dwarf plants grow with a single stem after such treatment.

Gibberellins increase the size of many young fruits, such as grape and fig. The fact that exogenous gibberellins can increase the berry size of seedless grape two- or threefold is the basis of an important commercial practice. In such plants as grasses and celery, gibberellin application results in greater yield increases than that produced by untreated plants.

FIGURE 4-6

Effect of spraying gibberellin at a concentration of 100 ppm on growth of dwarf-5 corn. *Left,* untreated dwarf corn; *center,* normal corn sprayed with gibberellin; *right,* dwarf corn sprayed with gibberellin. Photographed 6 weeks after spraying.

Some plants can be stunted as a result of virus disease. In some of these diseases, such as sour cherry yellows, the effects of the virus can be overcome by gibberellin.

Mechanism of Action

GERMINATING SEEDS. One of the best-known examples of enzyme induction by hormones is the gibberellin-induced production of α-amylase in barley aleurone. GA_3 can replace an α-amylase-inducing factor that is produced by germinating barley seeds. A naturally occurring gibberellin is produced by the barley embryo and translocated into the aleurone layers of the endosperm where syntheses of enzymes occur. These enzymes, including amylases, proteases, and lipases, rapidly break down the cell walls of the endosperm and subsequently hydrolyze the starches and proteins, thus liberating the nutrients and energy needed for embryo development. It has been demonstrated that GA_3 causes the de novo synthesis of α-amylase in the aleurone cells (Filner and Varner, 1967). Enzyme activity resulting from gibberellin thus is not caused by a liberation of enzymes from a bound form but rather by increased activity in the cells as a result of the formation of new enzymes (see Marcus, 1971). Jacobsen *et al.* (1970) found that the addition of GA_3 to isolated aleurone layers of barley resulted in the production of four amylases.

Gibberellin probably causes changes at the gene level that, in turn, stimulate syntheses of enzymes in the cell (Osborne, 1965). Gibberellin results in the stimulation of RNA synthesis in the aleurone layers, which may be required for expression of the gibberellin effect (Varner and Chandra, 1964). One theory is that gibberellin is related to the DNA-directed synthesis of the messenger RNA in the nucleus. It is now believed that gibberellin modifies the RNA produced in nuclei and thus can exert its control over cell elongation as well as other growth and developmental activities in plants.

GIBBERELLINS AND CELL ELONGATION.   The role of gibberellins in cell elongation is still obscure, but several attractive theories have been advanced. Gibberellin may cause elongation by the induction of enzymes that weaken the cell walls (MacLeod and Millar, 1962). Treatment with gibberellin induces formation of proteolytic enzymes that would be expected to release tryptophan, a precursor of IAA (van Overbeek, 1966). Gibberellin frequently increases auxin content. Gibberellins may also transport auxins to their site of action in plants (Kuraishi and Muir, 1963). Another mechanism by which gibberellins might stimulate cell elongation is that the hydrolysis of starch resulting from the production of gibberellin-induced α-amylase might increase the concentration of sugar, thus raising the osmotic pressure in the cell sap so that water enters the cell and tends to stretch it. Another hypothesis is that gibberellins stimulate the biosynthesis of polyhydroxycinnamic acids (Kögl and Elema, 1960). The latter compounds then inhibit IAA oxidase, thus promoting auxin-mediated processes in plants by reducing the amount of auxin destroyed by the enzyme.

CYTOKININS

Biological Effects

Soon after cytokinins were discovered in the 1950's, their numerous diverse effects became apparent. Two striking effects of cytokinins are the induction of cell division and the regulation of differentiation in excised tissues. Extremely low concentrations (5×10^{-11} M) of the cytokinin zeatin induce cell division in tobacco pith and carrot-phloem explants (Letham, 1969); without zeatin or another cytokinin, only slight growth of these tissues occurs. Cytokinin is required for both initiation and continuation of cell division. Availability of cytokinin has made possible the culturing of many tissues; some of these tissues grow actively in vitro with the addition of only a few required organic compounds—sucrose, thiamine, myoinositol, and auxin, as

FIGURE 4-7

A radish seedling. The cotyledon on the right was painted with a solution of the synthetic cytokinin PBA (100 mg/liter). Stimulation of cotyledon development was caused mainly by promotion of cell enlargement. (After D. S. Letham, 1969.)

well as a cytokinin (Letham, 1967; Linsmaier and Skoog, 1965). In tissue-culture bioassays for cytokinin, auxin must be added to the basal medium because these hormones exert a synergistic action on the induction of cell division and undifferentiated growth in tissue cultures.

In addition to promoting cell division, cytokinins influence the differentiation of the cultures. Cytokinins interact with auxins to give different expressions of growth. Skoog and Miller (1957) showed, in vitro, how a change in the cytokinin-auxin balance can affect growth expression. They found that tobacco pith culture requires both a cytokinin and an auxin for active growth. When the ratio of cytokinin to auxin is low, root development occurs, but when the ratio is high, both buds and shoots develop. When the ratio is intermediate, undifferentiated callus tissue develops. Such results strongly suggest that cytokinins may be important in the control of plant form as well as in the control of cell division.

Cytokinins also cause enlargement of some leaves (Fig. 4-7) and elongation of segments of etiolated stems. These responses are largely caused by cell enlargement (Letham, 1969).

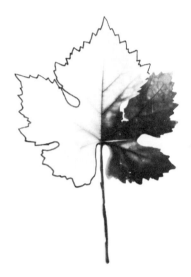

FIGURE 4-8

Autoradiograph showing the distribution of imported ^{14}C within a mature darkened grape leaf following application of 4.4×10^{-3} M BA to right half of the leaf. Left half was treated with water containing wetter. The shoot base was ringed and each of the two leaves immediately above the treated one was fed 25 μc $^{14}CO_2$. (After J. D. Quinlan and R. J. Weaver, 1969.)

Another effect of cytokinins is to delay senescence in plant tissues. If cocklebur leaves are floated on water, they lose their chlorophyll and after several days turn yellow. However, the addition of kinetin at concentrations of 1 to 5 ppm maintains the fresh appearance of the leaves by preserving much of the chlorophyll (Richmond and Lang, 1957). Cytokinins apparently have an antisenescence effect that is a result of a maintenance of protein and nucleic acid synthesis in the dark if they are first treated with kinetin (Miller, 1956). Figure 4–9 illustrates the effect of kinetin on termination of rest in cocklebur solution.

Cytokinins are also important in the mobilization phenomena of plants. When one portion of a leaf is treated with cytokinin, amino acids and other nutrients are attracted to the treated portion (see Letham, 1969). Similar results have been obtained with intact plants. Quinlan and Weaver (1969) showed that if a portion of a grape leaf is treated with cytokinin, translocation patterns in the vine are so altered that photosynthetic products move to the treated area (Fig. 4–8). Gibberellins and auxins, as well as cytokinins, also can often induce movement of assimilates to the treated area.

Cytokinins sometimes have an effect on germination. Seeds of certain varieties of lettuce that require light for germination germinate in the dark if they are first treated with kinetin (Miller, 1956). Figure 4–9 illustrates the effect of kinetin on termination of rest in cocklebur seeds.

The application of cytokinins to axillary buds of apple (Fig. 4–10) and shoots of apricot overcomes apical dominance (Chvojka *et al.,*

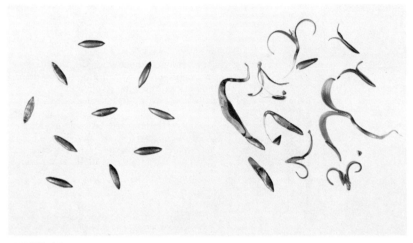

FIGURE 4-9

Application of kinetin solution to small, dormant seeds of cocklebur on right broke dormancy, allowing them to germinate. Untreated controls are on left. (After A. A. Khan, 1967a.)

1961; Williams and Stahly, 1968). Axillary buds on actively growing shoots produce spurs and lateral branches if treated with cytokinins. Rest of buds of apple (Williams and Stahly, 1968) and *Vitis vinifera* grape (Weaver, 1963) can be terminated by treatment with cytokinin.

Tuber formation on potato stolons is stimulated by cytokinins (Smith and Palmer, 1970). Kinetin-treated stolons showed more starch accumulation than untreated stolons before either showed any visible sign of tuber formation. Esashi and Leopold (1968) demonstrated that tuber development in *Begonia evansiana* is stimulated by BA.

Mechanism of Action

Cytokinins probably act at the molecular or gene level, but their mechanism of action is still unknown. It is now known that cytokinins may be incorporated into nucleic acids in the cell (Hall, 1968; Kovoor and Klämbt, 1968). The fact that many cytokinins have been isolated from RNA preparations indicates that cytokinins are involved in some way with the nucleic acids. They may act as gene derepressors (see Letham, 1969).

There is some evidence that in the transfer RNA for the amino acid serine, the cytokinin 6-(γ,γ-dimethylallylamino) purine is an "odd"

FIGURE 4-10
Spur development resulted when the lower 5 buds of the apple shoot on the right were treated with PBA. Shoot on the left is the control. (After M. W. Williams and E. A. Stahly, 1968.)

base immediately adjacent to the anticodon (the sequence of bases that carries the matching code for the messenger RNA, which, in turn, specifies the proper place for the amino acid in the new protein) (see Helgeson, 1968). Cytokinins have also been detected in serine transfer RNA from liver, yeast, and *Escherichia coli* RNA (Skoog *et al.*, 1966). The same base has been found adjacent to the anticodon of a tyrosine transfer RNA, but preparations of arginine, phenylalanine, glycine, valine, and alanine transfer RNA were found to contain no cytokinin activity. When the base sequence of alanine transfer RNA was determined, no cytokinin moiety was detected. This finding agrees with the lack of cytokinin activity in alanine transfer RNA (Holley *et al.*, 1965).

INHIBITORS

Inhibitors are a very diverse group of compounds and therefore have different biological effects on plants. Many theories have been advanced to explain their mechanisms of action.

Biological Effects

ABSCISIC ACID. This inhibitory hormone, which is widespread in the plant kingdom, interacts with growth promoters and thus has an important effect on growth phenomena. Since the availability of ABA is relatively limited, present knowledge of its diverse biological effects on crops is incomplete.

ABA seems to act as a general inducer of senescence, and foliage applications of ABA often cause senescent color changes in leaves (see Addicott and Lyon, 1969). Treatment of 'Temple' orange with ABA hastens coloring and onset of senesence (Cooper et al., 1968); treated potatoes became soft and senescent (El-Antably et al., 1967). In addition, early work showed that ABA accelerates abscission of petiole stumps of explants of various species. Subsequently it was shown that ABA promotes abscission of leaves in such intact plants as citrus (Cooper et al., 1968) and olive (Hartmann et al., 1968), as well as abscission of flowers and fruit in such plants as grape (Weaver and Pool, 1969).

ABA inhibits the growth of many plants and plant parts, as has been demonstrated in coleoptiles, seedlings, leaf disks, root sections, hypocotyls, and radicles (see Addicott and Lyon, 1969). Inhibition of shoot and leaf growth has also been shown to occur, but often several treatments with ABA are required because its effect persists for only a relatively short time.

The response of various species and cultivars to application of ABA shows great variation. In a comparative study using thirty-four soybean cultivars, marked differences occurred both in amount of inhibition of stem elongation and in senescence (Sloger and Caldwell, 1970).

Another biological effect of ABA is to prolong the rest of many seeds, such as cress and lettuce. The compound also inhibits the germination of seeds whose rest period has terminated, but this effect usually is ended if the seeds are washed in water to remove the ABA. Application of ABA also induces rest in the buds of certain species, including some deciduous fruit trees, citrus, and potato.

The finding that the amount of ABA increases in leaves during short days has stimulated interest in the effects of ABA on flowering. Flower initiation in some long-day plants grown during a noninductive photoperiod is enhanced by gibberellin; however, ABA has been shown to counteract this effect. One hypothesis is that ABA acts as a flowering inhibitor in leaves of long-day plants growing during short days. ABA can also induce flowering in some short-day plants growing under noninductive conditions. Some of these effects may be explained on the basis of growth retardation, which decreases the competition of vegetative parts so that more floral induction can occur.

PLANT GROWTH RETARDANTS. These compounds, which include SADH, CCC, Phosfon-D, and Amo-1618, retard stem elongation by preventing cell division in the subapical meristem, usually without similarly affecting the apical meristem (Sachs et al., 1960). However, inhibition of apical meristematic activity in apple as a result of SADH application has been demonstrated (Wilde and Edgerton, 1969). Plant growth retardants can induce plants having a tall growth habit to develop with a rosette type of growth. The effect of growth retardants is often just the opposite of that of gibberellins (Fig. 4–3).

Sachs and Kofranek (1963) have shown that Amo-1618, CCC, and Phosfon-D all inhibit subapical cell expansion and division in *Chrysanthemum morifolium*. 'Indianapolis Yellow' (Fig. 4–5). Compounds were added to the nutrient solution. GA_3 prevented inhibition of stem elongation in the presence of a growth retardant and maintained subapical meristematic activity at levels that were normal or greater than normal. None of the retardants inhibited transverse stem growth; rather they stimulated transverse cell expansion and division in the subapical tissues.

The time of flower initiation of many woody plants is greatly hastened by growth retardants. Large increases in the number of flowers and fruits have also been obtained in many herbaceous plants.

The effect of plant growth retardants on inhibition varies greatly with the chemical compounds and plant species. In a stud f fifty-five species of agronomic or horticultural interest, Amo-161c retarded the growth of only six species (Cathey and Stuart, 1961). Phosfon-D inhibited the growth of all species affected by Amo-1618, as well as twelve additional species, whereas almost all were affected by CCC.

Application of Amo-1618 does not stimulate flower initiation and in some plants causes a delay in flowering. However, the high cost of the compound limits the extent to which it can be tested on woody plants. The carvacrol isomer of Amo-1618 (Cardavan) is more effective in retarding the growth of poinsettias than is Amo-1618 (Cathey, 1959b).

Cathey (1969) tested the effectiveness of sixteen analogues of

SADH in controlling the growth of nineteen plant species. The most active compounds were SADH and N-pyrrolidinosuccinamic acid. The latter compound possessed greater activity than SADH during the summer, but the activity of the two was about equal during the winter. Active compounds were also produced when the two unsaturated dibasic acids, maleic and fumaric, were combined with the correct hydrazines.

CCC induces swollen root tips on seedlings of *Vitis vinifera* grape and on other plants grown from cuttings. It is probable that CCC acts on the root meristem to increase cytokinin production and that the latter compound (or compounds) cause the root tips to swell (Skene, 1970b). Increased amounts of cytokinin are also found in the plant sap as a result of application of CCC to the roots.

MALEIC HYDRAZIDE.   MH, one of the first inhibitors to be used commercially, has herbicidal effects when used at high concentrations. Because it is a general inhibitor of meristematic activity, it retards stem elongation and prevents leaf and flower initiation, as well as fruit set and enlargement. MH is used to prevent suckering of tobacco and sprouting of potato tubers and onions in storage. It has also been used to control excessive growth in trees, shrubs, and grasses.

MORPHACTINS.   Several morphactins are available, all of which are derivatives of fluorene-9-carboxylic acid, which has a fluorene nucleus. At high concentrations they are useful as weed killers. These compounds frequently are useful to control the growth of woody plants, but they have been known to cause foliar distortions, retardation of stem elongation, and the breaking of axillary buds.

Morphactins also have several interesting nonherbicidal properties. Khan (1967) demonstrated that applications of morphactin enable a plant to defy the effects of gravity and light. The roots and shoots of vertically growing plants usually orient themselves with respect to gravity: The roots are positively geotropic and the shoots are negatively geotropic. Higher plants respond to unilateral light by displaying positive phototropism. However, when seeds are soaked in morphactin at a concentration of 6×10^{-5} M before planting, both roots and shoots lose their capacity to respond to either gravity or light stimuli (Figs. 4–11, 4–12).

Morphactins also stimulate abscission of flowers and fruits. Because the firmness of attachment of fruit to the plant can be decreased by application of morphactins (Weaver and Pool, 1968),

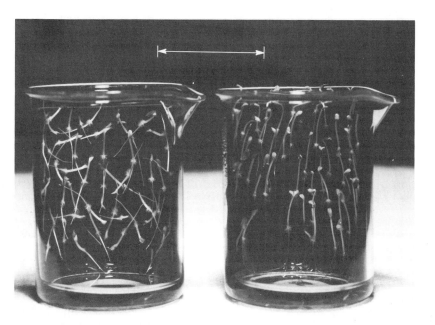

FIGURE 4-11

Lettuce seeds were germinated on blotter that was rolled and placed inside beaker and kept moist by water in the bottom of the beaker. (Seeds were placed between glass and blotter.) The roots of the beaker on the right are positively geotropic; the hypocotyls are negatively geotropic. When a morphactin solution was substituted for water in the beaker on the left, the root and hypocotyl grew in all directions irrespective of the force of gravity. The direction of growth depended on the orientation of the seeds on .the blotter. (After A. A. Khan, 1967b.)

they may become useful aids in the mechanical harvesting of grapes and tree fruits.

The various physiological responses of many crops to morphactins are the subject of many intensive studies that are yielding new and exciting possibilities (Schneider, 1970).

CHEMICAL PRUNING AGENTS. Some aromatic fractions from petroleum distillations that are referred to as "naphthalene base oils" are useful to control the growth and flowering of certain kinds of ornamental plants. Another group of compounds, the fatty acid derivatives (methyl nonanoate and methyl decanoate, for example), kill terminal shoots of azalea but cause no damage to axillary buds or foliage (Stuart, 1967). Chemical pruning is thus induced and new growth is stimulated.

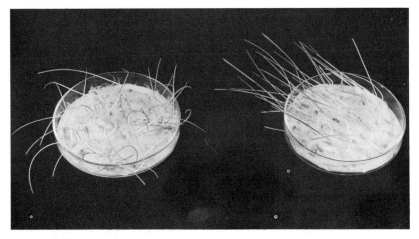

FIGURE 4-12

Oat seedlings were grown on filter paper disks soaked in a morphactin solution (*left*) and water (*right*). Four-day-old seedlings were illuminated unilaterally. Seedlings grown in water curve toward the source of the light; those grown in the morphactin solution show no definite response to light. (After A. A. Khan, 1967b.)

The effect of chemical pruning agents on the growth of terminal and lateral buds is dependent on the surfactant used to emulsify the oil, the type of cultivar, and the stage of plant development at the time of application. Treatment of vegetative plants of chrysanthemum, for example, kills the apical meristem and no flowers are produced. However, spray applications can abort the lateral flower buds if treatment is applied after terminal flower buds have completed flower initiation but before lateral floral buds have initiated all the florets.

Mechanism of Action

ABSCISIC ACID. This compound is widely distributed in plants, and evidently tissues of varying ages can synthesize the compound. Many of its physiological effects are antagonistic to those of such growth-promoting hormones as auxins, gibberellins, and cytokinins; for example, it has been demonstrated that GA_3-induced synthesis of a-amylase by barley grains is strongly inhibited by ABA (Chrispeels and Varner, 1967). Several investigators have noted that cytokinins reverse the inhibitory effects of ABA (see van Staden and Bornman,

1970). Growth and other responses of a plant may be the result of an interaction or a balance between ABA and the growth promoters (Addicott and Lyon, 1969). Perhaps the inhibitory effect of ABA on growth is due to its inhibition of hydrolytic enzymes, which are essential for plant metabolism.

The actual site of action of ABA is unknown. However, its interaction with promotive hormones suggests that it plays a role in nucleic acid metabolism and protein synthesis. The action of ABA may result from the inhibition of the synthesis of enzyme-specific molecules of RNA or from the prevention of their incorporation into an active enzyme-synthesizing unit.

PLANT GROWTH RETARDANTS. The mechanism of action of growth retardants is still vague. Since the effect of these compounds on plants (stem inhibition) is often precisely opposite that of gibberellin (stem acceleration), it seems logical to believe that retardants act as anti-gibberellins. Lang and his colleagues have demonstrated this hypothesis to be true with CCC and Amo-1618 in the fungus *Fusarium moniliforme* (Kende *et al.*, 1963) and with Amo-1618 (Baldev *et al.*, 1965) and CCC (Harada and Lang, 1965) in higher plants. In these experiments gibberellin synthesis was blocked, but the gibberellin already present in the tissues was not affected.

The mechanism of action of SADH may be based on the hydrolysis of the compound to UDMH, which subsequently inhibits diamine oxidase from converting tryptamine to IAA (Reed *et al.*, 1965). However, Ryugo and Sachs (1969) concluded from their in vitro and in vivo studies that the UDMH moiety is not the active portion of SADH and that the primary effect of SADH is to inhibit IAA synthesis.

MALEIC HYDRAZIDE. There are conflicting theories as to the mechanisms of action of MH at the molecular level. The compound is incorporated into the nucleolus as a part of the RNA fraction and probably is an analogue of uracil (see the summaries compiled by Zukel for a detailed review of MH action). Noodén (1970) found that MH is bound to macromolecules by an energy-requiring process in the older parts of corn seedling roots but not in the meristematic region or the coleoptiles; he believes that the physiological role of the binding process may be to provide a mechanism for inactivating or detoxifying MH.

MORPHACTINS. These compounds may exert their effect by influencing the plant's auxin metabolism (Tognoni *et al.*, 1967), thus causing alterations in hormonal control and such consequent responses as loss of apical dominance. In this regard it is interesting that applied auxin counteracts the abscission-promoting effect of morphactin (Weaver and Pool, 1968). Morphactins apparently do not interfere with the normal synthesis and action of gibberellin (Mann *et al.*, 1966). The action of morphactin at the genetic level is unknown, although Varner (see Mann *et al.*, 1966) found that morphactins do not affect amylase induction.

CHEMICAL PRUNING AGENTS. The mechanism of action of these agents includes the destruction of the meristematic and differentiating cells of certain species rather than the inhibition of some metabolic process. The effect is localized to the area of application, the first visible response occurring within fifteen minutes after the plants are sprayed with an emulsion of fatty acid esters or alcohols (Cathey *et al.*, 1966). The meristems initially turn black and become flaccid thirty to sixty minutes later. Within three to five days after treatment the tissues assume a tan color.

ETHYLENE

In its chemical structure, ethylene, a natural product of plant metabolism, is the simplest plant growth hormone. Such other volatile compounds as acetylene and propylene can produce ethylene-like effects, but ethylene is sixty to one hundred times more active than propylene, the next most effective compound of the group (see Pratt and Goeschl, 1969). Ethylene is also the only compound of the group of volatile compounds that is produced in any appreciable amount in plant tissues.

Biological Effects

One of the first effects of ethylene to be noted was the stimulation of sprouting and germination. Resting potato tubers are stimulated to sprout following application of ethylene at brief intervals; however, treatments of longer duration suppress sprouting. Growth of various corms, bulbs, hardwood cuttings, and roots, as well as seed germination in some species, is also stimulated if the gas is applied only as a short pretreatment—that is, if exposure is limited to a few hours or days before sprouting or during imbibition of seeds. Further treatments after sprouting or germination inhibit the growth of shoots and leaves.

Another effect of ethylene is to induce premature abscission of leaves, young fruits, and other organs. It is probable that the defoliation effects produced by 2,4-D, NAA, morphactins, and other compounds occur as a result of the induction of ethylene production by these compounds.

Ethylene, long known to be a fruit-ripening compound, has been applied to hasten the ripening of such harvested fruits as banana, mango, and honeydew melon and to degreen citrus fruits before marketing. Fruits that are mature but not yet ripe will respond to ethylene application before they produce their own ethylene.

It has also been known for many years that an application of 2,4,5-T during phase II of the growth pattern of fig (*Ficus carica*) accelerates growth (Crane and Blondeau, 1949); subsequent to this discovery, 2,4,5-T was found to induce ethylene formation (Maxie and Crane, 1968).

Ethylene can also induce flowering; for example, the compound enhances pistillate flower formation in cucurbits. A common technique, which has been the main method used in Hawaii to induce flowering in pineapple, is to spray the plant with ethylene adsorbed onto a bentonite suspension in water.

ETHEPHON.   Treatment of field-grown plants with ethylene gas is not practical because it dissipates too rapidly. However, the new product ethephon exerts its effect by gradually releasing ethylene as a decomposition product close to the site of action in plant tissue (Yang, 1969). Ethephon thus offers a means of treating field-grown plants with ethylene because its effects are often similar to those exerted by ethylene on flowering, fruit ripening, and abscission.

Levy and Kedar (1970) found that ethephon induces swelling of leaf bases and initiates bulbing in onion during noninductive day lengths (Fig. 4–13). Normally, day lengths of twelve to sixteen hours are required for bulb initiation. Other effects of ethephon are described in subsequent chapters.

Mechanism of Action

Man's knowledge of the role of ethylene in plant growth and development is still rudimentary, but recent discoveries have elevated the importance of its role as a hormone.

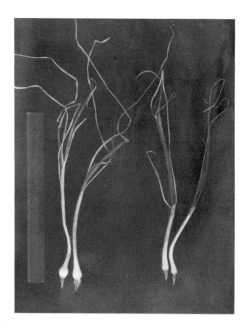

FIGURE 4-13
Effect of ethephon on bulb
initiation of onion, cv.
'Riverside.' *Left,* plants
grown in a solution containing
ethephon at a concentration
of 1,000 ppm and 1 percent
sucrose. *Right,* control plants
grown in 1 percent sucrose.
(After Levy and Kedar,
1970.)

STIMULATION OF GERMINATION AND SPROUTING. There are several possible mechanisms by which ethylene might stimulate germination and sprouting. For example, it might stimulate the movement of hydrolytic enzymes in storage tissues. Application of ethylene to isolated barley aleurone layers has been shown to stimulate the release of gibberellin-induced α-amylase from the aleurone cells into the endosperm, a storage tissue (Jones, 1968b). It is also possible that the ethylene produced during the growth of a bud may serve to control the mobilization of food reserves in the surrounding tissues.

ABSCISSION. The precise role of ethylene in the abscission process is still unknown, partly because the physiology of abscission is very complex in general. More information is needed on whether ethylene is produced in or near the abscission zone and whether it is produced before, during, or after abscission.

GROWTH REGULATION. Ethylene probably plays a role in the transcription and translation of the genetic code from DNA to RNA to protein and may be incorporated into the RNA in the same manner as are some of the other hormones. If so, it would also contribute to the regulation of other developmental phenomena, such as flowering, abscission, and fruit ripening. There is considerable evidence that when tissues are treated with ethylene, increases in various enzymes, especially peroxidases and other hydrolytic enzymes that are important in the growth process, follow (see Pratt and Goeschl, 1969). One hypothesis is that ethylene regulates growth by altering the transport or metabolism of auxin. Another possibility is that ethylene stimulates important enzyme systems associated with the cell membranes, thus aiding in the excretion from cells of certain enzymes important to growth.

FRUIT RIPENING.   The mechanism by which ethylene induces ripening is still unknown. One theory is that ethylene changes the physical state of the cells or membranes, thus allowing reactions to occur that previously had been prevented. Ethylene may be a causal agent of the changes in cell permeability that occur during the maturation and ripening of fruits. Ethylene stimulates respiration and protein synthesis in certain immature fruits, which may trigger a chain of biochemical events required for ripening, since the production of enzyme protein occurs early during the ripening process.

INTERACTION OF HORMONES

Each hormone produces many physiological responses, and frequently the responses overlap. Also, hormones probably do not act alone but rather interact with each other. The plant response or expression is often the result of a balance between growth promoters and growth inhibitors. Some of these interactions are discussed in the following paragraphs.

Hormonal Regulation of Plant Development

Plant growth processes can be controlled by the combined action of several hormones. The amount of cell division and cell enlargement is controlled by the varying levels of promotive and inhibitory hormones that regulate plant growth. Nonmobile inhibitors present in such organs as buds restrict their growth, whereas the other organs grow normally. Such environmental conditions as light and temperature alter growth by changing the amounts of different hormones present in the tissues. These variables can affect synthesis, transport, and inactivation of hormones.

Relative Levels of Promotive Hormones

Plant growth appears to be controlled mainly by the relative amounts of promotive hormones present. Apical dominance of the growing shoot is maintained by the downward movement of auxin produced in the apical bud, which keeps the lateral buds from growing. (In some species gibberellin may also contribute to apical dominance.) Production of ethylene in the lateral buds may block commencement of their growth (Burg and Burg, 1968), but bud growth is stimulated by the application of kinetin (Wickson and Thimann, 1958).

Gibberellin and auxin often act synergistically to accelerate plant growth. The presence of gibberellin usually increases the level of auxin in a plant—possibly through an auxin-sparing mechanism, since gibberellin has been noted to decrease the activities of certain presumedly auxin-destroying systems in some plants and to increase the auxin levels in others (Galston and Purves, 1960). The levels of auxin and gibberellin have also been found to influence development of the secondary vascular system (see Galston and Davies, 1969).

Auxin and Ethylene

Much research has been done on the interaction of these two hormones, especially since the finding that auxin application can trigger ethylene production in plants. It is a general rule that relatively low concentrations of auxin stimulate growth and that higher concentrations inhibit growth, although the optimal concentration depends on species and type of tissue. At certain critical concentrations of auxin, ethylene is produced. In etiolated pea stems, the start of ethylene production coincides with application of the concentration of auxin that causes maximum growth promotion in the tissue (see Galston and Davies, 1969). Although such data suggest a close relationship between auxin and ethylene, the role of auxin-induced ethylene production still is not clear.

Abscisic Acid and Promotive Hormones

In many systems the effects of the promotive hormones (auxins, gibberellins, and cytokinins) are counteracted by ABA. It would seem logical that in a growth system there should be such a brake; a push-pull system should be a more effective growth system than one in which no inhibitor is present. ABA has been shown to influence such processes as abscission of fruits and leaves, induction and prolongation of dormancy in the shoots of deciduous trees and tubers, inhibition of germination (by prolonging seed dormancy), and inhibition of flowering of long-day plants that are restricted to short days (see Addicott and Lyon, 1969). Apparently ABA interacts with different promotive hormones in different plant systems. A logical hypothesis is that each of the interactions is caused by one master promotive hormone and that ABA acts as the braking hormone.

Interaction of Other Hormones

Hormones probably interact to control the various physiological processes that occur during growth. The abscission phenomenon is a result of a complex interaction between auxin, ethylene, and ABA, and possibly other compounds as well. Even less is known regarding the interaction of hormones in fruit set, fruit development, and certain other physiological processes.

STIMULATION OF CAMBIAL ACTIVITY AND XYLEM DEVELOPMENT

The initiation of cambial activity in deciduous species during the spring depends on the presence of expanding buds. The initiation of cambial divisions beneath the expanding buds is followed by a wave of cambial divisions downward to the branches. The stimulus appears to travel basipetally only. For many years the commonly held theory was that endogenous auxin in the apexes of the growing shoots stimulates cambial division and xylem differentiation in the stem below (Wareing et al., 1964b). However, the finding that application of auxin to disbudded twigs causes cambial divisions and production of new xylem for only a relatively short distance suggested that other hormones might also affect in cambial activity (see Wareing et al., 1964b).

After the gibberellins had been discovered, Wareing and his colleagues (1964) showed that joint application of auxin and gibberellin to disbudded stems of sycamore, poplar, and ash results in xylem and phloem development that resembles the normal tissue more than that occurring if auxin is applied alone. Subsequently, promotion of xylem differentiation and development by the application of GA_3

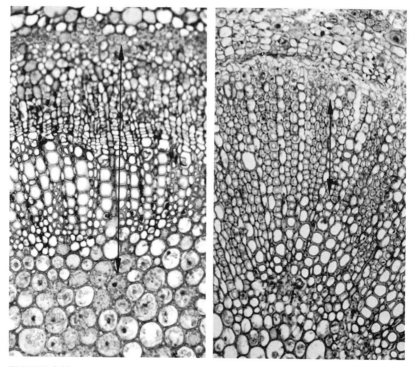

FIGURE 4-14

Transsections of the first internode of young olive stems formed after treatment with water (*left*) and GA₃ at a concentration of 100 ppm (*right*), showing effect of GA₃ on xylem development. Vertical lines indicate radial width of xylem. Magnified 200 times. (After S. A. Badr, M. V. Bradley, and H. T. Hartmann, 1970.)

was reported in many plants (see Badr *et al.*, 1970); for example, application of GA₃ at concentrations of 100, 250, or 500 ppm significantly promotes xylem differentiation in newly developed regions of olive shoots (Fig. 4–14). Application of GA₃ plus IAA, each at concentrations of 250 or 500 ppm, has a synergistic effect on xylem production, but application of auxin alone has no such effect in olive.

Research done in Poland reveals that cytokinins are also important in xylem production (Hejnowicz and Tomaszewski, 1969). These investigators found that in pine, IAA substitutes only partly for the substances produced by the growing apex, but if both GA₃ and cytokinin are used in addition to the auxin, normal wood formation is induced. They therefore hypothesized that, in pine at least, auxin is the principal limiting factor inducing xylem formation, whereas gibberellin and cytokinin augment the distribution pattern of auxin within the stem.

The present evidence suggests that a certain balance between gibberellin and auxin is required for maximum xylem differentiation and that, in some plants, cytokinin also may stimulate xylem differentiation.

Supplementary readings on the general topic of biological effects and mechanism of action of plant growth substances are listed in the Bibliography: Galston, A. W., and Davies, P. J., 1969, 1970; Leopold, A. C., 1964a; van Overbeek, J., 1966; and Wilkins, M. B., ed., 1969.

Chapter 5

ROOTING AND
PROPAGATION

Stimulation of cell division was the first role of auxin to be discovered. Stimulation of root initiation, the second, was the first practical use to be made of growth regulators. Nurserymen presently make widespread use of growth regulators to stimulate rootings of cuttings.

The portion of a plant that is severed from the parent plant for the purpose of propagation is called a cutting. Cuttings may be taken from a stem, a root, or a leaf and are called stem cuttings, root cuttings, and leaf cuttings, respectively. Cuttings taken from deciduous species during the dormant season are referred to as "hardwood cuttings," whereas those taken during the growing season while the plants are succulent or have only partially matured wood are referred to as "leafy," "softwood," or "semihardwood" cuttings. The ability of many plants to form roots on cuttings that are placed in favorable growing conditions is of great value in plant propagation.

ANATOMICAL DEVELOPMENT OF ROOTS
IN STEM CUTTINGS

Most adventitious roots of herbaceous stem cuttings arise from groups of thin-walled, living parenchyma cells that are capable of becoming

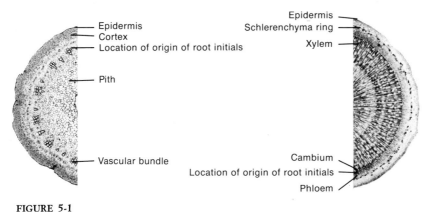

FIGURE 5-1

Stem cross sections showing usual location of origin of adventitious roots. *Left,* young herbaceous, dicotyledonous plant; *right,* young woody perennial plant. (After H. T. Hartmann and D. E. Kester, *Plant propagation: Principles and practices,* 2d ed., 1968. By permission of Prentice-Hall, Inc., Englewood Cliffs, New Jersey.)

meristematic. In herbaceous cuttings these cells are located just outside and between the vascular bundles (Fig. 5–1). In woody perennial plants, in which one or more layers of secondary xylem and phloem are present, adventitious roots in stem cuttings usually originate in the young secondary phloem tissue, although such roots also arise from other tissues such as cambium, vascular rays, or pith.

The root initials are small groups of meristematic cells that continue to divide and form groups composed of many small cells that develop further into new, recognizable root primordia. Cell division continues and soon each group of cells begins to form a root tip structure. A vascular system develops in the new root primordium and becomes connected with the adjacent vascular bundle. The root tip grows outward through the cortex and epidermis and emerges from the stem.

The formation of adventitious roots in most plant species occurs after the cutting is made. However, in some plants preformed root initials, whose location is generally the same as that of nonpreformed root initials, form during stem development and usually remain dormant until cuttings are made and are placed under favorable environmental conditions, after which they grow and develop as adventitious roots. There are a number of easy-to-root genera, such as willow (*Salix*), gooseberry and currant (*Ribes*), and poplar (*Populus*) that commonly develop preformed root initials.

Preformed root initials are not essential for rapid rooting. Stem cuttings of most varieties of grape, as well as most softwood and hardwood cuttings of many ornamental species, root easily even though preformed root initials are not present.

Roots that arise following application of plant growth regulators are similar in origin to those normally produced. However, the characteristics of the roots and their arrangement on the stem may vary considerably. High concentrations of growth regulators can produce abnormalities in root formation and necrosis of tissue. It is likely that the observation that high concentrations of auxin stunted the growth of adventitious roots led to the idea that growth regulators would be of value as herbicides.

PHYSIOLOGICAL BASIS OF ROOT FORMATION IN CUTTINGS

Exogenous Plant Growth Substances and Rooting

Sachs in 1882 and Bouillenne and Went in 1933 postulated that there was a substance or substances in the leaves, buds, and cotyledons of plants that moved to the roots and stimulated rooting. This hormone substance, termed "rhizocaline" by Bouillenne and Went, is as yet only a hypothetical compound.

AUXINS. Before the discovery of auxin, several chemical compounds, such as permanganate (Curtis, 1918) and carbon monoxide (Zimmerman et al., 1933), were reported to increase rooting in cuttings. These materials probably achieve this result by affecting the relative auxin levels in the cuttings. In 1934 Thimann and Went reported that auxin generally exerts the primary control over root formation. For centuries Dutch gardeners have promoted root formation by imbedding grain seeds into cuttings. Such a result is now known to have been caused by the large amounts of auxin produced by the germinating grain. Similar practices have been carried on for centuries in Afghanistan. Fifteen or twenty barley seeds are planted with each cutting of grape, poplar, and some citrus species. Before planting deciduous fruit trees, farmers place approximately one-quarter pound of barley seed 50 cm deep in each hole (M. Ghufran, private communication). Growers believe that root formation is promoted and that tree vitality is enhanced. Barley seeds planted deep in the soil will germinate but will not grow.

GIBBERELLINS. These growth regulators are antagonistic to root initiation (Brian et al., 1960). Figure 5–2 shows the effect of GA_3 on mung

FIGURE 5-2

Rooting of mung bean cuttings 6 days after bases were placed in (*left*) water, (*center*) IBA solution at a concentration of 1 ppm, and (*right*) GA₃ at a concentration of 1 ppm. IBA stimulated rooting but GA₃ completely inhibited it.

bean cuttings. Gibberellin probably prevents the cell division in mature tissues that is a prerequisite for creation of the meristematic condition and formation of root initials. The effect of gibberellin may also be in part nutritional because shoot growth is stimulated and thus competes for the assimilates required for root initiation.

CYTOKININS. The differentiation of a meristem into a root primordium is the first step in root formation. Skoog and his colleagues (1951) showed that the type of differentiation that occurs in a meristem is dependent on the proportion of auxin to cytokinin or to other substances, such as adenine, that stimulate cell division. They showed that the meristem of tobacco stem sections tends to form bud and leaf primordia when the ratio of auxin to such plant constituents as adenine and cytokinin is low. However, when the ratio is high, root primordia form. When the ratio between auxin and cytokinin and adenine is intermediate, a simple callus is formed, but there is no differentiation. Thus the physiological basis for the control of meristematic differentiation is the balance between auxin and cytokinin and other compounds evidencing cytokinin activity.

Although cytokinins would not be expected to stimulate root development, since they usually stimulate shoot development and are antagonistic to rooting, there have been some reports that low concentrations of cytokinins do stimulate root initiation. Kinetin at a concentration of 0.1 ppm more than doubles the rooting of terminal cuttings of one clone of the difficult-to-root *Feijoa sellowiana* (Meredith *et al.*, 1970). Kinetin also increases rooting of cuttings of *Acer rubrum* when it is applied to the leaves, but it is inhibitory when applied at the base of cuttings (Bachelard and Stowe, 1963). These researchers found no evidence that the stimulatory effect of the foliar treatment is caused by a change in the balance between auxin and kinetin.

This author has observed that BA stimulates root initiation in grape cuttings, but the effect may have been an indirect one because bud rest was also terminated, thus allowing shoot growth to commence. The basipetal transport of auxin produced by the growing shoots may have been the primary reason for the root initiation.

INHIBITORS.   These compounds have diverse effects on root initiation. Many of them are inhibitory but some can be promotive; for example, catechol reacts synergistically with IAA to initiate roots in the mung bean bioassay (Hess, 1962).

ETHYLENE.   Many years ago ethylene, as well as propylene, acetylene, and carbon monoxide, were shown to be stimulators of root initiation (Zimmerman and Hitchcock, 1933). However, at that time ethylene and related compounds could not be used in propagation because gases were then believed to move readily throughout the plant and to induce rooting where roots were not desired. Such ethylene-generating compounds as ethephon may also prove to be of value for rooting.

Cofactors Necessary for Rooting

Successful rooting is dependent on the presence in the cuttings of a number of cofactors, which, in combination with auxin, enable cuttings to root. The source of these cofactors is usually the leaves. Plant propagators are well aware that loss of leaves from cuttings greatly reduces chances for successful rooting (Figs. 5–3, 5–4). The sugars and nitrogenous materials produced in the leaves probably are rooting cofactors. There is also evidence that such phenolic compounds as caffeic acid, catechol, and chlorogenic acid interact with auxin to induce root initiation (Tomaszewski, 1964; Hackett, 1970).

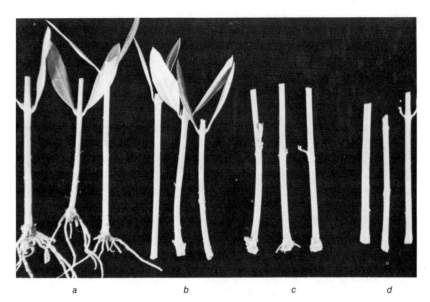

a   b   c   d

FIGURE 5-3

The presence of leaves and application of growth regulator result in best root production in olive (*Olea europea*). *a*, Cuttings with IBA-treated leaves; *b*, cuttings with untreated leaves; *c*, cuttings without leaves, given IBA treatment; *d*, untreated cuttings without leaves. (After H. T. Hartmann and D. E. Kester, *Plant propagation: Principles and practices*, 2d ed., 1968. By permission of Prentice-Hall, Inc., Englewood Cliffs, New Jersey.)

Experiments conducted by van Overbeek *et al.* (1946), using easily rooted cuttings of red hibiscus and cuttings of a difficult-to-root white variety, have provided considerable information on the cofactors required for root formation. Abundant roots were induced on leafy cuttings of the red variety as a result of applying IBA to the stem bases but no roots were formed on the IBA-treated white variety. Cuttings of the white variety would not root even though a scion of the easily rooted red variety was grafted onto them. But when such grafts were treated with IBA, abundant roots formed. These results show that two factors were necessary for root formation in the white variety: IBA, and an unknown cofactor or cofactors present in the leaves of the red hibiscus. The white hibiscus failed to root not only because it lacked auxin but also because its leaves failed to produce the other factor or factors that, in addition to auxin, were a prerequisite for root initiation.

These investigators then demonstrated that, by supplying the cuttings with sugars and nitrogenous compounds, they could entirely replace the promotive effect of the leaves on root initiation. They further showed that an increase in the number of leaves from zero to three

FIGURE 5-4

Effect of the presence of leaves on the rooting of cuttings of 'Lisbon' lemon. Both groups were rooted under intermittent mist and were treated with IBA at a concentration of 4,000 ppm by the concentrated-solution-dip method. (After H. T. Hartmann and D. E. Kester, *Plant propagation: Principles and practices,* 2d ed., 1968. By permission of Prentice-Hall, Inc., Englewood Cliffs, New Jersey.)

resulted in a quantitative increase in root initiation in red hibiscus and that the promotive effect of the leaves was completely replaced by a solution of sucrose plus ammonium sulfate. However, if IBA was not applied, practically no rooting occurred even when these nutrients were provided. Chemical analyses showed that the leaves actually did provide such nutritional substances.

In some species, thick cuttings having much reserve material stored in them do not require leaves for rooting, which indicates that sufficient cofactors are already present in the wood to stimulate root initiation.

Ryan *et al.* (1958) later repeated van Overbeek's experiment with the same varieties and obtained excellent rooting. The failure of van Overbeek to obtain rooting may be attributed to the fact that the red hibiscus scion lost its leaves early in his experiment. These studies of the rooting of many species have led to the conclusion that rooting capacity is not determined by the kind of leaves supplying the cutting but by the type of stem from which the roots arise. So long as healthy leaves were retained in these experiments, the ease or difficulty of rooting was not influenced by the rooting characteristics of the variety furnishing the leaves.

In *Rhododendron*, poor rooting of some clones can make propagation difficult. Lee *et al.* (1969) studied the level of root-promoting and root-inhibiting substances in clones of the easy-to-root *Rhododendron* 'Cunningham's White,' the difficult-to-root 'Dr. H. C. Dresselhuys,' and 'English Roseum,' which roots with a difficulty intermediate to the other two. The rooting cofactors were extracted with methanol, separated by paper chromatography, and measured using the mung bean rooting bioassay. The highest levels of four rooting cofactors in the stems and leaves in any season were found in 'Cunningham's White'; the lowest were found in 'Dr. H. C. Dresselhuys.' The intermediate 'English Roseum' contained an intermediate amount of the cofactors. An inhibitory material was found in all clones, but the variation in the levels of the rooting cofactors was believed to be responsible for the difference in the rooting response of the clones. The promoting activity in the clones increased in September and then decreased in November to the level found in the July extract.

Rooting of the difficult-to-root *Rhododendron* 'Dr. H. C. Dresselhuys' was significantly improved by grafting to it a leaf and bud scion of easy-to-root 'Cunningham's White.' Grafting of scions of the difficult-to-root 'Dr. H. C. Dresselhuys' to 'Cunningham's White' resulted in a decreased rooting of the cuttings. The results of these grafting trials performed on *Rhododendron* by Lee and his colleagues do not agree with the conclusion of Ryan *et al.* (1958) that rooting capacity is determined not by the kind of leaves retained but by the stem from which the roots arise. From the conflicting results discussed, it is clear that as yet there is no general agreement as to the contribution of leaves to the rooting of cuttings.

Thimann and Delisle (1939, 1942) demonstrated that some unknown factor other than auxin is involved in the root initiation of coniferous evergreen cuttings. They believed that this factor was present in large amounts in young plants and often was present in lower amounts in older plants. This hypothesis would explain why sections taken from young plants often root more readily than sections taken from older plants. The presence or absence of such cofactors could also explain why lemon cuttings treated with auxin root well but apple cuttings treated with auxin fail to root (Cooper, 1938).

That the presence of buds on a cutting is favorable to rooting can be demonstrated by removing the buds from a cutting or by making a girdle under the buds. The girdling experiment shows that some substances move downward through the phloem to the base of the cutting where root initiation is stimulated. The importance of buds in root initiation is also shown by the fact that cuttings root better after termination of bud rest (Howard, 1965). However, the presence of buds is not as essential in species with preformed root initials.

Hess (1964) studied differences in the rooting ability of the juvenile and mature forms of English ivy (*Hedera helix*) and two varieties of chrysanthemum. Cuttings of the juvenile form of *Hedera* and of *Chrysanthemum* 'Fred Shoesmith' are easy to root, and those of the mature form of *Hedera* and *Chrysanthemum* 'Salmon Shoesmith' are difficult to root. Both juvenile and mature forms of *Hedera* can be found on the same plant, so genetic differences can be eliminated. Differences in rooting capacity of *Hedera* were not caused by a lack of extractable auxin or the presence of an inhibitor. However, in both *Hedera* and *Chrysanthemum* there was a greater quantity of extractable root-promoting cofactors in the easy-to-root cuttings than in the difficult-to-root varieties. One of the extracts contained at least four active substances, which have been tentatively identified as oxygenated terpenoids (Gorter, 1958).

Kawase (1964) centrifuged cuttings of willow (*Salix alba*) and obtained an extract that induced root initiation. The active substance proved to be a nonauxin compound. Rooting cofactors have also been revealed in hardwood cuttings of 'Crab C' and 'E. M. 26' apple rootstocks (Challenger *et al.*, 1965).

Fadl and Hartmann (1967a) made extracts from the bases of cuttings of 'Old Home' pear, an easy-to-root variety, and of 'Bartlett,' a difficult-to-root variety. Approximately ten days after 'Old Home' cuttings were treated with IBA, a highly active root-promoting substance appeared in cuttings with buds, but no such substance appeared in 'Bartlett.' The promotive substance did not appear in 'Old Home' cuttings without buds, even when those cuttings were treated with IBA, nor in cuttings with buds that were taken when buds were in a resting condition. The exact chemical nature of this rooting factor is unknown, but ultraviolet and infrared spectrum analyses indicated that the compound had a high molecular weight and was possibly a condensation product that resulted from the combination of the phenolic substance with IBA.

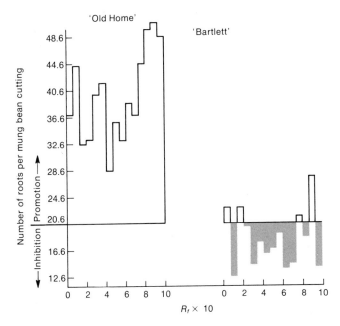

FIGURE 5-5

Comparison of the amounts of promoters and inhibitors, as determined by the mung bean bioassay, present in paper strips prepared from chromatograms loaded with extracts from bases of cuttings of 'Old Home' (*left*) and 'Bartlett' pear (*right*) 20 days after the cuttings were treated with IBA and placed in the rooting medium. Large amounts of promoters were present in the easily rooted 'Old Home' cuttings at all R_f levels. Bases of the difficult-to-root 'Bartlett' variety not only lacked promoters but showed strong inhibition at most R_f levels. (After H. T. Hartmann and D. E. Kester, *Plant propagation: Principles and practices,* 2d ed., 1968. By permission of Prentice-Hall, Inc., Englewood Cliffs, New Jersey.)

Endogenous Inhibitors

The preceding section discussed evidence in support of the theory that difficult-to-root cuttings fail to root because they lack the necessary cofactors. Another possibility is that such cuttings contain inhibitory substances in amounts high enough to mask the effect of the promotive substances present. When extracts of IBA-treated cuttings of easy-to-root 'Old Home' pear and difficult-to-root 'Bartlett' pear were fractionated by paper chromatography and the strips bioassayed by the mung bean bioassay, marked differences were apparent in the histograms (Fadl and Hartmann, 1967b). Much promotive activity was observed in the easy-to-root 'Old Home' cuttings and much inhibitory activity was found in the difficult-to-root 'Bartlett' cuttings (Fig. 5–5). A logical hypothesis seems to be that 'Bartlett' is a poor rooting variety because the inhibitors present are antagonistic to processes that lead to rooting.

Spiegel (1954, 1955) had previously discovered the presence of two inhibitors in grape. Water extracts from the difficult-to-root grape stock *Vitis berlandieri* had a deleterious effect on the rooting of the easily rooted *Vitis vinifera;* however, leaching of the difficult-to-root *Vitis berlandieri* cuttings enhanced their root initiation. Cuttings of the latter variety that were not leached had a high inhibitor content. Although Odom and Carpenter (1965) found inhibitors in stem cuttings of alternathera, coleus, chrysanthemum, geranium, and carnation, they could find no correlation between the presence of inhibitors and the ease of rooting of the cuttings.

Meredith *et al.* (1970) found an inhibitor in both moderately easy-to-root and difficult-to-root clones of *Feijoa sellowiana,* but the content was five times higher in the latter.

There is often considerable variation in the rooting of replicate cuttings. Profuseness of root production of *Acer rubrum* has been correlated positively with anthocyanin formation in the leaves (Bachelard and Stowe, 1963). These investigators speculate that some substance utilized in anthocyanin biosynthesis may also cause root formation.

USE OF GROWTH REGULATORS
TO STIMULATE ROOTING

One of the best and most commonly used rooting stimulators is the auxin IBA (Fig. 5–2). It has weak auxin activity and is destroyed relatively slowly by auxin-destroying enzyme systems. A chemical that is persistent is very effective as a root promotor. Because IBA translocates poorly, it is retained near the site of application. Growth regulators that readily translocate may cause undesirable growth effects in the propagated plant.

Another excellent auxin frequently used for root promotion is NAA. However, this compound is more toxic than IBA, and excessive concentrations of NAA must be avoided because of the danger of causing injury to the plant.

IBA and NAA are more effective in induction of rooting than IAA. IAA is very unstable in plants, and its decomposition occurs rapidly in unsterilized solutions, although it will remain active in sterile solutions for several months. Strong sunlight can destroy a 10 ppm solution of IAA in fifteen minutes.

The amides of IBA and NAA are also very effective rooting agents. The amide form of NAA is less toxic than NAA and thus is safer to use. Other homologues are effective rooting agents although none is superior to NAA.

Many phenoxy compounds promote root formation if used at low concentrations. When applied at high concentrations they tend to produce thick, stunted roots and their toxicity limit is near the optimal concentration for root initiation. 2,4-D promotes rooting in certain species but because it is potent and readily translocated, it usually tends to inhibit shoot development and cause injury to the shoot, especially if too much chemical is used. Several other phenoxy compounds have proved to be effective; 2,4,5-T, 2,4,5-TP, 2,4,5-TB, and 2,4-DB produce good rooting without injuring shoots if used in very low concentrations. Unfortunately, the effective range of concentrations of these compounds is quite narrow.

Growth regulators may alter the type of roots as well as the number that are produced. IBA produces a strong, fibrous root system, whereas the phenoxyacetic acids often produce a bushy, stunted root system consisting of bent, thick roots. Root-promoting substances frequently are more effective when used in combination (Hitchcock and Zimmerman, 1940); equal parts of IBA and NAA induce a higher percentage of cuttings to root in some species than either material used alone. These roots have some characteristics of root systems treated with either IBA or NAA.

In 1969 Read and Hoysler demonstrated that SADH is a practical aid in propagating geranium, chrysanthemum, and dahlia when applied to the basal portion of the cuttings as a fifteen-second aqueous dip. In 1971 the same investigators reported that this SADH treatment also promoted the rooting of carnation and poinsettia. Thirty days after SADH at a concentration of 5,000 ppm was applied to 'Red Alaska' carnation, the mean number of roots per cutting was 20.4, compared with 13.4 for the controls. The mean root lengths were 0.95 and 0.45 cm, respectively. Rooting of cuttings of 'Paul Mikkelson' and 'Ecke White' poinsettia was promoted when their bases were dipped in SADH at a concentration of 2,500 ppm. Foliar dips of 'Ecke White' cuttings in the same compound at the same concentration produced the same result. Treatments with ABA, SAPL, and EHPP had little effect, and GA_3, ethephon, and CCC retarded rooting.

The many lists that have been published summarizing the results of investigations of the use of auxins as rooting aids (Audus, 1959;

Thimann and Behnke-Rogers, 1950) are indicative of the tremendous amount of work that has been undertaken in this area.

METHODS OF APPLYING GROWTH REGULATORS TO STEM CUTTINGS

There are many methods of applying sufficient amounts of growth regulator to a stem cutting. However, at present the only three methods that have come into widespread practical use are the quick dip, prolonged soaking, and powder methods. Other methods, such as the use of lanolin paste, injection, and insertion of toothpicks soaked in auxin, are usually not used commercially because of their inconvenience.

Quick Dip Method

In this method the basal ends of the cuttings are dipped for approximately five seconds into a concentrated solution (500 to 10,000 ppm) of the chemical in alcohol. The chemical can be absorbed through intact tissue, leaf scars, wounds, or cut apical or basal ends of cuttings (McGuire et al., 1969). The cuttings are then placed immediately in the rooting medium. This method has the advantage that less equipment is necessary for soaking than in the prolonged soaking technique. The amount of auxin applied per unit surface area of cutting base is constant and is less dependent on external conditions than in the other two methods. The same solution can be used repeatedly, but it should be tightly sealed between use so that the alcohol will not evaporate.

A concentration of growth regulator just below the toxic point is optimal for root promotion. This concentration causes some swelling in the basal part of the stem, accompanied by profuse root production just above the base of the cutting. Too high a concentration may inhibit bud development or cause yellowing or dropping of leaves or even death of the cutting.

Prolonged Soaking Method

In this method a concentrated stock solution of auxin is prepared with 95 percent ethanol and is then diluted to the desired strength in water. Concentrations used vary from 20 ppm for the easily rooted species to 200 ppm for species that root with more difficulty. The cuttings (basal inch only) are soaked in the solution for twenty-four hours in a shaded location at room temperature and are then placed immediately in the rooting medium. The amount of chemical absorbed by each cutting depends on environmental conditions and on the species used. (Leafy softwood cuttings, for example, vary widely in the amount of solution they absorb.) Much more solution is absorbed in the transpiration stream in warm, dry conditions than in cool, moist conditions. Therefore, the cuttings should be kept in a humid atmosphere during the soaking period so that a slower but steady uptake is obtained. The amount of auxin required varies with the species, the time of year when the cuttings are taken, and the compound used. For leafy cuttings of woody species a one- to two-hour soak in a 100 ppm solution or a ten- to twenty-four-hour soak in a 5 ppm solution is optimal.

Powder Method

In this method the base of the cutting is treated with growth hormone mixed with a carrier (a fine inert powder, such as clay or talc). Approximately 200 to 1,000 ppm of the growth hormone should be used for softwood cuttings and five times this amount should be used for hardwood cuttings. Two main methods are employed to prepare the treating mixture. One is to grind the auxin crystals to a fine powder and then to mix this powder thoroughly with the carrier. The other is to soak the carrier in an alcoholic solution of a growth substance and then to evaporate the alcohol so that the carrier is left as powder.

It is often desirable to make fresh cuts at the base of the cuttings before treatment to facilitate absorption. The basal inch of the cuttings is then moistened with water and rolled in the powder. All excess powder should be shaken from the cuttings to prevent possible toxic effects. The cuttings are then planted immediately, using care taken not to rub off the thin layer of adhering powder. (To this end, a thick knife may be used to make a trench in the rooting media before the cuttings are inserted.) Uniform results may be difficult to obtain with this method because of the variability in the amount of material that adheres to the cutting.

Other Techniques

In addition to these three main methods for applying growth regulators, numerous other techniques have been used experimentally

FIGURE 5-6

Effect of site of application of 1 percent IBA on orientation of adventitious roots of *Ilex crenata* 'Convexa.' *Left,* untreated control; *center,* received basal dip; *right,* received terminal dip. (After J. J. McGuire, L. S. Albert, and V. G. Shutak, 1968.)

with varying degrees of success (see McGuire *et al.,* 1968). Foliar parts of cuttings have been dipped in auxin solutions, and foliage has been sprayed with such solutions either before or after the cuttings were removed from the stock plant. Auxin solutions have been injected into cuttings, and auxins have been forced into stems by vacuum infiltration. Auxins have been sprinkled on the soil to be absorbed by plants, and pegs soaked in auxin have been inserted into holes bored in the plant stems.

Recent data indicate that foliar or terminal applications of growth substances may effectively promote rooting of cuttings. McGuire *et al.* (1968), using fourteen different species of woody ornamental plants, found that root initiation is stimulated as a result of a foliar application of 1 percent IBA to cuttings. The concentration of IBA required for successful foliar application must be higher than that normally used for basal application. Early researchers failed to obtain successful results with foliar applications—probably because the concentrations used were too low. Terminal applications to *Ilex crenata* 'Convexa' and some of the other species used resulted in a different root orientation from that obtained with untreated plants or plants receiving basal applications. Basal applications inhibited rooting and root development at the lowest part of the cutting. Both untreated cuttings and terminally treated cuttings produced roots at the base of the cutting, but roots were also produced in the upper region of the basal area of terminally tested cuttings (Fig. 5–6). The high concentrations of IBA used did not cause phytotoxic responses in the zone of root initiation, which indicates that a wide range of concentrations of growth substance can be used with the foliar application method.

Precautionary Measures to Be Used with Chemicals

The use of old chemical solutions of growth regulators to aid rooting of cuttings has sometimes produced negative results. Therefore, it is best to make just enough solution required for the cuttings and then to discard it. New mixtures should be made from the stock solution each time. Dilute solutions of IBA may lose their activity within a few days. However, growth regulator preparations dispersed in talc will keep their activity for several months, and concentrated solutions that contain a high percentage of alcohol, such as are used in the quick dip method, retain their activity almost indefinitely.

USE OF GROWTH SUBSTANCES IN TRANSPLANTING

Nursery-grown trees usually grow only slightly during the first year or two after planting, mainly because they have difficulty in reestablishing a root system. Many transplanted trees have become weakened and died because they could not establish such a system. The application of auxin to hasten the recovery of the root system has generally yielded unsatisfactory results, although there have been several exceptions.

Romberg and Smith (1938) demonstrated that when toothpicks containing 4 mg of IBA were inserted into the roots of five- to seven-year-old trees, there was an increase in root development compared with that of the controls. Application of the auxin at 0.5 percent with lanolin or wheat flour dough as an inert carrier was also effective but was less economical in terms of material and time. In one experiment using the toothpick treatment, the average numbers of new roots that developed on control and treated trees were 36 and 66, respectively. The total lengths of new roots per tree were 219 and 586 inches, respectively. Romberg and Smith (1938) also showed that ten-year-old nursery pecan trees can be transplanted successfully and economically when the roots are first treated with IBA-impregnated toothpicks. A more fully developed root system results in a rapidly developing tree that is not as readily affected by sun scald or as subject to attack by borers as are weakened trees that grow more slowly.

In the major fruit-growing areas of British Columbia, pear trees frequently fail to produce more than a few inches of terminal growth during the first one or two years after being planted in orchard soil because of poor root growth. Looney and McIntosh (1968) treated roots of one-year-old 'Bartlett' pear trees that had been grafted to 'Bartlett' seedling roots with IBA before planting and found that the treatment stimulated the production of new roots. The two most effective methods of inducing development of new roots were found to be (1) dusting the cut end of the roots with a commercial rooting power containing 0.8 percent IBA and (2) inserting toothpicks impregnated with 1 mg of IBA into the roots. Trees treated with IBA developed larger root systems than did the controls (Fig. 5–7). Weight of new growth minus the top growth after the first growing season following treatment was: for control, 58.6 g; for trees treated with IBA-impregnated toothpicks (roots not cut), 66.3 g; for trees treated with IBA-impregnated toothpicks (roots cut 2.0 cm beyond point of insertion), 87.4 g; for trees treated with IBA dust, 76.5 g.

RESPONSE OF CUTTINGS TO GROWTH REGULATORS

The objective of treating cuttings with growth regulators is to increase the "take," or the percentage of cuttings that grow vigorously in the nursery or field. The favorable effects of such treatment are (1) stimulation of root initiation, (2) an increase in the percentage of cuttings that form roots, and (3) acceleration of the time of rooting. These effects lead to a saving in labor and a more rapid turnover of greenhouse space. Growth substances stimulate rooting of easy-to-root species most effectively but may not induce rooting on species that

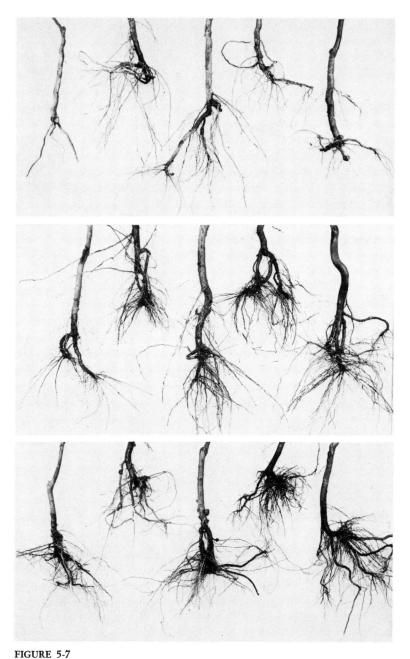

FIGURE 5-7

Roots of trees of 'Bartlett' pear that received various preplanting treatments with IBA. *Top*, untreated control; *middle*, treated with IBA dust; *bottom*, treated with IBA-impregnated toothpicks (roots not cut). (After N. E. Looney and D. L. McIntosh, 1968.)

normally fail to root. If cuttings root easily without treatment, there is no need to undergo the added expense required by such treatment.

Herbaceous Plants

Most herbaceous plants, including chrysanthemum, geranium, verbena, carnation, begonia, poinsettia, African violet, and English ivy, respond well to growth regulator treatment. Initial toxic effects, such as stem bending and injured roots, sometimes occur, but recovery is rapid and few cuttings are lost.

Deciduous Trees, Woody Plants, and Shrubs

More favorable results have been obtained by treating softwood cuttings during the active growing period than during the dormant season. The application of growth regulators to such deciduous plants as plum, cherry, apple, mulberry, pear, beech, birch, apricot, grape, pecan, black locust, and filbert can accelerate rooting by two to three weeks.

Coniferous Trees and Shrubs

Special care must be exercised to take cuttings from these plants at the proper time of year and to use the correct concentration of hormone. Pine, yew, fir, arborvitae, hemlock, and juniper are some of the plants that respond well to growth regulators under proper conditions.

Broad-leaved Evergreens

Many plants of this group respond well to growth regulator treatment. Cuttings from holly, rhododendron, gardenia, oleander, pyracantha, escallonia, camellia, holly, and other ornamental plants give good results. In addition, cuttings from orange, grapefruit, lemon, magnolia, and olive root well as a result of growth regulator treatment.

Tropical Plants

CACAO. A mixture of IBA and NAA is used in the propagation of cacao in West Africa and the West Indian Islands (van Overbeek, 1961). This practice has become a standard method in the propagation of cacao from cuttings (Evans, 1951).

RUBBER.   A mixture of IBA and NAA is useful in the propagation of rubber (Templeman, 1955).

AUXIN TREATMENT OF
DIFFICULT-TO-ROOT SPECIES

Application of growth regulator treatments to some plant species has proved to be unsuccessful. Hardwood cuttings of many forest trees and deciduous fruit trees, nut trees, and such tropical or subtropical fruit trees as avocado and mango have responded poorly to growth regulators. If cuttings are taken from the juvenile plant (less than two years old) of these types, they often can be rooted, but cuttings from older trees usually fail to respond. Auxin-type chemicals generally have been more successful when applied to herbaceous greenwood cuttings than to hardwood cuttings.

When attempting to root difficult-to-root cuttings, close attention should be paid to environmental conditions. Adequate light must be provided. Leafy cuttings must be prevented from wilting until a root system develops. One of the most effective techniques is an intermittent water spray (Hartmann and Kester, 1968). Use of IBA treatment and mist sprays separately will not always induce root initiation, but a good root system usually results when the two are used together.

USE OF CUTTINGS OTHER THAN
STEM CUTTINGS

Almost any plant organ, such as stem, leaf, stolon, root, and even flower and fruit, is capable of producing roots. Stems are generally the rooting structure of choice because they usually have sufficient undifferentiated tissue to permit differentiation of root primordia; furthermore, their buds have already formed. The presence of buds on propagation pieces is desirable because auxin treatments that promote rooting do not favor shoot development.

Leaf Cuttings

Either the leaf blade alone or the petiole plus the leaf blade are utilized to start a new plant. Leaf cuttings require the differentiation of both root and bud primordia. Adventitious roots and adventitious shoots usually form at the base of the cutting, and the original leaf itself does not become a part of the new plant. Although bud formation is often inhibited by auxin treatment, growth regulators are useful for inducing root initiation of leaf cuttings of such plants as begonia, India rubber plant, and African violet. The scale leaves of Easter lily root more readily after treatment with auxin.

When large veins of some plants, such as begonia, are severed, new plants arise at the site of vein severance. The endogenous hormones that are trapped on the apical side of the cut veins doubtlessly stimulate root initiation.

Leaves can be laid flat on moist earth or sand. In some plants (African violet, for example) the petiole can be placed in water or on moist sand or earth. Growth regulator treatment is not generally necessary for leaf cuttings because most of them root readily in a short time without such treatment.

Leaf Bud Cuttings

These cuttings consist of a leaf blade, a petiole, and a short piece of stem with the attached axillary bud. Such cuttings are of particular value for plants that are able to initiate roots, but not shoots, from detached leaves and are also valuable when rapid propagation is desired because each node can be used as a cutting. Rooting has been induced in a number of species by application of auxin to the stem portion (Fig. 5–8).

Root Cuttings

This type of cutting has often been used satisfactorily to produce new plants. The likelihood of good results is increased if the cuttings are taken in late winter or early spring before new growth starts and when abundant food is stored in the roots. Root cuttings have been used successfully in the propagation and cultivation of the Russian dandelion (*Taraxacum koksaghyz*), which was grown experimentally during World War II as a possible source of latex for rubber production. Auxins definitely enhance the rooting of this species. Root cuttings of blueberry can also be used successfully in propagation. However, auxins have generally hindered rather than improved propagation of root cuttings since no buds are present on the roots and since exogenous auxin does not favor bud formation.

FIGURE 5-8

Boysenberry leaf bud cuttings. *Top*, untreated control; *bottom*, basal ends were treated with IBA at a concentration of 4,000 ppm by the concentrated-solution-dip method. (After H. T. Hartmann and D. E. Kester, *Plant propagation: Principles and practices*, 2d ed., 1968. By permission of Prentice-Hall, Inc., Englewood Cliffs, New Jersey.)

LAYERING

Layering is often used to propagate species that fail to root or that root with extreme difficulty. In this method a root develops on a stem that is usually pegged down, except for the apex, in a shallow trench and covered with soil while it is still attached to the parent plant. After the layered stem has developed a root system, it is detached from the parent plant and then grows independently.

Root initiation in layered stems may be facilitated either by girdling or by tightly ringing the stem with wire. (In the latter method, the stem is "girdled" with wire as it grows in diameter.) By either technique the phloem is interrupted and the downward movement of carbohydrate assimilates and hormones is stopped so that they accumulate above the girdle and stimulate rooting. Application of exogenous auxins has sometimes been reported to stimulate rooting further.

Layering is frequently used to replace a missing grapevine in a vineyard. A rooted cutting that is directly planted to replace a missing vine usually fails to develop because there is too much competition from the surrounding vines. If the vine that arises from the layered shoot is not detached for one to two years, the young vine may draw on the parent vine for nutrients until it is established sufficiently to compete successfully with older vines.

Another type of layering is termed "marcottage" or "aerial layering." Slits or girdles are made on an aerial portion of the stem where rooting is desired. Growth regulators are sometimes applied, after which the area is enclosed in a wet, moisture-holding material, such as sphagnum moss. Polyethylene film is then wrapped around the moss to prevent evaporation of water. Both the naturally occurring growth regulators trapped above the girdles and the applied hormones stimulate root initiation. After a good root system has developed, the stem is severed below the root system and the severed portions are planted.

In guava the success of aerial layering can be improved by topping the shoot and applying IBA at a concentration of 5,000 ppm. Rooting is then 100 percent and survival is approximately 80 percent, compared with 68 percent survival of noninvigorated shoots (Mukherjee, 1967).

Because a clonal rootstock for pecan is not available for commercial use, seedling rootstocks must be employed. Aerial layering of the cultivar 'Stuart' has been improved by applying IBA (Sparks and Chapman, 1970). One-year-old nonjuvenile branches were girdled, treated with IBA at a concentration of 3 percent, and air-layered fifty days after bud break; they were left on the tree for approximately 5.5 months. Air layers were grown in the soil, and maximum survival occurred after one season's growth.

STOOLING

Stooling is a propagation technique, used commonly with apple trees, in which soil is mounded around the bases of young shoots to stimulate rooting. Stooling sometimes fails to result in initiation of roots unless plant regulators are also used.

Mango

Stooling of mango has been successful in India but only when ringing and plant regulators are used. The method of Mukherjee and Majumder (1963) is to head back three-year-old seedling plants to 6 to 8 cm above the ground level in March. When the sprouts are 25 cm long, soil is heaped around the shoot bases to a height of 10 cm, which causes the shoots to become etiolated. During July, the soil is removed from the base of the sprouts and IBA at a concentration of 5,000 ppm in lanolin is applied to a girdle ring. After eight to ten days, when root initials are visible, the shoot bases are again covered with soil. In September the rooted shoots are removed. In these experiments the survival rate after planting was 93 percent for shoots that are girdled and treated with IBA. Shoots that were not girdled or treated with IBA failed to survive.

FACTORS THAT INFLUENCE ROOTING

The primary factors that affect rooting of cuttings have been thoroughly discussed by Hartmann and Kester (1968). The use of growth regulators does not obviate the necessity for other recommended propagation practices, such as selection of good cutting material (including wood of proper size and age), use of a good rooting medium, maintenance of sufficient moisture, and choice of suitable light, aeration, temperature, and humidity, all of which are prerequisites for optimal root initiation.

Etiolation

Cuttings usually root more readily if they are initially grown in the dark so that the stems are in an etiolated condition. Herman and Hess (1963) found that etiolated 'Red Kidney' bean and Hibiscus stems rooted readily after application of IBA and that a slightly higher auxin content was present in etiolated tissues than in nonetiolated ones. During the propagation of stem cuttings the bases are darkened by the rooting medium, and it is possible that rooting is stimulated by etiolation effects. During the period of rooting, the etiolated site of the cutting has a higher auxin content than do portions in the light (Kawase, 1964).

Air layering of mango has not been adopted by nurserymen because the percentage of survival of the air layers after planting has often been low, even though rooting on the tree was high. This technique is

142

FIGURE 5-9

Etiolation of air layers of 'Bombay Green' mango increases amount of rooting resulting from application of a mixture of IBA at a concentration of 5,000 ppm and NAA at a concentration of 5,000 ppm. *Left*, nonetiolated; *right*, etiolated. (Photograph courtesy of S. K. Mukherjee.)

of value, however, for the propagation of mango clones from mature mango trees. In India Mukherjee and Bid (1965) found that etiolation plus growth regulator treatment results in profuse rooting and that air layers can be successfully established after transplanting (Fig. 5–9). In March the terminal portions of shoots are placed in tubes of black paper containing moist vermiculite. After fifteen to twenty days, shoots emerge from the tops of the tubes and grow rapidly. Approximately three months later, etiolated portions of shoots are treated with growth regulator and are air layered. These investigators found that application of IBA at a concentration of 10,000 ppm or NAA at a concentration of 5,000 ppm to the surface of a girdle induces 100 percent rooting in the etiolated shoots of mango air layers. When the etiolated shoots were cut off and grown under mist, the survival rate, after one year, of shoots treated with IBA was 95 percent, compared with a rate of 90 percent for shoots treated with NAA. Air layers of etiolated shoots that were not treated with growth regulator produced roots, but to a lesser extent than those that were treated. However, none of these layers survived after planting.

Vitamin B_1 (Thiamine Chloride)

Some experiments conducted in the late 1930's led to the belief that vitamin B_1 is an effective root stimulator. However, subsequent experiments produced erratic and usually unsuccessful results. Evidently most plants have a sufficient supply of this vitamin, which is necessary for root growth, so no response is obtained from exogenous application.

Mineral Nutrients

Nitrogenous compounds, such as amino acids and ammonium sulfate, have stimulated rooting in a number of plants, possibly because of an increased synthesis of auxin; but in general no stimulation of rooting results. Boron seems to have a synergistic effect with IBA on the rooting of cuttings of English holly (Weiser and Blaney, 1960). Application of boron alone often promotes root growth, although it does not affect root initiation.

Girdling

The value of using girdling in conjunction with layering to enhance root initiation has already been discussed. Girdling of stems usually results in an increase in the production of auxin above the girdle for about ten days, after which there is a gradual decrease, often correlated with a cessation or retardation of shoot growth (Kato and Ito, 1962). In pear, girdling of stems results in greater root initiation than does application of auxin, which indicates that girdling does something more than merely increase the auxin content (Higdon and Westwood, 1963). An explanation of this phenomenon may be found in some studies made by Stolz and Hess (1966) on the effect of girdling easy-to-root and difficult-to-root varieties of *Hibiscus*. These studies showed that girdling has a different effect on the rooting cofactors in the two types of plants. The cofactors that these investigators numbered 1 and 2 increased above the girdle in both varieties during the first ten days, but cofactor 4 increased in the tissues above the girdle of the easy-to-root variety only.

Fungicides

Cuttings are subject to attacks by various fungi during the rooting period. However, treatment of cuttings with fungicides frequently results in survival of the cutting and improved root quality. It is frequently unclear whether fungicides affect root initiation directly or whether they merely protect the cutting from fungus attack. Several fungicides, including Ferbam and Phygon XL, have been reported to improve root quality (Doran, 1953; White, 1946).

Several investigators have noted that Captan results in increased rooting (Van Doesburg, 1962; Wells, 1963; and Hansen and Hartmann, 1968). Hansen and Hartmann (1968), working under the mild winter conditions of central California, investigated the effects of IBA, Captan, and mixtures of the two on the rooting and survival of hardwood cuttings of a clonal peach rootstock and a clonal peach-almond hybrid rootstock, both of which were resistant to root-knot nematode. Treatment with IBA was generally found to be required for satisfactory rooting, and application of Captan (25 percent wettable powder) was found to increase the percentage of cuttings that survived. In one experiment cuttings were taken on December 12 and were planted in nursery rows. Survival counts were made on June 21 of the following year. Percentages of survival were as follows: for controls, 1.3; for cuttings dipped briefly in IBA at a concentration of 4,000 ppm, 49.0; for cuttings dipped briefly in IBA at a concentration of 4,000 ppm and then briefly in 25 percent Captan, 88.7; and for cuttings dipped only in 25 percent Captan, 12.0. The survival rates for the controls and for the cuttings that received only Captan were not significantly different, but the rates for the other treatments were significantly improved. The investigators concluded that the higher survival rate that resulted from Captan treatment is probably due to the protection it provides against attacks by soil-borne organisms during the rooting period rather than to direct stimulation of root formation or development.

INCREASING THE TAKE OF GRAFTS BY USING GROWTH REGULATORS

Because auxins often stimulate cambial activity, they are sometimes used to improve the take of grafts. In Hungary NAA is used widely to increase the percentage take of grape grafts: More than six million rootstock cuttings received such a treatment in 1964. Rootstocks approximately 40 cm long are soaked in water for several days and are then placed upside down for sixty hours in a solution of NAA at a concentration of 10 ppm (Fig. 5–10). The scions are not treated. A short whip graft is then made by hand. The grafts are packed in sawdust and are placed in a cool room until calluses form. The NAA treatment increases the take by 10 to 20 percent.

The take of bud grafts is sometimes increased by pretreating the budwood with growth substances. Samish and Gur (1962), experimenting with budding avocado, immersed the budwood for twenty-four hours in a solution of IAA at a concentration of 25 ppm. This procedure improved the take of bud grafts, particularly when older

FIGURE 5-10

Rootstocks are soaked in concrete tanks, first in water for several days and then, upside down, in NAA at a concentration of 10 ppm. This procedure is believed to increase the graft takes. (Photograph courtesy of J. Eiffert.)

rootstocks were used. The percentage success of bud grafts performed on three-year-old stocks grown in rows in the nursery was increased 40 percent as a result of IAA budwood pretreatment. The immersion of budwood in distilled water alone reduced the take of the bud grafts, as compared with that of the nonimmersed controls, but application of IAA overcame this deleterious effect (Samish and Gur, 1962).

Supplementary readings on the general topic of rooting and propagation are listed in the Bibliography: Audus, L. J., 1959; Hartmann, H. T., and Kester, D. E., 1968; and Leopold, A. C., 1955.

Chapter 6

DORMANCY

The growth activity of higher plants usually undergoes marked seasonal changes. In temperate-zone perennials, periods of visible growth alternate with periods of little or no growth, whereas in annuals death of the entire plant, except for the seeds, occurs. Actively growing plants show little resistance to such unfavorable external conditions as frost, heat, and drought, whereas dormant plants exhibit very high resistance. Thus the onset of the dormant state before unfavorable environmental conditions begin usually ensures plant survival. Except for certain tropical species, almost all plants undergo a period of dormancy.

Dormancy is a difficult term to define, and many definitions have been formulated (Amen, 1968). There are two general causes of dormancy. Growth may be stopped by such external conditions as unfavorable water supply or temperature or by internal factors that prevent growth even though environmental conditions are favorable. The former type of dormancy is termed *quiescence* and is under *exogenous* control; the latter type is termed *rest* and is under *endogenous* control (Samish, 1954; Dennis and Edgerton, 1961).

Dormancy is observable in many plant parts, including seeds, buds, and bulbs. Most studies have focused on seeds, probably because of their easy availability. (For articles on dormancy, see Amen, 1968; Lang, 1965; Rappaport and Wolf, 1969).

CAUSES OF SEED DORMANCY

The failure of apparently ripe seeds to germinate can result from one factor or from a combination of factors (Amen, 1968; Bonner, 1965). The main causes of seed dormancy are: (1) rudimentary embryos, (2) physiologically immature embryos, (3) mechanically resistant seed coats, (4) impermeable seed coats, and (5) presence of germination inhibitors. Dormancy caused by rudimentary embryos is found in orchids, ginkgo, and holly. In these species the embryos are still imperfectly developed when the seeds are shed, and no germination can occur until embryo development is complete.

In lettuce, barley, and many trees, for example, the embryos are completely developed when the seeds are shed, but they fail to germinate when placed under environmental conditions favorable for growth. Such seeds germinate only after a period of after-ripening. In nature, seeds of many plants after-ripen during the low temperatures of winter; sometimes several years are required. Man subjects seeds to a moist condition and low temperature for several weeks or months, depending on the species, to after-ripen the seeds. During this period physiological changes occur in the embryo that allow germination to occur.

Seeds of some plants, including many leguminous species, have such resistant seed coats that the embryos are unable to expand and develop. Under natural conditions the structural strength of such seed coats is gradually broken down by freezing and thawing, by leaching, by passage through an animal's digestive tract, or by certain light and temperature conditions (Varner, 1965). Man can reduce or eliminate the seed coat barrier by mechanical scarification procedures or by other treatments that so weaken or rupture the seed coats enough that water can easily enter the seed. Seed coats can also be made permeable by treatment with strong mineral acids, but care must be taken not to injure the embryo.

Some seeds, such as those of cocklebur and many legumes, fail to develop because their seed coats are impermeable to oxygen and water. These seeds will germinate only if the seed coat is ruptured. The intensity of dormancy of intact seeds probably diminishes as the seed coats gradually become more permeable to oxygen.

The presence of germination inhibitors in the fruit tissues prevents the germination of some seeds while they are still in the fruit. An interaction of inhibitors and promoters is part of the process that determines the onset and termination of rest. In recent years research has, for the most part, been concentrated on this interaction and its relation to certain light and temperature conditions.

CAUSES OF ONSET AND TERMINATION OF SEED REST

According to Amen (1968), seed dormancy can be divided into four relatively distinct developmental phases: (1) *induction*, characterized by a striking decrease in hormone levels; (2) *maintenance*, a period of partial metabolic arrest; (3) *trigger*, a time when the seed is especially sensitive to environmental cues, and (4) *germination*, characterized by increased hormonal and enzymatic activity followed by growth of the latent embryonic axis (see Amen, 1968, for a more detailed description than that presented here).

Induction Phase

Processes occur during the maturation of the seed that lead to the onset of rest. These processes may be triggered or affected by changes in light, temperature, chemicals, and other environmental factors. The light and temperature conditions that prevail during seed development affect the later germinability of lettuce seeds (Koller, 1962), indicating that rest is preconditioned. Pretreatment of certain seeds with varying temperatures and chemicals also has a noticeable effect on rest. Pretreatment of peanut seeds at 40°–50° C may shorten their rest period from forty days to fifteen (Bailey et al., 1958).

Dormancy and germination are among the many plant growth responses that are probably controlled by the balance of growth inhibitors and promoters. This balance seems to be shifted in favor of the inhibitory substances during seed maturation, resulting in a resting condition. For example, in the early development of seeds of mazzard cherry (*Prunus avium*), high concentrations of promotive substances are present, but as the fruit ripens and embryo growth is arrested, these substances rapidly decline, resulting in a higher ratio of inhibitory to promotive materials (Pillay, 1966). Auxins, gibberellins, and cytokinins can serve as promoters; ABA is probably an important part of the inhibitor complex.

The development of impermeable seed coats may indirectly affect rest by altering the environment of the interior of the seed. For example, impermeable seed coats can reduce the oxygen content inside the seed. Under anaerobic conditions the seed may synthesize growth-inhibiting substances (Roberts, 1964).

Maintenance Phase

During this period of *rest*, general metabolism is very low and the inhibitor-promoter balance is still weighted in favor of the inhibitor. The maintenance of seed rest results from the presence of certain endogenous inhibitors that cause partial and/or specific metabolic blocks (Amen, 1968). The concept of an inhibitor-promoter balance as the regulatory mechanism in many types of seed dormancy is supported by the effects of exogenous growth substances on rest. Specific inhibitors can impose rest on a seed that would otherwise germinate, and applications of promoters can sometimes terminate rest. For example, application of GA_3 can terminate rest in seeds of grape (Yeou-Der *et al.*, 1968), cherry (Fogle and McCrory, 1960), and many woody species (Frankland, 1961). On the other hand, rest in ash seeds can be extended by application of ABA (Sondheimer and Galson, 1966).

Trigger Phase

In this phase some *triggering agent* shifts the promoter-inhibitor balance in favor of the promoter. In seeds requiring light, removal of inhibitor by leaching or scarification, or a thermochemical reaction such as that occurring in stratification (see p. 150) and after-ripening, the triggering agent may be photochemical in nature. The actual *germination agent* is usually a hormone. The action of this mechanism is determined by the type of hormones or their concentrations.

LIGHT AND TEMPERATURE AS TRIGGER MECHANISMS. Light is essential for the germination of many seeds and encourages germination in others; for example, light greatly increases the percentage of germination in blueberry, strawberry, blackberry, and raspberry (Scott and Draper, 1967).

A photochemically reversible reaction caused by the pigment *phytochrome* and its response to light of different wave lengths may sometimes control germination. Borthwick *et al.* (1952) demonstrated the repeated reversibility of this photoreaction using lettuce seed:

Type of Light:	red	far-red
Form of Pigment:	phytochrome 660 ⇌	phytochrome 730
	(red-absorbing)	(far-red-absorbing)
Response:	germination	germination
		inhibited

Phytochrome probably occurs in all higher plants. In the dark it is in the stable form with absorption maximum at 660 mμ, but in red light it is converted back to the far-red form with absorption maximum at 730 mμ. In far-red light the far-red form is instantly converted back to the red form; in the dark, in the presence of oxygen and at favorably low temperatures, this reaction occurs only slowly. Under natural light, red wave lengths dominate the far-red so that phytochrome is converted to the active far-red form. The latter form is believed to cause a chain of reactions that ultimately result in germination.

The role of light in the termination of rest is not always clear since the plant response varies with the kind of seed and the environmental conditions under which the seed is grown. Light induces the production of specific enzymes that are essential for growth. In some seeds, the light requirement can be partly or fully replaced by such compounds as gibberellin, cytokinin, and thiourea. GA$_3$ stimulates the synthesis of ribonuclease, amylase, and proteases in the barley endosperm (Varner, 1967). Thus the hydrolytic systems of dormant seeds appear to be under the control of endogenous hormones.

Many seeds require exposure to low temperatures (0°–10°C) and moist conditions for the termination of rest. This treatment is called *seed stratification*. The amount of time required for termination of seed rest varies from a few weeks to several months; some seeds require even longer periods. During stratification there is usually a decline in inhibitory materials and an increase in promoters in the embryo. In some seeds, daily alteration of low and high temperatures are more effective for termination of rest than are constant temperatures.

Germination Phase

During the early stage of germination the dry seed has imbibed water, seed coats have softened, and hydration of the protoplasm has occurred. After rest is terminated, the seed completes the germination process if external environmental conditions are favorable and if there are no other limiting factors, such as hard seed coats. Metabolic activity increases and there is a corresponding increase in enzyme activity and respiratory rate. Gibberellin plays an important role in increasing the metabolic activity. In cereal grains gibberellin appears in the embryo and is translocated to the aleurone layer (the outer layer of endosperm, one or two cells thick) where it activates enzymes. One of these enzymes, α-amylase, is secreted into the endosperm where it converts starch to sugar. Complex insoluble food reserves, including fats, carbohydrates, and usually proteins, are digested to soluble forms that are translocated to the growing areas. Assimilation of these substances in the meristems provides energy for cellular activity and growth. The seedling develops by division, enlargement, and differentiation of cells at the growing point and is dependent on its own food reserves until green leaves develop and actively produce assimilates for it.

SPECIAL TYPES OF SEED DORMANCY

Sometimes the failure of seedlings to develop from seeds is not a result of any process that occurs during seed dormancy. In some species the epicotyls push through the seed coat but then remain dormant. Young root systems can also emerge from the seed only to revert to a dormant condition. Both of these types of dormancy can usually be terminated by exposure to low temperatures (0–10°C) or by treatment with growth substances.

BUD REST AND DORMANCY

The buds of such woody plants as grape and tree fruits undergo a period of dormancy each year, the length of which varies with the species. The mechanisms that determine bud rest and dormancy are believed to be similar to those in seeds, and therefore it is not surprising that rest in buds can be altered by light, temperature, and application of plant growth substances.

Light controls dormancy of some trees. Both the onset and termination of rest can be controlled by photoperiod or day length. Wareing (1953) has demonstrated that the buds of some woody trees perceive the photoperiodic stimuli and that the presence of leaves is unnecessary for this function. Erez et al. (1966) have shown that leafless dormant shoots of peach are light receptors and that light can terminate rest of the buds.

Low temperatures often terminate bud rest; temperatures just above freezing are usually optimal. Heat treatment can also terminate

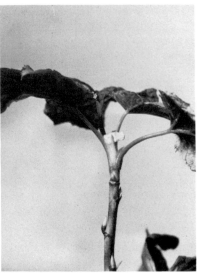

FIGURE 6-1

Effect of ABA on growth of European black currant. Apex of shoot of untreated plant on left grows actively under long days, but dormancy has been induced by ABA in plant on right, also growing under long days. (Photographs courtesy of P. F. Wareing.)

rest in some species; soaking resting grape canes in water at 30°C for twenty-four hours has been found effective (Weaver *et al.*, 1961). Loomis and Evans (1928) provided a more classic example, breaking rest in gladiolus corms by soaking them in water at 39°C for two weeks.

Plant growth substances are important in bud dormancy and rest. From the results of his work on dormant potato buds (1949a) and ash buds (1949b), Hemberg advanced the theory that rest is controlled by the inhibitor content of the buds. He found that a high inhibitor content exists in resting buds but that it declines rapidly at the termination of rest. The present view is that dormancy in buds, as in seeds, is controlled by a change in the inhibitor-promoter balance.

Treatment with exogenous growth substances can also terminate or prolong bud rest, which gives further support for the importance of hormones in the control of rest. GA_3 has been found to terminate the rest of buds of peach (Donoho and Walker, 1957), potato (Brian *et al.*, 1955; Rappaport, 1956), some forest trees (Larson, 1960), and certain other plants. Gibberellic acid also substitutes in part for the chilling requirement in pear (Brown *et al.*, 1960). However, gibberellins delay the termination of rest in buds of grape (Weaver *et al.*, 1961) and

FIGURE 6-2
Budbreak of *Fraxinus americana* is delayed by ABA treatment. *Left to right:* plants growing in ABA solution at concentrations of 0, 0.4, 2.0, and 10.0 ppm. (After C. H. A. Little and D. C. Eidt, 1968.)

cherry when applied the previous year (Brian *et al.*, 1959b). Application of auxins to dormant buds has delayed bud break in peach (Marth *et al.*, 1947) and grape (Nigond, 1957). Cytokinins have been shown to terminate rest in fruit trees (Benes *et al.*, 1965) and grape (Weaver, 1963).

Application of ABA has been found to induce dormancy in sycamore and black currant (El-Antably *et al.*, 1967). The effect of ABA on European black currant is illustrated in Figure 6–1. When ABA is applied through cut bases of stems of *Salix viminalis* and *Ribes nigrum*, it retards bud development (El-Antably *et al.*, 1967). Little and Eidt (1968) obtained similar results with *Acer rubrum* and *Fraxinus americana* (Fig. 6–2). ABA almost completely inhibits sprouting of potato buds (Blumenthal-Goldschmidt and Rappaport, 1965). Phillips (1962) demonstrated that naringenin, an inhibitor found in peach buds, competitively antagonizes gibberellin's effect of breaking the rest of peach buds.

EFFECT OF ENDOGENOUS GROWTH SUBSTANCES ON REST

The hypothesis that the balance between promoters and inhibitors determines whether a seed or bud is in a resting or quiescent state is based mainly on the effects of applied growth substances. Before the truth of this theory can be assumed, however, the processes that occur in the seed or bud must first be ascertained. Even though experiments with endogenous growth substances have sometimes produced conflicting results, there are considerable data to indicate that hormonal control of bud rest sometimes operates through the inhibitor-promoter balance. Some of the experimental results will now be discussed.

Inhibitors

ABA is probably one of the most potent physiologically active inhibitors found in resting buds and seeds. ABA or abscisic-like substances have been shown to decline during stratification in several seeds including walnut (Martin et al., 1969), peach (Lipe and Crane, 1966), and apple (Rudnicki, 1969). Chilled seeds of Fraxinus americana were found to contain approximately one-third as much ABA as nonchilled seeds (Sondheimer et al., 1968).

Phillips and Wareing (1958) found that a correlation between inhibitor levels and winter rest exists in species of Acer and that short days in the autumn increase the inhibitor level in buds and leaves of such species.

Gibberellins

Many investigators have attempted to demonstrate an increase in growth promoter near the termination of the rest period, and some have noted an increase in gibberellin. Wareing (1969) found that when shoots of European black currant are stored at 2°C, gibberellin levels increase during the period of chilling. Chilling of hazel seeds results in formation of gibberellins, but this process occurs after transfer from the chilling temperature to 20°C (Ross and Bradbeer, 1968).

Interaction of Gibberellins and ABA

An interesting theory is that an interaction between gibberellins and ABA occurs during dormancy. According to this theory, induction of rest is accompanied by high levels of ABA and low levels of gibberellin, but at termination of rest the reverse is true. Wareing (1969) has described various ways that gibberellins and ABA interact in bud dormancy. If birch seeds are induced to form buds by application of ABA, they will rapidly renew growth after treatment with GA_3. If quiescent shoots of birch are collected in the spring and placed under conditions favorable for growth, bud break is inhibited by standing the shoots in solutions of ABA, but if gibberellin is mixed with the ABA, bud break occurs.

Cytokinins

Domanski and Kozlowski (1967) noted that after rest was terminated in buds of *Betula* and *Populus*, cytokinin-like activity increased until shortly before the buds opened and decreased thereafter, indicating that cytokinins may interact with other growth promoters in termination of rest.

EVALUATION OF MECHANISMS
AFFECTING DORMANCY

Plant responses obtained from exogenous applications of growth substances indicate that dormancy is controlled by a hormonal mechanism. Correlations of the endogenous promoters and inhibitors (especially gibberellin and ABA) occurring in some buds and seeds support the hypothesis that the inhibitor-promotor balance determines the beginning and termination of rest. High inhibitor-to-promoter ratios induce rest, and high promoter-to-inhibitor ratios terminate rest. However, proof of this mechanism depends on knowledge of the mode of action of the hormones at the molecular level. Among the other hypotheses that have been advanced are that (1) the restricted gaseous exchange rate, especially of oxygen, by bud scales or seed coats induces dormancy (Vegis, 1964) and (2) repression and depression of DNA occur during dormancy (Bonner, 1965). All theories may prove to be valid for some plants.

HASTENING OF SEEDLING EMERGENCE
FROM QUIESCENT SEEDS

Early emergence and rapid growth of seedlings may be of considerable advantage in that they enable young plants to avoid many of the insect and disease hazards and crusting of soil surface that frequently accompany their germination and early growth. Soaking seeds in gibberellin or coating them with a slurry containing the growth regulator frequently accelerates germination. Hayashi (1940) found

156

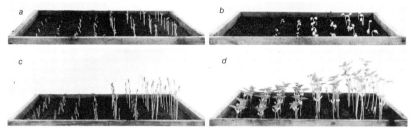

FIGURE 6-3

Effect of incorporating gibberellin in slurry seed treatments on emergence of seedlings of 'Alaska' pea (*a* and *c*) and 'Sanilac' bean (*b* and *d*), at 13°C (*a* and *b*) and 18°C (*c* and *d*). Paired rows in *a*, *b*, *c*, and *d*: *left*, treated with seed protectant slurry (Delsan AD) only; *right*, treated with seed protectant slurry containing gibberellin at concentrations of 500, 2,500, and 10,000 ppm. In similar experiments, application of gibberellin at concentrations of 2,500 and 10,000 ppm often resulted in excessive stem elongation and severe seedling abnormalities. (After S. H. Wittwer and M. J. Bukovac, 1958.)

that application of gibberellin to barley and rice results in more rapid growth and that germination is hastened as the concentration is increased. More recently, Wittwer and Bukovac (1957d, 1958) studied the effect of gibberellin treatments on seeds of pea, bean, and sweet corn. When gibberellin was incorporated with a slurry seed protectant and applied to the seed coats of pea and bean, emergence was hastened in both greenhouse and field plantings (Fig. 6–3). The optimal concentrations of compounds were from 500 to 1,000 ppm, and gibberellin-treated seedlings usually emerged three or four days before the control seeds. However, there was no effect on final germination percentage. Treating seeds of sweet corn with gibberellin did not yield favorable results (Wittwer and Bukovac, 1958).

In California the fruit of 'Duke' avocado is of mediocre quality, but growers utilize it as a rootstock because of its tolerance to the avocado root rot fungus, *Phytophthora cinnamomi*. However, nurserymen have not used 'Duke' avocado extensively because germination of its seeds is usually slower and less uniform than that of other rootstocks. Tests made at Riverside, California, showed that seeds soaked in GA_3 exhibited relatively rapid germination and growth (Burns *et al.*, 1966). Seeds were obtained in November and stored at 3°C. In December each seed was cut at top and bottom and the seed coats were removed. The seeds were soaked for twenty-four hours in solutions of KGA_3 ranging from 0 to 10,000 ppm and were then planted. Within two weeks it was evident that seedlings soaked in gibberellin at concentrations of 1,000 and 10,000 ppm germinated earlier and were taller. The plants were moved from the greenhouse in June and were allowed to grow in partial shade. On July 20, average heights in inches for plants whose seeds were soaked in gibberellin at concentrations of 0, 1, 10, 100, 1,000, and 10,000 ppm were 16.6, 16.0, 15.5, 17.3, 21.7, and 24.9, respectively. Plants whose seeds were soaked in gibberellin at concentrations of 1,000 and 10,000 ppm were significantly larger than those whose seeds were soaked in weaker concentrations.

TERMINATION OF SEED REST BY APPLICATION OF GROWTH SUBSTANCES

Seeds of many plants often require an extended period of after-ripening at low temperatures before they will germinate. Consequently, some means of terminating rest would definitely be of value to accelerate breeding programs.

Early Work on Effect of Gibberellin on Woody Species

Soon after the discovery of gibberellin, the hormone was applied experimentally to the seeds of several woody species that grow in temperate regions (Frankland, 1961). These seeds are in a resting condition when harvested and require a period of chilling (stratification) before they are capable of germination. Seeds of European hazel (*Corylus avellana*) and European beech (*Fagus sylvatica*) were sown on filter paper or cotton soaked with water or GA₃ solution. Almost no unchilled seeds of either variety germinated in water, even after removal of pericarp and testa. However, application of gibberellin at a concentration of 100 ppm caused germination of both varieties within three weeks. Beech embryos responded to gibberellin at a concentration of 1 ppm. Nuts of both beech and hazel with pericarp and testa intact failed to germinate even in the presence of gibberellin at a concentration of 500 ppm.

These tests also showed that gibberellin stimulates germination of birch seeds but has no effect on dormant sycamore seeds and that delayed epicotyl extension in acorns of *Quercus petraea* can be overcome by gibberellin treatment.

In all seeds used, coat structures were found to prevent a response to gibberellin; lack of response could be caused by failure of the compound to penetrate the coat structures or failure of the embryos treated with gibberellin to overcome the retarding effect of the coat structures. If the latter were the cause, then gibberellin only partially replaced the chilling requirement in these tests, since fully stratified seeds germinated in spite of the presence of coat structure.

Subsequent research has revealed that gibberellins stimulate germination in a wide range of plants. Gibberellins have been found to vary in their effect on seed germination; GA_4 is often more effective in terminating rest than is GA_3 (Thompson, 1969).

Citrus

The sweet orange, *Citrus sinensis*, is a valuable rootstock but its seed is slow to germinate. Burns and Coggins (1969) demonstrated that soaking seeds of sweet orange in water for twenty-four hours hastens germination and that soaking them in gibberellin at a concentration of 1,000 ppm is even more effective. Approximately two months after the seeds were planted, their rates of germination were as follows: seeds that were not soaked, 60 percent; seeds that were soaked in water, 66 percent; seeds that were soaked in gibberellin at a concentration of 1,000 ppm, 79 percent. Eleven months after planting, seedlings that had been soaked in gibberellin were somewhat larger and more uniform.

Camellia

The effects of various gibberellin treatments on the germinability of seeds of several varieties of camellia were investigated by Furuta (1961). He discovered that seeds sown after harvest germinate slowly but readily, and usually a tap root begins to develop. The epicotyl does not begin to develop until some time later. A solution of gibberellin at a concentration of 100 ppm was generally found to be the most effective in hastening germination. For example, when seeds of camellia 'Sasanqua Texas Star' were presoaked for twenty-four hours in gibberellin at concentrations of 0, 50, 100, and 500 ppm before planting, germination rates fifty days after sowing were 3, 44, 60, and 52 percent, respectively. Epicotyl growth of treated seeds was much more rapid. Furuta found gibberellin dusts to be quite ineffective in stimulating germination.

Grape

Grape seeds are in a resting condition at fruit maturity. The usual method of terminating rest is to stratify the seeds at a low temperature

FIGURE 6-4

Seeds of 'Tokay' grape 22 days after soaking for 20 hours in KGA at concentrations of (*left to right*) 0, 100, 1,000, and 8,000 ppm. (After K. Yeou-Der, R. J. Weaver, and R. M. Pool, 1968.)

(approximately 5°C) for three months. Randhawa and Negi (1964) showed that the required stratification period can be shortened by application of gibberellin. More recently, Yeou-Der *et al.* (1968) demonstrated that soaking seeds of 'Tokay' grape in high concentrations of gibberellin (8,000 ppm) for twenty hours can completely replace the cold requirement (Fig. 6–4). Application of the compound at a low concentration of 10 ppm terminated the rest in 'Tokay' seeds if they had first been scarified with a rotating sandpaper disk. This result indicates that the seed coat does hinder, to a degree, the penetration of gibberellin into the interior of the seed; it also explains why high concentrations must be used on unscarified seeds. It may be that the seeds are in a state of rest due to a deficiency of gibberellins and that this deficiency can be overcome by application of exogenous gibberellin. In contrast to gibberellin, application of the cytokinin BA at a concentration of 8,000 ppm did not replace the cold requirement.

Apple

For commercial production of apple seedlings, the seeds are usually stratified under moist conditions at a temperature of 5°C for seventy to eighty days. However, when ruptured seeds or excised embryos from nonstratified seeds are exposed to temperatures favorable for germination, a few embryos do grow but they fail to elongate normally. Such plants are termed "physiological dwarfs."

Gibberellin was the first chemical shown to stimulate the growth of physiological dwarfs. In 1956 Barton applied lanolin preparations or aqueous solutions of GA_3 to physiological dwarfs from embryos of *Malus arnoldiana* that had not been after-ripened. Treatment resulted in extension of the seedlings' internodes and consequent elimination of the dwarfed condition.

Although gibberellin accelerates the germination of dormant rosaceous seeds when the seed coat has been removed, it fails to have any effect when the coat structures are intact (Nekrasova, 1960). This observation was confirmed by Frankland (1961). Using seeds of apple (*Malus*) cultivars 'Red Stoke' and 'Medaille d'Or,' he found that gibberellin stimulates isolated embryos to germinate but has no effect if the endosperm is left intact, even if the testa is first removed.

Subsequently, a cytokinin was found to have effects somewhat similar to those of gibberellin. Badizadegan and Carlson (1967) showed that soaking excised embryos from mature seeds of 'McIntosh' and 'Wealthy' apple in BA at concentrations of 5 to 25 ppm significantly stimulates rate and percentage of germination. Twenty-eight percent of the nontreated excised embryos germinated, but only a few of them showed elongation of the radicles and epicotyls. Treated embryos germinated rapidly, and application of BA at a concentration of 25 ppm resulted in 66 and 50 percent germination of the mature 'McIntosh' and 'Wealthy' embryos, respectively. These embryos were planted in soil and soon grew into small seedlings, but they were dwarfed and had rosetted leaves and short internodes. Most of the 'Wealthy' seedlings failed to survive.

Badizadegan and Carlson (1967) also studied the effect of BA on embryos that had been excised from dry stored seeds after having been soaked in water for one to five days. After one day of soaking there was no difference in germination between treated and untreated embryos. However, a three- to four-day soaking in water, followed by a twenty-four-hour soaking in BA at a concentration of 10 ppm, greatly increased germination. These investigators suggest that since cytokinins stimulate cell division and plant growth, the inhibitor in the embryo was partly diluted by growth, thus enhancing embryonic development.

Peach

The standard amount of time required for stratification of peach seeds is sixty to one hundred days at 5°C. Gibberellin treatment can replace at least some of the low-temperature requirement. Donoho and Walker (1957) tested the effect of gibberellin on seeds of 'Elberta' peach that had received only thirty-five days of stratification, during which they were placed in a moist medium and stored at a near-freezing temperature. Before planting, the seeds were soaked for twenty-four hours in solutions of GA_3 at concentrations of 20 to 1,000 ppm. After sixteen days the germination rates for controls and for seeds treated with gibberellin at concentrations of 0, 20, 100, 200, 500, and 1,000 ppm were 30, 50, 80, 70, 40, and 30 percent, respectively. Concentrations of 100 or 200 ppm increased germination, but concentrations higher than 200 ppm resulted in poor germination. Twenty days after planting, plants grown from seed that had been treated with a concentration of 100 ppm had 48 percent more top growth than the untreated control plants. The roots of plants grown from seeds that had been treated with concentrations of 100 or 200 ppm were much larger than those of the untreated controls. Seeds that had been soaked in the 100 ppm solutions produced plants with root systems of 56 percent greater length and 80 percent greater fresh weight than did the untreated seed.

There is probably a varietal difference in the response of peach to growth regulators. Seeds of peach (*Prunus persica*) cultivars 'Amber Gem,' 'Suncling,' and 'Rutgers Red-Leaved,' produced either negative or poor results after treatment with GA_3 at a concentration of 100 ppm (Carlson and Badizadegan, 1967). However, application of BA at 10 or 20 ppm was effective in stimulating germination of the excised embryos. Seeds were removed from the pits, soaked in solutions of thiourea, GA_3, or BA for sixteen hours, and placed in petri dishes. Three days after treatment, the BA-treated embryos had turned green and the cotyledons had separated; by the fifth day embryo growth was evident and the cotyledons were much greener than those of the controls. When planted in soil, seedlings treated with BA continued to grow more vigorously than seedlings that had been treated with other compounds.

Kola

Seeds of kola (*Cola nitida*) require a period of dormancy or after-ripening before they begin effective germination and vigorous seedling growth. Following a seven-month storage period, kola seeds usually germinate within three to four months after planting; freshly harvested kola seeds require from three to nine months to germinate after planting (Ashiru, 1969). Ashiru found that kinetin, as well as thiourea and thiourea dioxide, increases the percentage of germination. The average numbers of freshly harvested seeds of a twelve-seed sample that germinated after they had been soaked for twenty-four hours in distilled water, kinetin at a concentration of 100 ppm, thiourea at 1,000 ppm, and thiourea oxide at 1,000 ppm, were 4.93, 6.68,

6.75, and 7.12, respectively. None of the chemicals completely substituted for the postharvest storage requirement of freshly harvested kola seeds, but the chemicals' facilitation of germination reduced the storage time required after planting and produced more uniform seedlings.

Cherry

Poor germination of seeds of cherry (*Prunus avium*), particularly those of early maturing varieties of sweet cherry, has been a major problem for the breeder, as well as for the nurseryman using mazzard seedlings as rootstocks for cherry varieties. Seeds of sweet cherry require an after-ripening period of at least six months at 3°C for germination. Gibberellin can substitute for part of this after-ripening period. Soaking the seeds for twenty-four hours in gibberellin at a concentration of 100 ppm immediately after collection substitutes for two or three months of the after-ripening period (Fogle, 1958). A combination treatment of two to four months of after-ripening followed by a gibberellin soak seems to offer the most promise for reducing storage losses and facilitating germination during the season when the seed is produced. However, when seeds are treated with gibberellin but are not after-ripened, some of the seedlings produce terminal buds and rosette, a condition that is typical of plants that receive insufficient after-ripening. The condition can be alleviated by application of foliar sprays of gibberellin at a concentration of 100 ppm at three-week intervals.

More recently, Pillay *et al.* (1965) experimented with mazzard and mahaleb cherry, two species that differ distinctly in endocarp thickness. Mazzard seeds were found to require between 120 and 150 days of after-ripening at 7°C in a moist medium; mahaleb seeds require 79 to 90 days. Soaking seeds in gibberellin at a concentration of 100 ppm increases the rate of germination and partially replaces the chilling requirements. Removal of the endocarp does not hasten germination or curtail the necessary chilling requirements. The investigators found that chilling seeds for twenty-four and thirty-four days at 7°C and then soaking them in gibberellin at a concentration of 100 ppm results in 75 to 100 percent germination. It was also found that seeds whose endocarp was left intact produced better germination than seeds whose endocarp had been removed and that were subsequently treated with gibberellin. Even excised seeds seem to require a certain amount of after-ripening in a moist medium.

Endive

Seedlings of unvernalized endive (*Cichorium endivia*) grow rapidly after treatment with gibberellin. Harrington *et al.* (1957) placed seeds of 'Fulheart No. 5' in an environment favorable for growth. When the seedlings emerged, 0.5 ml of an aqueous solution of gibberellin at a concentration of 1,000 ppm was applied to the stem apex of each. Some plants were treated once and others were treated weekly with 50 μg of the solution until first anthesis. Ten days after the first application of gibberellin, stem elongation was visible on all treated plants irrespective of vernalization. Repeated applications resulted in increased stem elongation. Flowering occurred on vernalized plants of all treatments, and weekly applications of gibberellin were found to induce flowering in nonvernalized plants.

Lettuce

In the desert areas of Arizona and California, the winter lettuce crop is often planted when soil temperatures are extremely high, and as a result much of the seed remains dormant. A soil temperature of 27°C usually induces such dormancy in seeds (Borthwick and Robbins, 1928). Kinetin and thiourea are the two main chemicals that have been found effective in overcoming high-temperature dormancy. In California, Smith *et al.* (1968) dipped three varieties of lettuce seed in kinetin at a concentration of 100 ppm for three minutes and obtained a striking increase in the percentage of germination at a soil temperature of 35°C. When this procedure was utilized in a total of eight tests performed with cultivar 'Great Lakes-659,' an average germination rate of 67 percent was obtained whereas none of the untreated controls germinated. Both kinetin and BA were effective at a soil temperature of 30°C, but at 35°C only kinetin had a marked effect on germination (Fig. 6–5). There was a good germination response when seeds were submerged in kinetin and then planted in soil at a temperature of 30°C for sixteen hours, but the percentage of germination was much higher after forty-eight hours. Kinetin also reduced the incubation time at low temperatures that was necessary for successful germination at high temperatures.

Other studies have shown that kinetin is effective in overcoming the inhibition of lettuce seeds to germination caused by salt concentration (Odebaro and Smith, 1969). Seeds of 'Great Lakes-Phoenix'

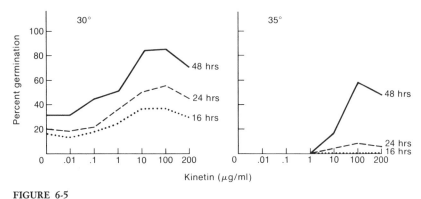

FIGURE 6-5

Germination response of seeds of 'GL-659' lettuce to a 3-minute dip in a series of concentrations of kinetin at 30°C and 35°C. (After O. E. Smith, W. W. L. Yen, and J. L. Lyons, The effects of kinetin in overcoming high-temperature dormancy of lettuce seed. In *Proc. Amer. Soc. Hort. Sci.* 93:444–453, 1968).

were dipped for three minutes in kinetin at a concentration of 10 ppm and were then planted on filter paper that had been soaked in sodium chloride solutions that varied from 0.02 to 0.10 M. Kinetin was found to increase germination at most concentrations of salt used, but its effectiveness was dependent on the temperature maintained during germination. As the temperature increased from 20°C to 25°C and 30°C, the percentage of germination of kinetin-treated seeds as compared with the controls increased. This increase can be interpreted to mean that kinetin treatment overcomes both salt-induced and high-temperature-induced dormancy. Seedling growth was also found to be stimulated by soaking the seeds in kinetin. In the irrigated valleys of the southwestern United States, lettuce seeds are often planted in soil that is somewhat saline; thus treating seeds with kinetin may prove to be commercially useful.

TERMINATION OF BUD REST BY APPLICATION OF GROWTH REGULATORS

Peach

A period of chilling is necessary to terminate the rest period of peach buds. The 'Elberta' peach requires 950 hours of chilling at temperatures below 7°C. Donoho and Walker (1957) demonstrated that gibberellin is effective in terminating rest in this cultivar. In November two-year-old trees that had dropped their leaves and that had been exposed to temperatures below 7°C for a maximum of 164 hours were transplanted from the orchard to large cans and placed in a storage room at 18°C. On February 23, ninety-five days after the trees had been placed in the storage room, they were transferred to a green-house and sprayed with GA₃ at concentrations of 50 to 4,000 ppm. On March 8, at the time of a second application of GA₃, a large per-centage of the trees that had received the applications of GA₃ at concentrations of 1,000 and 4,000 ppm had produced buds and small green leaves, and on March 29, these trees were growing rapidly. Shoot length increased with higher concentrations of gibberellin. Trees that had received gibberellin at concentrations of less than 100 ppm remained dormant.

Application of the cytokinin PBA at concentrations of 100 or 200 ppm has also been shown to terminate rest in four varieties of peach (Weinberger, 1969). However, since the growth substance is effec-tive only when the chilling requirement is nearly satisfied, the chemi-cal compensates for only a small amount of chilling.

Grape

Much interest has been shown in forecasting crop yields of grape. If crop yields can be predicted early enough, then the degree of dormant pruning can be adjusted to result in the desired crop level. In Aus-tralia and South Africa, crop prediction for 'Thompson Seedless' is based on an annual study of dissected buds. Because this method is tedious, efforts are being made to replace it with a forcing method. If dormant buds can be forced to grow, then the clusters can easily be counted.

Early work showed that soaking cuttings with resting buds in warm water, ethylene chlorohydrin, thiourea, or rindite hastens termination of rest in 'Thompson Seedless' (Weaver et al., 1961). However, these investigators found that gibberellin delays termination of rest and that NAA usually causes some delay. On the other hand, exogenous ap-plications of BA at a concentration of 1,000 ppm hasten termination of rest (Weaver, 1963). Large quantities of inhibitor disappear from treated buds within ten days after such treatment, suggesting that, this compound may terminate rest by exerting a destructive effect on inhibitor concentration in the bud (Weaver et al., 1968).

The fact that gibberellin delays termination of rest in buds of grape does not support the theory that an inhibitor-promoter balance controls rest. Since application of an exogenous cytokinin terminates rest, a cytokinin may be the promoter of dormancy in grape buds.

PROLONGATION OF BUD REST BY APPLICATION OF GROWTH SUBSTANCES

Grape

Delay of bud break in the spring frequently would be of value because it would enable the plant to escape damage from spring frost. For some plants a delay of a few days would be very helpful, and for others a delay of up to a month would be desirable. Bud break of seeded varieties of grape in the spring is delayed if gibberellin is sprayed on the plants during the previous growing season (Weaver, 1959; Weaver et al., 1961). However, there is a danger of killing buds of seeded varieties if concentrations of gibberellin are too high. Interestingly, the same treatment fails to retard bud break of 'Thompson Seedless' and probably would produce the same result on other seedless varieties.

In France, Nigond (1957) demonstrated that spraying vines of 'Aramon' grape in February or March with NAA at concentrations of 750 to 1,000 ppm retards bud break by sixteen to twenty-seven days. The delaying effect of gibberellin on bud break and flowering of certain tree fruits is discussed in Chapter 7. Increased concentrations of this growth regulator may reduce the number of fruitful buds produced and thus be of value in the year following treatment.

Citrus

A quiescent state occurs in citrus during the winter when temperatures of the night air fall below the minimum temperatures necessary for growth. In sweet orange this temperature is from 0° to 13°C (Webber, 1943). However, when the weather becomes warm, buds will start to grow if soil moisture is sufficient. Shoot growth at this time of year is undesirable because frost injury may subsequently occur. Cooper and his colleagues (1969) sprayed several varieties of citrus with growth substances to determine their effect on bud growth. They found that spraying during the winter with gibberellin and BA induced bud break and growth, whereas application of ABA, SADH, and CCC frequently delayed bud break.

In Florida the 'Temple' orange often grows sporadically all winter.

When trees of this variety were sprayed once with SADH at a concentration of 500 ppm (on December 29) or twice with ABA at a concentration of 500 ppm (on December 29 and January 29), winter bud growth was greatly inhibited (Cooper et al., 1969). For example, on April 5 the numbers of new shoots per tree on controls and on trees sprayed with SADH or ABA were 706, 170, and 27, respectively. However, the number of shoots that developed on each tree in the spring flush in early April was not influenced by either compound. When applied to trees of 'Pineapple' orange, SADH was found to be a more effective spring-flush inhibitor than CCC. These investigators believe that tree responses are affected by timing of the sprays and varietal sensitivity to various growth inhibitors.

In studies made by Young and Cooper (1969), 'Redblush' grapefruit was placed in a greenhouse at a day temperature of 21°C and a night temperature of 10°C and was sprayed one to five times with either ABA at concentrations of 100 to 1,000 ppm or CCC at concentrations of 500 to 3,000 ppm. Both chemicals delayed bud growth of leafy and defoliated seedlings, but ABA was more effective. Weekly sprays were found to be much more effective than single sprays in preventing bud growth. Some concentrations of these compounds were toxic, especially to defoliated shoots, a result that Young and Cooper ascribed to absorption of toxic amounts through the leaf scars.

FORCING OF RHUBARB

The rest period in rhubarb begins in the fall when the foliage dies. About two months of chilling or of cold weather in the field are required to terminate the rest period (Tompkins, 1965). Gibberellin can replace the cold requirement for rhubarb. In western Washington, Tompkins (1965) dug unchilled crowns and sprayed or poured solutions of GA_3 onto the crown buds and surrounding tissue before forcing at 13°C. Application of gibberellin at concentrations of 125 to 1,000 ppm consistently resulted in greater production of all cultivars treated at various stages of their rest period. Crowns that were not exposed to chilling before forcing yielded well when treated with sufficient gibberellin. Production is reduced if crowns are forced before the rest period is completely broken, and untreated, unchilled plants produced few or no marketable petioles. Production was increased by as much as 40 percent when crowns that had been exposed to sufficient chilling were treated with gibberellin. These crowns had more large petioles of extra fancy grade and produced them earlier in the season than did the untreated controls. For best results gibberellin should be applied within one or two days after the crowns are brought into the forcing house. In general, when used in conjunction with partial chilling, application of gibberellin at a concentration of 250 ppm is adequate, but a concentration of 500 ppm is required if there has been no chilling.

If the temperature of the forcing house exceeds 16°C for several days, as it frequently does in the spring, petioles with poor color may develop. When sucrose at 1.2 percent is added to the GA$_3$ solution, petioles with good color develop even when forcing is done at 18°C (Tompkins, 1966).

TERMINATION OF REST IN POTATO

Soon after gibberellin became commercially available, the discovery was made that it can be used to break rest in potato (Brian and Hemming, 1955; Rappaport, 1956). In California the year-round production of potatoes frequently necessitates the use of relatively immature resting tubers for seed pieces. To assure uniformity of sprouting and maximum stand of these plants, a chemical that can break the rest period is desirable. At one time, ethylene chlorohydrin was used for terminating the rest period of potato tubers, gladiolus corms, and dormant buds of deciduous trees and shrubs, but it is no longer commercially available for this purpose. Gibberellin is very effective for breaking rest of potato tubers (Fig. 6–6). Rappaport et al. (1957) found that the rest period of buds of newly dug 'White Rose,' 'Kennebec,' and 'Russet Burbank' potato is terminated by five- to ninety-minute dip treatments of GA$_3$ at concentrations of 50 to 2,000 ppm. There was also a two- to three-week acceleration of sprouting.

Studies made with both dormant and sprouted potatoes have shown that emergence of shoots is more rapid from gibberellin-treated seed pieces than from untreated seed pieces (Timm et al., 1962). However, vigorously sprouting seed pieces are less affected by gibberellin than dormant seed pieces. Concentrations of gibberellin as high as 5 ppm have no material effect on total yield of tubers, but higher concentrations reduce yield and change tuber shape. Gibberellin treatment frequently results in a more uniform size of U.S. No. 1 grade tubers.

A common practice is to dip seed pieces in gibberellin at a con-

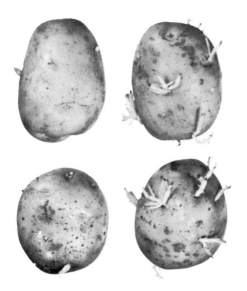

Effect of gibberellin on sprouting of 'Katahdin' potato. Untreated control is on left. More uniform and rapid development of sprouts in treated tubers resulted in earlier and more uniform crop emergence. (After S. H. Wittwer and M. J. Bukovac, 1958.)

centration of 1 ppm. Resting seed pieces that have received this treatment sprout more rapidly and uniformly when planted in cold soil than do untreated pieces.

Hemberg (1970) has demonstrated that the cytokinins kinetin and zeatin can terminate rest in potato tubers of the cultivar 'Majestic,' which has a long rest period. Sprouting of tuber pieces occurs within two or three days after treatment when these compounds are applied at concentrations of 10 or 100 ppm.

TERMINATION OF REST IN LILY BULBS

Florists have noted that during some seasons, bulbs of the Easter lily (*Lilium longiflorum*) are immature or in a resting condition that is evidenced by a delay in sprouting. Wang and Roberts (1970) found that harvested 'Ace' bulbs that are stored at 21°C for six weeks require sixty-six days for sprouting. Storage for six weeks at 4°C reduces the time required for sprouting to thirty-seven days. However, treatment with GA_3 at a concentration of 2,500 ppm, followed by six weeks of storage at 21°C, reduces the time required for sprouting to ten days; treatment with GA_3 followed by six weeks of storage at 4°C does not result in more rapid sprouting.

INHIBITION OF SPROUTING OF TUBERS AND OTHER PLANT PARTS IN STORAGE

There are many situations in agriculture in which it is useful to delay the onset of sprouting of shoots by inducing or prolonging dormancy. For example, early sprouting of potatoes and other vegetables in storage results in rapid loss of weight and eventual loss of marketability. Similarly, nursery trees often fail to remain dormant until they are ready for sale.

Potato

Guthrie (1938, 1939) showed that soaking the cut bases of a tuber in IAA at concentrations of 250 to 1,000 ppm prolongs dormancy and that MENA, applied as a vapor, is more effective than IAA. An excellent method for applying the growth regulator is to mix NAA-impregnated paper strips with the tubers in boxes or bins. (Approximately 100 mg of MENA per kilogram of tubers is optimal.) Treatment should not be given near susceptible plants, which can be severely damaged by the highly volatile vapor of MENA. MENA can also be applied in combination with an inert material, such as talc, or sprayed directly on the tubers in a solution of 95 percent ethanol.

The time of treatment in relation to the dose of MENA is important. Heavy dosages are necessary when the tubers are in the early stages of rest, but much less is required when treatment is given just before growth of the tubers commences. Storage temperatures are also important. Denny et al. (1942) showed that treatment of tubers with MENA before storing them at 10°C prevents shrinkage and sprouting for a year or more. Storage life was found to decrease as temperatures are elevated. Storage at 23°C reduces storage life to three to six months.

Preharvest applications of growth regulators to potatoes grown in the field have also successfully induced dormancy. There are two advantages of this method: more rapid and efficient translocation of growth regulators to the growing meristems of the tubers, and ease of application of the compound. Application of 2,4,5-T or related compounds at a concentration of 50 ppm is more effective than treatment with the sodium salt of NAA (Ellison and Smith, 1949), and MH has been found to be very effective on some varieties of potato (Rao and Wittwer, 1955). The optimal time for spraying 2,4,5-T is just before flowering; MH should be sprayed just after flower shedding.

Rhodes *et al.* (1950) have pointed out that IPC is effective for suppressing sprouting of tubers in storage. The compound CIPC also effectively prevents sprouting in potato. Hruschka and his colleagues

$$\text{IPC} \qquad \qquad \text{CIPC}$$

(1965) found that no sprouts developed on tubers that were stored for five months at 16° C after they were dipped in a 0.5 percent suspension of CIPC in water (Fig. 6–7). An aerosol treatment of 1.5 g of CIPC per hundredweight of tubers applied twice at a two-week interval reduced sprouting, especially at storage temperatures of 13° and 16°C. A disadvantage of CIPC is that, at low concentrations and under certain storage conditions, it stimulates development of internal sprouts (Fig. 6–7), which are regarded as commercially serious defects. Davis (1961) noted that such sprouts may develop when the apical dominance of the exposed sprouts is destroyed without inhibition of development of the unexposed buds. Internal sprouts are often produced in potatoes that are stored at some distance from treated tubers containing trace amounts of CIPC (Hruschka *et al.*, 1965).

Onion

Although NAA and 2,4,5-T retard the growth of sprouts in stored potatoes and in some vegetable root crops, they have no apparent inhibiting effect on the sprouting of onions (Wittwer and Sharma, 1950). A probable reason is that the growing points in the onion are so firmly enclosed in layers of leaf bases that the chemicals fail to penetrate to the meristems.

In Michigan, Wittwer and Sharma sprayed field-grown yellow sweet Spanish onion with five different growth regulators on August 15 (approximately two weeks before harvest). By treating the onions at a preharvest stage, they hoped to induce translocation of the growth regulators to the meristematic regions (Wittwer and Sharma, 1950). After harvesting, the onions were cured at 29°C for two weeks and were then stored for one month at 2°C and for four months at 13°C.

FIGURE 6-7

External sprouting of potato tubers after storage for 5 months at 16°C and 85 percent relative humidity. *Left,* untreated tubers stored in potato cellar produced long sprouts. *Upper right,* CIPC-proximity-treated tubers and tubers stored in laboratory produced appressed rosette-type sprouts. *Lower right,* CIPC-dip-treated tubers stored in potato cellar produced no sprouts. (After H. W. Hruschka, P. C. Marth, and P. H. Heinze, 1965.)

After the five months of storage no sprouting was evidenced by the bulbs of plants previously sprayed with MH at a concentration of 2,500 ppm and there was a significant reduction of sprouting of plants previously sprayed with MH at a concentration of 500 ppm (Fig. 6–8). The other chemicals used, NAA and BOA (Fig. 3–1, XXIII), did not inhibit sprouting. Deterioration in quality during storage was also decreased as a result of MH treatment.

Sweetpotato

Sweetpotatoes are usually cured for approximately one week after harvest at 29°C at a high relative humidity (Kushman, 1969). If this period exceeds one week or if the storage temperature after curing cannot be maintained at 16°C, sprouting often occurs. The use of

FIGURE 6-8

Results of the original experiment (August 1950) to test the effectiveness of a preharvest spray of MH as an inhibitor of sprouting of yellow sweet Spanish onion while in storage. *Left,* untreated control; *center,* plant sprayed with MH at a concentration of 500 ppm; *right,* plant sprayed with the compound at a concentration of 2,500 ppm. Photographed after. one month of storage at 2°C and 4 months of storage at 13°C. (After S. H. Wittwer and R. C. Sharma, 1950.)

MH, MENA, and 2,4,5-T has not received commercial acceptance. However, CIPC has been reported to cause a sizable reduction in sprouting (Kushman, 1969).

Commercial tests have been performed using the 'Centennial' and 'Goldrush' varieties of sweetpotato (Kushman, 1969). In one test 0.015 pound of CIPC per bushel of sweetpotatoes, applied as an aerosol spray, was found to reduce sprouting whether the roots were stored at temperatures of 13° to 18°C or 21° to 27°C. A second and third application were required to maintain the sprouting of roots stored at higher temperatures at the same level as the sprouting induced by one application to roots stored at the lower temperatures. In a second experiment a fog applicator was used to spray CIPC at a rate of three pounds per 1,000 boxes of roots. CIPC reduced sprout growth from 55 to 85 percent of that of the controls and maintained such growth at less than 0.75 inch except for sprouts that were longer than this before treatment. Most sprouting normally occurs at the top of the bin because it is often warmer than the bottom. When untreated roots were stored in bins at a temperature of 13° to 16°C, the weight of sprouts growing on ten roots, taken from the top of the bin, was 16.2 g. The weight of sprouts growing on ten roots, taken from the bottom of the bin, was 0.2 g after five months of storage.

Similar samples of roots that received the three-pound treatment of CIPC weighed 2.4 g and 1.2 g, respectively.

Root Crops Other than Sweetpotato

MENA reduces sprouting of stored carrots and other root crops. However, callus is stimulated and lateral roots are produced.

INHIBITION OF SHOOT GROWTH OF SHRUBS AND TREES IN STORAGE

After nursery trees are dug in the winter, they are often stored for several weeks or months, during which shoots frequently begin growth. These shoots are spindly and etiolated and consume food reserves, thus weakening the whole plant. Therefore, a means of retarding shoot growth would be advantageous to nurserymen.

In the early 1940's it was found that NAA and its derivatives delay sprouting of shoots of stored rose bushes (Marth, 1943). Methyl and ethyl esters and nitrile compounds were found to be the most effective derivatives. Marth sprayed the tops of plants before storage with a wax-emulsion solution of MENA at concentrations of 50 to 100 ppm. Marth found that another effective method was to apply the vapors of the esters to the entire plant. The plants are placed in an air-tight space equipped with a fan to ensure equal distribution of the vapor; 0.3 to 0.4 g of ester are used per 1,000 cubic feet of space. At higher storage temperatures, treatment time must be decreased; for example, the treatments given at 0°, 4°, and 21°C should last approximately sixteen hours, four hours, and one hour, respectively. Bud growth of some plants is delayed for forty to sixty days.

Experiments have shown that shoot elongation can also be retarded in peach, cherry, apple, pear, and persimmon. Experiments with forest trees have produced more erratic results than those obtained with roses.

EFFECT OF STORING LILY BULBS IN MENA

Storing lily bulbs in MENA affects their growth after planting. At the University of Chicago, Acker (1949) treated Croft lily and estate lily bulbs with the compound at concentrations of 0.0, 0.062, 0.62, and 6.2 g per cubic foot of storage space. Bulbs were stored at 1°C for

54, 91, and 129 days and were then planted in greenhouse pots. Plants from bulbs treated with the two lower concentrations of MENA flowered two weeks earlier at all plantings. In the second and third plantings, these same plants produced larger flowers and shoots having greater dry weights than those of the controls. More bulblets were produced on Croft lily plants previously treated with 0.62 g or 6.2 g of the compound than were produced on untreated plants or plants that received 0.062 g.

The large size of the plants and flowers that were produced following some of the treatments may have been an indirect result of the more extensive root system possessed by the treated plants. The growth regulator also may have affected the carbohydrate metabolism of the bulb in such a manner that growth of the young shoots was stimulated.

When plantings were made outdoors, previous treatment of bulbs with MENA failed to stimulate shoot growth and flower size. Plant growth was stunted by the highest concentration of compound.

Supplementary readings on the general topic of dormancy are listed in the Bibliography: Amen, R. D., 1968; Kefeli, V. I., and Kadyrov, C., 1971; Vegis, A., 1964; Wareing, P. F., and Saunders, P. F., 1971; and Woolhouse, H. W., ed., 1969.

Chapter 7

FLOWERING

VARIABLES THAT AFFECT FLORAL INITIATION

Floral initiation, like other physiological processes, is determined by genotype. In some plants this factor seems to be the only determining one, whereas in others genotype must interact with specific environmental conditions to induce floral initiation. The two most important conditions are low temperature and a specific range of light. The role of certain growth regulators to induce floral initiation in some plants and inhibit it in others is also important.

Once the plant reaches the physiological stage of readiness for flower initiation, the first noticeable morphological change indicating the transition of a vegetative meristem to a reproductive one is enhanced cell division in the central zone immediately below the apical part of the vegetative meristem. This division results in a group of undifferentiated parenchymatous cells surrounded by meristematic cells, which, in turn, give rise to flower primordia.

Vernalization

The initiation of flower formation by subjecting seeds, seedlings, or bulbs to low temperature is known as vernalization. Plants requiring vernalization generally grow in temperate regions and are classified as biennials or perennials. In nature they receive their quota of cold treatment during the winter months when they remain small, the majority of them in a rosette condition, and when metabolic activity is minimal. When spring arrives and temperatures are favorable, vegetative growth resumes, and flowering occurs in spring and early summer. The temperature requirement for the flowering of these vernalized plants varies. Low temperatures followed by relatively high temperatures are essential to induce flowering in many of them; in others, initiation of flower primordia occurs at low temperatures. Some plants have a dual requirement of low temperature and proper photoperiod. Certain plants, subjected to low temperature for a specific duration and subsequently given favorable conditions of light and temperature, can be induced experimentally to flower at any time of the year.

Photoperiodism

Plants whose flower formation is determined exclusively by genotype and that have no specific light requirement are known as "dayneutral" plants. Such plants flower when they reach a certain stage of vegetative development. Other plants require specific periods of light before they can flower, a phenomenon termed *photoperiodism*. Depending on the nature of the day-length requirement, such plants have been classified as "short-day plants" or "long-day plants." The former plants will flower only when the period of illumination remains shorter than a certain critical period, whereas flowering occurs in the latter only when a certain minimal day length is exceeded. There are yet other categories of plants that will not flower if maintained exclusively under either short-day or long-day conditions, for example, there are long-short-day plants and short-long-day plants.

The fact that perception of day length occurs in the leaves whereas the response is manifested in the bud indicates that a substance is produced in the leaf and is then translocated to the site of action, the apex. This is the basis for the theory that a flower hormone controls photoperiodism, an idea that has been substantiated by grafting experiments conducted between induced and noninduced plants. A "receptor" organ growing under noninductive conditions can be caused to initiate flowering if a leaf is grafted onto it from a plant that has been grown under inductive conditions. Further, transmission of the stimulus occurs not only within groups of plants that require the same photoperiodic treatment but also between plants that differ in response (that is, between short-day plants and long-day plants and vice versa). These results indicate that the flower-forming hormones of both short-day and long-day plants are identical physiologically and probably also chemically. However, efforts to isolate the responsible substance have met with partial success because such extracts have been found to produce flowers in only a few plants. The significance of these results is not yet entirely clear.

Effects of Plant Regulators

GIBBERELLINS. These compounds are the only chemical substances capable of inducing flower formation in plants that are representative of well-defined physiological classes when such plants are grown under experimental conditions in which they otherwise would stay completely vegetative. Gibberellin thus appears to be capable of replacing certain specific environmental conditions that control flower formation. Application of GA_3 induces the majority of long-day and cold-requiring plants to form flowers; it also induces flower formation in certain long-short-day plants when it is substituted for the long-day requirement. Gibberellin usually enhances flowering in short-day plants growing under inductive conditions, but its effect is generally negative when these plants are grown under noninductive conditions. The flower formation of long-day or long-short-day plants can be controlled by regulating the endogenous level of gibberellin-like substances through the use of such growth retardants as CCC that inhibit gibberellin synthesis.

In many short-day plants or in plants that do not require variations in light for flowering, gibberellin application usually delays flower initiation or blocks it entirely. This delay may be caused by rapid shoot growth, which results in much competition between vegetative growth and floral development.

AUXINS. The role of auxins in flower formation is not clear, and it seems uncertain whether these substances play a decisive role in photoinduction. When auxins are applied to some plants at an appropriate time, the flowering response is modified in a consistent manner, and consequently certain auxins have been used for commercial purposes. Auxins inhibit flowering in some plants and stimulate flower induction in others, but the effects are only slight.

INHIBITORS. Stimulation of shoot growth usually retards floral initiation; inhibition of shoot growth often enhances floral initiation. Such plant growth retardants as CCC and SADH are especially effective. When shoot growth is retarded, there is less competition for the nutrients required for floral bud development. Inhibitors sometimes may directly induce such development.

A significant finding was made by Professor Wareing and his colleagues when they noted that ABA, when applied under long-day conditions, promotes flowering in the short-day plants *Pharbitis nil,* *Ribes nigrum,* and *Fragaria* (El-Antably *et al.,* 1967). They found that ABA fails to induce flowering in certain other typical short-day plants and inhibits flowering in some long-day plants. Cathey (1968) found that repeated doses of ABA delay flowering of the long-day 'Pink Cascade' petunia (*Petunia hybrida*) and the short-day 'Double Eagle' marigold (*Tagetes erecta*), whether grown under short or long photoperiods (Fig. 7–1). Stem elongation was usually less in the ABA-treated plants than in the controls during the same photoperiod.

When used in very high concentrations, some of the growth retardants, such as CCC and Amo-1618, inhibit rather than promote flower formation in some plant species. As these compounds are known to inhibit endogenous synthesis of gibberellin-like substances, the delay in flower formation may result from reduction of the endogenous gibberellin content.

Importance of Growth Regulators in Flowering

It is probable that naturally occurring hormones play an important role in the process of flower bud induction. This view is strengthened by the fact that exogenous growth regulators frequently induce or promote flowering or prevent or delay flowering. Several reviews have been published concerning the hormonal mechanisms that promote or delay flowering (Chailakhian, 1968; Hillman, 1962; Lang, 1965b; Searle, 1965; and Evans, 1971).

Chemical treatments have been developed that will induce certain plants to flower at desired times. The remainder of this chapter is concerned mainly with the effect of exogenous plant regulators on the flowering habits of agricultural plants.

180

FIGURE 7-1

Plants of 'Pink Cascade' petunia, a long-day plant (*top*), and 'Double
Eagle' marigold, a short-day plant (*bottom*), grown at 16°C under (*left to
right*) 8-, 12-, 14-, and 16-hour photoperiods. Plants were sprayed 3
times a week for 8 weeks with ABA at concentrations of 0 (*back row*),
500 (*center row*), and 1,000 ppm (*front row*). Photographed 10 weeks
after start of chemical treatments. (After H. M. Cathey, 1968b.)

FLOWER INDUCTION AND PROMOTION OF FLOWERING

Pineapple

Pineapple ceases to grow when the temperature drops below 4°C. Inhabitants of the Azores erected tents over rows of pineapple and built fires to prevent frost damage; the smoke forced early flowering. In Puerto Rico, brush fires adjacent to pineapple fields also stimulated flowering (Rodriguez, 1932). These observations led to the discovery that smoke, or such unsaturated gases as ethylene that are found in smoke, brings about flower initiation. Subsequent experiments showed that acetylene gas has a similar effect; it was used commercially in Hawaii in 1935. Clark and Kerns (1942, 1943) later showed that auxin can force flower initiation in pineapple.

Although the main growing points of the pineapple plant produce only one fruit, there are often three or four suckers that produce small but quality fruits the following year. The latter fruits are called the ratoon crop. After the ratoon crop is harvested the plants are destroyed. Usually two suckers are used for the ratoon crop; the rest are utilized for propagation, as are the slips that arise from lateral buds on the upper part of the peduncle just under the fruit.

Flower initiation in pineapple usually begins in November and continues throughout the winter. The plant can produce marketable fruit of good quality approximately eighteen months after flowering. However, the early and late fruits produced during one season are usually picked months apart. The practice of forcing by chemical means results in fruits that are ready for harvesting at the same time six months after flower formation. Application of small amounts of auxin to vegetative pineapple plants induces flowering, whereas treatment with larger amounts inhibits flowering of plants that would normally be expected to flower (Clark and Kerns, 1942).

Some plant scientists believe that flowering in pineapple is brought about by the accumulation of auxin at the stem apex. They contend that if the plant is placed horizontally, the auxin level is increased on the lower side of the fruit and the plant is induced to flower. However, the fact that ethylene causes no increase in auxin content of the stem tip raises a doubt as to the validity of this theory (see Leopold, 1955). Another suggestion is that ethylene makes the tissues of the vegetative apex more sensitive to the naturally occurring auxin. Gowing (1956) hypothesized that NAA acts by competitively lowering the effective native auxin level in the stem tip. Both NAA and 2,4-D are effective but IAA is not, probably because of the large amounts of auxin-destroying enzymes in pineapple tissue and also because light destroys IAA. GA_3 neither promotes nor inhibits flowering and is relatively inactive when applied to pineapple plants that have progressed beyond the seedling stage (Gowing and Leeper, 1961a).

In Hawaii many pineapple fields are sprayed with the sodium salt of NAA at a concentration of 25 ppm (approximately 20 g in 200 gallons of water per acre). This compound was the original forcing agent meant to be used in industry and it is still commonly used, especially with ratoon crops (J. B. Smith, private communication). In Puerto Rico sprays of 2,4-D at concentrations of 5 to 10 ppm are commonly used. However, ethylene gas, applied as a saturated water solution, gives more consistent results and is used more widely than any of the other agents. Acetylene has also been applied. A simple method is to drop 1 g of dry calcium carbide into the heart of each pineapple plant. Acetylene gas is liberated by reaction of the calcium carbide that occurs in the reservoir of rain water usually existing in each plant. BOH at a rate of 200 g in 200 gallons per acre has also been used. Gowing and Leeper (1961b) have reported on results obtained with thirty-nine hydrazine derivatives. The only hydrazine derivatives tested by them that forced pineapple plants to flower were ethylhydrazine, sym-diethylhydrazine, BOH, unsym-bis-(β-hydroxyethyl)hydrazine, 2-hydroxyethyl N-(2-hydroxyethyl)carbazinate, and 2-(2-hydroxyethyl)semicarbazide. The last two compounds were not found to be active in all tests.

The new group of growth substances, the 2-haloethane-phosphonic acids, have been experimentally demonstrated to be very effective in forcing flowering of pineapple (Cooke and Randall, 1968). In field tests, spraying of ethephon at rates of one, two, and four pounds per acre on smooth Cayenne pineapple plants produced 100 percent flower induction; the control plants remained vegetative. The higher levels were also found to hasten flowering; plants treated with the compound at a rate of four pounds per acre matured two to three weeks earlier than did those receiving a rate of one pound per acre. The four-pound treatment retarded vegetative growth somewhat, but slip and sucker production was normal on all treated plants.

Chemical forcing of flowering has several advantages. First, all fruits are ready for picking at the same time, thus eliminating the need for several harvests. Second, sprays can be so timed that fields are ready to be picked at different times, and peak loads in the cannery can thus be avoided. Finally, higher yields per acre can be obtained since many plants fail to produce fruit under normal conditions.

Conifers

In 1958 Kato *et al.* demonstrated that GA_3 induces flowering in *Cryptomeria japonica*. Subsequently other Japanese reports have shown that gibberellins can induce other conifers in the families Cupressaceae and Taxodiaceae to flower. GA_3 induced flowering of Arizona cypress (*Cupressus arizonica*) as early as eighty-eight days after germination (Pharis *et al.*, 1965); this precocious flowering response as a result of GA_3 treatment has also been observed in Portuguese cypress (*Cupressus lusitanica*), pigmy cypress (*Cupressus pygmaeae*), and western red cedar (*Thuja plicata*) (Pharis and Morf, 1967). However, use of gibberellin to induce flowering in members of the family Pinaceae has been unsuccessful.

Pharis and Morf (1968) tested the efficiency of seven gibberellins on flower induction in four members of the family Cupressaceae: Arizona cypress, Portuguese cypress, pigmy cypress, and western red cedar. The criteria used to measure flower induction were (1) the number of flowering meristems produced after ten weeks of gibberellin spray and (2) the number of days to first flower. Seedlings were sprayed twice weekly with gibberellin compounds at concentrations of 140 to 150 ppm; 100 ppm of GA_1 was also used. The results showed that GA_3, GA_9, and GA_4 are approximately equal in their ability to induce flowering and that GA_1, GA_7, and a mixture of GA_4 plus GA_7 are slightly less effective. Gibberellin A_{13} has a relatively low activity and GA_5 is inactive. The experimental seedlings were grown under long-day conditions because under short-day conditions, the response to gibberellin was decreased fivefold.

The coastal redwood (*Sequoia sempervirens*) and giant redwood (*Sequoia gigantea*) usually do not start to flower until they are twenty and seventy years old, respectively (see Pharis and Morf, 1969). Repeated sprays of GA_3 or of mixtures of GA_4 plus GA_7 applied to one- to two-year-old coastal redwood trees over a period of six months failed to induce flowering. Larger amounts of gibberellin were introduced into the plants either by injection or by external application to the main stem of 5-μl quantities of gibberellin compounds in 75 percent ethanol. Dosages of 50 μg to 125 μg were applied either once or at weekly intervals. Most plants of each specie flowered as a result of extended applications, some of which lasted as long as four months.

A total of 1,500 μg of gibberellin was required to induce flowering in a giant redwood less than one year old. The effectiveness of GA_3, the $GA_{4,7}$ mixture, and GA_{13} in producing strobili was approximately the same. From 2,500 to 85,000 times more gibberellin is required to produce a strobilus in the coastal and giant redwoods than in Arizona cypress, which emphasizes the tremendous difference among conifers in their response to gibberellin.

Tree Fruits

LITCHI. This subtropical tree can be girdled to control flowering. The same effect can also be achieved by application of auxin. Few flower primordia are produced during periods of active vegetative growth. (In Hawaii the time of flower initiation in litchi coincides with a flush of vegetative growth.) Application of NAA retards this growth and promotes flowering. Since girdling causes flowering, Nakata (1955) has suggested that NAA causes a mobilization of assimilates in the tree.

PEACH. Working in New York State during the month of July, Edgerton (1966) found that application of SADH at a concentration of 2,000 ppm stops terminal elongation of peach shoots but does not prevent breaking of lateral buds at the shoot apex. Formation of flower buds was found to increase slightly.

APPLE. An increase in flower induction in apple as a result of inhibitor application has frequently been demonstrated. Such treatment shows promise as a means to correct alternate bearing as well as to control fruit tree size.

Batjer et al. (1964) found that spraying apple trees in the northwestern United States with SADH at concentrations of 500 to 2,000 ppm at ten-day intervals beginning fifteen to seventeen days after full bloom reduces shoot growth. One of the most spectacular responses to sprays of SADH was found to be the great increase in number of blossoms during the year following treatment (Fig. 7–2). Treated trees bore two to twelve times as many flowers as did untreated trees. Some spray treatments were found to retard anthesis. Application of SADH at a concentration of 3,000 ppm delayed full bloom of trees of 'Starking' apple by six days, whereas application of SADH at a concentration of 1,000 ppm resulted in a two-day delay.

The shoot growth of trees on EM 2 stock bearing 'Red Delicious' apples is sharply inhibited by SADH at concentrations of 2,000 to

FIGURE 7-2

Effect of application of SADH at a concentration of 2,000 ppm on return bloom of apple trees in the season subsequent to spraying. Unsprayed control is on left. (Photographs courtesy of M. W. Williams and G. C. Martin.)

5,000 ppm (Edgerton and Hoffman, 1965; and Greenhalgh and Edgerton, 1967). Trees sprayed in mid-June showed a reduction in terminal growth of approximately 40 percent compared with untreated trees. Simultaneous single applications of SADH were found to be as effective in promoting flowering as were repeat applications of TIBA, MH, and 2,4,5-T, which promoted flowering but did not reduce growth as effectively as SADH. No evidence of phytotoxicity was found, even after three applications of SADH at a concentration of 5,000 ppm. Treatments other than SADH caused some epinasty as well as the formation of deformed leaves near the shoot apex after the sprays were applied.

In an experiment conducted in Maryland, young trees of 'Royal Red

Delicious' apple were sprayed with SADH near the time of blossoming for four successive years (Rogers and Thompson, 1968). In one treatment, a spray of SADH at a concentration of 1,200 ppm was applied to some trees two weeks after bloom. Other trees were similarly sprayed with SADH at a concentration of 600 ppm during the first year and 400 ppm during each of the three succeeding years. The latter group received three applications: the first during bloom and the second and third after bloom, at seven- to ten-day intervals. Reduction in terminal growth of shoots occurred only during the first and second years of the experiment. At the end of the experiment sprayed trees had a smaller number of secondary laterals making extended growth from one-year-old wood and consequently had a "thin" look. The number of blossoms was increased only during the season following the first-year application, indicating that the single spray was the most effective.

In Canada, the work of Fisher and Looney (1967) showed that the increase in flower bud induction as a result of application of SADH differed among five varieties of apple. In 1965 the cultivars 'Delicious,' 'Golden Delicious,' 'Spartan,' and 'Winesap' were sprayed with SADH at concentrations of 500 to 2,000 ppm ten to eighteen days after bloom. The average bloom ratings (0 = no bloom, 10 = maximum possible bloom) of these four cultivars the following year were 8.5, 7.0, 6.2, and 8.6. Corresponding figures for control trees were 7.6, 4.8, 1.9, and 8.0. The most marked effects were observed on the 'Spartan' and 'Golden Delicious' varieties, which, in Canada, have a greater tendency toward biennial production than do the 'Delicious' and 'Winesap' varieties.

In one experiment with 'Starking' apple, in which only the lower one-third of the tree was sprayed with SADH, shoot growth was not arrested in the upper part of the tree (Batjer et al., 1964), yet the return bloom over the entire tree increased. It may be that SADH is translocated throughout the tree in a quantity sufficient to increase flower initiation but not enough to suppress vegetative growth.

PEAR.   Flower formation and suppression of shoot growth of 'Bartlett' pear have been shown to increase strikingly with increased concentrations of SADH (ranging from 1,000 to 3,000 ppm) Griggs and Iwakiri, 1968). Double spray applications caused production of 2.8 to 3.1 times as many flower buds as produced by unsprayed controls. Fruit set, however, is reduced, probably because some of the nutrients used in excessive flower bud formation are diverted from the developing fruitlets. Whether the effect of inhibitors on increased flower initiation is direct or indirect is a matter of speculation. The greater leaf area resulting from SADH application, together with the early cessation of terminal growth, could indirectly account for the increased flowering. However, the fact that an increase in the amount of bloom is sometimes obtained without an appreciable effect on shoot growth would seem to indicate that the chemical is a direct causal factor.

Dennis (1968), working in New York State, also found that SADH appears to promote the flowering of pear seedlings grown from seedling 52915 even when there is no vegetative response. Various pear seedlings sprayed with SADH at a concentration of 2,500 ppm in June 1965 and June 1966 evidenced a significant increase in flowering in both 1966 and 1967. The percentages of control trees that flowered in 1966 and 1967 were 8 and 58, respectively; the percentages for SADH-treated trees were 33 and 75, respectively. Application of SADH at a concentration of 500 ppm increased flowering only during the first year of the experiment.

Trials conducted in Maryland with trees of 'Magness' pear demonstrated that application of SADH at a concentration of 1,200 or 2,400 ppm reduces terminal shoot growth during the first year only; however, no inhibition of growth occurred when the experiment was extended for another two years (Rogers and Thompson, 1968). Sprayed trees evidenced increased fruit bud initiation and consequently produced greater yields as compared with unsprayed trees.

LEMON. Plant growth retardants and the auxin BOA have been found to induce flowering in 'Eureka' lemon. A common practice in Israel is to retard growth by withholding water from lemon trees for two months beginning in mid-August in order to produce blossoms in November and a fruit crop during the following summer. However, because this practice has the disadvantage of possible injury to the trees, inhibition of growth by chemical means is more desirable.

In Israel trees of 'Eureka' lemon were sprayed with CCC at a concentration of 1,000 ppm, SADH at a concentration of 2,500 ppm, or BOA at a concentration of 25 ppm (Monselise and Halevy, 1964; Monselise et al., 1966). Five treatments were made at three-day intervals, from the last week of August through the first week of September, the period when flower induction may be obtained by withholding irrigation. All three compounds considerably increased flowering and production of lemons. BOA was found to be relatively more effective on branches that were older than six months, and the two growth retardants were found to be more effective on branches that were six months old or younger. Spraying only twice with the chemicals was also effective; however, unsprayed branches usually failed to develop flowers.

ORANGE. In Israel shoots that are part of the summer flush usually bear few flowers during the following spring. However, certain antimetabolites of nucleic acid and protein synthesis increase flower bud formation and fruit set when applied to summer shoots during the flower induction period. Goren and Monselise (1969) found chloramphenicol succinate, 5-fluorodeoxyuridine, and 5-bromo-3-*sec*-butyl-6-methyluracil to be effective. These results are interesting since nucleic acid and protein synthesis are believed to be necessary for flower induction (see Lang, 1965). The increased bloom was accompanied by the shortening of internodes that is characteristic of flowering branchlets. The investigators suggest that perhaps the compounds stimulate flower formation by weakening the branches and decreasing vegetative vigor.

PAPAYA. The cultivar *Carica papaya* 'Solo' flowers from axillary buds that arise along the central stem. All stages of fruit development from flower to mature fruit are present in papaya at any one time. Dedolph (1962), working in Hawaii, showed that applications of BOA can hasten flowering, produce flowers at a lower node, and increase early fruit yield. Applications of NAA, TIBA, and IAA were found to be less effective.

In one experiment conducted by Dedolph in mid-October, four-month-old papaya plants were sprayed with BOA at concentrations of 0, 10, 20, 30, 40, and 50 ppm. Tween 20, a surfactant, was added to each solution at the rate of 1 ml per gallon to aid penetration of BOA into the plants. Dedolph found that application of BOA increased flower production, especially at concentrations of 30, 40, and 50 ppm. One month after treatment the percentages of trees flowering at concentrations of 0, 10, 20, 30, 40, and 50 ppm were 8.3, 16.7, 8.3, 58.3, 50.0, and 66.7, respectively. The number of nodes to the first flower was generally decreased by approximately one node per 10 ppm increase in concentration. During the first twenty-two weeks of harvest, plants sprayed with BOA yielded a greater total weight and number of fruits per tree during any given period than did the untreated plants. In the early part of the harvesting season, the yield of trees sprayed with the chemical at a concentration of 50 ppm was two to three times greater than the yield of the unsprayed trees.

The primary effect of the chemical was to force earlier yields; the average total weight of fruit was not materially altered. One spray of BOA at a concentration of 30 ppm was found to increase yield during the first six weeks of harvest by approximately 1,400 pounds per acre. Another advantage of spraying papaya with BOA is that early maturation of the fruit eliminates many of the field hazards that might spoil the crop.

Blueberry

SADH retards shoot elongation and increases flower bud formation in the 'Coville' highbush blueberry (Shutak, 1968). Bushes were cut back in the fall leaving only stubs two to three inches long, and on July 31 of the subsequent season the vigorous shoots produced were sprayed with SADH at a concentration of 5,000 ppm. After leaf fall in the autumn following the spray applications, the number of flower buds per five inches of shoot growth was calculated. During the first flush, no flower buds developed on either control or treated plants. During the second flush, there were 3.5 buds per five inches of the control plants and 7.9 buds per five inches of the SADH-treated shoots.

This work was then extended to highbush blueberries 'Collins' and 'Bluecrop' (Hapitan et al., 1969). Plants were sprayed with SADH at a concentration of 5,000 ppm either on July 18 or August 22 or on both dates. The repeat sprays reduced the shoot growth of 'Collins' by 31 percent and the shoot growth of 'Bluecrop' by 34 percent; single sprays were found to be less effective in reducing shoot growth. SADH increased the number of flower buds on both cultivars; the greatest increase resulted from the repeated spray applications. In 'Collins' there were 1.20 flower buds per five inches of new shoot growth of the unsprayed plants. The corresponding number for plants given the repeated spray was 4.89. The figures in the same order for 'Bluecrop' were 1.58 and 4.73. There was no difference in the effect of the two single sprays on flower bud formation in either variety.

Vegetables

SEED STALK PRODUCTION IN HEAD LETTUCE. The greatest expense of growing a seed crop of compact heading varieties of lettuce is the cost of deheading. Removal of the tight head means that the seed-stock is free to bolt instead of being trapped within the compact

a   b

FIGURE 7-3

Gibberellin-induced flowering of lettuce grown under a 9- to 11-hour photo-period and at night temperatures of 10° to 13°C. *a,* 'Great Lakes' variety of *Lactuca sativa: left,* untreated control; *right,* plant whose stem apex received 60 μg of gibberellin in 3 20-μg treatments applied at 4-week intervals. *b, Lactuca dentata: left,* untreated control; *right,* plant that received an aqueous spray of gibberellin at a concentration of 100 ppm 4 weeks before photograph was made. (After S. H. Wittwer and M. J. Bukovac, 1957.)

head where it often breaks and rots. Mechanical methods of removing heads often severely injure the plants and thus reduce yields.

Gibberellin application stimulates bolting before head formation and thus eliminates the need to dehead (Fig. 7-3). Spraying of 'Great Lakes' lettuce with gibberellin at concentrations of 3 to 10 ppm at the four- and eight-leaf stages of growth significantly increases seed yield

(Harrington, 1960). Seeds of the treated plants were found to mature two weeks earlier than did those of the untreated plants, and the uniformity of maturation was improved. Germination and subsequent plant growth of the seed produced by gibberellin-treated plants were normal.

ACCELERATION OF GROWTH OF GLOBE ARTICHOKE. In the United States the globe artichoke is produced mainly along the coast of California. The peak of production occurs between October and April, but yields are often poor during the winter months because of low temperatures. If production could be hastened, artichokes would be available in November and December when prices are high and when other vegetable crops are very expensive.

In tests conducted in California, single applications of gibberellin at concentrations of 25 or 50 ppm made in September were effective in accelerating bud production, both in number and weight as compared with nontreated plants, between January and March. In some tests harvest was advanced by six weeks or more. At the termination of the harvest period, total accumulated yields were, in general, nearly the same (Snyder *et al.*, 1970).

In southern Italy application of gibberellin advanced ripening of artichokes of the variety 'Molese' (Casilli, 1969). Plants sprayed at the six-leaf stage or earlier with the compound at a concentration of 50 ppm could be harvested approximately sixty days earlier than the controls.

CONTROL OF FLOWERING IN SWEETPOTATO. The frequent failure of the Jersey type of sweetpotato to flower in the continental United States is disadvantageous to the plant breeder because the uniform shape and size of the Jersey type could otherwise be profitably incorporated with the advantages of disease resistance, high vitamin content, vigor, and adaptability of the moist-fleshed varieties.

Sprays of the sodium salt of 2,4-D at concentrations of 100 to 500 ppm were found to induce flowering in plants of 'Porto Rico' and 'Yellow Jersey' sweetpotato (Howell and Wittwer, 1955). Plants growing in the ground bed of a greenhouse were sprayed in September when the fleshy roots had begun to enlarge. 'Yellow Jersey' produced the greatest number of flowers following treatment with 2,4-D at a concentration of 100 ppm; higher concentrations were injurious. 'Porto Rico' showed greater tolerance to 2,4-D, and its flower number was highest following treatment with 2,4-D at a concentration of 500 ppm.

INDUCTION OF FLOWERING IN BIENNIALS. Many vegetables that require cold treatment or long days to flower respond to gibberellin. Lang (1956) was the first to show that gibberellin can replace a biennial's low-temperature vernalization requirement for flowering.

In Michigan several biennials, including carrot, beet, cabbage, kale, collards, and rutabaga, which were grown at temperatures slightly above the critical temperature for flower formation, were induced to flower by gibberellin (Bukovac and Wittwer, 1957). These results suggest that the cold-temperature requirement for flowering in biennials may be partially or even completely replaced by one or several foliar sprays of gibberellin at concentrations of 100 to 1,000 ppm. All plants used by Bukovac and Wittwer, with the exception of 'Golden Acre' cabbage and two varieties of collards and turnip, remained vegetative when grown at a night temprature of 10° to 13°C. At a night temperature of 15° to 18°C, gibberellin induced flowering in carrot. Of the other biennials grown at this temperature, the beet, turnip, rutabaga, and kale plants did not flower as a result of the gibberellin treatment; only a small percentage of the celery, cabbage, and collard plants produced flowers. Stems of gibberellin-treated plants were greatly elongated; considerable stem elongation also preceded the appearance of flower buds in the plants that flowered.

These results suggest that the normal cold-temperature requirement for flowering of biennials may be partially or, in a few instances, completely replaced by the application of one or several foliar sprays of gibberellin or by weekly applications of gibberellin to the stem apices. Wittwer and Bukovac (1957) suggest that such a means of controlling reproduction in biennials, previously regulated by temperature, offers the possibility of extending the geographical boundaries wherein many flower and seed crops may be grown commercially.

INDUCTION OF FLOWERING IN LONG-DAY ANNUALS GROWN UNDER SHORT-DAY CONDITIONS. Before the discovery of gibberellin, flower induction in long-day plants grown under short-day conditions was generally not subject to chemical regulation. However, many long-day annuals grown in environments that are not conducive to flowering can now be made to flower by the use of gibberellin. Wittwer and Bukovac (1957b) grew several genera and species of long-day plants at temperatures of 10° to 13°C. The plants were grown under noninductive, short photoperiods (nine to eleven hours). Application of gibberellin stimulated stem elongation and induced flower and seed production. Several crops, including leaf and head lettuce, endive, radish, mustard, spinach, and dill, showed a positive response. No plant that was sensitive to long days failed to flower when grown under a short photoperiod and appropriately treated with gibberellin.

Gibberellin was found to stimulate earlier flowering in such long-day plants as lettuce, endive, radish, and mustard that were maintained under an inductive (eighteen-hour) photoperiod (Wittwer and Bukovac, 1957). However, other plants, including spinach and dill, were not similarly affected. A single foliar spray of gibberellin at concentrations of 100 or 1,000 ppm applied to 'Grand Rapids' and 'Tendergreen' lettuce, radish, and dill at varying stages of seedling growth induced flowering under noninductive, short-day conditions. Repeated application of the compound was necessary to induce flowering of other crops, including 'Bibb' and 'Great Lakes' lettuce, endive, mustard, and spinach.

TOMATO.   Several plant growth retardants, such as CCC and the related compounds (2-bromoethyl)trimethylammonium bron'de and (2,3-n-propylene)trimethylammonium bromide, when appliec to the roots of tomato plants at concentrations of 10^{-3} to 10^{-7} M, alter 'rowth and promote earlier flowering (Wittwer and Tolbert, 1960). 'Teated plants were found to develop thick stems and dark green .eaves, changes that are similar to those produced by exposure of the plants to intense light and low temperatures. Treatment with these retardants partially duplicates the effects of cold exposure of seedlings, such as a decrease in the number of leaves subtending the first inflorescence (Wittwer and Teubner, 1956) and complements the effects of N-m-tolylphthalamic acid in increasing flower numbers (Teubner and Wittwer, 1957). These researchers also found that the stem-elongation effect of high temperature and low light intensity was counteracted by the application of a 10^{-3} M solution of (2,3-n-propylene)trimethylammonium bromide to the soil root medium.

The control of growth and flowering by the use of these chemicals offers several practical possibilities for their utilization in tomato culture. Plants having thick leaves and stems, short internodes, and dark green foliage and that flower early may be grown in greenhouses during midwinter. Seedlings produced for field transplanting would have shorter stems, stronger laterals, and heavier roots, and they might flower and fruit earlier.

Certain mutant tomato plants obtained by treating seeds with ^{32}P produce only an occasional flower at a later date after seeding than the original variety. Experiments with mutants of the 'Canary Export' variety showed that of several growth substances tested, only the cytokinin PBA improved flower bud retention and development (Coggins and Lesley, 1968). Young inflorescences from a rooted mutant cutting were treated daily for sixteen days with PBA. Flowering and bud development were enhanced; many flowers were produced on the treated plant, but only one small flower appeared on the control plant. The other growth substances tested, IAA, TIBA, and GA_3, failed to give consistently favorable results.

BEAN.   Spraying with TIBA has been found to increase the number of flowers of soybean, a short-day plant. In Ames, Iowa, spraying of soybeans (*Glycine max*, cv. 'Hawkeye') with TIBA at a concentration of 15 ppm between the stage at which the plants had two trifoliate leaves and the stage at which they had five or six trifoliate leaves was found to increase the number of flowers and seed yield, although yield increases in these field experiments were moderate (5 to 15 percent). The spray also decreased the rank vegetative growth and the tendency to lodge (Anderson *et al.*, 1965). In greenhouse experiments TIBA has stimulated pod formation tenfold, but very few seeds were formed (Greer, 1964).

Anderson *et al.* (1965) obtained best results when soybean plants were grown in narrow rows (twenty inches or less). The increase in seed yield was caused by both the narrowness of the row and the hormone. Application of TIBA sprays to plants growing in wider rows (forty inches), especially the late or midseason varieties, also increased yield when environmental conditions were favorable to growth.

Some photoperiodic-sensitive bean plants are poorly adapted to areas with short growing seasons because of late flowering and maturation. A technique to promote early flowering in certain late-maturing bean types would therefore be useful. In Nebraska Coyne (1969) studied the effect of CCC, SADH, and GA_7 on time of flowering of the short-day field bean *Phaseolus vulgaris* cv. 'Great Northern Nebraska No. 1 sel. 27' when grown under long and short photoperiods. This variety is resistant to the common blight bacterium (*Xanthomonas phaseoli*), but because of late flowering and maturation is not adapted to western Nebraska.

Sprays of CCC at a concentration of 500 ppm, applied when the first trifoliate leaf was emerging, promoted earlier flowering under long-day conditions; the flowering occurred as early as that of control plants grown under short-day conditions. Under long days the number of days required to induce flowering averaged 48.6 for controls and 37.9 for plants treated with CCC. Plants sprayed with SADH at a concentration of 1,500 ppm or GA_7 at a concentration of 100 ppm and grown under both long and short days flowered at approximately the same time as the respective control plants.

Floral and Decorative Plants

BEGONIA. Solutions of CCC at 8×10^{-2} M, applied as a soil drench to plants of the cultivar 'Mörk Marina' grown at 21°C in a twenty-four-hour photoperiod, induced much earlier and richer flowering than that produced by the controls (Heide, 1969b). Treated plants required 64.0 days to anthesis; control plants required 102.5 days. Treatment with a soil drench of Phosfon-D had no effect on time of flowering, whereas spraying with SADH delayed flowering somewhat.

In a second experiment with 'Mörk Marina' Heide (1969b) showed that when the CCC treatment was applied at 21°C, the appearance of visible buds was accelerated by one week, but at 18°C there was no significant effect. No flowering was obtained at 24°C either with or without application of CCC. The effect of CCC on growth of lateral shoots was found to be dependent on temperature. At 24° and 21°C, only the highest concentration of CCC retarded shoot growth, whereas application of CCC at 5×10^{-3} M at these temperatures stimulated growth. However, shoot growth was retarded by all concentrations when applied at 18°C.

BOUGAINVILLEA. Flowering in Bougainvillea is favored by short days, moderate temperatures (24°C average), intense light (greater than 4,000 foot-candles), and increasing plant age (Hackett and Sachs, 1967). However, rooted cuttings of the 'San Diego Red' cultivar do not flower until the twentieth or thirtieth node even under favorable conditions. Since these conditions are difficult to obtain in most greenhouses, acceptable flower displays are not usually possible during the first year of propagation. The use of growth retardants to promote earlier bloom and to reduce shoot elongation in the nursery offers a means of developing Bougainvillea as an ornamental pot plant.

Treatment with CCC accelerates flowering under short days but not under long days (Hackett and Sachs, 1967). There is a 20 to 25 percent inhibition of stem elongation under both long- and short-day conditions. In one experiment CCC at the rate of 2 g per six-inch pot was applied as a soil drench and the plants were placed under short (eight-hour) days. At the time of treatment, two-month-old plants were cut back so that there were five leaves and a bud in the axil of the uppermost leaf, to develop a single strong shoot.

The first inflorescence of the short-day control arose at the twenty-fifth node; the first flowering of the CCC-treated plant arose at the fourth node.

Foliar sprays of SADH at a concentration of 5,000 ppm also promote flowering (Hackett and Sachs, 1967). Such sprays inhibit shoot elongation to approximately the same degree as treatment with CCC. The average numbers of nodes from the base of the stem to the first infloresence for control plants and for SADH-treated plants were found to be 24.0 and 9.4, respectively.

Application of gibberellin greatly delays flowering when plants are subjected to short-day conditions, and it increases shoot elongation and node number under both long and short-day conditions. In addition, CCC fails to induce flowering when 10 μg of gibberellin is applied weekly to the tip of the shoot. Since gibberellin prevents the CCC-induced promotion of flowering, the retardant effects on flowering may be explained by a reduction in the gibberellin level within the treated plants (Ninnemann et al., 1964). Hackett and Sachs (1967) found that the inhibition of vegetative growth by CCC is similar under long and short days, but only under short days does the compound promote flowering, which shows that CCC does not replace the effect of day length but does increase the effectiveness of short-day treatment.

HOLLY (ILEX). Several plant growth retardants stimulate flower initiation in holly. In a greenhouse experiment conducted in Beltsville, Maryland, two types of holly, *Ilex cornuta* 'Burfordi,' and a hybrid, *Ilex aquifolium* × *I. cornuta* 'Nellie R. Stevens,' were grown in six-inch soil plots and treated with growth retardants CCC or Phosfon-D (Marth, 1963). The 'Burfordi' plants were also treated with CO-11, an analogue of SADH (Riddell et al., 1962). The chemicals were applied at the rate of 150 mg per plant as an aqueous drench in July, just after the plants had completed their initial vegetative flush. By October no flowers or fruits had been produced on the untreated plants, but all plants treated with Phosfon-D had flowered and produced many mature fruits. CCC and CO-11 were found to be somewhat less effective than Phosfon-D in stimulating berry formation. The 'Nellie R. Stevens' plants treated with either Phosfon-D or CCC were approximately 40 percent shorter than the controls, but the vegetative growth of 'Burfordi' holly was not affected.

Flowers of the 'Burfordi' plants treated with Phosfon-D or CO-11 persisted and developed into mature fruits, whereas flowers of the CCC-treated plants abscissed and fruiting was sparse. This result suggests that certain growth-retardant chemicals may have a direct effect on flower abscission.

IRIS.   Most experiments testing the effect of GA_3 on bulbous plants have produced no positive results. However, Halevy and Shoub (1964) have reported that injection of bulbs with 50 or 500 μg of GA_3 can, under suitable conditions, accelerate flowering by up to nineteen days. Bulbs of Iris cv. 'Wedgewood' and 'Prof. Blaauw' were stored at 10°C for eighteen or thirty-five days. After planting, bulbs of both varieties previously stored at 10°C emerged sooner than those previously stored at a warmer temperature in common storage. Treatment with GA_3 had no effect on bulbs that had been held in common storage but further accelerated the emergence of bulbs held at a low temperature. The most favorable results were obtained with 'Wedgewood' and 'Prof. Blaauw' after eighteen and thirty-five days of cooling, respectively. The investigators believe that the most opportune time to apply GA_3 to accelerate emergence and flowering is soon after flower initiation. GA_3 probably has no direct effect on flower production and thus cannot substitute for low-temperature storage.

In one experiment the plants were sprayed with GA_3 at a concentration of 3,500 ppm seven times, from the time of emergence until a few days before anthesis. Flowering of 'Wedgewood' was accelerated by ten days; flowering of 'Prof. Blaauw' was accelerated by fifteen days.

AZALEA (RHODODENDRON SPP.).   In 1961 Stuart discovered that the growth retardants Phosfon-D, CCC, and SADH suppress vegetative growth and cause rapid initiation of flower buds in several cultivars of evergreen azalea (Fig. 7-4). Soon after publication of this report such treatments were widely tested both experimentally and commercially (Stuart, 1965; Furuta, 1965).

Azalea plants treated with growth retardants were found to require several months to develop flower buds, indicating that treatment must be made early (Cathey, 1965). Stuart (1965) has recommended the following technique for treating plants and manipulating the environment. Plants should be pinched or sheared six or seven months before bloom is desired. After three to six weeks, depending on the variety and growing conditions, new shoots develop, and a growth retardant is then applied or short days are begun. Cathey (1967) found an electric aerosol machine that applies SADH in very fine droplets to be efficient. Plants to which chemical treatment is applied should be approximately the same size at time of treatment as that desired at the time of bloom, since treated plants produce little additional vegeta-

FIGURE 7-4

Effect of growth retardants on flowering of rhododendron. *Left to right:* untreated control; recipients of applications of Phosfon-D (soil drench), CCC (foliar spray), and SADH (foliar spray). (Photograph courtesy of H. M. Cathey.)

tive growth before budding and flowering. Such late varieties as 'Sweetheart Supreme' and 'Sun Valley' are not suitable choices when early flowering is desired.

SADH and CCC are the most effective compounds for retarding growth of azalea. One application of SADH at a concentration of 2,500 ppm or two sprays at a concentration of 1,500 ppm one week apart are effective. A single spray of CCC at concentrations of 1,844 to 2,305 ppm can be applied, or two applications one week apart are also satisfactory (Stuart, 1965). After chemical treatment is applied, at least eight weeks should elapse before the plants are placed in cool storage, which usually consists of four weeks of long days and four weeks of short days.

Earlier work demonstrated that application of gibberellin at a concentration of 1,000 ppm resulted in uniform flowering of azalea cultivars 'Hexe' and 'Sweetheart Supreme' that had been grown to forcing size at a minimum temperature of 16°C and that had not received any cold treatment. The time required to force with gibberellin was approximately the same as that required for storing plants at low temperature and then forcing them in the greenhouse. However, application of gibberellin eliminated the low-temperature requirement (Boodley and Mastalerz, 1959).

Furuta and Straiton (1966), working with unchilled azalea (cv. 'Red Wing'), noted that a combination of gibberelin at concentrations of 100 or 500 ppm plus kinetin at a concentration of 100 ppm, applied at four-day intervals, hastens flowering. Kinetin alone did not hasten flowering, but kinetin acted synergistically with gibberellin in breaking dormancy.

Information on the time of flower induction and rate of bud differentiation is very important because these considerations affect the response of buds to growth substances. Criley (1969) made an anatomical study of the effect of photoperiod and growth regulators on the early stages of flower bud initiation and development using the azalea cultivar 'Hexe.' Flower initiation was most rapid under an eight-hour photoperiod, and a change in the shape of the apex was discernible four or five weeks from the start of short days. When a soil drench of CCC was combined with short-day treatment, floral initiation was more rapid than that occurring under short days alone. When GA_3 was applied to plants three or four weeks after the CCC treatment and the start of short days, flower initiation was delayed; application of GA_3 four to eight weeks afterward prevented flower initiation. McDowell and Larson (1966) have also pointed out that azalea plants cv. 'Red Wing' that were treated with CCC and SADH evidenced earlier flower induction and more rapid bud development than plants that were not chemically treated, whereas Ballantyne (1966) has noted that GA_3 does not promote earlier flowering.

CAMELLIA. Applications of Phosfon-D and CCC induce plants of *Camellia japonica* to form flower buds within one year of propagation from cuttings. Warm temperatures and long photoperiods favor the process. Treated plants flower but untreated ones remain vegetative (Stuart, 1962; Gill and Stuart, 1961).

GERANIUM. The flowering of F_1 geranium 'Carefree Scarlet' was advanced eight to sixteen days over that of the controls when a soil drench of CCC at a concentration of 5,000 ppm was applied thirty-one days after sowing (White, 1970). Mature plants treated with CCC were 8 to 10 cm shorter than mature plants that received no CCC. Plants that received no CCC but were manually pinched were also short and well branched, but there was a delay in flowering that was commercially undesirable.

HYDRANGEA.   When hydrangeas cease terminal growth in autumn, they do not resume growth for several weeks or months unless they receive cold-temperature treatment (six weeks at 4° to 5°C). Better results are obtained when plants are defoliated and when gibberellin application replaces part of the cold requirement. Old leaves inhibit shoot growth, and their removal increases the effectiveness of gibberellin application (Stuart and Cathey, 1962). In one experiment plants of hydrangea cv. 'Sainte Therese' were sprayed twice a week for one or two weeks after defoliation with gibberellin at concentrations of 10 or 50 ppm. Other plants of this cultivar were unsprayed but were stored at 4° to 5°C for two, four, or six weeks and were then returned to the greenhouse, defoliated, and sprayed with gibberellin. Cold storage accelerated stem growth of these plants and hastened flowering. Gibberellin applications produced similar effects irrespective of type of storage.

GLADIOLUS.   Gladiolus is one of the few plants whose stem length is stimulated by application of CCC. Halevy and Shilo (1970) applied CCC at a concentration of 8,000 ppm three times as a soil drench to potted 'Sans Souci' corms. The first application was made immediately after planting and the second was made four weeks later; the third was made three weeks after the second application, approximately twenty-five days before flowering. Increased stem growth and a greater number of flowers per spike resulted. These investigators state that successful field trials have been made and recommend CCC for practical use in gladiolus fields.

CHRYSANTHEMUM (JAPANESE).   Japanese varieties are not sensitive to photoperiod but require cold treatment in order to flower. The varieties used by Harada and Nitsch (1959), 'Shuokan,' 'Kinkozan,' and 'Shinisno,' required a cold treatment of three to four weeks at a temperature of approximately 1°C before flowering occurred. Application of 10 μg of GA_3 to the growing point of these plants induced bolting and flowering (Fig. 7–5).

KALANCHOE.   High temperatures delay flowering in *Kalanchoe*, and this may be a problem in greenhouses where adequate temperature control is not available during the summer months (Zawawi and Irving, 1968). Experimentation with *Kalanchoe blossfeldiana* cv. 'Red Glow' revealed that neither five months of short days nor a combination of short days plus application of TIBA could induce flowering at a relatively high temperature (approximately 24°C). However,

FIGURE 7-5
Induction of bolting and flowering in Japanese
chrysanthemum, cv. 'Shuokan,' 19 weeks after
application of GA₃ (*left*) and at the beginning of
a 4-week cold treatment (*right*). The control
(*center*) remained rosetted and vegetative. (After
H. Harada and J. P. Nitsch, 1959.)

these investigators found that when temperatures were maintained
below 24°C, applications of TIBA spray at a concentration of 250
ppm combined with short-day treatment almost doubled the amount
of flowering.

On August 18 Zawawi and Irving sprayed five-week-old plants that
were growing under short days (daytime temperature ranging from
21° to 24°C, nighttime temperature ranging from 18° to 21°C) with
TIBA at a concentration of 250 ppm. By December 7 the average
number of flowers per plant had increased from 2.6 on control plants
to 4.7 on treated plants. Application of higher concentrations of TIBA
or repeat application of the compound at the same concentration

failed to increase flowering, and ring fasciations, which enclosed the stem and occasionally the terminal bud in a tubule-like structure, often developed.

In contrast to the effect produced by TIBA, multiple applications of IAA at a concentration of 250 ppm to plants growing under short days at temperatures below 24°C delayed flowering by approximately two weeks and decreased the number of flowers produced.

PREVENTION OR DELAY OF FLOWERING

Tree Fruits

PEACH. Sprays of gibberellin promote vegetative growth, but the invigoration is accompanied by a reduced number of flc‑‑‑‑‑ buds, delayed differentiation, and production of smaller buds. In New York State, trees of 'Redhaven' and 'Golden Jubilee' were sprayed with potassium salt of GA_3 at concentrations of 50 or 200 ppm in late July when the length of the shoots averaged nine inches (Edgerton, 1966). Examination of sections of buds in September showed that the gibberellin treatment had delayed the initiation of flower buds at the shoot tips and retarded flower bud development toward the bases of the shoots. There was a reduction in both the number of flowers per foot of shoot and the percentage of buds that were fruit-ful. 'Redhaven' controls had eighteen flower buds per foot of growth, whereas shoots treated with gibberellin at a concentration of 50 ppm had only five.

Gibberellin delayed full bloom of 'Elberta' peach up to six days in trials conducted in Washington State (Proebsting and Mills, 1964). A delay in bloom is likely to be greater if cold temperatures prevail when the flowers start to open. The gibberellin treatment also increased the cold hardiness of the wood. Twigs were collected from gibberellin-treated trees in November and February and were subjected to freezing tests. The results showed increases in the hardiness of the treated flower buds, which was probably caused by the delayed differentiation of the flower primordia (Edgerton, 1966). Only the small buds with partially developed primordia survived the coldest temperature (−26°C). Buds of trees of 'Elberta' peach growing in the Northwest that were sprayed with gibberellin in August also showed increased cold hardiness (Proebsting and Mills, 1964).

A high percentage of the flowers on gibberellin-treated trees usually set fruit. The high set is probably a result of the sharp reduction in the number of flower buds per shoot. Application of gibberellin at a concentration of 50 ppm produced a nearly optimal set of fruit (Edgerton, 1966). If too high a concentration of gibberellin is used, no fruitful buds are formed (Hull and Lewis, 1959).

In experiments in which trees of 'Redskin' peach were sprayed with gibberellin after bud differentiation but before leaf fall, bud development was retarded and some buds were killed (Stembridge and La Rue, 1969). In one experiment trees were sprayed with potassium salt of GA_3 at a concentration of 50 ppm on six dates from July 31 to October 21; observations were made on the following March 24. The spray applied on August 30 resulted in production of the least number of flowers per linear foot of fruiting wood, whereas the October 3 treatment retarded bud growth most. The reason why bud mortality response did not peak at the same time that maximum retardation was induced is obscure. However, all treatments that killed buds also retarded bloom.

APRICOT, ALMOND, AND PLUM.   Spraying apricot cv. 'Royal,' almond cv. 'Jordanolo,' and plum cv. 'President' during the floral initiation period with GA_3 at concentrations of 100 to 1,000 ppm at the initiation of the pit-hardening stage of the fruit growth that generally occurs during this period inhibits development of both floral and vegetative buds (Bradley and Crane, 1960). In this California experiment, apricot flower buds collected from control trees on August 24 contained well-defined primordia of sepals, petals, and stamens. However, the growing points of branches sprayed with compound at a concentration of 50 ppm were rounded rather than flattened, indicating that the buds had not attained the initial phase of floral differentiation. At bloom during the following spring there were almost no flowers on any treated branches except at the tips of some long shoots, where the fruitful nodes probably developed after the gibberellin treatment. Growth of vegetative buds was also generally inhibited. Higher dosages of compound were required to block vegetative growth than to block floral bud growth. Bradley and Crane obtained similar results with plum. Two applications of gibberellin at a concentration of 50 ppm caused complete inhibition of floral bud development.

Blossoming in almond may be retarded for several days by application of gibberellin the preceding fall in order to reduce the frost hazard. Another advantage of gibberellin treatment is that it facilitates better cross-pollination among varieties having nonsynchronous bloom periods. In California, application of gibberellin to the 'Peerless,' 'Nonpareil,' and 'Mission' varieties in August or September either

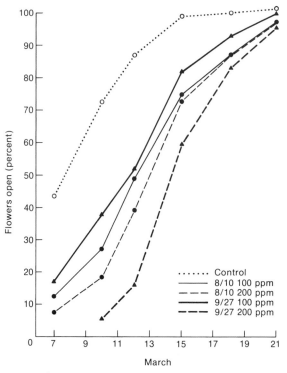

FIGURE 7-6

Effect of 2 applications of GA₃ (on August 10 and September 27, 1965) on time of bloom of 'Peerless' almond in the spring of 1966. (After J. R. Hicks and J. C. Crane, The effect of gibberellin on almond flower bud growth, time of bloom, and yield. *Proc. Amer. Soc. Hort. Sci.* 92:1–6, 1968).

before or after flower bud differentiation retarded bloom the following season by four to seven, four to six, and two to five days, respectively (Hicks and Crane, 1968). More delay resulted with early blooming varieties. ('Peerless' is the earliest-blooming variety; 'Mission' is the last to bloom.) Figure 7–6 shows the effect of August and September applications of GA₃ on the time of bloom of 'Peerless' the following spring. Application of gibberellin at a concentration of 200 ppm generally delayed bloom one to two days more than did a concentration of 100 ppm. Application in September retarded bloom one to two days more than did application in August, probably because by September the crop had been removed. Before harvest some gibberellin may have translocated into the fruits, leaving relatively lesser amounts in the branches. The sprays may also reduce yields by inhibiting fruit development and causing abscission of part of the flower buds, so the economic benefits of the treatments must be carefully weighed.

The immediate effect of gibberellin on the trees studied was an inhibition of cell division, which led to the restriction of lateral bud development. Gibberellin seems to interfere with the processes that occur during floral initiation, differentiation, and floral development (Hicks and Crane, 1968).

SADH can also effectively delay flowering. Applications of the compound at concentrations of 2,000 or 4,000 ppm to almond trees in June, September, or October were found to delay opening of blossoms the following spring (Ryugo et al., 1970). The compound was effective in proportion to the rest requirement of the variety. The full-bloom date of the early blooming variety 'Jordanolo' was only slightly affected by the treatment, whereas that of a midseason bloomer 'Nonpareil' was delayed for two to four days. Blossoming of the late blooming variety 'Ne Plus Ultra' was delayed by approximately one week. SADH also inhibited shoot growth and reduced the average nut and kernel weights of mature trees.

CHERRY.   Postbloom applications of gibberellin at a concentration of 100 ppm to branches of mature, bearing trees of 'Montmorency' cherry was found to result in weak flowering the following year (Hull and Lewis, 1959). Flowers appeared on the distal portion of the shoots of treated branches. Spurs on the treated trees were entirely vegetative, but many nontreated trees were fruitful. One postbloom application of gibberellin at a concentration of 500 ppm or two applications at a concentration of 250 ppm caused complete inhibition of flower bud development in sweet cherry cv. 'Bing' (Bradley and Crane, 1960).

These results suggest some practical future applications. Flowering of young cherry trees might be inhibited by applications of gibberellin during the preceding growing season. As a result, the trees should attain a larger size at an earlier age. Also, trees might be sprayed with gibberellin during one season to cause a thinning of fruit the following season.

CITRUS.   Gibberellin retards flower induction in citrus trees. Experiments in Israel have shown that application of gibberellin at a concentration of 200 ppm at two-week intervals, each treatment consisting of three, four, five, or six sprays, from November until the end of January, inhibits flower induction in trees of 'Shamouti' orange (Monselise and Halevy, 1964). Three sprays resulted in some delay in blossoming the following spring. In February, subsequent to three spray applications, there were only 426 flowers per treated branch compared with 1,782 on untreated branches. If more than three sprays were made, the effect of the gibberellin persisted longer; the flower inductive period elapsed before the tree was able to differentiate flower buds, and few or no flowers developed. Flowers on treated branches were progressively younger the longer the gibberellin influence persisted, indicating a delay in the differentiation of the few flowers present.

Appropriate concentrations of gibberellin also prevent flower induction in lemon. In Israel, branches of trees of 'Eureka' lemon that were sprayed twice in August with GA_3 at a concentration of 500 ppm developed no flowers. On the other hand, growth-retarding chemicals, such as CCC, SADH, and the auxin BOA, stimulate flowering in lemon.

PEAR.   In California, September sprays of gibberellin at concentrations of 10 to 500 ppm failed to effect flowering in the following spring. However, higher concentrations applied in March when flower bud scales were separating reduced flower bud formation. The greatest reduction in flower buds resulted when highest concentrations were applied at full bloom or petal fall (Griggs and Iwakiri, 1961).

PECAN.   Sprays of KGA_3 applied to mature trees of pecan (*Carya illinoensis*) were found to inhibit catkin formation during the spring following treatment (Sparks, 1967). The growth substance at a concentration of 200 ppm was applied one, two, or three times at twenty-three-day intervals either early in the growing season (beginning on June 1) or late (beginning on August 8). Final cumulative concentrations of spray applied during both parts of the growing season were 0, 200, 400, and 600 ppm, respectively. In the spring following the early application, the control trees had 3.43 catkins per terminal; the trees treated with KGA_3 at concentrations of 200, 400, and 600 ppm had 1.09, .07, and .26 catkins, respectively. Only a small decrease in the number of catkins resulted from the late treatment. KGA_3 probably affects catkin initiation or differentiation, and since differentiation in the pecan is usually completed in late summer, there was no adverse effect on the number of catkins formed after the August–September applications.

Other effects of the sprays were (1) termination of rest of some buds as a result of early applications and (2) an increase in total weight per nut as a result of the late sprays.

APPLE.   Many horticulturists have noted that although gibberellin stimulates shoot growth, relatively fewer flowers are produced. Stimulation of the vegetative development of the tree is probably antagonistic to flower formation.

In Scotland, Guttridge (1962) demonstrated that GA_3 can inhibit fruit bud formation in apple without affecting the bursting of buds in the following spring. Branches of six varieties were sprayed with gibberellin at concentrations of 10 or 50 ppm at weekly intervals from May 21 until August 29. By the following spring the percentage of spurs bearing blossom clusters had been reduced from 40 percent on unsprayed branches to 14.7 percent on sprayed branches. Similar results were obtained in France when sprays of gibberellin at concentrations of 100 and 300 ppm were applied to the 'Golden Delicious' variety (Marcelle and Sironval, 1963). These findings suggest that gibberellin sprays can be used to restrict flower bud formation during the nonfruiting year of biennially bearing apple trees.

Research on the effect of seed formation on subsequent flowering in apple indicates that alternate bearing, frequently a characteristic of apple trees that bear seeded fruits, in which a crop is obtained only in alternate years, may be hormonally controlled. Using varieties 'Spencer Seedless' and 'Ohio 3,' which bear annual crops of parthenocarpic fruits but which produce seeded fruits following hand pollination, Chan and Cain (1967) have shown that flower bud formation in apple is inhibited by seed development. Removal of seeded fruits after bloom showed that 65 percent of the inhibition occurred within the first three weeks following pollination. Production of seedless fruits did not have any effect on subsequent flowering. Thus, seed formation appears to be the controlling factor in flower initiation. Since appropriate growth regulators can either hasten or inhibit flowering, the correlative inhibition of flowering by seeds may be hormonal in nature. In general, the inhibition of flower bud formation by abundant fruiting in apples has previously been explained by nutritional competition between developing fruits and formation of flower buds.

Similar results have been obtained in pear (Griggs et al., 1970), indicating that the inhibitory effect of seeds on flower bud formation may occur in other tree fruits and plants as well.

TUNG. The size of the tung crop produced in the belt located in the southern United States is frequently reduced by spring cold. When the inflorescences have just projected beyond the bud scales they are very susceptible to frost. If anthesis could be delayed by one or two weeks, most cold injury could be avoided.

The efficacy of many growth regulators and chemicals has been tested, but promising results were not obtained until the 1960's. In Louisiana, Sitton *et al.* (1968) tested nearly 200 chemicals, both in laboratory experiments (excised shoots were dipped in the compounds) and in trials conducted in the tung orchard (trees were sprayed or individual buds were dipped). In the laboratory experiments, conducted in a humid chamber at 27°C, applications of thiothymine, thioadenine sulfate, and dichloropyrimidine at concentrations of 0.02 g/liter retarded bud break by approximately seven days. In the orchard trials best results were obtained by applying thiouracil combined with either glycerol or propylene glycol in October or early March. In one orchard, spraying of trees in March with thiouracil at concentrations of 0.5, 1.5, and 3.5 g/liter retarded bud development so that attainment of the loose-cluster stage was delayed by two, three, and six days, respectively. Fruit set per terminal bud was reduced by 8, 16, and 15 percent, respectively. A repeat spray of thiouracil at 1.5 or 3.5 g/liter on February 21 and March 7 retarded bloom by ten days.

Flowers

CARNATION. Flowering in carnation, a facultative long-day plant, is delayed by repeated applications of ABA. The effect of the compound is dependent on length of photoperiod (Cathey, 1968). When carnation plants were sprayed once daily for fifteen consecutive days with ABA at a concentration of 1,000 ppm, the flowering of plants grown under long days was delayed by several weeks, compared with unsprayed plants grown under long days. A weekly spray of ABA at a concentration of 1,000 ppm for eight weeks delayed flowering of plants grown under long days more than it did flowering of untreated plants grown under short days. Also, ABA delayed flowering of plants grown under short photoperiods (eight hours) considerably more than it did flowering of plants grown under long photoperiods (twelve and twenty-four hours). Cathey suggests that a stoichiometric relationship exists between length of photoperiod and frequency of ABA application required to delay flowering. In his experiments, retardation of stem elongation as a result of ABA treatment was accompanied by a delay in flowering. Both untreated and treated plants were taller when grown under longer photoperiods. Interestingly, when ABA treatments were discontinued, normal growth characteristics immediately resumed, indicating that the effect of the hormone is very transitory.

Cathey (1968) extended his work on carnation to many other ornamental plants. Some results with a long-day petunia (*Petunia hybrida*) and a short-day marigold (*Tagetes erecta*) are shown in Figure 7–1.

FUCHSIA.   Many varieties of *Fuchsia hybrida* are long-day plants. Although applications of gibberellin replaces the long-day requirement for flower initiation in many plants, it inhibits flower induction in *Fuchsia hybrida*. Detailed studies made with the 'Lord Byron' fuchsia have shown that at least four days with a day length in excess of twelve hours are required for flower development. However, when 'Lord Byron' plants were sprayed with gibberellin at concentrations of 10 to 100 ppm, the long-day induction of flowering was prevented (Sachs and Bretz, 1962). The plants were maintained under long-day conditions for twelve, twenty, and forty days and were sprayed every second day with gibberellin. Whether the primary effect of gibberellin was on long-day induction in the leaves or on the differentiation of flower primordia in the axillary buds is not known.

POINSETTIA.   The formation of flowers and decorative bracts in poinsettia (*Euphorbia pulcherrima*) is induced by short-day conditions; during long days the plants remain vegetative. In Scotland, Guttridge (1963) showed that spraying poinsettia plants with gibberellin during the short-day inductive period delays as well as reduces the amount of flowering and also causes poor coloration of bracts. In one greenhouse experiment plants were sprayed weekly with GA_3 at concentrations of 2.5 to 40.0 ppm during a six-week inductive period commencing on October 25. By the end of February the induced control plants and those sprayed with GA_3 at a concentration of 10 ppm or less were flowering, but those sprayed with higher concentrations of compound were not flowering. By mid-May some plants treated with GA_3 at a concentration of 40 ppm were still not in flower.

HYDRANGEA.   Hydrangea ceases terminal growth and differentiates flower buds in the fall. If plants are treated with gibberellin in the summer they elongate rapidly, but flower bud differentiation is greatly delayed. In Maryland, Stuart and Cathey (1962) sprayed potted hydrangea plants cv. 'Sainte Therese' in August with gibberellin at concentrations of 0, 0.01, 1.0, or 10.0 ppm. Plants were sprayed five times with the 10 ppm solution and ten times with each of the other concentrations. Flower bud differentiation was greatly delayed by all treatments.

CHRYSANTHEMUM.   Auxin causes a slight delay of flowering in certain varieties of chrysanthemum. Applications of IAA at concentrations of 25 to 400 ppm to *Chrysanthemum morifolium* varieties 'Forty-Niner' and 'Iceberg' under greenhouse conditions (with a nine-hour photoperiod) increased the number of short days until first visible flower color and greatly restricted the appearance of the reproductive buds (Lindstrom and Asen, 1967). Auxin inhibits flower induction in some species and stimulates it in others, but both effects are only slight.

Ten weeks of short photoperiods are required to induce flowering in 'Indianapolis Yellow' chrysanthemum. Cathey (1959b) found that five applications of gibberellin at a concentration of 100 ppm, made during the fourth week of the photoperiod, reduced the number of inflorescences produced. Plants not treated with gibberellin averaged 9.2 lateral inflorescences; treated plants averaged only 4.2. The remaining flower primordia in the gibberellin-treated plants failed to develop.

Sprays of gibberellin at a concentration of 100 ppm, applied on five consecutive days, elongated shoots most when applied during the third week of short photoperiods, after initiation of the inflorescences but before extensive floret development. On the other hand, application of the growth retardant Amo-1618 at a concentration of 500 ppm to 'Yellow Lace' on three alternate days inhibited stem growth. Maximum inhibition of elongation occurred during the first week of short photoperiods, and the treated plants flowered fourteen to eighteen days later than the controls. Simultaneous application of gibberellin and Amo-1618 during the first two weeks of short photoperiods neutralized the growth-regulating properties of these compounds, and consequently the growth and flowering of treated and untreated plants were similar.

Some flower growers in Missouri have reported that during the winter when there was poor ventilation in their tightly constructed plastic houses, chrysanthemums grew abnormally and failed to flower even under conditions that were favorable for flowering (including short days) (Tija *et al.*, 1969). The causal agent was shown to be the ethylene contained in the exhaust gases of gas or oil heaters and open-flame burners used for heating purposes. Many years ago it was demonstrated that the presence of ethylene in the air in concentrations as low as 0.1 ppm causes injury to carnation (Crocker and Knight, 1908). Tija and colleagues subjected plants of chrysanthemum '#3 Indianapolis White' continuously to atmospheres containing ethylene in concentrations of 1 to 4 ppm. Plants maintained under photoinductive conditions failed to initiate or develop flower buds. The vegetative development of these plants also showed changes typical of ethylene injury: Internodes were shortened, stems were thickened, and leaves were epinastic. When plants were placed alternatively in normal atmospheres and in atmospheres containing ethylene, flowering usually was also prevented.

MARIGOLD.   Plants of 'Sovereign,' an F_1 cultivar of *Tagetes erecta* (marigold), that were growing under both short and long days were given weekly foliar sprays of SADH at concentrations of 500, 1,000, or 2,000 ppm (McConnell and Struckmeyer, 1970). Short-day control plants were shorter, flowered earlier, and had shorter leaves than long-day control plants. SADH delayed flowering of treated plants up to eight days in each photoperiod compared with the respective controls. However, the terminal flowers of the short-day plants were the same size as those of the controls. SADH sprays also resulted in shorter plants, denser foliage, and wider leaflets.

STRAWBERRY.   Flowers of strawberry are usually initiated only under short-day conditions. Under long days runners are formed in the leaf axils and the petioles grow long and upright. In Scotland, Thompson and Guttridge (1959) sprayed plants of 'Talisman' growing under natural short days with gibberellin at concentrations of 0 to 100 ppm. Plants were kept under short days from September 4 until November 19. The increase in plant height, measured from the node of the leaf emerging at the start of treatment to the tenth node above it, was directly proportional to the concentration of the gibberellin applied. Dissection of plants showed that the average numbers of inflorescences per crown of plants treated with gibberellin at concentrations of 0.0, 12.5, 25.0, 50.0, and 100.0 ppm were 1.00, 0.58, 0.25, 0.00, and 0.00, respectively.

In November runners that had failed to emerge were found in the axils of the leaves of some of the treated plants. The plants were then placed in a warmer and better-illuminated greenhouse; after twelve days the numbers of runners per crown of control plants and plants treated with the same concentrations of gibberellin (in the same order) were 0.00, 0.25, 1.16, 1.22, and 1.59, respectively. All the runners had been initiated under natural short-day conditions.

Since the initial work of Thompson and Guttridge (1959), an effort has been made to increase runner production in everbearing varieties of strawberry. The fact that flower bud development continues throughout most of the growing season but that runner production is low creates a problem for the nurseryman. Runner production can be encouraged by removing flowers during the summer to prevent fruiting, but this operation is very laborious. In New York State, Dennis and Bennett (1969) applied single sprays of gibberellin at a concentration of 550 ppm to one-year-old plantings of the everbearing variety 'Geneva' in June; double sprays were applied in June and July. The sprays stimulated runner production of runners and/or increased the numbers of marketable runner plants per mother plant. Increases in runner plants among the various experiments ranged from 111 to 1,600 percent. In another experiment defloration increased subsequent flowering and yields of runner plants but had no marked effect on runner initiation. Defloration combined with gibberellin application resulted in the highest yield of runner plants. Gibberellin inhibited subsequent flowering of most plants; this phenomenon was most noticeable in deflorated plants because they produced many flowers (Dennis and Bennett, 1969).

REGULATION OF SEX EXPRESSION

It is well known that application of auxin shifts the balance of sex expression from maleness to femaleness in certain flowering plants (Heslop-Harrison, 1959) and that application of gibberellin induces formation of staminate flowers in certain flowering plants (Galun, 1959). Consequently, such growth regulators have been used to facilitate vegetable seed production and as an aid in breeding several species.

Cucumber

In their classical work on the modification of sex expression by hormones, Laibach and Kribben (1950) noted that application of auxin stimulates the development of female flowers rather than male flowers in some cucurbitaceous plants.

MONOECIOUS CUCUMBER. Treatment of cucumber plants with gibberellin increases the number of staminate flowers produced in mo-

FIGURE 7-7

Female flower (*left*) from cucumber plant treated with ethephon at a concentration of 250 ppm and male flower (*right*) from untreated plant. (After R. W. Robinson, H. Wilczynski, M. D. de la Guardia, and S. Shannon, 1970.)

noecious cucumber (Bukovac and Wittwer, 1961). These investigators applied gibberellin at a concentration of 100 ppm to young pickling-type cucumber seedlings ('Wisconsin SMR-12') during two weeks of short-day exposure (nine hours daily). Pistillate flowers are usually formed earlier (that is, at a lower node) when cucumber plants are grown under a short rather than long photoperiod. The effect of the short photoperiod of hastening pistillate flower formation was noticeably reduced by the growth regulator. In other words, the effect of gibberellin was opposite to that of a short photoperiod and simulated the effects of a long photoperiod.

Iwahori *et al.* (1970) have shown that a mixture of GA_4 and GA_7 causes maleness in the cultivar 'Improved Long Green.'

Soon after ethephon was made available for experimentation, the compound was observed to shift the sex expression of cucumber to femaleness. These effects are illustrated in Figures 7–7 and 7–8. Mc-Murray and Miller (1969) also made this observation in greenhouse and plot trials with pickling cucumber. *Cucumis sativus* cultivars 'Model,' 'Chipper,' 'SC 19,' and 'SC 23' were given single and multiple spray applications of ethephon at concentrations of 120, 180, or 240 ppm. These investigators noted that the usual flowering pattern for 'SC 23,' when grown in the greenhouse, was that nodes 3, 9, and 16 produced pistillate flowers, whereas nodes 17 and 20 produced stami-

214

FIGURE 7-8

Cucumber plant treated with ethephon at a concentration of 250 ppm (*left*) has only female flowers; untreated plant (*right*) has only male flowers. (After R. W. Robinson, H. Wilczynski, M. D. de la Guardia, and S. Shannon, 1970.)

nate flowers. The staminate-to-pistillate flower ratio was approximately 10:1. However, ethephon-treated plants usually produced pistillate flowers at nodes 1 to 16; the staminate-to-pistillate flower ratio ranged from 1:6 to 1:14, depending on the concentration of ethephon used.

In these experiments, two field applications of ethephon at a concentration of 240 ppm to 'Model' resulted in exclusively pistillate nodes for the first 2.5 weeks of the harvest season. Application of ethephon to 'Model,' 'SC 23,' and 'Chipper' resulted in significant yield increases as well as an increase in value per acre as a result of early yield.

The results of research conducted in California on the effects of ethephon on pickling cucumber cultivar 'SMR-58' were similar to those obtained by McMurray and Miller (Sims and Gledhill, 1969). 'SMR-58' usually produces male flowers at the first five or six nodes; above this point there is a ratio of female-to-male flowers that is dependent on environmental conditions. In a field trial ethephon produced complete femaleness. Treated plants were smaller and matured earlier. This last result is important because it makes possible a single harvesting of cucumbers for pickling.

Results obtained in Israel from application of ethephon to 'Beit Alpha' were generally similar to those previously described (Rudich et al., 1969).

There is apparently considerable latitude in choosing the stage of plant development at which successful results can be obtained. Treatment of plants of the cultivar 'Galaxy' with ethephon at a concentration of 120 ppm at the one-leaf stage or at subsequent leaf stages through the stage at which seedlings had twelve true leaves increased pistillate flower formation (Lower et al., 1970).

Earlier work demonstrated that auxins, especially NAA, also increase the femaleness of monoecious cucurbits (Wittwer and Hillyer, 1954).

GYNOECIOUS CUCUMBER. For mechanical harvesting of cucumber, new hybrids derived from gynoecious lines are becoming increasingly important. Maintenance and production of gynoecious plants is a problem because of insufficient male flowers and viable pollen. Gibberellin has been found to induce male flower formation in gynoecious cucumber (Peterson and Anhder, 1960). Commercially one row out of three rows of gynoecious cucumber plants, such as 'MSU 713-5,' is sprayed three times a week with gibberellin at a concentration of 1,000 ppm beginning at the time of expansion of the first true leaf. The induced male flowers serve as the source of pollen for the unsprayed females, and the seed sold is the F_1 hybrid. Hybrids are more uniform in size and more productive; however, there have been some instances of poor seed yields, which may result from the inhibiting effects of the hormones on pollen germination.

Clark and Kenney (1969) found that GA_3, GA_4, GA_7, GA_{13}, and a mixture of GA_4 and GA_7 were equally effective in producing staminate flowers in the gynoecious cucumber 'MSU-713-5.' The activity of a mixture of equal parts of GA_4 and GA_7 was nearly equal to that of pure GA_7, but GA_{13} was much less active than GA_7 or a mixture of GA_4 and GA_7. Gibberellins A_3 and GA_4 and the mixture of GA_4 and and GA_7 showed the most activity as judged by the number of staminate flowers produced per plant.

True gynoecious plants produce only female flowers, but when several of the new hybrids thought to be gynoecious were grown under commercial field conditions in California, they produced up to 50 percent male flowers. This result was probably caused by variable environmental conditions. Application of ethephon at concentrations of 50 to 250 ppm at the stage at which seedlings had one fully expanded true leaf induced femaleness in the hybrid 'Piccadilly' and reduced the size of the plant by shortening the internodes (Sims and Gledhill, 1969).

Begonia

The genus *Begonia* is monoecious, bearing male (staminate) and female (pistillate) flowers on the same inflorescence. The variety *B.* × *cheimantha* is an exception to the rule that gibberellins usually enhance male sex expression. Heide (1969a) sprayed plants of the cultivar 'Mörk Marina' with GA$_3$ or IAA and found that both compounds increased femaleness. For example, when IAA at a concentration of 10^{-5} M was sprayed daily on plants growing under eight-hour days, the number of female flowers increased nearly sixfold compared with the nontreated controls; the number of male flowers was unaffected. Application of GA$_3$ also increased the femaleness of the plants but to a lesser degree than did application of IAA. Heide suggests that the effect of GA$_3$ on sex expression may result from its general inhibition of flowering or from the fact that the compound seems to augment auxin biosynthesis in *Begonia* as it does in a variety of other plants (Paleg, 1965). Other research conducted by Heide (1969b) showed that the growth retardant CCC, which is an inhibitor of gibberellin biosynthesis, reduces extractable auxin and favors male rather than female sex expression in *Begonia*.

Short days and low temperatures increase maleness in *Begonia* (Heide, 1969a). However, most monoecious plants, under these conditions, increase femaleness (Nitsch, 1965). Heide believes that auxin is the mediating compound through which temperature and photoperiod effect sex balance in *Begonia*.

Hops

Gibberellin sprays increase the number of female inflorescences in hops, although vegetative growth remains normal. This increase often results in the production of more fruits (cones) and an increased yield.

In Oregon, Zimmerman *et al.* (1964) found that application of gibberellin at a concentration of 5 ppm to five-foot-high seeded 'Fuggle' hop plants resulted in a 25 percent increase in the number of cones, which amounted to a yield increase of approximately 300 pounds per acre. The response was caused by the presence of a large set of small cones, but there was no reduction in hop quality. The response was correlated with stage of plant growth, and the action of the hormone may be caused by its influence on endogenous hormone activity at the time of floral differentiation.

The effects of gibberellin on hops vary with variety and probably with climate. In Australia, Nash and Mullaney (1960) obtained a 40 percent increase in yield by applying gibberellin at a concentration of 12.5 ppm to hops when the stigmas of the female flowers were .125 inch long or longer. This increase in yield was probably a result of increased cone size. In England, Roberts and Stevens (1962), using the same procedure, noted a slight reduction in yield of hops.

Grape

Nearly all cultivated varieties of grape are hermaphroditic and self-fruitful. However, the indigenous forms of the *vinifera* grape are dioecious. Professor H. P. Olmo of the University of California at Davis collected seeds from wild vines of *Vitis vinifera* growing in northern Iran. On rare occasions one of the progeny vines classified as male produced hermaphroditic flowers and mature fruits. Of the many hormones tested, only the cytokinin PBA changed the sex of a cluster from male to hermaphrodite (Negi and Olmo, 1966). When the flower clusters of the male vine were dipped in an aqueous solution of PBA at a concentration of 1,000 ppm three weeks before anthesis, they produced typical hermaphroditic flowers instead of the usual male flowers, and normal fruit setting ensued (Fig. 7–9). The control clusters and clusters treated with other chemical solutions had predominantly male flowers, which characteristically dried up a few weeks after anthesis. Subsequently these researchers obtained similar results with male vines of other *Vitis* species, and Moore (1970) successfully converted genetically staminate plants of *Vitis* to functional hermaphrodites by applying cytokinin to eight of a total of fifteen clones.

It may be that the presence of a high level of endogenous cytokinin in flower buds during development causes or mediates a conversion of the sex of a male vine to hermaphrodite. The ability to convert a male vine to a hermaphroditic one should be of great value in plant breeding because it means that the male vine can be utilized as a female parent.

Muskmelon

Application of the growth retardant SADH to an andromonoecious cultivar of muskmelon (*Cucumis melo* cv. 'Ananas PMR') shifts sex

FIGURE 7-9

Effect of treating clusters of a wild male *Vitis vinifera* grape with the cytokinin PBA at a concentration of 1,000 ppm. *Left,* untreated control; *center,* cluster treated with PBA; *right,* untreated control with poor set that sometimes occurs in nature. Photographed 3.5 months after treatment. (Photograph courtesy of H. P. Olmo and S. S. Negi.)

expression toward femaleness (Halevy and Rudich, 1967). In this plant flowers are borne on very short branches in leaf axils growing from the main shoot. Anthesis occurs in regular ascending order as the main shoot develops. The lower nodes produce only staminate flowers. A mixed phase subsequently begins during which both male and hermaphrodite flowers are formed. As the plant develops the ratio of hermophrodite flowers to male flowers steadily increases, and this increase can be used as measure of the plant's sex tendency.

SADH at a concentration of 0.5 percent was applied to the muskmelon variety 'Ananas PMR' either by soaking the seeds, before they were sown, in the chemical solution for twenty-four hours or by applying a foliar spray at one or all three developmental stages (Halevy and Rudich, 1967). The growth retardant changed the plant from a climbing to a bushy type and effectively shifted the sex expression toward femaleness. The most effective treatment was a combination of the seed application plus three repeated foliage sprays, which resulted in a ratio of hermaphrodite to male flowers of 1:1.4 compared with 1:6.2 for control plants.

In earlier research Laibach and Kribben (1950) and Brantley and Warren (1960) demonstrated that auxin treatments also increase femaleness.

Application of ethephon also increases femaleness in the muskmelon cultivar 'Ananas PMR' (Rudich et al., 1969). When seedlings were sprayed at the one- or three-leaf stages with the compound at a concentration of 500 ppm, hermaphroditic flowers developed on the lower nodes of the main axis as well as on the short side branches where such flowers are normally formed in this variety.

Squash

The staminate phase of the main axis of the butternut squash (Cucurbita moschata) terminates after production of thirteen to sixteen staminate nodes (Hopp, 1962). In field trials Hopp and Rochester (1967) found that applications of SADH at concentrations of 1,000 or 5,000 ppm when seedlings had one true leaf and approximately six weeks later extended the staminate phase. Less than half the plants treated with the compound at a concentration of 5,000 ppm produced fewer than sixteen staminate nodes below the first pistillate node. The remaining 52.5 percent of the plants produced 19.5 nodes during the staminate phase. When SADH was applied at a concentration of 1,000 ppm, the extension was less pronounced. However, SADH treatments did not affect the normal 2:1 ratio of staminate buds to pistillate buds in the subsequent mixed phase. The extension of the staminate flower phase by several nodes could be of value in plant-breeding studies.

Rudich et al. (1969) made the interesting finding that application of ethephon at a concentration of 500 ppm to squash (Cucurbita pepo cv. 'Spotted Zucchini') at the three-leaf stage completely prevents the formation of male flowers during the first three weeks of flowering and increases the number of pistillate flowers. Application of ethephon to 'New Hampshire Butternut' squash results in increased pistillate flower and fruit production but decreased fruit weight (Coyne, 1970).

Pumpkin

Application of ethephon to plants of pumpkin (Cucurbita moschata cv. 'Dickinson Field') results in increased production of female flowers,

220

FIGURE 7-10

Effect of treatment with GA₃ on heterostylism in tomato. *Left,* untreated control; *right,* plants following treatment with GA₃ (note extended stigmas). (Photographs courtesy of M. J. Bukovac and S. Honma.)

shorter internodes, and earlier fruit set, but application of GA₃ results in more male flowers, longer internodes, and later fruit set (Splittstoesser, 1970). However, since most of the flowers induced by ethephon in these experiments aborted, there was only a slight increase in number of fruits per plant.

Tomato

HETEROSTYLISM. A heterostylous condition is one in which the stigma is extended beyond the stamen cone before pollen maturation. The existence of such a condition might facilitate hybridization without the necessity of hand emasculation. Elongation of styles occurs in many plants when they are treated with gibberellin. A heterostylous condition can be induced in flowers of tomato (*Lycopersicum esculentum*) by spraying them with GA₃ at concentrations of 10^{-3} or 5×10^{-3} M four to six days before bloom (Bukovac and Honma, 1967). A single treatment was found to cause a significant elongation of styles, although there were no significant differences in ovary growth (Fig. 7–10). The ovaries formed were functional and produced viable seeds following hand pollination. The extended stigma was easily hand-pollinated, and there was no contamination as a result of self-pollination. Style elongation was induced in cultivars of indeterminate or semideterminate growth habit but not in cultivars of determinate growth habit.

Chemical control of style elongation by application of gibberellin may serve as an effective hybridization technique for certain tomato cultivars. The female parent can be maintained by hand-pollination because the pollen of gibberellin-treated plants is viable. Maintenance of the female parent can also be achieved by permitting flowers that develop later to self naturally, since the effect of the gibberellin is transient. Two disadvantages of the technique are: (1) the number of seeds produced per fruit following hand-pollination is low; and (2) only flower clusters of tomato cultivars with indeterminate or semideterminate growth habit are responsive.

ANTHER AND POLLEN DEVELOPMENT. Application of gibberellin induces anther and pollen development in stamenless tomato mutants (Phatak et al., 1966). Rooted cuttings were exposed to a solution of gibberellin at a concentration of 25 ppm. Within three to four weeks anthers developed in treated plants and viable pollen was produced. However, application of CCC was found to inhibit pollen formation in gibberellin-induced anthers. These results show that gibberellins are necessary for the development of male gametophytes in the tomato and that this influence is exerted after flower initiation.

Cotton

Male-sterile strains of cotton have been sought for many years because they would provide the means for controlled production of hybrid planting seed. Some promising results have been obtained with cotton by using a chemical as selective gametocide to produce male sterility. Plants of 'Empire' cotton produced no pollen after they had been sprayed with a 1.2 percent solution of FW-450 (Eaton, 1957). Several varieties of cotton react well to the compound. A spray of FW-450 at a concentration of 0.25 percent was found to result in good male sterility without a reduction in the number of bolls so long as there were enough bees to ensure good cross-pollination (Eaton, 1958–1959). Plants should be sprayed thoroughly approximately one week before the first buds are visible; the sprays should be repeated every three weeks during the period of bud development and boll setting. The disadvantage of using this method for cotton and for some other plants is that 100 percent male sterility is difficult to obtain.

Supplementary readings on the general topic of flowering are listed in the Bibliography: Chailakhian, M. K., 1968; Evans, L. T., ed., 1969; Evans, L. T., 1971; Hillman, W. S., 1962; Lang, A., 1965b; Leopold, A. C., 1958; and Searle, N. E., 1965.

Chapter 8

FRUIT SET AND DEVELOPMENT

Methods of increasing or decreasing fruit set (see p. 250) and altering fruit size and development are utilized to much advantage in agriculture. The control of these processes through the application of plant growth regulators can also produce favorable effects on rate of maturation, induction of seedlessness, and fruit quality.

DEFINITION OF FRUIT

A fruit is a mature ovary with associated parts. Some fruits consist only of the ovary; they may be fleshy or dry, dehiscent or indehiscent. In other fruits such flower parts as petals, stamens, sepals, peduncles, and receptacles develop at the same time as the ovary. From a physiological rather than morphological point of view, a fruit may be defined as the structural entity that results from the development of the tissues that support the ovules of a plant (Nitsch, 1952, 1965). This definition encompasses such dissimilar organs as the floral axis of pineapple, receptacle of strawberry and apple, and syconium of fig, in that an ovule is present in all these fruits. Since ovules are initially present but fail to develop in seedless fruits, the definition can also be correctly applied to them.

PHYSIOLOGY OF FRUIT SET

Fruit set is the rapid growth of the ovary that usually follows pollination and fertilization. Other changes, such as wilting of petals and stamens, usually occur simultaneously. In many plants including grape, fruit set is accompanied by abscission of many of the flowers and fruits that fail to set. In some species fruit set is induced parthenocarpically.

Pollination and Fertilization

Pollination is the transfer of pollen from anther to stigma. The pollen grains develop pollen tubes that grow down through the style into the ovary (Fig. 8–1). On reaching the ovary the pollen tube usually enters the micropyle and penetrates through the nucellar tissues into the embryo sac. The sperm nuclei break out of the pollen tube and one of them fuses with the egg cell. "Fertilization" is defined as the fusion of the male and female gamete to form the zygote. The other sperm nucleus fuses with the two sperm nuclei in the embryo sac to form the primary endosperm nucleus. The zygote has double the number of chromosomes as the sperm or egg, and the endosperm nucleus has triple the number of chromosomes as the sperm or egg.

Arrest of Growth of Unpollinated Ovaries

The ovary fails to enlarge following anthesis unless fertilization occurs and unless parthenocarpy is initiated. In many species the whole flower abscisses if there is no pollination. However, in some plants, such as gherkin, the ovaries remain attached even though pollination has not taken place. Unpollinated ovaries cease growth one day following bloom, whereas pollinated fruits develop into large fruits (Nitsch, 1952). Thus an ovary needs more than a simple connection with the parent plant in order to develop into a fruit. The presence of an ovary in a plant is only a prerequisite, not a cause, of fruit set and growth.

Auxins and Pollen

Both pollination alone and growth of pollen tubes in the style can stimulate the growth of the ovary, even if fertilization does not occur.

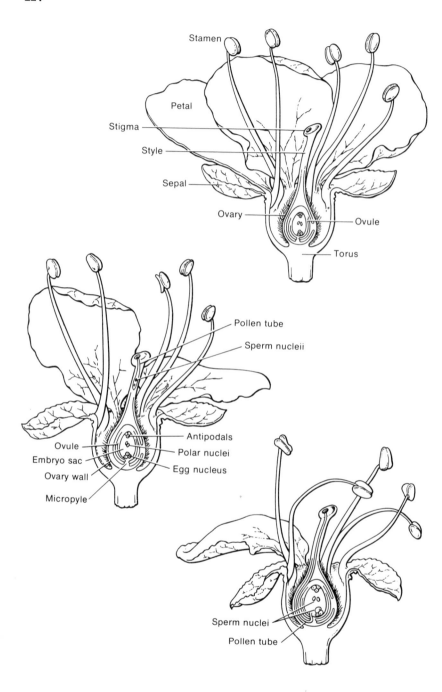

Stamen

Petal

Stigma

Style

Sepal

Ovary

Ovule

Torus

Pollen tube

Sperm nucleii

Ovule

Embryo sac

Ovary wall

Micropyle

Antipodals

Polar nuclei

Egg nucleus

Sperm nuclei

Pollen tube

Fitting (1909) showed that application of aqueous pollen extracts to orchid flowers stimulates ovary growth, indicating that the stimulation is caused by a chemical substance that fits the definition of a hormone. Fitting first worked with orchids, but he subsequently demonstrated that pollen of *Hibiscus* can produce set in orchid ovaries. Other investigators (Yasuda, 1934; Gustafson, 1937) later showed that pollen extracts of many unrelated genera can stimulate ovaries to develop fruit. Other investigators subsequently showed that pollen extracts cause curvature in the *Avena* coleoptile tests, proof that an auxin-type compound is present in pollen. Gustafson (1936) proved that auxins are responsible for the stimulus of pollination by demonstrating that synthetic auxins can induce unpollinated ovaries to develop into full-sized fruits. However, the hypothesis that auxin plays some part in pollen-produced fruit set cannot be accepted as a general explanation of fruit set because the amounts of pollen contributed during normal pollination are far too small to account for the relative increase in the auxin content of the ovary at the time of fruit set. One theory is that an activator contributed by the pollen results in an enzymatic release of auxin in the tissues of the ovary (Muir, 1947). It may also be that the process of fertilization itself is the main stimulus that results in the high concentrations of auxin found in the fertilized ovary.

Parthenocarpy

Parthenocarpy is the development of fruit without fertilization of the ovule. Two types of parthenocarpy occur in nature. In *vegetative parthenocarpy*, fruits develop without pollination. Examples of such fruits are pineapple, 'Washington Navel' orange, cultivated banana, oriental persimmon, and some varieties of fig and pear. In *stimulative parthenocarpy*, which occurs in such species as the 'Black Corinth' ('Zante' currant) grape, fruit set results from the stimulus of pollen alone without subsequent fertilization.

In his attempt to determine the cause of natural parthenocarpy, Gustafson (1939a) compared the auxin content of the ovaries of seeded

FIGURE 8-1

Reproduction of fruit in apricot flower. *Top,* one pollen grain from an open stamen has fallen on the pistil. *Middle,* the pollen tube, bearing the sperm nuclei, has begun to grow down the style. Fertilization takes place when the pollen tube grows through the micropyle and enters the ovule and when a sperm nucleus unites with an egg nucleus to produce a new seed and fruit. *Bottom,* the ovule has begun to grow. By that time some petals have fallen and the stamens have withered. (After J. B. Biale, 1954.)

and parthenocarpic varieties of grape, orange, and lemon and found that seeded varieties have a lower auxin content at anthesis than do parthenocarpic varieties. He concluded that the relatively high auxin content of varieties that develop by parthenocarpy allows further development of the fruit, even though there is no fertilization or seed development. In stimulative parthenocarpy the liberation of auxin by enzymes in the pollen and pollen tubes is sufficient to cause fruit set and growth.

Parthenocarpy may be viewed as the ultimate stage of a sequence in which fruit development becomes progressively independent of seed development (Luckwill, 1957). Some fruits are entirely dependent on their seeds. In strawberry, removal of the achenes (commonly called seeds) at any stage of development leads to the cessation of fruit growth. There is no free auxin in the receptacular tissue. At the other end of the scale, such plants as 'Washington Navel' orange and 'Black Corinth' grape develop even though they are not fertilized and have no seeds, and thus they are completely independent of them. Some other species are intermediate between these two extremes. In apple, seeds are necessary for fruit growth until fruit is approximately one-third full size, but the receptacular tissue subsequently synthesizes sufficient hormone for fruit growth.

Auxins and Fruit Set

Many synthetic auxins set fruit in plants. Best set is usually obtained with 4-CPA or BNOA. IAA is usually ineffective, probably because it is unstable in light and is rapidly destroyed in the plant by oxidative processes. Auxins are most effective on fruits with many ovules, such as fig, strawberry, squash, tomato, tobacco, rose, and eggplant. Auxins are ineffective on peach, plum, cherry, and other stone fruits. In all, only 20 percent of the horticultural crops have responded to auxins, indicating that hormones other than auxins must affect fruit set.

Gibberellins and Fruit Set

Many fruits that can be set with auxins can also be set with gibberellins. However, gibberellins have also been effective for setting fruit in several species that do not respond to auxins. Gibberellins produce a good set in tomato but the fruit remains small.

Attempts to induce parthenocarpy in apple by applying auxins have produced rather poor results, but gibberellins have yielded positive results (Luckwill, 1959c; Dennis and Edgerton, 1966). However, the

induced fruit set has been generally low, and application of gibberellins to open-pollinated trees at petal fall has resulted in reduced
yields.

There is a considerable specificity among the gibberellins in promoting parthenocarpy. Bukovac and Nakagawa (1967) treated emasculated flowers of 'Wealthy' apple with gibberellins GA_1 through GA_{10}
and GA_{14} by applying lanolin paste at 5×10^{-3} M to the cut style and
adjacent receptacle tissue. Nonpollinated controls did not enlarge and
abscissed within two or three weeks after treatment. After four weeks
GA_4 and GA_7 showed the best activity; the growth rate of fruit treated
with them was equal to that of pollinated controls. GA_5, GA_6, and
GA_8 showed the least activity and GA_1, GA_3, GA_4, GA_{10}, GA_{13}, and
GA_{14} evidenced intermediate activity. At maturity all persisting parthenocarpic fruits were approximately the same size as seeded fruits,
except for those in which parthenocarpy had been induced by GA_5
and GA_{14}. It is not surprising that GA_4 and GA_7 are very active since
they have been found to be present in immature seeds of apple
(Dennis and Nitsch, 1966).

A mixture of $GA_{4,7}$ at a concentration of 400 ppm also increased
fruit set of the emasculated 'Sturmer' variety of apple in New Zealand
and the 'Red Delicious' variety in Washington State, but not the
'Golden Delicious' variety (Williams and Letham, 1969). The $GA_{4,7}$
acted synergistically with the cytokinin PBA at a concentration of 400
ppm in increasing set in 'Sturmer.' Cytokinins alone enhance parthenocarpic fruit set in some varieties of apple but are less effective than
gibberellins.

Dennis (1967) used the 'Wealthy' variety of apple because it has a
tendency to set by parthenocarpy and because its unpollinated flowers
show a definite response to gibberellins. He extracted seeds from
young fruits of this variety and obtained substances that evidenced
gibberellin activity. When he applied these extracts to unpollinated
blossoms of the same variety, he obtained mature seedless fruits (Fig.
8–2). His conclusion was that the gibberellins that are produced in the
ovule after fertilization are responsible for fruit set in apple. However,
the role of gibberellins in the early stages of fruit development is not
yet clear because Luckwill and his colleagues (1969) failed to find
gibberellin activity until five weeks after pollination.

GA_3 usually has little effect on fruit set in seeded commercial
varieties of apple unless pollination is prevented. Dennis (1970)
treated the blossoms of six apetalous clones with gibberellins to
ascertain their response. He found that GA_3 was effective in increasing
fruit set in only two of the six clones, and he confirmed that GA_7 and

228

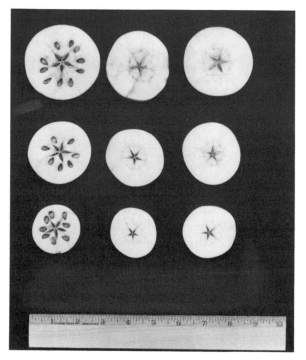

FIGURE 8-2

Fruits of 'Wealthy' apple produced by (*left*) pollination, (*center*) application of lanolin paste of KGA₃ (5 × 10⁻³ M), (*right*) application of lanolin paste containing apple seed extract (1 kilogram equivalent per gram of paste). Blossoms were treated on May 26, 1966, fruits were photographed on September 7, 1966. (Seeds were removed from pollinated fruits and placed on cut surface.) (After F. G. Dennis, Jr., 1967.)

the mixture $GA_{4,7}$ were much more potent than GA_3 in inducing fruit set. However, Dennis (1970) concluded that seedlessness alone is not correlated with responsiveness to gibberellin application.

Auxins are ineffective in setting citrus, but gibberellins have shown promising results. Coggins *et al.* (1966) demonstrated the possibility of increasing fruit set by applying gibberellin to lemon, lime, and other varieties of citrus, but unfortunately application of sprays to whole trees caused phytotoxic effects. Soost and Burnett (1961) obtained an increased set in the self-compatible 'Clementine' mandarin (*Citrus reticulata*). In India, Randhawa *et al.* (1964) found that application of gibberellin increased set in four varieties of emasculated grapefruit ('Marsh,' 'Excelsior,' 'Duncan,' and 'Thompson Seedless')

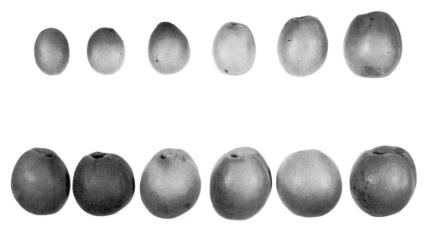

FIGURE 8-3

Promotion of parthenocarpy in emasculated blossoms of 'Royal' apricot (*top row*) resulting from two applications of gibberellin at a concentration of 500 ppm. Fruits range in size from smallest to largest. *Bottom row*, open-pollinated fruits of the same variety. (After J. C. Crane, P. E. Primer, and R. C. Campbell, 1960.)

and three emasculated varieties of mandarin ('Kaula,' 'Lahore Local,' and 'Nagpuri'). All mandarin fruits dropped before maturity. These investigators suggest that, in spite of the labor involved in emasculation, it should be possible to produce parthenocarpic grapefruits, at least on a restricted scale, in home gardens.

Both auxins and gibberellins effectively increase set in 'Bartlett' pear (Griggs and Iwakiri, 1961), fig (Crane, 1965), apple, currant, *Rosa* spp., *Zephyranthes,* tomato, cucumber, eggplant, and pepper.

Auxins fail to set fruit in *Rosa arvensis,* but GA_3 is very effective; it has been shown to induce a 71 percent parthenocarpic set compared with a 45 percent set of pollinated blossoms (Prosser and Jackson, 1959). In *Rosa spinosissima* a synergistic action occurs between GA_3 and NAAm that promotes parthenocarpy (Jackson and Prosser, 1959). The two growth regulators together resulted in a 94 percent set. GA_3 alone and NAAm alone resulted in sets of 70 and 65 percent, respectively.

Gibberellin has effectively stimulated parthenocarpy in several stone fruits, including apricot (Fig. 8–3), peach, and almond (Crane *et al.,* 1960, 1961), fruits in which negative or, at best, poor results were obtained with auxin. The size of mature parthenocarpic fruits was found to be approximately the same as that of open-pollinated fruits. Parthenocarpy has also been induced in the 'Sultan' plum by application of GA_3 and a mixture of GA_4 and GA_7 (Jackson, 1968);

the mixture was found to be more effective. Although parthenocarpic fruits of plum grew more rapidly early in the season, their final diameters were only about 60 percent of the controls. Parthenocarpic development of fruits of sweet cherry resulted from application of gibberellin only when applied in conjunction with 2,4-dichlorophenoxyacetyl methionine. When both compounds were applied to fruits of the self-incompatible 'Bing' cherry, a parthenocarpic set of 27 to 39 percent resulted, compared with 44 percent for open-pollinated blossoms (Rebeiz and Crane, 1961). Application of gibberellin in concentrations of 200 to 250 ppm, in combination with such auxins as 2,4-D, 4-CPA, 2,4,5-T, NAA, and picloram, has also been shown to be effective in increasing fruit set in 'Bing' cherry but no commercial advantage has resulted because undesirable vegetative responses and inhibition of flower bud production were produced with all concentrations and combinations (Crane and Hicks, 1968).

Gibberellin induces parthenocarpy in emasculated clusters of both seedless and seeded varieties of Vitis vinifera grape, although the final berry size is somewhat smaller than that of open-pollinated clusters (Weaver and Sachs, 1968). The set of open-pollinated clusters is usually decreased (see Chapter 10).

In some trials conducted with Vitis labrusca, cv. 'Concord,' application of gibberellin either during or shortly after anthesis has resulted in a moderate increase in set (Bukovac et al., 1960; Shaulis, 1959). This result contrasts with results obtained with 'Carignane,' a seeded variety of Vitis vinifera, in which set was sharply reduced by bloom-time applications (Christodoulou et al., 1968).

Thompson (1967) successfully induced parthenocarpic development in the pistillate flower of strawberry varieties 'Freya' and 'Tardive de Leopold.' Both NAA and IBA were usually effective, especially when applied in lanolin emulsions or agar gels. Application of GA_3 alone promoted development in 'Freya' only. Application of GA_3 combined with either auxin produced fruit as large as or nearly as large as that of pollinated controls.

Mango is a very difficult plant in which to induce fruit set. Chacko and Singh (1969), working in India, successfully induced fruit set in mango cv. 'Dashehari' by applying BA at a concentration of 250 ppm at anthesis, followed by biweekly applications of a combination of BNOA at a concentration of 10 ppm and GA_3 at a concentration of 250 ppm. Parthenocarpy was induced neither by application of the auxin, cytokinin, or gibberellin alone nor by application of the compounds in pairs.

Cytokinins and Fruit Set

Application of BA and PBA was found to be effective in increasing set in open-pollinated clusters of two seedless varieties ('Black Corinth' and 'Thompson Seedless') and three seeded varieties ('Muscat of Alexandria,' 'Tokay,' and 'Almeria') of *Vitis vinifera* (Weaver *et al.*, 1966). However, the more soluble and mobile PBA was always more effective than BA. Both auxin and gibberellin induced parthenocarpic set in the 'Calimyrna' ('Lob Ingir') fig, and Crane (1965) demonstrated that PBA is also effective. Treatment with BA is also useful for increasing fruit set in muskmelon (Jones, 1965). It is well known that cytokinins increase DNA, RNA, and protein synthesis and that they can mobilize metabolites to the area of application of a compound. Jones believes that the action of BA in muskmelon is to increase the competitive ability of the treated fruits. Cytokinins are also effective for setting fruit in emasculated flowers of some varieties of apple, although they are generally less effective than gibberellins (Williams and Letham, 1969).

Inhibitors and Fruit Set

In Australia, Coombe (1965) treated four varieties of *Vitis vinifera* grape ('Zante' currant, 'Sultanina,' 'Muscat of Alexandria,' and 'Doradillo') with CCC or Phosfon-D and found that when treatments were made two to three weeks before anthesis, the number of berries that set was considerably increased in all cultivars except 'Doradillo.' This was the first time an increase in set of seeded berries ('Muscat of Alexandria') had been induced chemically in grape. In no instance was set increased significantly by treatment with these substances either at or after anthesis.

Pre- and postbloom applications of SADH at a concentration of 500 ppm tend to increase fruit set in apple (McDonnell and Edgerton, 1970). However, treatment with CCC tends to reduce set.

ABSCISIC ACID. Since application of ABA usually causes abscission of flower parts, its effect on fruit set is negative. ABA causes abscission of both flowers and young berries of grape (Weaver and Pool, 1969).

Ethylene and Fruit Set

Ethylene usually causes abscission of flowers and young fruits, so its effect on fruit set is frequently a negative one.

PHYSIOLOGY OF FRUIT DEVELOPMENT

Fruit Enlargement

CELL DIVISION. The increase in volume that is associated with fruit growth is largely a result of cell division or cell enlargement or both. In addition, in some fruits, such as apple, expansion of intercellular spaces may also contribute to the growth of the fruit, especially during the later stages. Generally growth by cell division predominates in the early stages of growth, whereas growth by cell expansion predominates during the later stages, but there is much varietal variation. The cell-division stage usually overlaps the cell-enlargement stage. In one species of tomato (*Lycopersicum pimpinellifolium*), some cell division continues even to maturity, whereas in *Lycopersicum esculentum*, division ceases at anthesis (Houghtaling, 1935). After anthesis in this species, all fruit growth is a result of cell expansion. In other fruits the duration of cell division after anthesis varies. In apple and peach, division ceases three or four weeks after bloom; in avocado and strawberry it persists to maturity. More-complicated patterns of development occur in some fruits in which cell division ceases at different times in different parts of the fruit.

INCREASE IN SIZE. Large increases in size are characteristic of fruit growth. The European black currant undergoes a hundredfold increase in a period of ten weeks, and the apple may increase volume 6,000 times during a twenty-week growth period (Luckwill, 1957, 1959b). The period of fruit growth varies from one or two weeks to several years. However, fruits usually are initiated and mature within several months.

Two distinct types of fruit growth curves are observable when the increase in such variables as the volume, fresh weight, dry weight, and diameter of the fruit is plotted as a function of time after anthesis. Many plants, including apple, pear, tomato, cucumber, and strawberry, as well as many plant organs, have a smooth sigmoid curve. The growth of other fruits, such as fig, currant, grape, blueberry, and many stone fruits, including cherry, olive, apricot, peach, and plum, is characterized by a double sigmoid curve. In this type two periods of rapid growth are separated by an intermediate period when either less growth or no increase in volume occurs (Fig. 8–4). A double sigmoid curve can be viewed as two successive sigmoid curves. There are thus three clearly defined stages of growth. In the first (cell-division)

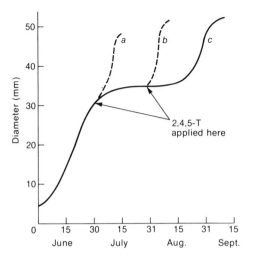

FIGURE 8-4

Curves of growth, as measured by diameter, of (*a*) unpollinated and (*b*) pollinated fruits of 'Calimyrna' fig sprayed with 2,4,5-T at a concentration of 25 ppm, compared with (*c*) pollinated but unsprayed control fruits. (After J. C. Crane and R. Blondeau, Controlled growth of fig fruits by synthetic hormone application. *Proc. Amer. Soc. Hort. Sci.* 54:102–108, 1949).

stage, the ovary and its contents grow rapidly except for the embryo and endosperm. Stage two is characterized by rapid growth of embryo and endosperm, lignification of endocarp, and slight growth of the ovary wall. In stage three, rapid growth of the mesocarp occurs, causing the final swell of the fruit, which is followed by maturation.

Auxins and Fruit Growth

Increase in fruit size is caused mainly by cell enlargement. It is then not surprising that since auxins control cell extension, they have been thought to play a dominant part in determining the growth patterns of fruits. Two main lines of evidence support this hypothesis. First, there is a correlation between seed development and final size and shape of fruit; second, application of auxin to certain fruits at particular stages of their development induces a growth response.

SEED DEVELOPMENT IN RELATION TO SIZE AND SHAPE OF FRUIT. Fruit size and shape are closely correlated with seed number and seed distribution in many fruits. Müller-Thurgau (1898) noted that in grape there is a direct correlation between size of berry and number of seeds, varying from 0 to 4. The endosperm and embryo in the seed produce auxin, which moves outward and stimulates growth of the endosperm development (Nitsch, 1950). In apple the embryo also of developed achenes and weight of receptacle has also been demonstrated in strawberry (Nitsch, 1950).

The location of seeds also greatly influences or controls fruit shape.

FIGURE 8-5
Effect of seeds on shape of fruit. A seed-less pear, cv. 'Conference' (*right*), and a normal seeded fruit of the same variety (*left*). (After L. C. Luckwill, 1959b.)

This phenomenon can be best demonstrated with strawberry because the achenes are borne on the outside of a fleshy receptacle and are easy to remove. Each achene induces growth of the receptacular tissue around itself, and removal of achenes from young fruits causes fruits of varying shapes to be produced (Nitsch, 1950). There is often in-sufficient pollination in strawberry, resulting in development of only a small number of fertile achenes and extremely distorted fruits. The effect of seed distribution on shape of fruit can also be demonstrated with certain varieties of pear because many of its seeds abort before reaching maturity. An uneven distribution of seeds results in asym-metrical fruit growth (Fig. 8–5).

CHANGES IN PATTERNS OF FRUIT GROWTH CAUSED BY AUXIN APPLICATION. The hypothesis that auxins control fruit growth is strengthened by the fact that applications of synthetic auxins can enlarge many fruits or change their pattern of growth. The effect of auxin application on increase in fruit size has been used to commercial advantage with several plants, such as blackberry, grape, strawberry, and orange. A striking example of the change that can be effected in the fruit growth pattern of fig was demonstrated by Crane (1949). When 2,4,5-T was applied to the fruit of fig (which has a double sigmoid curve) at the beginning of the second stage of growth, the fruit continued to grow and ripened in 60 days instead of the normal 120 days (Fig. 8–4).

INFLUENCE OF SEEDS ON FRUIT GROWTH. Developing seeds usually have an influence on fruit growth. Convincing evidence for this fact has been presented by Nitsch (1950), who experimented with culti-vated strawberry (*Fragaria* spp.). When achenes were removed from

a   b   c   d

FIGURE 8-6

Effect of removal of fertilized achenes and application of auxin on growth of strawberry receptacle. *Left to right: a,* control; *b,* all achenes were removed and replaced with lanolin paste alone; *c,* all achenes were removed, but a lanolin paste containing BNOA at a concentration of 100 ppm was added to receptacle; *d,* growth was induced by 3 fertilized achenes. Magnified 3 times. (After J. P. Nitsch, 1950.)

the area of a receptacle, growth ceased in that area (Fig. 8–6). Growth occurred around areas where achenes were retained; however, unfertilized achenes apparently produced no auxin because their removal had no effect on the growth of surrounding tissues. The auxin-producing effect of the ovules is continuous since growth of the receptacle is stopped by their removal even as late as twenty-one days after pollination. If BNOA or another auxin is applied to replace the effect of the achenes, full-sized fruits are obtained.

In other fruits seeds are required for growth during only part of the growth period. For example, fruit growth in apple is dependent on the presence of growing seeds up to the end of the June drop (Abbott, 1959). After this stage removal of the seeds no longer results in a cessation of fruit growth. In later stages of fruit development, tissues other than those in the seeds probably synthesize auxin. Berries of the 'Thompson Seedless' ('Sultanina') grape are dependent on seeds for an even shorter period. In this variety of *Vitis vinifera* grape, seeds abort while they are still very immature. Subsequent to this stage the fruit tissues apparently are able to synthesize sufficient auxin for growth. In the seedless 'Washington Navel' orange and 'Black Corinth' grape, fruit development has become completely independent of seeds.

PRODUCTION OF AUXIN IN SEEDS. Ovules and young seeds are the main sources of auxin. Gustafson (1939b) demonstrated that the concentration of auxin in the seeds of immature fruits of tomato was higher

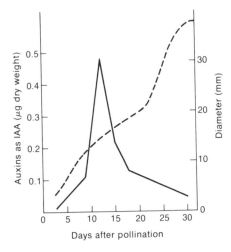

FIGURE 8-7

Changes in auxin concentration (*solid line*) and diameter of fruit (*broken line*) of 'Marshall' strawberry. (After J. P. Nitsch, 1952, 1955.)

than that in the placenta and that the concentration of auxin in the central axis and in the partitions of the fruit was in turn higher than that in the carpel wall. These data would seem to indicate that auxin is produced in the seeds and then diffuses outward along a concentration gradient to other parts of the fruit.

Auxin production in strawberry correlates closely with endosperm development, reaching a peak approximately at the time of maximum endosperm development (Nitsch, 1950). In apple the embryo also contains a considerable amount of auxin although not as much as the endosperm (Luckwill, 1948).

CORRELATION BETWEEN AUXIN CONTENT OF SEEDS AND FRUIT GROWTH. Although close correlations frequently exist between number of seeds and final fruit size and between seed distribution and shape of fruit, there is usually no close correlation between the total amount of auxin produced in the seeds and fruit growth. Careful studies have been made in which amount of seed auxin and fruit size have been followed through the growth cycle of the fruit in fig, grape, apple, strawberry, and peach, and although the auxin level was usually correlated with endosperm development, it was not correlated with fruit growth. The typical results that were obtained by Nitsch (1952, 1955) with the strawberry are shown in Figure 8–7.

Most studies of the correlation between auxin and fruit growth have focused on the total free auxin of the seeds. It could be hypothesized that growth is not related to the total auxin complex but may be controlled by one or more specific auxins in the complex. Wright (1956) found that two of four auxins correlated well with fruit

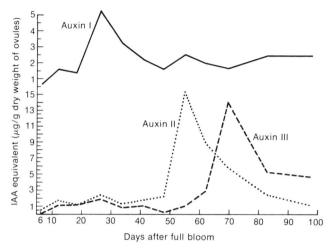

FIGURE 8-8

Levels of auxins I, II, and III in ovules of 'Halehaven' peach. (After E. A. Stahly and A. H. Thompson, 1959.)

growth in the European black currant. To date, however, there has been little evidence in support of this idea.

The slow growth of stone fruits in stage two is as yet unexplained. Stahly and Thompson (1959) found three auxins in peach seeds during growth. Auxin I was present in very small quantities and showed no correlation with growth. However, auxins II and III, which were produced by the endosperm and embryo, respectively, were present in largest amounts during the period of slow growth of the fruit (Fig. 8–8). Stahly and Thompson (1959) suggest that the slow growth period was caused by auxin concentrations that were high enough to inhibit cell enlargement. Similar results were obtained by Collins *et al.* (1966) with fruit of the lowbush blueberry (*Vaccinium angustifolium*), although these investigators believe that the decline in growth and the accompanying peak in auxin concentration were correlated with a peak in embryo growth.

The general lack of correlation between endogenous auxins and fruit growth suggests that, for some species, auxins do not control fruit growth. The conclusion follows that other classes of plant regulators also play a part.

Gibberellins and Fruit Growth

PRESENCE OF GIBBERELLINS IN SEEDS AND FRUITS. Young seeds are rich sources of gibberellin-like substances. Such substances have been isolated from the seeds of plum, apricot, almond, and young *Phaseolus,* for example. More gibberellin is usually present when seed growth is rapid. Bound gibberellins have also been found in some seeds, and often several gibberellins are present. MacMillan *et al.* (1961) identified GA_1, GA_5, GA_6, and GA_8 in immature seeds of runner bean. Another indication that seeds produce gibberellin is the fact that varieties of seeded grape are richer sources of gibberellin than are seedless varieties (Iwahori *et al.,* 1968).

CORRELATION BETWEEN ENDOGENOUS GIBBERELLIN AND FRUIT GROWTH. As with auxin, little correlation has been found between the gibberellin content of fruit and fruit growth. The concentration of gibberellin-like substances in the seed, endocarp, and mesocarp of apricot correlates closely with the growth rates of these tissues between anthesis and maturity but not with general fruit growth (Jackson and Coombe, 1966); see Figure 8–9. Iwahori *et al.* (1968) studied the gibberellin activity in berries of two cultivars of *Vitis vinifera* grape—'Tokay,' a seeded variety, and 'Seedless Tokay.' The seedless variety developed as a mutant of the seeded plant. These investigators found that gibberellin-like activity begins shortly after fertilization, when berries are young and rapidly growing. In 'Seedless Tokay' seed development ceases at an early stage. The earlier decrease of gibberellin content in 'Seedless Tokay' is probably caused by the abortion of the embryo and cessation of seed development. A close correlation was found between the gibberellin-like activity and berry growth during stages one and two but not in stage three.

A study of the variations in gibberellin content of the seeds of three cultivars of apple during the growing season showed that GA_4 and GA_7 appeared in the seeds approximately five weeks after bloom (Luckwill *et al.,* 1969). By the ninth week the gibberellin content had increased to a maximum concentration; it then decreased, completely disappearing by the time of seed maturity. Again, although the presence of the hormones was correlated with embryo and endosperm development, there was little correlation with fruit growth.

Research done by Jackson (1968b) in New Zealand prompted him

FIGURE 8-9

Relation of fruit growth to gibberellin-like activity of apricot fruit. *Top,* rate of volume increase of nucellus, endosperm, and embryo, which are the principal tissues in the seed. Graph shows growth rate as measured by fruit diameter. *Bottom,* concentrations of gibberellin-like substances in methanol extracts of seed, endocarp, and mesocarp. (After D. I. Jackson and B. G. Coombe, 1966.)

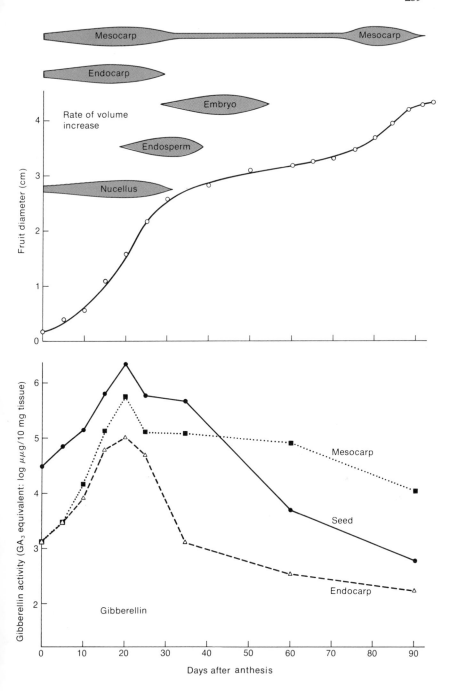

to conclude that the gibberellin concentration in peach tissues is closely correlated with the rate of cell expansion in each tissue but not with cell division or general fruit growth.

Using a fluorometric technique, Wiltbank and Krezdorn (1969) studied changes in the amount of GA_3 plus GA_1 in flowers and young fruits of navel orange. A positive correlation was found to exist between concentration of gibberellins and rate of fruit growth and also between total gibberellins per fruit and cumulative fruit growth. However, because this experiment was terminated nine weeks after bloom, the correlation between gibberellins and growth during the entire cycle of fruit growth is not known.

CHANGES IN GROWTH PATTERN INDUCED BY APPLICATION OF GIBBERELLIN. Experiments in which gibberellin was applied to parthenocarpic fruits yield further evidence against the theory that the auxin produced in the seeds controls fruit growth. Application of gibberellin, but not auxin, induced parthenocarpy in *Rosa arvensis* (Prosser and Jackson, 1959). Two acidic auxins and one neutral auxin were found in fertilized or normal fruits (hips). The growth pattern of the parthenocarpic fruits produced following application of GA_3 was similar to that of normal hips, even though the former contained none of the three. Similar results were obtained with the 'J. H. Hale' peach by Crane et al. (1960). The growth pattern of parthenocarpic (seedless) fruits produced by gibberellin closely paralleled that of the cross-pollinated fruits. An emasculated cluster of 'Tokay' grape treated with gibberellin produced large, seedless berries having the same truncated shape as that of the normal pollinated berries (Weaver and McCune, 1960).

Gibberellin applications enlarge the fruits of several species, including most seedless varieties of grape. The time of application also has a marked effect on berry shape (Christodoulou et al., 1968; Zuluaga, 1968). To date most work has been done on the 'Thompson Seedless' and 'Black Corinth' varieties. Figure 8–10 shows that gibberellin treatments applied at the time of bloom elongate the berries and that later applications produce larger berries (Christodoulou et al., 1968). The effect of gibberellin application on increase in size of seeded varieties of grape is usually minimal, although Lavee (1960) showed that applications of gibberellin to the seeded variety 'Queen of the Vineyard' after set partially compensated for the decrease in berry weight associated with a reduction in the number of seeds (Fig. 8–11). Application of gibberellin resulted in the greatest increase in berry size of seedless varieties; the amount of increase declined as seed

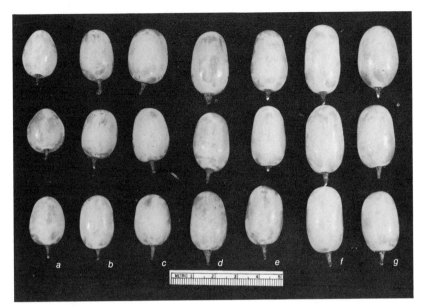

FIGURE 8-10

Effect of time of GA₃ treatment applied to berries of 'Thompson Seedless' grape. *a*, berries sprayed with GA₃ at a concentration of 20 ppm 5 days before bloom; *b*, berries sprayed at beginning of bloom; *c*, berries sprayed at 25 percent calyptra fall; *d*, berries sprayed at 50 percent calyptra fall; *e*, berries sprayed at 75 percent calyptra fall; *f*, berries sprayed at 100 percent calyptra fall; *g*, berries sprayed 2 days after bloom and just before shatter. Later sprays produced larger berries. (After A. J. Christodoulou, R. M. Pool, and R. J. Weaver, 1966.)

number increased from one to four. In India, postbloom applications of gibberellin resulted in an increase in the size of berries of the seeded varieties 'Bhokri,' 'Gros Colman,' and 'Anab-e-Shahi' (Dass and Randhawa, 1968).

Gibberellin and IAA have a synergistic effect on the growth of tomato fruits. Fruits more than twice as large as those obtained by application of either hormone alone were obtained (Luckwill, 1959a), indicating that both hormones affect fruit growth.

There is also evidence that the effect of gibberellin on fruit growth may be to increase the amount of auxin in the ovaries. Sastry and Muir (1963) treated emasculated tomato ovaries with gibberellin in lanolin at a concentration of 3×10^{-4} M at anthesis when there was no auxin present. Within twenty-eight hours after treatment there was an IAA concentration of 2.7×10^{-7} M. Development of the treated ovaries was similar to that of ovaries that were pollinated and fertilized, and both had approximately the same amount of diffusible auxin.

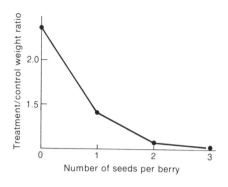

FIGURE 8-11

Effect of number of seeds on ratio between weight of gibberellin-treated berries of grape and that of controls. (After S. Lavee, 1960.)

Asymmetric fruit growth can be induced in seeded and partheno-carpic fruits of apple and Japanese pear following a localized application of GA_4 and GA_7 to one side of the fruits two weeks after bloom (Nakagawa et al., 1968). The number and size of cells of the treated sides increased; cells of the nontreated sides remained unchanged. However, application of GA_3 failed to induce asymmetric fruit in Japanese pear. Bukovac and Nakagawa (1968) also noted that localized lateral applications of GA_4 to seeded and parthenocarpic fruits of 'Wealthy' apple resulted in pronounced tissue enlargement (Fig. 8–12).

Seedless fruits of apple that are induced to set by gibberellins usually produce much more elongated fruit than do normal seeded fruits (Fig. 8–13). Similarly, fully seeded fruits of apple respond to application of GA_3 by producing longer fruit, just as do the partheno-carpic fruits set by the hormone (Westwood and Bjornstad, 1968). Application of GA_3 has also been reported to increase the depth of the stem cavity of 'McIntosh' apple (Webster and Crowe, 1969).

Application of gibberellin develops parthenocarpic fruits of peach and fig that are similar to those resulting from pollination (Crane, 1964; Stembridge and Gambrell, 1970). Application of gibberellin has also resulted in production of mature parthenocarpic pears (Griggs and Iwakiri, 1961).

Applications of GA_3 at a concentration of 500 ppm to trees of apricot cv. 'Moorpark' significantly increased the size of the fruit within seven days after treatment, and the increase in size over that of the controls was maintained until maturity (Jackson, 1968b).

Cytokinins and Fruit Growth

PRESENCE OF CYTOKININS IN SEEDS AND FRUITS. Developing fruits are rich sources of cytokinins, which are found in tissues where rapid

FIGURE 8-12

Typical form of fruits of 'Wealthy' apple at maturity as influenced by a localized application of GA₄ (*arrows*) 2 weeks after full bloom. *Left,* seeded nontreated; *center,* seeded GA₄-treated; *right,* parthenocarpic GA₄-treated. (After M. J. Bukovac and S. Nakagawa, 1968.)

cell divisions are occurring. The highest concentrations have been found in young fruits, particularly the seeds. Active extracts have been obtained from fruitlets of apple, quince, pear, plum, peach, and tomato (see Letham, 1967). Cytokinins play an important role in the regulation of cell division in fruits of apple and plum and probably in most other fruits as well.

EFFECTS OF SYNTHETIC CYTOKININS. Much of the research in this area has been done with BA and PBA. Crane (1965) treated 'Calimyrna' fig with PBA at a concentration of 500 ppm and obtained parthenocarpic figs similar to those obtained by application of an auxin or a gibberellin.

The berry size of the seedless 'Black Corinth' grape was increased

FIGURE 8-13

Effect of a prebloom spray of GA₃ at a concentration of 30 ppm on shape of 'Rome Beauty' apple. Untreated control is on left. Photographed 40 days after full bloom. (After M. N. Westwood and H. O. Bjornstad, 1968.)

FIGURE 8-14

Clusters of 'Almeria' ('Ohanez') grape (*bottom*), photographed 128 days after being dipped in PBA at a concentration of 1,000 ppm, have many small seedless berries. Untreated control clusters (*top*) are very loose. (After R. J. Weaver, J. van Overbeek, and R. M. Pool, 1966.)

by application of PBA at a concentration of 1,000 ppm (Weaver and van Overbeek, 1963), but the berry size of the 'Thompson Seedless' variety was only slightly increased (Weaver *et al.*, 1966). When applied to such seeded varieties as 'Almeria,' PBA produces many small berries that fail to develop (Fig. 8–14). BA is usually less effective

FIGURE 8-15

Fruits of 'Delicious' apple at harvest. *Top left,* treated
with *N*-(purin-6-yl)-*a*-phenylglycine; *top right,* treated
with PBA; *bottom left and right,* untreated controls.
(After M. W. Williams and E. A. Stahly, 1969.)

than the more soluble and mobile PBA. The results of the experiments performed with grape would seem to indicate that cytokinins exert most of their effect on fruit set, gibberellins affect mostly fruit growth, and auxins affect both set and growth.

Cytokinin treatment causes fruits of apple to enlarge and to have well-developed calyx lobes. When 'Delicious' apples were treated with *N*-(purin-6-yl)-*a*-phenylglycine, PBA, BA, and zeatin at concentrations of 100 to 500 ppm four days after full bloom, all compounds stimulated development of elongated fruits with prominent, well-developed calyx lobes (Fig. 8–15). This gave the fruit a knobby appearance and a more open calyx (Williams and Stahly, 1969). A different result was obtained by Letham (1968), working in Australia. Treatment of the 'Cox's Orange Pippin' apple with zeatin at a concentration of 400 ppm suppressed fruit elongation although fruit weight was not significantly changed. Letham concluded that shape of the apple at maturity probably depends on the balance between gibberellins and cytokinins in the fruitlets and that apple varieties differ in their response to these compounds.

Inhibitors and Fruit Growth

Inhibitors, especially those in the plant growth retardant group, usually decrease the size of most fruits, including apple, grape, and squash.

ABSCISIC ACID. Little is known concerning the effects of this compound on fruit growth. Davis (1968) discovered that a poor correlation exists between growth of cotton bolls and ABA content.

Ethylene and Fruit Growth

Ethylene is now considered to be one of the main plant hormones. Many plant responses that were formerly believed to result from the presence of auxins are now ascribed to induced ethylene production (Burg and Burg, 1966). It has been shown that fig fruits produce large amounts of ethylene when treated with 2,4,5-T and that this gas causes the growth and development of the auxin-treated fruits (Maxie and Crane, 1967, 1968).

Maxie and Crane (1968) applied ethylene to fruits and other plant parts of 'Kadota' and 'Mission' fig that were enclosed in polyethylene bags. Although the gas at a concentration of 5 ppm inhibited the growth of fig fruits in stage one (the cell-division stage), both growth and maturation were enhanced during the second and third (cell-enlargement) stages. Rapid growth of fruits commenced immediately after ethylene treatment began. Fruits treated for four, five, or six days were stimulated to grow to maturity and usually ripened by the seventh day (Fig. 8–16). Maturation was sometimes attained within six days after treatment. Fruits treated with ethylene for three or four days were producing ethylene when harvested and production of the gas steadily increased until rot terminated the experiment.

Typical flavor did not develop in fruits that were treated in the first half of stage two, but normal flavor developed in those treated during the last half of stage two or during stage three. Maxie and Crane hypothesized that ethylene initiates growth during stage three and stimulates maturation of fig fruits.

Crane et al. (1970) also demonstrated that application of ethephon hastens maturation of the fig (Fig. 8–17).

The research on fig prompted the testing of ethephon on peach. Byers et al. (1969) sprayed limbs of trees of 'Redskin' peach with ethephon at the midway point in stage two. Four days before harvest, fruits treated with the compound at a concentration of 100 ppm

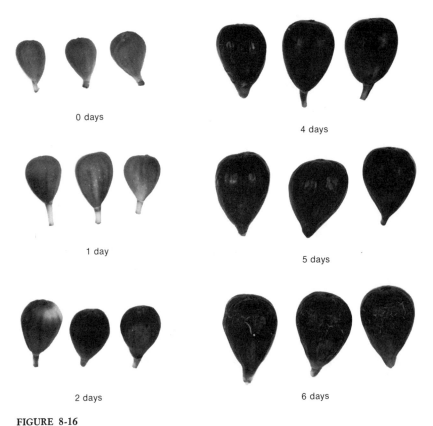

0 days

4 days

1 day

5 days

2 days

6 days

FIGURE 8-16
Typical fruits of 'Mission' fig exposed to ethylene for various periods. Photographed on the seventh day after treatments began. (After E. C. Maxie and J. C. Crane, 1968.)

measured 2.91 inches in diameter compared with 2.49 inches for the control fruits. Higher concentrations of the compound caused severe abscission of both fruit and leaves. Concentrations of 12.5 to 100.0 ppm also increased the carotenoid content of harvested fruits. Concentrations of 12.5 to 50.0 ppm had little effect on time of ripening, although a concentration of 100 ppm hastened ripening, as measured by the increase in ethylene production.

Research done with grape has emphasized the importance of the slow-growth stage (stage two) of berry development (Hale *et al.*, 1970). During this stage the berry ages and reverses its response to ethylene; although application of ethylene or ethephon in stage one (rapid-growth stage) inhibits ripening, application of these compounds in stage two hastens ripening. Application of the auxin BOA

248

FIGURE 8-17

Effect of application of ethephon at a concentration of 500 ppm on growth, maturation, and ripening of fruits of 'Mission' fig (*bottom*). Ethephon was applied on June 12, 1969, photograph was taken on June 19, 1969. Untreated control fruits (*top*) matured on July 3, 1969. (Photograph courtesy of J. C. Crane.)

during the second half of stage one or at the beginning of stage two retards ripening, but application of ethylene at this time partially reverses the effect. Hale and his colleagues suggest that an auxin-ethylene relationship determines the speed at which berries ripen.

Evaluation of the Role of Hormones in Controlling Fruit Set and Development

The role of hormones in control of fruit set and development remains to be determined. The evidence suggests that all classes of plant growth regulators probably play a part and that their influence is perhaps effected by changes in their balance or proportion.

GROWTH OF DOUBLE SIGMOID FRUIT. Many studies have been made of the development of varieties of fruit having a double sigmoid curve. The reason for the slow-growth period (stage two) that occurs between two periods of rapid growth in these species is still unexplained. Competition between the embryo and pericarp for nutrients was for many years thought to be the cause. However, this theory is not valid because parthenocarpic fruits show the same growth pattern as that of pollinated ones that contain seeds. In some species, the growth occurring during stage one correlates closely with auxin and gibberellin content. Coombe (1960), working with grape, hypothesized that the osmotic pressure resulting from accumulation of sugars in the berry effects the beginning of stage three, since the movement of water into the fruit causes cell enlargement and growth. Maxie and Crane (1968) later proposed that ethylene initiates the growth occurring in stage three. More research will be required to answer the questions that have been raised concerning fruit development.

MOBILIZATION. The movement of nutrients into flowers, fruits, tubers, and bulbs has been recognized for many years. Observations of such movements have led to the conclusion that organic and inorganic materials travel from source locations in plants, such as mature leaves, toward various sinks, such as shoot apexes and fruits. Application of auxin has long been known to increase the extent of translocation of organic material to the treated region. It has also been demonstrated (Mothes, 1960) that exogenous cytokinins are strong mobilization agents in leaves. Shindy and Weaver (1967) showed that gibberellins alter translocation patterns in grape shoots, and Kriedemann (1968) proved that cytokinins cause movement of photosynthate into the fruit of citrus. The fact that plant tissues that act as sinks are rich in hormones lends weight to the theory that hormones play a role in the mobilization of nutrients and may even be the controlling force.

Crane (1965) applied either an auxin, a gibberellin, or a cytokinin to 'Calimyrna' fig and obtained parthenocarpic fruits that were similar in gross morphology to pollinated figs. He concluded that fruit growth is not controlled by hormones emanating from the seeds but is caused instead by the capacity of these hormones to attract metabolites from other regions of the plant. Crane suggested that the fruit tissue surrounding the seeds receives hormones synthesized in the shoots and acts as a storage organ. In the seedless varieties of Vitis vinifera grape 'Thompson Seedless' and 'Black Corinth,' different growth regulators produce grapes of different size and shape. For example, gibberellin produces very elongated, large berries, whereas auxin produces oval or round berries and cytokinin results in small, round berries (Weaver and Sachs, 1968). For this reason the mobilization theory of fruit growth requires additional clarification when applied to grape. Application of different growth regulators to young fruits may result in differential attraction of various amino acids, organic acids, and sugars (Weaver et al., 1969).

However, the theory that the hormones found in such high concentrations in seeds mobilize essential metabolites and nutrients against the competition afforded by the developing shoots is widely held. Supporting evidence for this theory is that fruits with few or no seeds usually cannot survive shoot competition, but their development is enhanced if vegetative growth is suppressed (Abbott, 1960).

PRESENT AND POTENTIAL USES OF GROWTH SUBSTANCES TO CONTROL FRUIT DEVELOPMENT

The use of growth substances to control fruit set, size, and maturation has become increasingly important in agriculture. Auxin-type compounds are used most frequently to induce fruit set. A reduction in fruit set is also desirable in some situations, which are discussed in detail in Chapter 10. Under some environmental conditions fruit set is reduced or prevented because the production of growth regulators is low, but it can be raised to the desired level by application of synthetic hormones.

Fruit size is an important determinant of marketability. Growth substances can often be used to produce the larger, more attractive fruits preferred by most consumers. By hastening or delaying maturation, the grower can utilize peak demands for fruit, and earlier ripening may enable him to avoid unfavorable environmental conditions or extend the market period.

Increasing Apple Quality with SADH

Some of the effects of growth substances, when applied as preharvest drop sprays, on the maturation of apple are discussed in Chapter 10; the effects of SADH on induction of flowering were covered in Chapter 7. This section will discuss some effects of SADH and other growth substances on the quality of apple.

Sprays of SADH usually, but not always, have the disadvantage of decreasing fruit size, principally as a result of a decrease in number of cells per fruit (Martin et al., 1968). Typical data were obtained in the eastern United States by Southwick et al. (1968). The size of harvested fruit of 'McIntosh' apple decreased in proportion to the concentration of SADH applied in a single spray (1,000 to 5,000 ppm) and the earliness of application. Sprays applied on June 15 depressed

0.99   0.94   0.89

0.96   0.88   0.80

FIGURE 8-18

'Delicious' apples from trees sprayed with SADH at concentrations of 1,000 ppm (*center*) and 2,000 ppm (*right*) during the previous season, 85 days after full bloom (*top*) and 125 days after full bloom (*bottom*). Untreated controls are on left. Numbers represent length-to-diameter ratios. (After M. W. Williams, R. D. Bartram, and W. S. Carpenter, 1970.)

size the most and sprays applied on August 12 depressed size the least, whereas a spray applied on July 13 produced intermediate results. Apples sprayed with SADH at a concentration of 5,000 ppm on June 15 were approximately 0.14 inches smaller in diameter than the controls at harvest time.

A reduction in fruit size as a result of application of SADH was also demonstrated with 'McIntosh' by Forshey (1970), with 'McIntosh' and 'Delicious' by Looney *et al.* (1967), and with 'Delicious' by Sullivan (1968). Some flattening of apples also results from summer and fall applications of SADH. Williams *et al.* (1970) observed that when trees of 'Delicious' apple were sprayed with SADH at a concentration of 1,000 ppm 85 and 125 days after full bloom, flattened fruits were produced the following season (Fig. 8–18).

Application of SADH usually also delays maturation as evidenced by the slowing down of the rate of softening of the flesh (Southwick et al., 1968; Forshey, 1970; Fisher and Looney, 1967). Depending on the concentration of SADH used, delays in harvest of from two to four weeks can be obtained. Sometimes improvements in fruit color are obtained during the delay. The SADH-treated fruits are usually as firm or firmer than untreated fruits harvested at the normal time. Often there is no effect on the soluble solids content as compared with that of the controls, although some reports show an increase or decrease in the amount of soluble solids as a result of SADH treatment (Fisher and Looney, 1967). Sprays of SADH reduce fruit drop, which, of course, is often a prerequisite if maturation of fruit on the tree is to be retarded. The standard chemicals usually used to control drop, NAA or 2,4,5-T, hasten the rate of flesh softening.

The respiration of apples from trees sprayed with SADH, when such apples are placed in storage, is strikingly reduced, and this probably accounts for the improved storage and shelf life of fruit that has been reported by some investigators (Looney, 1967).

Midsummer applications of SADH, followed by application of ethephon and auxin just before harvest, may produce apples of desirable color and quality (Edgerton and Blanpied, 1970). The firmness and quality of the 'McIntosh' variety equals that of nontreated fruit harvested at the normal time. Development of red color may be much enhanced and the delay in ripening resulting from application of SADH may be minimized by subsequent treatment with ethephon plus NAA or ethephon plus 2,4,5-T. Addition of an auxin to the ethephon is necessary to counteract its tendency to induce abscission.

Increasing Fruit Set in Grape

AUXINS. 'Black Corinth' ('Zante' currant) grape is a variety of raisin from which commercially sold currants are made. Fruit set is induced by stimulative parthenocarpy and varies from year to year but the berries never grow to full size and are characteristically small. In some seasons environmental conditions apparently are not conducive to good pollen tube growth, and relatively few berries set. In other seasons, a fair percentage of ovaries may set. The practice of girdling grape vines at anthesis has been used for over a century to improve set and increase berry size. The effect of girdling this ancient variety was discovered in 1847 when a grower on the Island of Zante, Greece, tied his donkey to a vine trunk. The rope rubbed or tore bark from around the trunk and the alert grower later observed that many large berries had developed (U. X. Davidis, private communication).

In 1949 it was discovered that application of 4-CPA at concentrations of 2 to 10 ppm made three or four days following anthesis could replace girdling, and the industry rapidly adopted this method as a substitute (Weaver, 1956). In most varieties, application of 4-CPA induces set equal to or greater than that obtained by girdling and the berries produced are equal in size or larger. Since foliar application alone has little effect, the clusters themselves must be thoroughly sprayed. Clusters at the bloom stage have been shown to be very weak sinks for assimilates from the foliage (Hale and Weaver, 1962). One disadvantage of 4-CPA is that concentrations higher than those recommended are very injurious to the vine. In addition, when the growth regulator is applied at full bloom, there is a tendency for hard but empty seeds to form. 'Black Corinth' is used to produce the small, commercial variety of "currant" raisin and hard seeds in this cultivar are naturally very objectionable. This condition can be eliminated by spraying the clusters one to three days after 95 percent bloom.

Spraying with 4-CPA sometimes results in compact clusters that are subject to rotting, especially if the vines are planted in heavy soil. For this reason gibberellin has replaced 4-CPA as a girdling substitute for 'Black Corinth' in California. Application of gibberellin to this variety either has no effect or decreases set, as can be observed by comparing gibberellin-treated clusters with the more compact clusters produced by 4-CPA-treated vines.

Auxin spray induces set of "shot" berries (small seedless berries that have failed to enlarge normally) in such varieties as 'Muscat of Alexandria' and 'Thompson Seedless,' but the size of these berries renders them commercially worthless for use as either raisins or table grapes.

GIBBERELLINS. Bukovac et al. (1960) sprayed the 'Concord' variety of Vitis labrusca grape with gibberellin at a concentration of 100 ppm eleven days after full bloom. Fruit set was increased by approximately 16 percent, and there was an increase of 0.2 to 0.9 kilograms in the fresh weight of twenty-five basal clusters. These benefits are marginal but may have promise in vineyards where early berry fall is a problem.

INHIBITORS. SADH dramatically increases fruit set in the variety 'Himrod' (a seedless table grape resulting from a cross between 'On-

FIGURE 8-19

SADH treatment has stimulated set of berries of 'Himrod' grape on right; untreated cluster of the same variety is on left. (After L. D. Tukey and H. K. Fleming, 1967b.)

tario' and 'Thompson Seedless') and in *Vitis labrusca* cv. 'Concord' (Tukey and Fleming, 1967a). Application of SADH at a concentration of 2,000 ppm to 'Himrod' just before anthesis resulted in a 100 percent increase in set (Fig. 8–19). The treatment did not reduce berry size, but there was some inhibition of vegetative growth. Application of SADH at concentrations of 500 to 1,000 ppm to 'Concord' either before or at bloom, but not after bloom, increased fruit set and vine yield (Tukey and Fleming, 1968). Berry size was reduced, but yield was increased 16.4 percent in 1965 and 29.7 percent in 1966. The greater increase in yield in 1966 was a result of poor natural fruit set. The average numbers of berries on clusters treated with SADH at concentrations of 0, 500, 750, and 1,000 ppm were 44.0, 57.0, 58.4, and 62.1, respectively. Foliar sprays of SADH intended to increase berry set and yield of 'Concord' grape usually had little or no effect during the following year (Tukey and Fleming, 1970).

In New York State, Barritt (1970) showed that application of CCC

a   b   c   d

FIGURE 8-20

Clusters of 'Black Corinth' grape sprayed with gibberellin at a concentration of 5 ppm (c) are larger than girdled but unsprayed clusters (b), but those sprayed with the compound at a concentration of 20 ppm (d) are largest. a is the untreated control. Photographed 59 days after spraying on May 31. (After R. J. Weaver and S. B. McCune, 1959.)

at concentrations of 750 to 1,250 ppm to 'Himrod' increases fruit set and cluster weight.

Increasing Fruit Size of Seedless Grape

'BLACK CORINTH.' This ancient variety has been grown in Greece for thousands of years. The historical background and the advantages and disadvantages of using girdling and growth regulators to control berry development of this variety have been previously discussed.

In the early 1960's sprays of gibberellin at concentrations of 2.5 to 5 ppm replaced 4-CPA. Relatively loose clusters with berries of suitable size for raisin-making are produced, but care must be taken not to use too much gibberellin or berries too large for commercial use may result (Fig. 8–20).

In Australia, Antcliff (1967) conducted a five-year trial on 'Black Corinth,' during which chemical treatments were made at 90 percent capfall. Sprays of 4-CPA and 2,4-D were found to be somewhat less effective than girdling but sprays of GA_3 were more effective than girdling in increasing crop yield. Application of 4-CPA at a concentration of 20 ppm and 2,4-D at a concentration of 5 ppm produced equal effects. Application of GA_3 at a concentration of 10 ppm produced results similar to those obtained with girdling, but a concentration of 20 ppm was found to be more effective. Girdling plus application of gibberellin at a concentration of 10 ppm further increased yield. However, increased yields were a result of larger berries, rather than the relatively small berries from which are made the raisins most desired by the bakery trade.

Further research in Australia has indicated that results obtained from applying mixtures of different hormones are superior to those obtained from application of one compound. El-Zeftawi and Weste (1970a) found that although application of 4-CPA alone at a concentration of 20 ppm produced a lower fresh and dried weight yield than that obtained from girdling alone, the addition of GA_3 at a concentration of 0.5 ppm to the 4-CPA spray produced much smaller reductions in yield. These investigators (1970a) also reported that application of GA_3 at a concentration of 1 ppm plus CCC at a concentration of 100 ppm produced higher yields than the mixture of GA_3 and 4-CPA. The CCC evidently increased the set, whereas the GA_3 maintained berry size.

'THOMPSON SEEDLESS' AND OTHER SEEDLESS VARIETIES.   When vines of 'Thompson Seedless' grape are girdled at the fruit-set stage, a large increase in berry size usually results—sometimes as much as 100 percent. Because consumer preference dictates the desirability of producing large berries, vines of 'Thompson Seedless' grape have been girdled in California since the 1920's, and concurrent efforts have been made to increase the berry size of this variety by chemical means.

In the 1950's the plant regulator 4-CPA at concentrations of 5 to 15 ppm was applied to 'Thompson Seedless' on a limited scale to increase berry size. The compound was used either as a replacement for or as a supplement to girdling. However, growers did not accept use of this auxin on 'Thompson Seedless' as they did on 'Black Corinth' because it sometimes resulted in a delay in maturation, and it often failed to produce as uniform a berry size as did girdling.

Within three or four years after the first experiments in which gibberellin was applied to grape in 1957 (Weaver, 1957; Stewart et al., 1958), practically all grapes of the 'Thompson Seedless' variety that were intended for table use were being sprayed with gibberellin at a concentration of 20 to 40 ppm at the fruit-set stage to increase berry size. Some of these clusters were very compact, subject to bunch rot, and difficult to pack in shipping boxes, since sufficient berry thinning (removal of a portion of the cluster at fruit set) usually had not been performed to compensate for the greatly enlarged berries. Therefore, the method recommended by Christodoulou et al. (1968) was to use two applications of the compound. The results can be seen in Figure 8–21. The first application of gibberellin at concentrations of 5 to 20 ppm is made at anthesis when capfall (separation of calyptra from flower) is between 20 and 80 percent. This method thins the clusters by reducing berry set and also increases berry size. A second application at concentrations of 20 to 40 ppm is made on the same vines at the fruit-set stage to increase further the size of the berry. The bloom-time thinning spray tends to change the shape of the berry from the typical oval to a longer configuration.

Girdling is still performed on gibberellin-treated vines at the fruit-set stage to increase berry size and to make it more uniform. It has been demonstrated (R. J. Weaver, unpublished) that three gibberellin treatments (at bloom, at fruit set, and two weeks after fruit set) produce even larger berries than do two treatments, but this procedure is not recommended for commercial use because of the additional cost and labor required.

Sprays of gibberellin at time of bloom can also be used on grapes of the 'Thompson Seedless' variety that are utilized for raisins or canning when clusters are compact and subject to rotting. Higher concentrations of gibberellin may be required at fruit-set stage in desert regions because of the small size of berries growing in such environments.

The response to gibberellin of other seedless varieties, such as 'Black Monukka,' 'Perlette,' 'Delight,' 'Beauty Seedless,' and 'Seedless Concord' is generally similar to that of 'Thompson Seedless.' The proper timing of sprays is more critical with 'Perlette' than with 'Thompson Seedless' (Daris, 1966; Kasimatis et al., 1971).

Delaying Maturation of Vitis vinifera Grape

The auxin BOA is useful in viticulture because it strikingly delays maturation (Weaver, 1956, 1962). Application of this hormone can cause a delay in maturity of from a few days to several weeks in both seedless and seeded varieties of grape. The compound should be applied at concentrations of 5 to 40 ppm, depending on length of delay desired, four or five weeks after fruit set. Since few growers are presently interested in inducing late maturation of grape, there has been

258

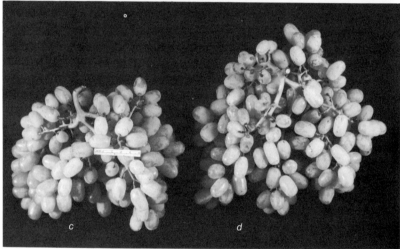

FIGURE 8-21

Response of clusters of 'Thompson Seedless' grape to bloomtime application of GA₃ at a concentration of 20 ppm (*b*), application of GA₃ at a concentration of 40 ppm at fruit-set stage (*c*), and application of GA₃ at concentrations of 20 ppm at bloom and 40 ppm at fruit set. Unsprayed cluster (*a*) is loose and has small berries. Bloom-sprayed clusters (*b, d*) are loose; cluster sprayed only at fruit set (*c*) is compact. Largest berries were produced by the repeated spray (*d*).

no commercial use of BOA. However, the compound offers possibilities for staggering harvest dates or meeting late market demands.

Promotion of Ripening of Grape

For many years grape growers have universally sought a chemical means for hastening ripening of grape. Application of auxins and other growth regulators has not been successful for this purpose. Some of the advantages of an earlier grape crop are that early market demand could be met, harvest dates could be staggered, and fall rains might be avoided, especially in regions with a Mediterranean climate such as California and Greece.

Hale and his colleagues (1970), working in Australia, were the first to demonstrate that ethephon hastens maturation. They found that dipping clusters of the cultivar 'Doradillo' in ethephon at the onset of ripening advanced maturation by six days and that a four-day advance could be obtained in ripening of 'Shiraz.' A concentration of 500 ppm was found to be more effective than a concentration of 1,200 ppm.

These researchers also found that two indexes of maturity are an acceleration in color development and an increase in the sugar-to-acid ratio. Studies made in California confirmed the theory that ethephon increases development of anthocyanin but found that the increased sugar-to-acid ratio is mainly a result of lower acidity (Weaver and Pool, 1971).

Induction of Seedlessness in 'Delaware' and Other Varieties of Grape

It has been demonstrated in Japan (Kishi and Tasaki, 1958) that gibberellin treatment can induce seedlessness and advance maturity in 'Delaware,' a seeded variety of *Vitis labrusca.* Clusters are dipped in gibberellin at a concentration of 100 ppm ten days before anthesis and again two weeks after bloom. The first dip results in seedlessness; the second induces berry enlargement. Treated clusters produce large seedless berries that are somewhat smaller than seeded berries but that color and ripen two to three weeks before untreated berries. 'Delaware' is the most important table grape in Japan and the use of gibberellin is standard commercial practice in that country. If this variety is treated at the proper stages of development, practically 100 percent seedlessness can be obtained. Similar results have been obtained in the northwestern United States (Clore, 1965).

Use of gibberellin to produce seedlessness in other varieties of grape has proved to be commercially unsuccessful. (A mixture of seedless and seeded berries reduces marketability.) In India application of gibberellin at a concentration of 50 ppm to the variety 'Anab-e-Shahi' seven to eight days before bloom decreased the bunch weight and the number of berries, causing a loosening of the cluster (Dass and Randhawa, 1968). Almost one-third of the treated flowers developed seedless berries. Nijjar and Bhatia (1969) obtained similar results with this variety by applying GA₃ at a concentration of 100 ppm. Application of 4-CPA or GA₃ to 'Tokay' at the prebloom stage has been found to induce formation of seedless berries (Zuluaga *et al.*, 1968).

Studies with the 'Delaware' grape indicate that the seedlessness induced by gibberellin is caused mainly by the compound's injurious effect on the ovules (Ito *et al.*, 1969) and not by its reduction of pollen germinability, an effect of gibberellin that has also been demonstrated (Weaver and McCune, 1960). A simple experiment in support of this theory shows that almost all berries on clusters that were dipped at the prebloom stage become seedless, whether or not emasculation or pollination occurs at time of bloom (Itakura *et al.*, 1965).

Berry Shrivel in the 'Emperor' and 'Calmeria' Varieties of Grape

In the San Joaquin Valley of California, the 'Emperor' and 'Calmeria' varieties are sometimes affected by a condition known as berry shrivel, in which the fruit loses turgidity approximately one month before harvest. If more than a few berries per cluster are affected, the cluster must be discarded. As a result, some growers suffer an appreciable crop loss every year. Jensen (1970) found that applications of gibberellin to 'Emperor' reduce the amount of berry shrivel in these varieties (Fig. 8–22). On further investigation he discovered that application of gibberellin at a concentration of 20 ppm to 'Emperor' one or two weeks following the fruit-set stage decreased the amount of berry shrivel. Another beneficial effect of this spray was that average berry size was somewhat increased, mainly as a result of increased size of seedless berries in the clusters. Jensen made similar applications to the 'Calmeria' variety and found that they reduced losses caused by berry shrivel but caused no increase in berry size.

Increasing Fruit Set in Tomato

GREENHOUSE TOMATOES.   Fruit set of tomatoes grown in greenhouses in the northern United States is poor during short, overcast winter days. Little or no pollen is produced under these conditions. In addition, the pistil often elongates and projects beyond the stamen cone, so that the structure of the flower is poorly adapted to self-pollination. The lack of wind currents in the greenhouse also is ad-

FIGURE 8-22

Berry shrivel in 'Calmeria' grape. *Left,* cluster from vine sprayed with gibberellin; *center and right,* clusters from vines not sprayed with gibberellin. (Photograph courtesy of F. L. Jensen.)

verse to pollination. Application of growth regulators has proved very effective for increasing fruit set and yields under these circumstances.

There are several effective methods of applying growth regulators, including water and aerosol sprays, vapors, and dusts; water sprays are the most common. Three useful compounds are 4-CPA at a concentration of 15 ppm, BNOA at a concentration of 50 ppm, and α-ortho-chloropropionic acid at a concentration of 40 ppm. A vigorous vibration of the flower clusters with an electric vibrator supplements these auxin sprays by improving normal self-pollination (Wittwer, 1949). After the first three or four flowers in a cluster have opened, flowers should be sprayed at intervals of seven to ten days.

Variations in results are caused by varietal differences among species and environmental differences among greenhouses, as well as varying cultural practices. Generally application of growth regulators ensures a good set of fruit, increases fruit size, and hastens fruit maturity. Some interesting results were obtained in Michigan with 'Spartan Hybrid,' an American globe type of forcing tomato, and with 'Improved Bay State,' an English forcing tomato, when both were treated with a mixture of BNOA at a concentration of 30 ppm and 4-CPA at a concentration of 10 ppm. Other plants received auxin and were also vibrated (Wittwer, 1951). Yields of the spring crop were three times greater than those of the fall crop, probably at least in part because Michigan experiences greater solar radiation in the spring than in the fall. The daily solar radiation received by the spring crop, in gram calories per square centimeter, was 354.2 during the fruit-set period and 437.3 during the harvest period. The figures for the fall crop, in the same order, were 183.2 and 102.6. Application of auxin resulted in a favorable response in both spring and fall crops of 'Spartan Hybrid,' but desired results were obtained with 'Improved Bay State' only in the fall. For example, the yields of 'Spartan Hybrid,' in pounds of fruit per plant treated in the spring, were 9.5 for control, 12.0 for flowers pollinated by vibration, and 14.0 for flowers pollinated by vibration plus application of hormone. Another experiment showed that significant differences were obtained in the fruit size of seven varieties grown in the spring, but significant yield increases were obtained only with American forcing types (Wittwer, 1949). Investigators working with other tomato varieties have generally confirmed these results (Hemphill, 1949; Randhawa and Thompson, 1949; Howlett, 1949).

In spite of these favorable experimental results, successful induction of fruit set in the greenhouse is still a major problem. The flower clusters must be vibrated periodically to ensure pollination and fertilization, and ultimately fruit development and maximum yield—a time-consuming and expensive operation. At first the use of exogenous growth substances appeared to be the solution to the fruit-set problem. However, because it frequently results in irregularly shaped (cat-faced) fruit with a short shelf life, this practice has been all but abandoned.

FIELD TOMATOES. A common complaint of tomato growers is that the early flower clusters fail to set fruit. Poor fruit set is caused by low temperatures, especially at night. The optimal range of night temperatures for fruit setting ranges from 15° to 20°C; fruit set does not occur at temperatures below 13°C (Went and Cosper, 1945; Wittwer et al., 1948). Pollen production, germination, and growth of pollen tubes are also poor under such conditions. There may be extended periods when night temperatures are too low for proper setting. For example, in areas of California where tomatoes are grown for the spring and summer market, after the danger of winter frost has passed there follows a period of six weeks to three months when day temperatures are suitable for good vegetative growth but night

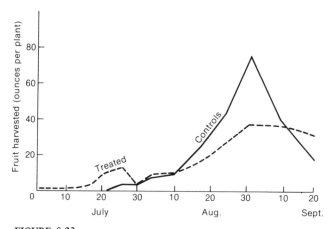

FIGURE 8-23

Comparative seasonal patterns of fruit production in hormone-treated and control tomato plants. The harvest period is extended and peaks of production are avoided when growth substances are used to overcome delayed fruit set induced by cold night temperatures. (After S. H. Wittwer, 1951.)

temperatures may drop too low to allow proper setting (Mann and Minges, 1949). Application of growth regulators will usually induce fruit set in spite of cool night temperatures. Very high temperatures also limit set, but again growth regulators can overcome this limitation.

Use of hormones to increase fruit set of field tomatoes has produced both positive and negative results. In California Mann and Minges (1949) performed twenty-nine experiments, in all but three of which set and size of early yield were increased. A favorable response was obtained with each of the five tomato varieties used. The most effective growth regulator was found to be 4-CPA at a concentration of 50 ppm, although BNOA and 2,4-D were also useful. To keep plant injury, such as occurred in these experiments, to a minimum Mann and Minges recommend that the chemical be confined as much as possible to the flower clusters. These investigators found that fruit size was consistently increased by growth regulator treatment, and in some tests much of the increase in the size of the early yield resulted from larger fruit size. The harvest season can be extended by treatment with growth regulators. A main effect is to shift the yield of fruit to an earlier period in the season rather than to increase the yielding capacity of the plants.

Wittwer (1951) obtained similar results in Michigan. 4-CPA at a concentration of 30 ppm was applied to the flower clusters of fourteen varieties. Figure 8–23 illustrates the averaged effect of such treatment, compared with results obtained with controls, in altering the production of field tomatoes. Leveling of production peaks is advantageous in that seasonal overloading of markets is lessened.

Promotion of Ripening of Tomato

EFFECTS OF ETHEPHON APPLICATION. Sprays of ethephon hasten the initiation and rate of ripening of tomato fruit. Trials conducted in New York State in which tomato plants cv. 'NY 903' were sprayed with ethephon at concentrations of 0, 1,000, 5,000, or 10,000 ppm two weeks before harvest showed that as the concentration of hormone increased, the proportion of ripe fruit increased and the proportion of green fruit decreased, but there was no change in the quantity of overripe fruit (Robinson et al., 1968). For example, at harvest the yields (in pounds per plot) of green fruit of plants sprayed with ethephon at concentrations of 0, 1,000, 5,000, or 10,000 ppm were 16.6, 10.0, 3.2, and 2.1. The corresponding yields of ripe fruit were 41.0, 39.4, 49.6, and 49.3. There were no significant differences in yield of overripe fruit or total yield among treatments. Although high concentrations of ethephon resulted in epinasty and chlorosis of foliage, fruit appearance was not affected. Application of ethephon can facilitate production of marketable fruit early in the season when the demand is high. Because of the weather conditions existing in the central and eastern United States, ethephon treatment might also be useful later in the season to stimulate ripening of fruit that would not otherwise ripen before frost in these areas. Robinson and his colleagues also point out that the efficiency of mechanical harvesting could be increased by application of ethephon since a larger proportion of the fruit would be ripe at harvest.

The timing of the ethephon application is not critical over a considerable period. Iwahori and Lyons (1970), working in California, applied ethephon at a concentration of 500 ppm to greenhouse tomatoes cv. 'VF 428' from fifteen to thirty-five days after anthesis. In all instances the time between anthesis and the first appearance of color was decreased by approximately seven days. Ethephon-treated fruit attained full color four days after the first appearance of color; control plants required five days. This acceleration resulted in an increase in early yield: The point at which 75 percent of the fruit was harvested was reached ten days earlier in the ethephon-treated field (Fig. 8–24). In Davis, California, the results obtained by Sims

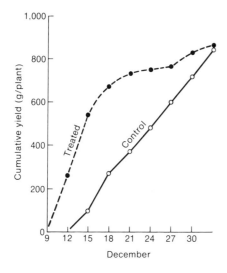

FIGURE 8-24

Effect of application of ethephon at a concentration of 500 ppm on the ripening of tomato. Dates shown are those on which fruit reached breaker stage (the stage at which fruits begin to color). Actual weight was obtained at a later stage of ripeness and was assigned to each date. Cumulative yields of the treated plots were significantly different from those of the controls at the 5 percent level except on December 30 and January 2. (After S. Iwahori and J. M. Lyons, Maturation and quality of tomatoes with preharvest treatments of 2-chloroethylphosphonic acid. *Jour. Amer. Soc. Hort. Sci.* 95:88–91, 1970).

(1969) were generally similar to those just described, but excessive defoliation sometimes occurred.

EFFECTS OF SADH APPLICATION. Sprays of SADH can be used to increase the yield of tomato plants and to cause the fruits to ripen more uniformly, thus allowing a shorter harvest period. The technique is different from that used with ethephon. Read and Fieldhouse (1970) applied SADH at a concentration of 2,500 ppm to field-grown 'Delaware 6553-2' tomato at the one- or four-leaf stage and found that both treatments increased yield. Yields in tons per acre for untreated controls and for plants treated with SADH at the one-leaf stage, four-leaf stage, or both, were 25.0, 31.9, 30.7, and 28.4, respectively. The yield increase resulted from the production of more flowers per cluster and the fact that a greater number of fruits were set per cluster.

Read and Fieldhouse (1970) found that to obtain more uniform ripening, a spray of SADH at a concentration of 5,000 ppm should be applied after the desired number of fruits have set. These sprays retard late vegetative growth, hasten fruit maturation, and prevent the formation of late fruit, thus encouraging uniform maturation and facilitating a single mechanical harvest. The hastening of maturation is probably a result of reduced competition of the vegetative parts for soil moisture and nutrients. Sprayed plants also have more resistance to the stresses of water and heat than do unsprayed controls.

Increasing Fruit Set in Bean

Snap beans frequently must be grown under environmental conditions that are unfavorable to set. Hot, dry weather during bloom is the most unfavorable condition; temperatures above 32°C cause the blossoms to drop. Many growth regulators have been tested, and such tests have shown 4-CPA generally to be most effective. This compound should be sprayed at a concentration of 2 ppm so that the plant is thoroughly wetted. Since flowering occurs over a period of several weeks, several applications may be necessary.

The greatest increases in fruit set occur when environmental conditions are not conducive to good set; for example, when hot, dry days occur during the normally short interval of flowering and pod formation. However, application of a growth regulator can offset the unfavorable effects of the weather and increase yields by 40 percent by inducing the production of more but smaller pods (Wittwer and Murneek, 1946). The optimal time for applying growth regulators to snap bean is when unfavorable weather conditions, normally conducive to poor set and low yields, prevail. Although the compounds are less effective when applied to plants growing under favorable environmental conditions, yield is often increased by 10 to 25 percent over that of the controls, mainly as a result of larger fruit size rather than increased set.

Increasing the Water-Holding Capacity of Bean Plants

There have been several reports that the water content of some plant tissues or plant parts is increased following treatment with auxin. When pods of 'Black Valentine' bean are sprayed with 4-CPA at a concentration of 1,000 ppm four days before normal harvest time, these treated fruits retain considerably more water than do the controls (Mitchell and Marth, 1950). The average fresh weights of treated fruits harvested one, two, and three weeks after treatment were approximately 10 percent greater than those of control fruits. The percentages of increase in moisture content of treated fruits harvested at these three intervals over the increase in moisture content of the controls harvested at comparable periods were 2.3, 8.7, and 14.0. Approximately 62 percent of the treated fruits that were left attached to the plant for three weeks were still green and turgid, but only 3 percent of the untreated ones were green; the remainder dried up and turned brown. If fruits are picked at the normal time and

FIGURE 8-25

Fruits of 'Black Valentine' bean plant. *Right,* sprayed 4 days before harvest with aqueous mixture containing 4-CPA at a concentration of 400 ppm and 0.5 percent Tween 20; *left,* unsprayed controls. Plants photographed after remaining 10 days on greenhouse bench. (After J. W. Mitchell and P. W. Marth, 1950.)

stored at room temperature, treated fruits remain fresh much longer than untreated ones. An example of this response is shown in Figure 8–25. In addition, a relatively high vitamin C content is maintained in treated bean pods following harvest (Mitchell *et al.,* 1949).

Increasing Fruit Set in Blueberry

Because inadequate fruit set is often a problem in blueberry production, development of a suitable growth regulator to increase set would be of value. Fruit set in plants of highbush blueberry developed from interspecific crosses was increased by application of NAAm (Darrow, 1956). Application of IAA, NAA, 2,4-D, and IBA to lowbush blueberry (*Vaccinium angustifolium*) resulted in a small increase in fruit set, but the developing berries were misshapen (Barker and Collins, 1965).

Early results obtained from field trials with gibberellin were erratic but more recent experiments with gibberellin have been promising. In one trial, sprays of GA_3 at a concentration of 100 ppm applied to 'Coville' plants resulted in a threefold increase in yield; fruit set was increased from 10 to 30 percent (Smith, 1960).

Mainland and Eck (1968) treated the style bases of flowers of

emasculated highbush blueberry with NAA and GA₃ at concentrations of 0, 5, 50, and 500 ppm. The concentration of 500 ppm resulted
in a 70 percent fruit set. A synergistic effect was observed to exist
between the two compounds. For example, at a concentration of 50
ppm, GA₃ and NAA alone produced sets of 28.5 and 19.0 percent, respectively, whereas in combination the compounds resulted in a set
of 72.5 percent. The concentration of 500 ppm produced a considerably higher set than that induced in the hand-pollinated fruit. Application of a cytokinin had no positive effect on set. Combinations of
NAA and GA₃ at higher concentrations promoted earlier ripening
and higher levels of soluble solids than those present in pollinated
fruit. As a result of application of a combination of GA₃ and NAA,
both at concentrations of 500 ppm, fruit ripened in seventy days,
seventeen days earlier than did the pollinated fruit.

Both greenhouse (Mainland and Eck, 1969a) and field experiments (Mainland and Eck, 1969b) conducted in New Jersey indicate that application of gibberellin increases fruit set in the 'Coville'
highbush blueberry. The increase in fruit set resulting from a 1966
field application of gibberellin at concentrations of 50 to 500 ppm
ranged from 78.2 percent for the controls to more than 90 percent
for treated blossoms (Mainland and Eck, 1969b). Ultimately the set
induced in controls would be expected to be far lower because the
environmental conditions were adverse to pollination. This result
could have been predicted from an experiment in which vines were
grown in cages to exclude pollinating insects and in which application of gibberellin at a concentration of 500 ppm resulted in increases in set as great as 83 percent. The parthenocarpic berries produced by the gibberellin were slightly smaller and ripened later than
the controls. However, the increased yields that were produced as
a result of gibberellin treatment in years when fruit set was low
would more than offset this disadvantage.

Promotion of Ripening of Blueberry

The harvest period for individual blueberry vines varies from three
to six weeks. Sprays of ethephon applied at the proper time hasten
ripening and allow more of the fruit to be harvested at a single time.
Eck (1970) applied ethephon to plants of 'Weymouth' and 'Blueray'
highbush blueberry two weeks before the anticipated harvest. An
increase in ripe fruit at the first picking of 'Weymouth' occurred at a
concentration as low as 240 ppm, but an application of 1,920 ppm
was required to produce the same effect at the same time in 'Blue-

ray.' The percentages of total crop yield obtained at first picking of 'Weymouth' as a result of application of ethephon at concentrations of 0, 240, 480, 960, 1,920, and 3,840 ppm were 18, 30, 32, 34, 40, and 46, respectively. However, the amount of fruit harvested from the controls at subsequent pickings was approximately the same as that obtained from treated bushes, but ethephon-treated berries were usually smaller and less acid than unsprayed berries.

Increasing Fruit Set in Cranberry

Fruit set in cranberry is often low, especially when weather conditions are unfavorable during blossoming. Application of NAA and 2,4,5-T at the fruit-set stage increases set, but berries are smaller so there is no increase in yield (Doughty, 1962). Treatment with 4-CPA increases flower initiation so that yield is increased the following year. Application of gibberellin has also been shown to increase fruit set and yield (Mainland and Eck, 1968), but results have been erratic. For example, application of GA_3 at concentrations of 50, 250, and 500 ppm at bloom increased fruit set and yield of the cultivar 'Early Black' in a year when natural set was high. The percent of set of control plants was 44.2; corresponding figures for plants treated with gibberellin at concentrations of 50, 250, and 500 ppm were 64.2, 77.4, and 76.8, respectively (Mainland and Eck, 1968). However, the gibberellin-treated fruit failed to size normally, and most of the fruits either were seedless or had low seed counts. Application of gibberellin reduced flower bud initiation, resulting in fewer flowers and a low yield the following year. After being sprayed with gibberellin for two years, treated plants developed uprights (shoots) that were more than twice as long as those of the controls, and few of these long uprights developed flower buds. Application of SADH at a concentration of 2,000 ppm in combination with gibberellin at a concentration of 500 ppm did not reduce the gibberellin-induced upright elongation. The commercial benefit that might be derived from the effect of gibberellin on fruit set is negated by the effect of the compound on the flowering and growth characteristics of cranberry.

Promotion of Ripening of Cranberry

Enhancement of concentration of anthocyanin pigments in cranberry by chemical means would be of value to the producer. Clarity and intensity of red color determine the quality of cranberry juice and thus have always been of prime importance in marketing. Eck (1969) sprayed plants of 'Early Black' cranberry at the preharvest stage with ethephon at a concentration of 600 ppm and found that the amount of pigment present in treated berries at harvest was significantly greater than that in the controls. The yield and size of berries were not altered by the ethephon treatment. Preharvest applications of 57 percent emulsifiable concentrate of malathion at a rate of 2.5 pounds per acre increased anthocyanin pigment somewhat, but application of SADH at a concentration of 4,000 ppm resulted in poor fruit coloration.

Increasing Fruit Set in Fig

Three of the four varieties of fig grown in California ('Mission,' 'Kadota,' and 'Adriatic') are the common type and produce their fruits by vegetative parthenocarpy. The 'Calimyrna' variety, which leads in production and is most popular with the consumer, belongs to the Smyrna type and has to be cross-pollinated either with the male or with the caprifig. The pollen-carrying agent is a tiny wasp (*Blastophaga*) that completes its life cycle in the male fig. The cost of purchasing caprifigs and distributing them in 'Calimyrna' orchards at the time of pollination is rather high. Growers of 'Calimyrna' figs were thus anxious to have a material that could induce the set of a satisfactory crop of fruit without cross-pollination.

In 1949 Crane and Blondeau found that more fruit was set in 'Calimyrna' by an application of 4-CPA at a concentration of 80 ppm than by caprification or cross-pollination. The parthenocarpic fruits developed to a normal size and had a suitable sugar content, but were completely seedless. The baking industry, which uses most of the 'Calimyrna' figs produced in the United States, did not want seedless figs because they lack the crunchy quality imparted by fig seeds. Subsequently, Crane (1952) found that BOA induces parthenocarpy and the formation of drupelets with hollow, sclerified endocarp, but the industry still has not adopted its use.

It has been known since the third century B.C. that growth and maturation of the fig fruit result and are advanced by a few days if a drop of olive oil is applied to the ostiole during the ten-day period following the time at which all drupelets in the fruit have turned red (see Saad *et al.*, 1969). This period brackets the transition from stage two to stage three of fruit growth. The stimulative agent is ethylene, which is produced as a breakdown product of olive oil, especially when the oil is exposed to solar radiation.

Increasing Fruit Set in Phalsa

Phalsa (*Grewia asiatica*) is a bushy tropical plant that bears small fruits containing large seeds. Frequently fruit set is low and small fruits are produced, so that an increase in the fleshy portion of the fruit is desirable. In India, Randhawa *et al.* (1959) found that sprays of gibberellin at a concentration of 10 ppm applied to phalsa at full bloom increased fruit set by 35 percent. Application of gibberellin at a concentration of 40 ppm increased fruit size slightly over that of the control, and the total yield was increased by 20 percent.

Increasing Fruit Set in Pear

Most varieties of pear are partially self-incompatible and require cross-pollination for fruit set. However, when the weather is cool and wet during the bloom period, a very light crop of 'Bartlett' pear can often be produced by parthenocarpy; when such conditions prevail a growth regulator spray to increase set would be of value.

AUXINS. Griggs *et al.* (1951), working in eight different 'Bartlett' orchards in California, found that spray application of either 2,4,5-TP or its alkanol amine salt at a concentration of 100 ppm at the pink-bud stage greatly increases the parthenocarpic set. Averages taken in the eight orchards showed that 66.7 fruits set for each one hundred flower clusters of the control plants; corresponding figures for the salt and the acid treatments were 87.7 and 93.1, respectively. Treated fruits in general appeared quite normal. Their size was slightly reduced, and many appeared to be somewhat apple-shaped, especially those growing in the less productive orchards. Since there is a danger of inducing injury to foliage, only vigorous trees capable of carrying heavy crops should be sprayed with growth regulator. Another result of spraying with auxins noted by these investigators was premature ripening and breakdown of tissues at the calyx end of the fruit, a type of injury most common in small fruits. Because such fruits are commercially unacceptable, few pears of the 'Bartlett' variety are sprayed with auxin to induce set.

GIBBERELLINS. Parthenocarpic fruit set of 'Bartlett' pear is also increased by gibberellin (Griggs and Iwakiri, 1961). Application of this growth regulator at concentrations of 10 to 50 ppm at the pink-bud stage, at full bloom, or at petal fall increases fruit set by approximately 25 percent. Because treated fruits and pedicels are long, the calyx ends of the fruits are distorted, and fruit-softening is hastened, commercial application of gibberellin has not yet become a reality. In Holland several varieties that set poorly, including 'Triomphe de Vienne' and 'Beurre Hardy,' are commercially sprayed with GA$_3$ at bloom to produce a good set (van Eijden, 1965). Other varieties, such as 'Doyenne du Comice,' do not respond.

INHIBITORS.   An indirect way to increase fruit set in 'Bartlett' pear by delaying bloom is to spray trees with SADH the preceding fall. Griggs et al. (1965), working in California, found that spraying SADH at a concentration of 4,000 ppm on September 9 increased set the following year by 31.1 percent compared with the unsprayed controls. However, it is not known whether SADH stimulated parthenocarpic fruit set or whether the increased set was caused by the delay of bloom, which reduced the loss of flowers because of the freezing temperatures that occurred in April. (The delay of bloom also might increase fruit set by allowing the fruit and flowers to develop under more favorable fruit-setting weather.) The most likely explanation for the increase is that the growth retardant reduced vegetative vigor and thus reduced the competition between the fruit and foliage. Pears that developed on the sprayed trees had shorter, thicker stems. No breakdown of fruit tissue occurred during the cold storage following harvest, and when fruits were placed in ripening rooms, pears with normal flavor developed.

Increasing Fruit Set in Muskmelon

Fruit set in hand-pollinated flowers of muskmelon is poor. The environmental conditions and the physiological condition of the plant influence the success of pollination. There is also competition among the vegetative portion of the plant, the fruit already on the plant, and the pollinated flowers. Fruit set can be enhanced by removing fruits that have already set or by pruning the stem tips at time of pollination to minimize vegetative competition. Muskmelon-breeding programs and production of hybrid seed would be greatly facilitated by a simple and effective means to increase set.

Auxins and gibberellins have not been found to be suitable for induction of fruit set. However, Jones (1965) demonstrated that the cytokinin BA at concentrations of 0.1 to 2.0 percent, applied in lanolin when the fruits of field-grown melons of the cultivars 'Supermarket

Hybrid' and 'Harvest Queen' were beginning to set, produced, on the average, a 37.5 percent set. The most effective concentration of BA, 0.2 percent, set more than 50 percent of the flowers. Untreated controls and melons treated with gibberellin and with the auxin NAAm each set less than 1 percent of the fruits. Further work showed that fruit set and number of seeds per pollination increased with higher levels of BA but that the average number of seeds per melon remained unchanged following application of BA. Seed quality, measured by germination or by percentage of seeds with empty seed coats, was not affected by the treatments.

Jones (1965) used BA routinely in a muskmelon-breeding program. He attributes the role of BA in inducing set to an increase in the ability of the young fruit to compete for assimilates with the rest of the plant. He found that senescence and formation of an abscission zone were prevented.

Promotion of Ripening of Cantaloupe

In the San Joaquin Valley of California, cantaloupes are often harvested ten to twenty times during one growing season. Therefore, any treatment that might enhance uniformity of ripening will lower harvesting costs by reducing the number of harvests.

Application of ethephon hastens apparent ripening of immature melons (Kasmire et al., 1970). Typical results have shown that melons treated with ethephon at a concentration of 1,000 ppm are more marketable than are control melons when picked two, three, four, and five days after treatment but are less marketable seven days after spraying. The yields, in crates per acre, of control melons when picked two, three, four, five, and seven days after treatment were 69, 100, 92, 121, and 136, respectively. Corresponding figures for ethephon-treated melons were 202, 262, 299, 232, and 96, respectively. However, the percentage of soluble solids was slightly lower in the treated melons whose apparent ripeness was the same as that of the controls (Kasmire et al., 1970).

Increasing Fruit Size of Blackberry

Sprays of a mixture of 4-CPA and BNOA have been shown to increase berry size in several varieties of blackberry grown in different parts of the United States. Berries should be sprayed when they are one-fourth to one-half their full size. Bringhurst et al. (1956) found that application of equal proportions of 4-CPA and BNOA at concentrations of 50 to 100 ppm produced the best results. A concentration of 100 ppm resulted in an 18 percent increase in berry weight in the 'Boysen' variety; the 'Olallie' variety showed a 16 percent increase. However, the 'Thornless Logan' variety showed only a 3 percent increase. These investigators found that two or three applications are generally necessary to obtain the greatest possible increase in yield. Since blackberry may bloom over a three- to four-week period, there is considerable variation in the stage of development of the fruits on a plant. After the first application is made, a second one is made approximately ten days later, and if the bloom period extends over several weeks, a third application may be necessary.

Increasing Fruit Size of Apricot

Branch girdling of apricot at the time of pit-hardening hastens fruit maturity by as much as two weeks. However, this procedure has not become standard commercial practice because girdling wounds are slow to heal. Apricot has a tendency to drop its fruits during the entire period from initiation of pit-hardening to harvest. Sometimes as much as three-fourths of the crop may drop before harvest.

Research by Crane and his associates (1955) has shown that the proper application of growth regulators to apricot can eliminate most of the drop, hasten maturation, and produce larger fruit. These researchers found that spray applications of 2,4,5-T at a concentration of 25 ppm to 'Tilton' apricot at the initiation of pit-hardening reduced preharvest drop to 6.6 percent of the total crop, compared with the figure of 52.9 percent recorded for the controls. Application of the regulator at a concentration of 50 ppm also increased the rate of fruit growth. Fruit diameter and fresh weight at harvest were increased 10.1 and 21.2 percent, respectively. Results obtained with the 'Royal' variety are shown in Figure 8–26. The increase in fruit size resulted from cell enlargement within the fleshy tissues. There was also a hastening of maturation of one to two weeks. Similar results have been obtained with 'Stewart,' a variety whose fruit drop is caused by seed abortion, and with other varieties.

Increasing Fruit Set in Citrus

Application of auxin has failed to induce fruit set in citrus. However, experiments conducted in California have shown that an increase in fruit set results when GA_3 is applied to flowers or to individual young fruits of trees of lemon, lime, and 'Washington Navel' orange

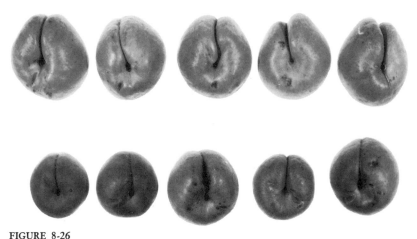

FIGURE 8-26

Representative mature fruits of 'Royal' apricot. *Top*, fruits sprayed with 2,4,5-T at a concentration of 100 ppm on April 13, 1951; *bottom*, unsprayed controls. (After J. C. Crane and R. M. Brooks, 1952.)

(Hield *et al.*, 1958). In other investigations, gibberellin sprays have been applied to mandarin (Soost and Burnett, 1961), tangelo, and 'Dream Navel' orange (Krezdorn and Cohen, 1962), 'Valencia' orange (Coggins *et al.*, 1960), and grapefruit (Coggins *et al.*, 1962). Although increased fruit set has sometimes been achieved, the undesirable effects of gibberellin treatment have outweighed the benefits.

Hield *et al.* (1965) found that a single spray application of gibberellin at concentrations of 46, 92, and 184 ppm to trees of 'Washington Navel' orange on four dates during flowering and early fruit set caused leaf drop and twig dieback, which were correlated with a reduction in set and decreased yield. On the other hand, in experiments conducted in California and Florida, gibberellin application that was restricted to flower clusters or to isolated small branches of citrus trees resulted in increases in fruit set. This finding indicates that if a technique is developed for applying gibberellin to the flower parts only, the compound will have a commercial potential for citrus.

Poor fruit set in citrus is a common problem in India and is often coupled with a high rate of fruit drop (Randhawa and Dhillon, 1965). Experiments conducted at the Indian Agricultural Research Institute in New Delhi showed that fruit set in sweet lime (*Citrus limettioides*) is increased by sprays of gibberellin at a concentration of 10 ppm. The auxins, 2,4-D at concentrations of 10 to 15 ppm and 2,4,5-T at concentrations of 10 to 15 ppm were reported to be effec-

TABLE 8–1

Application of 2,4-D to Orange and Grapefruit According
to Diameter of Fruit

	Amount of 2,4-D (ppm)	Diameter (inches)
Orange	12	.188–.250
	16	.250–.500
	20	.500–.625
	24	.625–.750
Grapefruit	12	.250–.375
	16	.375–.625
	20	.625–.750
	24	.750–1.00

SOURCE: Hield et al., 1964.

tive but to a lesser degree than gibberellin (Randhawa et al., 1959). Gibberellin was also found to be effective in increasing fruit set in 'Jaffa' and 'Pineapple' sweet orange (Randhawa and Sharma, 1962). However, some of the fruits that set abscissed during the June or summer drop. More recent studies showed that the fruit set of two varieties of sweet orange, 'Hamlin' and 'Valencia Late,' is increased by application of gibberellin at concentrations of 10 to 15 ppm. Some improvement in set also results from application of 2,4-D at a concentration of 5.0 ppm and 2,4,5-T at a concentration of 7.5 ppm (Sharma and Randhawa, 1967). At present growth regulator sprays are not used commercially in India to increase fruit set in citrus. The high cost of gibberellin in India has discouraged its use in that country.

Increasing Fruit Size of Citrus

'Valencia' oranges grown in southern California tend to be smaller than is desirable, and cultural practices that facilitated production of large fruits of better quality would be of value.

In the late 1940's various auxins were tested for their ability to control preharvest drop of navel orange. These tests showed that application of 2,4-D early during the fruit growth period increases fruit size (Stewart et al., 1952). The proper amounts of 2,4-D that should be applied to oranges and grapefruit of various sizes are given in Table 8–1. When regulator-treated fruit matures it usually is one commercial grade size larger than untreated fruit. In orange this increase is often accompanied by greater thickness and roughness of the rind. Therefore, if the rind of the fruit in one orchard is rougher than the average fruit in the district, 2,4-D should not be used. Sizing sprays cause a greater degree of roughness during years when the crop is light than during years when the crop is normal or heavy. Overthinning results if sprays are applied too early during the bloom period, but there is only a slight increase in size if treatment is delayed until three months after bloom.

Fruit size is a greater problem with the 'Valencia' orange than with the navel orange. Concentrations of compound higher than 24 ppm should not be applied to 'Valencia' orange because damage to foliage may result. As the fruit develops, its responsiveness to regulator spray rapidly declines. Grapefruit is less responsive than orange to regulator sprays. The use of sizing sprays on lemon is not recommended because results are extremely erratic.

Increasing Fruit Size and Hastening Ripening of Loquat

In northern India fruits of loquat (*Eriobotrya japonica*) are normally harvested in April, when the weather is quite hot and when severe sunburn of fruit often occurs. A means of advancing maturity of the fruit to avoid the sunburn would be of value. The compound 2,4,5-T at concentrations of 20 and 40 ppm was sprayed on branches of 'Improved Golden Yellow' and 'Pale Yellow' on February 10 when the fruit was approximately 1.5 cm in diameter and again on February 17 (Singh *et al.*, 1960). Application of 2,4,5-T at a concentration of 40 ppm to 'Improved Golden Yellow' caused a 63 percent increase in weight per fruit over that of the control; the 'Pale Yellow' variety showed a 41 percent increase. The fruit of both varieties matured approximately twenty-one days earlier than did the controls. Application of the compound at a concentration of 40 ppm increased the total yield of 'Improved Golden Yellow' by 74.2 percent; the total yield of 'Pale Yellow' was increased by 95.5 percent. Treatments with 2,4,5-T also resulted in a decrease in the percentage of dropped fruit.

Promotion of Ripening of Pepper

The pepper plant can be made to ripen more rapidly as a result of ethephon applications. In greenhouse trials conducted in California, chili peppers cv. 'California' were sprayed at breaker stage (the stage at which fruits begin to color) with ethephon at a concentration of

100 ppm. The fruits turned a ripe red color within eight days (Sims *et al.*, 1970). If the concentration of the compound was too high (250 to 500 ppm), the leaves and fruit abscissed within five days.

Relatively high concentrations of ethephon are required for field applications. In Davis, California, pimiento peppers cv. 'Pimiento select' were sprayed at breaker stage with ethephon at concentrations of 100, 250, and 500 ppm. Ripening varied proportionally with concentration of compound. Seventeen days after treatment, when the peppers were harvested, the percentages of red, ripe fruit that were harvested in the lots that had received ethephon at concentrations of 0, 100, 250, and 500 ppm were 15.3, 18.3, 24.3, and 61.1, respectively. The corresponding figures for green fruit were 77.4, 53.2, 48.6, and 11.9 (Sims *et al.*, 1970).

Promotion of Ripening of Strawberry

Gibberellin sprays can be used to increase the quantity of fruit harvested during the early part of the picking season. An investigation of the compound's effect on 'Sparkle' strawberry indicated that, for best results, gibberellin at a concentration of 10 ppm should be applied three times at weekly intervals starting in the autumn, when flowers are first initiated (Smith *et al.*, 1961). These investigators found that as a result of treatment, more flowers were produced at the beginning of the bloom period, and consequently the peak of the harvest was advanced. In one experiment 4,332 quarts of strawberries were picked per acre in the first three harvests, compared with 2,711 quarts of the controls. However, yield for the total of seven pickings was the same (approximately 10,000 quarts per acre) regardless of treatment. In the third harvest the weight of the treated fruit was somewhat less than that of the controls, but in the seventh harvest there was no significant difference in fruit size. These findings indicate that, as a result of gibberellin treatment, more berries can be picked in an early harvest; fruits grown in control plots ripen later in the season.

Promotion of Ripening of Coffee Berries

Experiments conducted in Kenya have demonstrated that application of ethephon to mature green coffee berries accelerates their ripening (Browning and Cannell, 1970). Berries growing on trees sprayed with ethephon at concentrations of 700 or 1,400 ppm begin to ripen two and four weeks earlier than untreated berries. Because quality coffee is picked by hand, controlling the length of the harvest period in this manner can reduce harvesting costs.

Application of ethephon at a concentration of 1,400 ppm also caused abscission of 60 to 65 percent of the young, expanding berries four to six days after spraying. According to these investigators, application of ethephon can replace hand-thinning of fruit during the first fruiting year and can be used in later years to prevent overcropping.

Promotion of Ripening of Peach

Application of SADH can hasten maturation of peach, as has been demonstrated with both 'Loring' and 'Redskin' varieties (Byers and Emerson, 1969). Trees of 'Loring' peach that were sprayed with SADH at concentrations of 1,000 to 8,000 ppm on four dates, April 22 (twenty-two days after full bloom), May 18, June 22, and July 2, ripened earlier than did untreated trees. In April and May the greatest effect on fruit maturation resulted from the high concentrations (4,000 and 8,000 ppm) (Fig. 8–27), but in June and July the greatest effect resulted from the two lower dosages (1,000 and 2,000 ppm). The accelerated rate of ripening caused by application of SADH was accompanied by an intensification of internal flesh color and red and yellow skin color, a slight decrease in the percentage of soluble solids, and a decrease in flesh firmness (Byers and Emerson, 1969).

Sansavini et al. (1970), working in California, obtained a six-day advance in maturation of 'Dixon' peach by applying SADH at the onset of pit-hardening. In addition, fruit growing on SADH-sprayed trees ripened more uniformly than did the fruit growing on unsprayed control trees.

Promotion of Ripening of Cherry

In Michigan it was shown that sprays of SADH at concentrations of 1,000 to 4,000 ppm applied to 'Windsor' sweet cherry two weeks after full bloom advance anthocyanin development by two weeks and sugar accumulation by one week (Chaplin and Kenworthy, 1970). Fruit firmness was not decreased, but there was a decrease in the size of treated fruits. These results show that it is possible to harvest SADH-treated fruit two weeks earlier than nontreated fruit for brining or canning purposes.

An experiment conducted in California showed that application of SADH at a concentration of 2,000 ppm in April enhances the biosynthesis of anthocyanins but does not measurably advance the

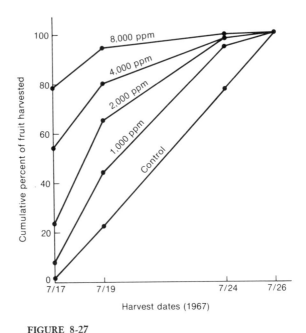

Harvest dates (1967)

FIGURE 8-27

Effect of application of SADH 21 days after full bloom on cumulative percent of fruit of 'Loring' peach harvested on April 22, 1967. (After R. E. Byers and F. H. Emerson, Effects of succinamic acids 2,2-dimethyl hydrazide (Alar) on peach fruit maturation and tree growth. *Jour. Amer. Soc. Hort. Sci.* 94:641–645, 1969b.)

physiological maturity of sweet cherry, cultivars 'Hative Burlat,' 'Bigarreau Moreau,' and 'Bing' (Ryugo, 1966). In the first two varieties, SADH treatment altered the ratio between the major anthocyanin, keracyanin, and the minor pigment, chrysanthemin, in favor of the latter.

Application of SADH at concentrations of 2,000 to 4,000 ppm to red tart cherry hastens coloration of fruit, increases fruit firmness, and makes fruit size more uniform (Unrath and Kenworthy, 1968). The force required to separate the fruit from its pedicel was found to be reduced, but the percentage of soluble solids was not affected. Treated trees ripened five to seven days earlier and bloomed two to three days earlier the following year. Shoot growth was depressed.

Preharvest application of ethephon at a concentration of 500 ppm to 'Montmorency' sour cherry accelerates fruit maturation without excessive defoliation (Anderson, 1969).

Prevention of Internal Browning of Fruit

PRUNE.   Early ripening varieties of prune grown in the northwestern United States are subject to internal browning, or the breakdown of the flesh next to the pit. Application of auxin usually aggravates this condition. However, experiments with several growth regulators have shown that a preharvest spray of gibberellin at a concentration of 100 ppm to 'Demaris Early Italian' and 'Richard Early Italian' substantially reduces internal browning and results in firmer fruit (Proebsting and Mills, 1966). The shelf life of the fruit is extended by approximately three days. Firmer fruit can be handled more easily, thus allowing an extension of the period during which fruit can be harvested. Such fruit is also better suited for mechanical harvesting.

A disadvantage of applying gibberellin to prune is that there is sometimes a delay in development of skin color and a reduction in the level of soluble solids. Experiments in which ethephon was applied both alone and in combination with gibberellin have indicated that the beneficial features of both chemicals can be combined to hasten maturation and decrease browning (Proebsting and Mills, 1969). In these investigations, application of ethephon alone at a concentration of 80 ppm 3.5 weeks before commercial harvest of untreated fruit advanced coloration and hastened development of soluble solids by two weeks and softening of fruit by one week. The amount of browning of ethephon-treated fruit was less than that of control fruit but more than that of gibberellin-treated fruit. Application of gibberellin alone at a concentration of 50 ppm 3.5 weeks before the maturity of control fruit delayed softening by one week and development of soluble solids by one or two days. When ethephon was combined with gibberellin, fruit with better color, a higher level of soluble solids and less internal browning than the controls was produced. The coloration and soluble solids level of this fruit were equal to those of the ethephon-treated fruit; amount of browning was equal to that of gibberellin-treated fruit.

PEACH.   Application of ethephon and GA_3 to peach trees has consistently prevented the browning of pureed and sliced peaches subsequently prepared from such fruit. Application of ethephon at a concentration of 50 ppm to 'Early Amber' when the seed was approximately 12 mm in length and application of GA_3 two weeks after petal fall were found to be effective (Buchanan et al., 1969). Peeled, macerated samples of 'Early Amber' taken from treated trees failed

FIGURE 8-28

Samples of cut slices and pureed fruit of 'Early Amber' peach from trees treated with ethephon (*left*) and from unsprayed trees (*right*). Fruit was kept at room temperature for 24 hours before being photographed. (After D. W. Buchanan and R. H. Biggs, 1969.)

to darken whether frozen, freshly harvested, or held at 25°C for seven days after harvest. Unsprayed fruits consistently darkened. It was found that pureed and sliced peaches could be held up to twenty-four hours without darkening (Fig. 8–28).

Hastening the Ripening of Detached Fruit

Many varieties of fruit are brought to early maturity in ripening rooms, and ethylene gas is well known as an effective ripening agent. This section will discuss some effects of other growth substances that hasten or retard the ripening process of detached fruits.

The auxin 2,4-D promotes the ripening of several fruits, including banana, apple, and pear (Mitchell and Marth, 1944). When green 'Fortuna' bananas were dipped in the compound at concentrations of

200 to 1,600 ppm, they turned yellow and softened within seventy-two hours, whereas untreated fruit was still hard and deep green in color. Five days after treatment, treated fruits were ripe and had an excellent flavor; untreated fruits were still light green in color, bitter, and hard. Starch hydrolysis proceeded at a more rapid rate in treated bananas than in untreated ones. These investigators treated apples and pears with 2,4-D at concentrations of 100 to 1,000 ppm and found that 'Yellow Newtown' apples ripened within two weeks of treatment but that untreated apples failed to ripen. They obtained similar results with 'Grimes Golden' and 'Rome Beauty' apples and with 'Kieffer' and 'Winter Bartlett' pears.

Most of the preceding discussion of growth regulators has focused on the hastening of ripening by adding compounds to the plant material. In contrast, ripening may be delayed and the storage life of some preclimacteric fruits can be extended by removal of an endogenous hormone, ethylene. If fruits are stored in sealed polyethylene bags containing an ethylene absorbent, the ripening effect of the ethylene in the air is largely avoided. For example, green bananas that were sealed for eight days in polyethylene bags containing the ethylene absorbent Purafil (alkaline potassium permanganate on a silica carrier) did not show a climacteric rise until ten days after removal from the bags. Fruits sealed in polyethylene bags without the ethylene absorbent evidenced a climacteric rise the day after they were removed from the bags (Liu, 1970).

The growth regulator ethephon, which probably acts by breaking down in the plant to produce ethylene, ripens green bananas at approximately the same rate as does ethylene (Fig. 8–29). The results of immersing green bananas for one hour in a solution of ethephon at room temperature are the same as those obtained from subjecting the fruit to ethylene gas for twenty-four hours (Russo et al., 1968). In contrast to ethephon, GA_3 retards ripening of banana. Fruit dipped in 10^{-4} M gibberellin maintained a green, unripe appearance for several days while the untreated controls began to ripen. However, some gibberellin-treated fruit was sprayed with ethephon seven days after dipping and ripened normally within two days of treatment. Techniques of retarding and initiating fruit ripening by gibberellin and ethephon offer a means to facilitate storage of bananas and certain other fruits.

Ethephon also hastens ripening of tomato (Russo et al., 1968) and gibberellin retards it (Abdel-Kader et al., 1966; Dostal and Leopold, 1967).

Robinson et al. (1968) obtained improvement in color of mature

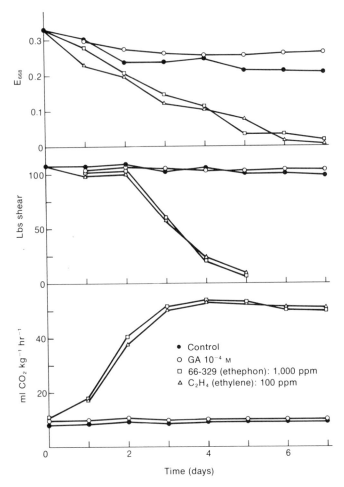

FIGURE 8-29

Changes in ripening of banana induced by application of gibberellin, ethephon, and ethylene gas. Ethephon and ethylene gas have equal effects on fruit ripening, as indexed by chlorophyll degradation (*top*), fruit firmness (*middle*), and respiration (*bottom*). (After L. Russo, Jr., H. C. Dostal, and A. C. Leopold, 1968.)

green fruits of tomato within two days after they were dipped in ethephon at a concentration of 10,000 ppm and stored at 24°C. When a concentration of only 1,000 ppm was used, there was no change in color until five days after dipping. During ripening the variability of color from fruit to fruit was strikingly reduced, a result that might be useful in cutting down the costly sorting of tomatoes in ripening

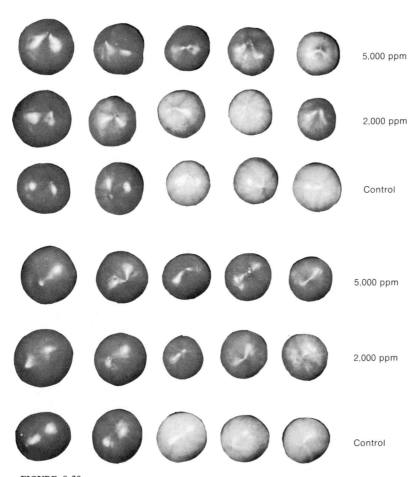

FIGURE 8-30

Ripening of mature green fruits of tomato after being dipped for 30 seconds in ethephon at concentrations of 0, 2,000, or 5,000 ppm. *Top,* 7 days after dipping; *bottom,* 12 days after dipping. (After W. L. Sims, 1969.)

rooms. Similar results were obtained by Sims (1969) and are depicted in Figure 8–30.

Experiments conducted in Israel have shown that maximum color development in tomato results when fruits are dipped in ethephon for one minute or less; longer immersion periods retard color development (Rabinowitch *et al.,* 1970) because they evidently increase the ethylene concentration in fruit above the optimal level.

Dennis *et al.* (1970) have noted that within two to five hours after green tomato fruits were dipped in ethephon at a concentration of

10,000 ppm, ethylene concentrations rose to a maximum of 30 to 70 ppm; however, they dropped to 5 ppm twenty-four hours later. The second increase in ethylene that occurred in both untreated and treated fruits accompanies normal fruit ripening of tomato. Untreated fruits ripened as a result of removal from the plant.

Abdel-Kader et al. (1966) found that ripening of detached fruits of tomato that were dipped into gibberellin at concentrations of 500 and 1,000 ppm at the mature green stage was retarded by eight and ten days, respectively, compared with control fruits. These investigators also found that dipping fruits in kinetin at concentrations of 10 and 100 ppm retarded ripening by five and seven days, respectively. Dostal and Leopold (1967) obtained similar results with gibberellin and concluded that the stimulation of color development by ethylene is prevented by treatment with gibberellin but that stimulation of respiration by ethylene is not.

Ripening of detached melons can also be hastened by treatment with ethephon. When fruits of Cucumis melo were picked thirty days after setting and immersed for ten minutes in ethephon at concentrations of 1,000 to 4,000 ppm, acceleration of ripening occurred at all concentrations but was most rapid at the higher concentrations. At a concentration of 4,000 ppm, 100 percent of the fruits reached full maturity within six days, compared with 37.5 percent of the control group (Rabinowitch et al., 1970).

Detached unripe fruits of 'Clapp Favorite' pear that were dipped for one minute in ethephon at a concentration of 25 ppm ripened in approximately twelve days, or at the same rate as fruits that were exposed for twenty-four hours to ethylene at a concentration of 100 ppm (Edgerton and Blanfield, 1968). Hastening of ripening of 'Anjou' pear as a result of an ethephon dip has also been demonstrated (Wang and Hansen, 1970).

In Canada, Looney (1968) sprayed fruits of 'McIntosh' apple two weeks after bloom with SADH at concentrations of 5 to 10,000 ppm. Application of the compound at the highest concentration resulted in a complete inhibition of ripening. From July 4 to September 20, fruits were harvested and placed in chambers containing ethylene at a concentration of 100 ppm. The delay of fruit softening that resulted from application of SADH was accompanied by a delay and inhibition of the climacteric. In this experiment, ethylene reversed the inhibitory effect of SADH on ripening, and even at the highest concentration of SADH, a climacteric was evident throughout the course of the experiment when apples were exposed to ethylene. Looney suggests that SADH retards maturation by suppressing any chemical action of ethylene within the fruit, including biosynthesis. It is interesting that another growth retardant, CCC, in contrast to SADH, fails to delay the climacteric in 'Tydeman's Early' apple (Looney, 1969). CCC either is ineffective in controlling ripening or does not persist in the fruit throughout the ripening period.

Harvested fruits of 'McIntosh' apple that were sprayed while on the tree with ethephon or with ethephon plus 2,4,5-TP ten days before harvest reached their climacteric five days before fruits treated with 2,4,5-TP alone and two weeks before the controls (Fig. 8–31). Ethephon stimulated fruit abscission but 2,4,5-TP nullified the effect (Edgerton and Blanfield, 1968).

Good color is a prerequisite for top quality of most fruits. Smock (1969) states that more than 200 chemicals have been screened in the laboratory for their effect on the color of apple. Although most compounds tested either had no effect or had an adverse effect on color, a number of carbonates, such as glycol carbonate and carbonate buffers, increased anthocyanin development. For example, fruits of 'McIntosh' apple that were dipped in ethylene carbonate at a concentration of .0016 M developed 44 percent more anthocyanins than did the controls. Other effective compounds were found to be $KHCO_3$, $CaCO_3$, quercetin-3-rutinoside (rutin), and 3-(3,4-dichlorophenyl)-1,1-dimethylurea (diuron). There was considerable variation in results among experiments because of time of sampling and other factors.

Citrus fruits often attain internal ripeness while the peel is still green. However, the green color is unacceptable to the consumer. The conventional degreening procedure is to use ethylene gas, even though ethephon is simpler to apply, and results obtained in Israel show that the effects induced by this compound on detached fruit are similar to those induced by ethylene (Fuchs and Cohen, 1969). Lemons dipped in ethephon at a concentration of 1,000 ppm attained a marketable yellow color seven days later, whereas control fruits were still only light yellow sixteen days later. Grapefruits treated with ethephon at a concentration of 50 ppm attained a marketable color seven days after treatment, whereas untreated controls required three weeks. Similar results were obtained with tangerines and oranges. Ethylene at a concentration of 50 ppm produced approximately the same advancement in degreening as did ethephon at a concentration of 1,000 ppm. Postharvest treatments with ethephon are preferable to preharvest treatments because the former eliminate the possibility of excessive defoliation or other phytotoxic effects.

288

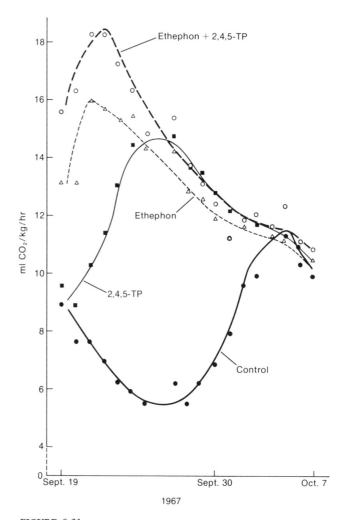

FIGURE 8-31

Effect of ethephon and 2,4,5-TP on respiration of 'McIntosh' apple. Compounds were applied as aqueous sprays to attached fruits and foliage on September 8, 1967; samples were harvested for study on September 18. (After L. J. Edgerton and G. D. Blanpied, 1968.)

Prevention of Frost Injury to Flowers and Fruits

One method of reducing danger to frost damage is to use growth regulators to delay the time of blossoming. However, this method has not been generally successful, and a delay in blossoming during a season in which frosts do not occur is usually not advantageous. A better solution would be to find a material that makes flowers and young fruits resistant to low temperatures or one that induces young fruits to grow in spite of frost injury.

APRICOT.   Frost causes many young apricot fruits to fall. The fruit that falls has dead seeds although the fruit tissue is still undamaged. This fact suggests that a synthetic hormone might replace the hormone normally produced by the seeds, thus allowing the fruit to develop. Crane (1954) increased the resistance of fruits of 'Royal' apricot to injury from low temperatures by applying 2,4,5-T at a concentration of 100 ppm fifteen hours before a frost occurred, which was forty days after full bloom. Sprayed trees dropped 83.9 percent fewer fruits that were injured by low temperature than did the unsprayed trees.

Crane also demonstrated that the effects of frost injury on the growth of apricot can be reduced by spraying the fruits with growth regulator subsequent to a frost. Severely frosted trees of 'Tilton' apricot were sprayed with 2,4,5-T at a concentration of 40 ppm two days after the frost occurred. In spite of the fact that the ovules had been killed and the endocarp tissue of some of the fruits was injured, treated fruits were induced to grow to a normal size as a result of growth regulator treatment.

PEAR.   Experiments conducted in England have shown that GA_3 can be used to improve the yield of pears in years when the styles and ovules of the flowers are killed by frost. In 1961 a frost at the Long Ashton Experiment Station killed approximately 78 percent of the blossoms on trees of 'Laxton's Superb' and 'Williams' pear. Luckwill (1961) sprayed the frosted trees at two different stages of development (white bud and full bloom), eight and eighteen days after the frost, respectively, with GA_3 at concentrations of 0, 20, 50, and 100 ppm. The most effective concentration for 'Williams,' 100 ppm, resulted in a fourfold increase in marketable crop. Application of GA_3 at a concentration of 50 ppm to 'Superb' at full bloom increased the marketable crop 2.5 times. Gibberellin treatment increased the proportion of 'Superb' fruits that were less than two inches in diameter but had no such effect on 'Williams.' Spraying frosted flowers of the cultivar 'Briston Cross' with the compound at a concentration of 10 ppm increased fruit set. When supraoptimal concentrations of gibberellin were applied, poorer fruit set and smaller, abnormally shaped fruits resulted.

An increase in set of frost-damaged 'Conference' pear was obtained by workers at the East Malling Research Station in England (Modlibowska, 1963) by applying GA₃ at a concentration of 50 ppm at different stages of development, from just before bloom (green cluster stage) to fruit set. The increase in production and in the number of marketable fruits resulted from an increased set of seedless fruits.

Griggs *et al.* (1956), working in California, found that 2,4,5-T and 2,4,5-TP sometimes reduce the loss of fruits of 'Bartlett' pear following exposure to freezing temperatures. The compounds should be applied at concentrations of 10 to 40 ppm either during a frost, at bloom, or within a few days following bloom.

Some reports from Holland have indicated that gibberellin sprays are very effective in preventing frost injury to pear. Favorable results were obtained with the 'Triomphe de Vienne' and 'Beurre Hardy' varieties. In one trial with the latter, gibberellin was applied eight days after a severe night frost that killed the flowers. Frost injury was reduced and a good crop was obtained.

Supplementary readings on the general topic of fruit set and development are listed in the Bibliography: Crane, J. C., 1964, 1969; Dilley, D. R., 1969; Gustafson, F. G., 1961; and Nitsch, J. P., 1965, 1970.

Chapter 9

SENESCENCE

Senescence has been defined as a general and increasing failure of many synthetic reactions that precede cell death (Osborne, 1967). Senescence, or aging, is the phase of plant growth that extends from full maturity to actual death and is characterized by an accumulation of metabolic products and a loss in dry weight, especially of the leaves and fruit. Senescence of leaves is evidenced by the yellowing and loss of chlorophyll that occurs before abscission or before the wilting and death of a leaf that does not abscise. In perennial plants senescence may also be an active protective process; assimilates are often exported from senescing leaves to the parent plant, which can afford greater protection to the plant from cold, drought, and other adverse conditions.

Addicott (1969) has suggested use of the term "phytogerontology," paralleling "gerontology" (the scientific study of the phenomena of aging and of the problems of the aged), to apply to the study of plant senescence and certain specialized aspects thereof, such as abscission.

NATURE OF SENESCENCE

The life-span of a leaf is determined by the conditions under which it grows and can be increased or decreased by several means. If only one leaf is left on a plant, that leaf can survive several times longer than the corresponding leaf on an intact plant (Osborne, 1967). For example, leaf cuttings or single leaves that are not in competition with other tissues and organs for nutrients and growth substances can live for long periods. On the other hand, length of life is decreased by the development of metabolic sinks in other parts of the plant, as well as by unfavorable photoperiods and lengthy exposure to shade.

Plant senescence is apparently under correlative control because the senescence of individual parts of the whole plant seems to be coordinated. The ability of developing fruits to mobilize nutrients from leaves and other plant parts might indicate that senescence in leaves and stems is a type of starvation (Molisch, 1938). In annual plants there is a synchronization between fruit maturation and the senescence of other plant parts.

Senescence is promoted by environmental factors that suppress plant growth, such as limitations imposed by insufficient soil nutrients, water, heat, and light (see Leopold, 1964a). The causal relationship between fruit development and plant senescence can be demonstrated by the removal of the plant's flowers and fruit, which delays the onset of senescence. However, different results have been obtained with some perennial plants. For example, the removal of fruit from apple trees at harvest hastens leaf senescence (G. S. Martin, private communication).

HORMONES AND SENESCENCE

The functional life-span of leaf cells can be extended or curtailed by hormone treatment. This regulation is probably effected by changes in the synthesis of protein and RNA (Osborne, 1967). Natural senescence has been shown to be accompanied by a general failure of the synthesis of nucleic acid and protein. When senescence is delayed by application of exogenous growth substances, the retardation is caused by either the maintenance of or an increase in the rate of such synthesis (see Osborne, 1967). On the other hand, ABA and other senescence stimulators cause a decrease in synthetic activity.

The balance of senescence promoters and senescence inhibitors present may be what determines the stage of senescence of a leaf or plant (Leopold, 1967). Thus, leaf longevity is affected by the levels of hormone concentrations in the metabolic sinks of the plant and also by the fluctuating levels of the hormones present in the leaves themselves.

Cytokinins

Soon after studies on the biological activity of cytokinins were begun, it was learned that cytokinins can control senescence. Richmond and Lang (1957) demonstrated that cytokinin treatment of detached leaves delays the onset of senescence. The next step was to study the interaction of treated and untreated parts. In 1959 Mothes *et al.* noted that if part of a tobacco leaf is treated with kinetin, that area remains green while the untreated portion turns yellow. That nutrients are mobilized from the untreated into the treated area was evidenced by the movement into kinetin-treated areas of radioactive glycine that was applied to one part of the leaf. Leopold (1964b), using bean cuttings with two primary leaves, treated one leaf with BA and found that although the cytokinin-treated leaf continued active photosynthesis, the untreated leaf underwent senescence at a more rapid rate than did a primary leaf from an untreated cutting. Such findings suggest that the treated leaf mobilizes assimilates from the untreated leaf, whose senescence is thus accelerated. Several investigators, using radioactive tracers, have demonstrated that cytokinin application does increase the movement of assimilates from untreated leaves to treated areas of the plant.

Leopold (1964b) distinguishes two types of deteriorative processes that occur in the leaves. The first consists of progressive deteriorations, such as a slowdown in photosynthesis and respiration and a decline in the level of RNA. The second consists of the simultaneous abrupt deterioration of the chlorophyll and protein content, which occurs near the time of the death of the leaf. Leopold (1964b) suggests that the second process may be triggered by the first.

Cytokinins are strong retarders of the leaf senescence of most species and probably act by retarding the terminal changes in chlorophyll (Leopold, 1964b) and protein content (Richmond and Lang, 1957). Osborne (1967) has suggested that the senescence-retarding effect of kinetin may be a result of its action on the production of nucleic acids and protein synthesis.

One hypothesis is that the senescence of leaves on plants with de-

veloping fruits results from the diversion from the leaves to the developing seeds of the movement of a naturally occurring cytokinin
that is formed in the roots and that is necessary for protein synthesis
(Wareing and Seth, 1967). There is some support for this theory,
but the lower levels of naturally occurring cytokinin found in such
leaves may be only a result of the initiation of senescence and not
its cause.

AUXINS, GIBBERELLINS, AND ACCELERATING SUBSTANCES.   The senescence
of disks cut from leaves of vegetative plants of common cocklebur
(*Xanthium pennsylvanicum*) is slightly accelerated by both IAA and
gibberellin but is strongly retarded by cytokinin (Osborne, 1967).
Some auxins, such as 2,4-D, often delay senescence of leaves of woody
species.

Senescence-accelerating substances are present in leaves that are
not classified as either auxins or gibberellins. ABA is one of these
compounds. The senescence of leaves is accelerated when there is an
absolute increase in the amount of endogenous ABA or when there is
a relative increase caused by a decrease in other hormones (Addicott
and Lyon, 1969).

ABA treatment often induces senescent effects in plants. Color
changes in leaves similar to those that occur during senescence commonly result from foliage application of ABA (Smith *et al.*, 1968).
ABA has stimulated the loss of chlorophyll in isolated disks of most
species that were examined (see Addicott and Lyon, 1969; see also
El-Antably *et al.*, 1967) and has accelerated the loss of chlorophyll in
some varieties of orange (Cooper *et al.*, 1968b). Potato tubers treated
with ABA became soft and senescent (El-Antably *et al.*, 1967).

CONTROL OF SENESCENCE BY APPLICATION OF EXOGENOUS PLANT REGULATORS

Cytokinins (BA and PBA), plant growth retardants (CCC, SADH,
and Phosfon-D), and auxins (2,4-D) are the compounds that have
been utilized most frequently to delay senescence in plants. Gibberellins are used to delay senescence in some species and especially
that of citrus rinds. Gibberellins also rejuvenate tissues. (For example,
they stimulate chlorophyll synthesis in citrus rinds.) Benzimidazole
delays senescence in some plants, but not enough is known about this
chemical to fit it into any of these classifications. Cytokinins act by
maintaining a high level of protein synthesis, delaying degradation

of protein and chlorophyll, slowing down rate of respiration and, in general, maintaining cell vigor (Wittwer *et al.*, 1962). Plant growth retardants probably affect the metabolism of protein and nucleic acid in the same way as do the cytokinins (Halevy, 1967).

There is a great variation in the response of plants to growth substances. SADH and CCC delay senescence of broccoli but BA is ineffective. The deterioration and discoloration of mushrooms is inhibited by SADH but BA and CCC have no observable effect. Such variability of response suggests dissimilar modes of action or a complex interaction of different chemicals with endogenous growth substances.

Use of plant regulators to delay senescence would not always be a viable alternative to present practices, but it would be useful as a supplement to rapid, careful handling and proper storage conditions. Vegetables treated with the appropriate growth regulator frequently retain their freshness for some time after harvest, in contrast to the untreated produce, which readily spoils.

BA and other synthetic cytokinins have not yet received clearance by the Food and Drug Administration for use on edible crops. When naturally occurring cytokinins such as zeatin become commercially available, they may receive a clearance. However, synthetic cytokinins can be used practically on such nonedible crops as ornamental shoots and cut flowers.

SENESCENCE OF DETACHED LEAVES OR LEAVES OF SEVERED STEMS

Excised leaves or leaves on detached stems usually senesce rapidly. They quickly turn yellow as a consequence of the degradation of chlorophyll and protein. However, senescence is usually retarded to some extent when the excised leaves are treated with cytokinin (Richmond and Lang, 1957; Osborne and McCalla, 1961; Miller, 1961). The amount of chlorophyll retained in excised leaves has been used as a bioassay for cytokinin.

Leaves of gibberellin-treated plants usually become paler than nontreated ones. This result has generally been attributed to pigment dilution caused by the failure of chlorophyll synthesis to keep pace with the increased cell and organ expansion, rather than to any direct effect of gibberellin on chlorophyll metabolism (Stuart and Cathey, 1961). However, Halevy and Wittwer (1965a) found that SADH, CCC, and BA delay chlorophyll degradation in bean plants (*Phaseolus*

FIGURE 9-1

Effect of immersing bases of detached primary leaves of bean plant in growth retardants for 20 hours. *Left to right:* leaves from cuttings immersed in CCC, SADH, water, and GA₃. Leaves were kept in tap water for 16 days after treatment before being photographed. (After A. H. Halevy and S. H. Wittwer, 1965a.)

vulgaris, cv. 'Contender'), whereas gibberellin induces an opposite effect and apparently interferes directly with chlorophyll metabolism. These investigators demonstrated this result by placing the bases of cuttings with two primary leaves in solutions of these chemicals. Twenty hours later the cuttings were transferred to water (Fig. 9–1) and the chlorophyll content was determined by reflectance measurement according to the procedure outlined by Leopold and Kawase (1964). Halevy and Wittwer (1965a) also performed experiments in which plants were sprayed with the chemicals. Gibberellin enhanced senescence of leaves when applied either through a cut stem base or in the form of a foliar spray. BA was found to be the most effective chemical for delaying chlorophyll degradation when applied as a spray. CCC and SADH were very effective in delaying senescence when they were absorbed through the cut stem bases. Direct interference of gibberellin in chlorophyll metabolism was also detected, corroborating the results of earlier experiments with leaves of citrus (Monselise and Halevy, 1962).

RETARDATION OF SENESCENCE OF VEGETABLES

Green vegetables usually deteriorate rapidly after harvest. Protein and chlorophyll content decrease in close proportion. A common result of senescence in such vegetables is the storage rot caused by the growth of bacteria and fungi on amino acids and other nutritive substances that are lost from the aging cells. The inevitable loss of quality that results is delayed or lessened by rapid, careful handling and proper storage conditions. Cytokinins and plant growth retardants often delay senescence in vegetables and offer an additional means of forestalling loss of quality of perishable crops. Application of these compounds to green vegetables may delay the visual manifestations of senescence that appear during storage.

Cabbage

Application of BA to cabbage (*Brassica oleracea capitata*) retards yellowing of the leaves (Tsujita and Andrew, 1966). In one experiment heads of 'Early Marvel' were dipped in BA immediately after harvest and stored at 5°C. Visible differences among treatments became evident twenty-one days after the storage period began. Chlorophyll determinations made after forty-five days of storage showed that application of the growth substance at a concentration of 30 ppm resulted in a chlorophyll content that was four times greater than that of the control heads. In a second experiment preharvest sprays of BA at concentrations of 20 and 40 ppm were applied to 'Glory of Enkhuizen' cabbage. After three weeks of storage, heads previously treated with BA at a concentration of 40 ppm were greener than those treated with a concentration of 20 ppm. It made no difference in this experiment whether heads were harvested immediately after spraying or twenty-four hours after. In all experiments, treated heads maintained high levels of chlorophyll and therefore appeared greener and more attractive during storage.

Lettuce

Like most green leafy vegetables, lettuce shows its age by a progressive yellowing of its leaves after harvest. Yellowing is one sign of degenerative breakdown of the tissues and is usually followed by the formation of slime rots and other diseases that live on dying tissues.

   Vacuum cooling and refrigeration are means used most frequently to delay the onset of senescence. However, these facilities are not available in many areas of the world, and a growth regulator treatment to delay senescence would be of value. Preharvest field application of BA to lettuce has been shown to delay senescence for several days. In Arizona, Bessey (1960) found that preharvest field sprays of BA at concentrations of 5 to 10 ppm helped lettuce to retain its fresh, green condition for an extra three to five days after packing. However, application of the compound more than three or four days before harvest was essentially ineffective (Zink, 1961). These results suggest that the effect of BA disappears rapidly under field conditions.

Lipton and Ceponis (1962) sprayed head lettuce just before harvest with BA at a concentration of 10 ppm. The heads were stored at a low temperature for one week and were then stored at 20°C. After three or four days control heads became yellow but treated heads remained green. Only the leaves exposed to the spray responded. Since the outer leaves are usually trimmed off before retail marketing, the percentage of marketable heads was approximately the same for treated and untreated heads. Thus postharvest treatment of trimmed heads is more useful than field treatment before trimming. Cytokinin treatment is most beneficial when the heads are stored at high holding temperatures for long periods. In this experiment, the delay of senescence was accompanied by an increased respiration rate.

Postharvest treatments with BA are also effective in delaying senescence. Zink (1961) obtained satisfactory results by applying BA at concentrations of 2.5 to 10.0 ppm to head lettuce one day after harvest. Zink (1961) conducted other experiments in which heads of 'Grand Rapids' leaf lettuce (*Lactuca sativa*) were held for two, four, six, and eight days at 4°C, after which the leaves were sprayed with BA at a concentration of 5 ppm and the heads held at 21°C. Treatment almost consistently resulted in a higher percentage of marketable plants, and the compound was usually more effective when the heads were stored for eight days. For example, five days after treatment only 20 percent of the control plants were marketable, compared with 70 percent of the treated plants. The results suggest possible use of this senescence inhibitor as a postharvest treatment at the terminal market. Treatment with BA has also been found to increase the storage life of varieties of butterhead lettuce, a relatively poor shipper that deteriorates more rapidly than head lettuce.

Vegetables handled through packing sheds can be treated with dips and washes. Compounds at concentrations of 2.5 to 10.0 ppm are best because higher dosages may cause phytotoxicity.

Plant growth retardants are sometimes much more effective than cytokinins in delaying senescence. In an experiment conducted in Michigan, leaves of greenhouse-grown 'Grand Rapids' lettuce were dipped or their cut stem bases were immersed overnight in SADH or CCC at concentrations of 5 to 1,000 ppm and in BA at concentrations of 5 to 200 ppm (Halevy and Wittwer, 1966). Dipping was found to be more effective than overnight immersion of the bases. After the leaves and stems were stored for five days at 22°C, the most effective treatments for delaying senescence were found to be

FIGURE 9-2

Effect of postharvest dip in growth regulator on longevity of 'Grand Rapids' leaf lettuce after 5 days of storage at 22°C. *Left to right:* leaves dipped in BA at a concentration of 25 ppm, in CCC at a concentration of 25 ppm, in SADH at a concentration of 25 ppm, and in water. (After A. H. Halevy and S. H. Wittwer, 1966.)

CCC at a concentration of 60 ppm and SADH at a concentration of 120 ppm (Fig. 9–2). After eight to ten days of storage, 10 ppm was found to be the optimal concentration for both chemicals. The investigators found BA to be ineffective, irrespective of concentration or method of application. In a subsequent experiment conducted by Halevy and Wittwer (1966), leaves of 'Grand Rapids' lettuce were dipped momentarily in the growth regulator solutions at concentrations of 10 to 100 ppm, and the leaves were then stored at 8, 15, and 22°C. BA did not preserve the quality of lettuce at 8° and 15°C and accelerated its deterioration at 22°C. Both SADH and CCC greatly increased the longevity at all these temperatures and concentrations. The lower concentrations were optimal at 15° and 22°C and the higher concentrations were optimal at 8°C. Halevy and Wittwer (1966) confirmed these results in a second experiment using storage temperatures of 10, 15, and 20°C.

Cauliflower

Yellowing and abscission of the jacket leaves of cauliflower limit the storage life of the head. Growth regulator application can retard leaf senescence. In New York State, Kaufmann and Ringel (1961) treated heads of Long Island-grown cauliflower with BA at a concentration of 10 ppm, 2,4-D at a concentration of 50 ppm, and a combination of BA at a concentration of 10 ppm plus 2,4-D at a concentration of 50 ppm, using a tap water control. A wetting agent, sodium lauryl sulfate at a concentration of 50 ppm, was added to each solution and to the tap water. The chemicals were applied as a continuous spray until the solution dripped from the cauliflower heads. The heads were stored at 9°C at a relative humidity of 90 percent. Application of BA in combination with 2,4-D prolonged the marketability of most heads by eighteen days, at the end of which 91 percent of the outer leaves of the treated heads were still completely green. Treatments with BA alone and 2,4-D alone were also effective. After eighteen days, 71 percent of the BA-treated leaves and 66 percent of the 2,4-D-treated leaves were still green. Only 34 percent of the control leaves were still green. The mixture of chemicals prevented leaf abscission and retarded yellowing during twenty-eight days of storage at 9°C. Retardation of senescence would benefit the grower by enabling him to market his cauliflower later than other producers.

Asparagus

Both cytokinin and plant growth retardants can delay senescence. Spears of asparagus (*Asparagus officinalis* cv. 'Mary Washington') were trimmed to 16 cm and were either dipped in BA at a concentration of 25 ppm for ten minutes or immersed (bases only) in a similar solution for eighteen hours (Halevy and Wittwer, 1966). Ten days later, changes in the color that existed at harvest were determined by reflectance as measured with a Bausch and Lomb "Spectrononic-20" spectrophotometer having a reflection attachment. A portion of the spear 8 cm from the tip was measured at 665 mμ. The percentages of reflectance, as measured ten days after treatment, of dipped spears and of controls (dipped in water) were 26 and 48, respectively. The percentages, in the same order, resulting from immersion of bases were 39 and 51. These figures indicate that both types of treatment delay degreening. Both SADH and CCC were found to reduce the degreening of asparagus spears although neither compound applied as a dip produced results equal to those obtained by dipping in BA at a concentration of 25 ppm. Immersion of bases in either SADH or CCC at concentrations of 125 and 500 ppm was found to retard degreening just as effectively as immersion in BA.

Broccoli

Experiments to test the effectiveness of 4-CPA, 2,4-D, and 2,4,5-T have shown that these chemicals can retard loss of green color by the sepals of broccoli flowerets (Marth, 1952). The most beneficial effect was obtained when 2,4,5-T was applied in the field either a few days before harvest or immediately after harvest. In one experiment heads treated with 2,4,5-T at a concentration of 1,000 ppm remained green and attractive for four to five days; control heads yellowed within three days after treatment.

BA is effective in delaying senescence of broccoli, especially when application is followed by low-temperature storage. Dedolph *et al.* (1962) dipped freshly harvested heads of broccoli (*Brassica oleracea* 'Spartan Early') in BA at a concentration of 10 ppm and then stored them at temperatures of 4°, 10°, 15°, and 21°C. The growth regulator was found to be effective in slowing the degeneration of the marketable appearance of broccoli at all storage temperatures. Senescence occurred much more rapidly at higher temperatures so that differences in acceptability between treated and untreated heads were visible earlier than at lower temperatures. However, only at 4°C were the weight losses of broccoli significantly less as a result of treatment with BA.

Other experiments with 'Spartan Early' broccoli, in which freshly harvested heads were given a ten-minute dip in BA at a concentration of 20 ppm and cut stem bases were immersed overnight for eighteen hours in compounds at a wide range of concentrations, demonstrated that BA is effective in retarding senescence but that SADH and CCC are ineffective (Halevy and Wittwer, 1966).

Celery

BA effectively delays senescence of celery. In Michigan, Wittwer *et al.* (1962) applied BA at a concentration of 10 ppm as a postharvest dip to freshly harvested stalks of green ('Utah 52–70') and golden ('Cornell 19') celery and found that the treatment extended the duration of visual freshness, leaf coloration, and market acceptability of both varieties. Evaluations at two-week intervals of celery held at 4°, 10°, 16°, and 21°C revealed that the appearance of the BA-treated celery was superior to that of the controls (Fig. 9–3). After forty days of storage at 4°C, 50 percent of the control celery was judged unacceptable but none of the BA-treated celery was unacceptable. It is

FIGURE 9-3

'Utah 52–70' celery when removed from respirometers after 22 days at 21°C. *Left*, controls, dipped in water; *right*, celery dipped in an aqueous solution of BA at a concentration of 10 ppm. (After S. H. Wittwer, R. R. Dedolph, V. Tuli, and D. Gilbart, 1962.)

interesting that the total evolution of CO_2 in BA-treated celery held at 21°C was reduced by 27 percent during a twenty-two-day storage period. This suppression of respiration rate was accompanied by a reduction in fresh weight of 47 percent compared with nontreated samples. The effectiveness of BA for preserving freshness in celery appears to be a consequence of respiration suppression.

Consumer acceptance tests conducted by Dedolph *et al.* (1963) showed that poststorage flavor of celery is improved by treatment with BA. Trimming losses of treated celery were cut fourfold compared with those of controls. This fact is important because celery is often packaged in transparent plastic before being displayed, a practice that prevents subsequent retrimming. Treated celery, both fresh and cooked, was preferable to the controls because of its higher chlorophyll retention.

Brussels Sprouts

Brussels sprouts dipped at harvest in BA at a concentration of 10 ppm were shown to have a much longer storage life than untreated controls (van Overbeek and Loeffler, 1962). Eleven days after storage for eleven days at 21°C all untreated produce was totally unsalable, but 90 percent of the BA-treated Brussels sprouts were in salable condition. The senescence-retarding effect was attributed to the cytokinin's maintenance of protein synthesis and retardation of protein degradation. According to these researchers, the cytokinin maintained a pattern of metabolism after harvest similar to that existing before harvest.

In another experiment Thomas (1968) demonstrated that a mixture of BA and NAA was more effective in delaying senescence of freshly harvested sprouts of the early, quickly frozen variety 'Jade Cross' and an early hybrid 'Avoncross' than BA applied alone. Freshly harvested sprouts were dipped in BA at a concentration of 10 ppm both with and without an added 25 ppm of NAA and were then stored at 18° to 20°C. The outer leaves of untreated 'Jade Cross' sprouts turned flaccid and yellow and the sprouts deteriorated in quality within four days. The mixture of hormones maintained all sprouts in a marketable condition for twelve days after treatment. Treatment with BA alone or in combination with NAA was less effective when applied to 'Avoncross' sprouts, which normally have a relatively long storage life. The controls showed little loss of color nineteen days after treatment. Preharvest application of BA or of a mixture of BA and NAA delayed postharvest yellowing of sprouts of 'Jade Cross' when they were harvested two days after treatment.

Eaves and Forsyth (1968) have provided some evidence that benzimidazole effectively retards senescence of Brussels sprouts. In a laboratory test Brussels sprouts cv. 'Jade Cross' were immersed in containers of benzimidazole at a concentration of 200 ppm and were kept for six days at 21°C in three different atmospheres (7 percent CO_2, air, and a combination of O_2-free air), both in light and in darkness. Controls were immersed in water. The sprouts that were kept in light, CO_2, and benzimidazole retained 35 percent more chlorophyll than the controls.

Other Vegetables

BA is effective in maintaining the green color and fresh appearance of endive, escarole, mustard greens, spinach, radish and carrot tops, parsley, and green onion (Zink, 1961; van Overbeek and Loeffler, 1962). Artichoke and snap and lima bean evidenced no response to the compound.

Carrots and radishes are often marketed with their tops attached to indicate freshness. Zink (1961) found that application of BA extends the storage life of the tops for two to three days over that of the controls.

He also found that the response of endive, escarole, spinach, and mustard greens to BA was striking (Zink, 1961). Sixty percent of the heads of escarole treated with BA at a concentration of 5 ppm and held for six days at 21°C were in a salable condition, whereas none of the controls were marketable. These results are typical of the response of this group of leafy vegetables to BA. Loss of quality by these and other leafy vegetables is characterized by soft rot and by a rapid disappearance of chlorophyll from the older leaves.

RETARDATION OF SENESCENCE OF FRUITS

Cherry and Strawberry

The marketability of some horticultural products is often determined by the inclusion of a nonedible portion of the commodity. The presence of green pedicels of sweet cherry and calyxes of strawberry favorably influences marketability by enhancing the fruit's fresh appearance.

Tuli et al. (1962) dipped freshly harvested sweet cherries (Prunus avium cv. 'Vernon') in an aqueous solution of BA at a concentration of 10 ppm and then held them for seven days at 21°C. At the end of that time the pedicels of treated cherries still appeared fresh and green and had an average chlorophyll content that was 35 percent greater than that of the control samples, which had withered and turned brown. These data suggest that treatment of sweet cherry with BA preserves the green color of the pedicel and fresh appearance of the fruit and reduces weight loss during storage.

In a similar experiment Tuli et al. (1962) dipped freshly harvested strawberries (Fragaria vesca cv. 'Robinson') in BA at a concentration of 10 ppm and found that there was no delay in senescence; in fact the percentage of decay was somewhat increased by the treatment. Nor was there any increase in the amount of chlorophyll retained in the fruit calyxes. On the other hand, Dayawon and Shutak (1967) obtained indications that BA may prolong the shelf life of strawberry. They noted that the growth substance decreased the respiration rate of the 'Geneva' variety.

Navel Orange

As the navel orange approaches maturity the rind color changes from green to orange because the concentrations of chlorophyll in the

flavedo (the outer, colored part of the rind) decrease and the carotenoid pigments increase. During this period the rind softens, first at
a rapid rate and then at a slower rate (Coggins and Lewis, 1965).
Coloration and softening precede fruit maturity; thus rind maturation
begins before harvest and continues as long as the fruit remains on the
tree. Under weather conditions existing in California, as long as eight
months may elapse from the time the rind begins to soften until
harvest (Coggins, 1969). A soft rind makes the fruit susceptible to
several physiological disorders that become apparent late in the
harvest season and that reduce storage quality and market value of the
fruit.

BIOCHEMICAL CHANGES OCCURRING IN NAVEL ORANGE DURING MATURA-
TION. During the softening period the ability of the rind tissue to
utilize glucose decreases and sugars accumulate to approximately
twice the initial concentration (Lewis et al., 1967). There is also an
increase in the ratio of potassium to calcium plus magnesium. Other
factors that lead to senescence of the rind of navel orange include
changes in pigment and in carbohydrate metabolism, but these can
be so modified by application of gibberellin that a retardation of
senescence occurs. Gibberellin retards the accumulation of rind caro-
tenoids and simultaneously slows the net loss of chlorophylls a and b
(Lewis and Coggins, 1964). Studies utilizing the electron microscope
have revealed that chromoplasts can revert to chloroplasts in regreening 'Valencia' orange (Thomson et al., 1967). Coggins et al. (1969)
also demonstrated that certain essential oils of the flavedo, linalool
and geraniol, decrease in a linear fashion during senescence but that
gibberellin treatment minimizes these changes.

ANATOMICAL CHANGES OCCURRING IN NAVEL ORANGE DURING MATURA-
TION. Striking anatomical changes also occur in the navel orange
before and during senescence. Cells of the flavedo and albedo (the
inner, white or nearly colorless part of the rind) usually enlarge,
become highly vacuolated, and change in shape during development,
maturation, and senescence (Coggins, 1969). Many intercellular
spaces develop and cell walls weaken and break. Senescent rinds are
structurally weak, and the cells contain small amounts of cytoplasm.

   Some of the rind disorders that may develop during senescence of
the rind are rind-staining, water-spotting, increased susceptibility to
decay, puffiness, and accumulation of sticky exudate. These disorders,
which are described more fully in the following paragraphs, are
obviated or alleviated by commercial application of gibberellin, which

delays senescence and thus maintains a firmer rind (Coggins and Lewis, 1965). During the 1965–1966 season approximately 10,000 acres of navel orange were so treated in California.

The brownish discolorations and shallow pits that frequently develop in the rind during the postharvest period (Fig. 9–4) are caused by a softening of the rind that makes it susceptible to mechanical abrasion. Rind-staining often occurs during harvesting, washing, waxing, drying, and packing. Unfortunately, this disorder often is not detected until the fruits arrive at the market since symptoms are not apparent until twelve to twenty-four-hours after the injury occurs (Coggins et al., 1963). Rind-staining results in a definite reduction in the market value of the fruit.

To prevent rind-staining, gibberellin at concentrations of 5 to 20 ppm (usually 10 ppm is sufficient) should be applied so that the outside leaves and fruit are wet thoroughly (Coggins et al., 1969). In California October or November sprays usually provide better protection from rind disorders than later sprays. However, fall sprays cause more of a delay in rind color development, and in some years the color delay can be sufficient to make the fruit unacceptable to the consumer if fruit is harvested before mid-March. The most acceptable fruits are those that are completely degreened at harvest. There is less of a delay in color if gibberellin is applied in December or January, but January applications sometimes result in lower production the following year. When harvest in California is scheduled for later than mid-March, a general rule is to apply gibberellin in October or November. When late coloration is expected to occur naturally, a later spray is more suitable. When a minimum effect on rind color is desired, the spray should be applied just after 90 percent of the green color has disappeared, preferably no later than December.

FIGURE 9-4

Gradations of rind-staining of naval orange. *Left to right:* none, slight, moderate, severe. (After I. L. Eaks, 1964.)

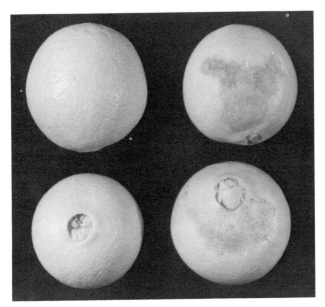

FIGURE 9-5
Mold-infected water spots on navel orange (*right*). Healthy oranges are on left. (Photograph courtesy of C. W. Coggins, Jr.)

Water spotting is a preharvest disorder that results from absorption of surface deposits of water by local areas of the rind. Affected fruits have no value on the fresh fruit market. Frequently the condition is worsened by secondary infections caused by blue and green mold (Fig. 9–5). A pesticide spray oil is used for insect control in citrus, but it has the disadvantage of increasing the susceptibility of the navel orange to water spotting. Application of gibberellin in early winter to 'Washington Navel' orange, if preceded by treatment with oil sprays, affords protection from water spotting (Riehl *et al.*, 1966).

Citrus fruits generally become more susceptible to decay as they become older. Coggins (1969) states that this is a result of both a reduction in resistance to entry of the organisms and changes in the rind that make it a better medium for growth of the organisms. Gibberellin treatment reduces the amount of decay by producing a tougher rind; such reduction may be accompanied by a reduced rate of sugar accumulation in the treated rind (Lewis *et al.*, 1967). In one experiment fruit from trees previously sprayed with gibberellin was

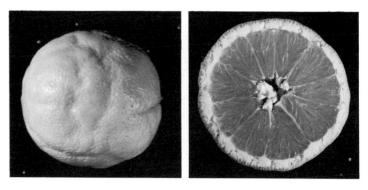

FIGURE 9-6

Puffiness in rind of navel orange. (Photographs courtesy of C. W. Coggins, Jr.)

stored at room temperature and the incidence of decay was observed. Typical results showed that treated fruits evidenced 60 percent less decay than control fruits (Coggins, 1969).

When puffiness occurs, the mesocarp separates from the endocarp in irregular areas and the affected areas become elevated above the remainder of the surface of the fruit (Fig. 9–6). Such rinds are structurally weak and have little eye appeal. Gibberellin application frequently reduces puffiness, although the results obtained have been inconsistent.

Sticky exudate (Fig. 9–7) usually begins to appear before harvest and accumulates more profusely subsequent to harvest. Its presence

FIGURE 9-7

Sticky rind exudate on unwashed fruit of navel orange (*right*), not present on control (*left*). (Photograph courtesy of C. W. Coggins, Jr.)

indicates that the rind is senescent and that the fruit will have a short shelf life. However, the problem is substantially reduced or eliminated by October or November sprays of gibberellin (Coggins, 1969; Coggins *et al.*, 1965). For example, 85 percent of the control fruits in one California orchard that was sprayed in early January with gibberellin at a concentration of 10 ppm had sticky rinds, whereas only 25 percent of the sprayed fruits were affected (Coggins, 1969).

Delay of Maturation of Lemon

A delay in maturation of trees of 'Lisbon' lemon occurs when trees are sprayed in the spring with high concentrations of gibberellin (Coggins and Hield, 1968). Lemons from trees sprayed with gibberellin at a concentration of 40 ppm were observed to be greener than controls as early as one month and as long as seven months after treatment. Sprayed lemons also developed color less rapidly during storage than did the controls. However, production was decreased during the year of treatment and was reduced the following year.

Coggins and Hield (1968) report that gibberellin application results in delayed yellowing, no matter what time of year it is made. According to these investigators, this response is caused by a delay in overall maturity rather than to a specific delay in rind maturity, as apparently occurs in navel orange. Low concentrations of gibberellin produce few undesirable effects. Fall applications of the compound at concentrations of 5 to 10 ppm induce the most response and are of most economic benefit. In addition to the delay in maturation and coloring that results during the first year, the harvest pattern is favorably affected in the year following application, probably because of the effect of gibberellin on flowering. Although less fruit is produced early in the season, yield is high during the summer when market demand is high. If sprays are applied during two successive years, the seasonal pattern is altered even more, since to the effect of the first-year spray on flowering is added the delay in coloring caused by the second-year spray. Other advantageous effects of the gibberellin sprays are a decrease in the number of small, yellow lemons produced and a longer storage life.

Delay of Maturation of Lime

In Southern California trees of 'Bearss' lime bear some fruit much of the year, but because most of the crop colors and ripens in autumn and winter, it must be picked when market demand is at a minimum.

Gibberellin delays the maturation of the lime fruit, thus providing the same advantages obtained from its application to lemon (Burns *et al.*, 1964). Fruit from trees sprayed with a concentration of 10 ppm is larger and remains green longer than untreated fruit, and these color differences continue during cold storage.

Delay of Maturation of Grapefruit and 'Valencia' Orange

Gibberellin delays senescence in the rind of 'Valencia' orange (Coggins *et al.*, 1960) and grapefruit (Coggins *et al.*, 1962), but the adverse effects that result outweigh the benefits. If gibberellin is applied when the fruit is fully colored, regreening of the rind occurs, which decreases crop value. Furthermore, severe phytotoxic injury, including leaf drop and twig dieback, as well as reduced production, often occur.

The response of citrus fruits to GA_3 varies with the variety. Application of the compound to 'Valencia' orange and grapefruit causes regreening of fully ripened fruit on the tree, but regreening does not occur in 'Washington Navel' orange (Coggins and Lewis, 1962; Coggins *et al.*, 1962). In Florida, Ismail *et al.* (1967) found that GA_3 delays the breakdown of chlorophyll in the rind of detached, mature-green fruits of 'Valencia' and 'Parson Brown' orange but does not promote regreening of detached, fully ripened fruits of 'Parson Brown' or 'Pineapple' orange.

Delay of Maturation of Persimmon

The oriental or 'Kaki' persimmon (*Diospyros kaki*) ripens very rapidly and soon becomes overripe, a characteristic that prevents it from becoming a fruit of wide commercial importance. This characteristic is easily perceived in the Japanese variety 'Hiratanenashi,' one of the most popular of the astringent- or puckery-type cultivars. In Japan, Kitagawa *et al.* (1966) applied various growth substances to trees of 'Hiratanenashi' and 'Fuyu,' a nonastringent type, in an attempt to prolong storage life and delay senescence. Gibberellin proved to be the most effective compound. Trees were sprayed with gibberellin at concentrations of 50, 100, and 200 ppm in October when the fruit was mature and ready to harvest. Fruit of 'Hiratanenashi' was harvested three, ten, and seventeen days after spraying; fruit of 'Fuyu' was harvested three and ten days after spraying. After harvest fruits of 'Fuyu' were stored at 7° to 15° C, but fruits of 'Hirantanenashi' were stored for one week in air-tight casks containing alcohol to remove astringency. Fruit firmness was subsequently determined with a pressure tester.

Sprayed fruits remained firm on the tree for three or four weeks longer than unsprayed fruits. Fruit enlargement and coloration of sprayed fruits were also retarded. For example, fruit of 'Hiratanenashi' harvested three days after being sprayed with gibberellin at a concentration of 200 ppm required 11.4 days of storage for the fruit firmness to decrease to 0.45 kg, but only 3.5 days were required for controls. Lower concentrations resulted in less retardation of ripening, and similar results were obtained when picking of fruit from the trees was delayed. 'Fuyu' fruits senesced less rapidly; gibberellin caused a delay of senescence in this variety, but not as great a delay as occurred in 'Hiratanenashi. It is interesting that gibberellin sprays also strikingly delayed autumn leaf fall of the trees bearing these varieties.

DEGREENING OF TANGERINE

Postharvest degreening of citrus fruit can be accomplished by ethylene treatment, but a high incidence of decay sometimes accompanies this process. In Florida degreening of stored fruits of 'Robinson' and 'Lee' tangerine was accelerated by preharvest sprays of ethephon at concentrations of 50 to 200 ppm (Young et al., 1970). Fruit of 'Lee' tangerine that was sprayed on November 5 with ethephon at a concentration of 200 ppm required twenty-four hours less time for degreening three weeks after treatment and had 21 percent less decay after three weeks of storage than did untreated fruit.

RETARDATION OF SENESCENCE OF MUSHROOM

SADH effectively retards senescence of mushroom. Halevy and Wittwer (1966) soaked freshly harvested mushrooms (Agaricus campestris) for ten minutes in solutions of SADH and CCC at concentrations of 10 to 4,000 ppm and in BA at concentrations of 6 to 400 ppm. They were then dried for two hours, wrapped in Saran and held for eight days at 5° and 22°C. BA and CCC were ineffective and even detrimental, whereas SADH, applied at a wide range of concentrations, significantly delayed deterioration (Fig. 9–8). Bisulfite was included in one experiment but retarded discoloration only slightly, if at all. In another experiment disks were cut from the center part of the mushroom cap and were held for three days at 20°C, four days at 10°C, and eight days at 5°C. SADH at concentrations of 10 to 1,000 ppm effectively reduced browning at all temperatures.

A subsequent experiment demonstrated that application of SADH

FIGURE 9-8

Effect of treatment with BA, SADH, and CCC on longevity of mushrooms stored at 5°C for 8 days after being harvested. *Left to right:* mushrooms dipped for 10 minutes in SADH at a concentration of 100 ppm, CCC at a concentration of 100 ppm, BA at a concentration of 100 ppm, and water. (After A. H. Halevy and S. H. Wittwer, 1966.)

at concentrations of 10 or 100 ppm to mushrooms held at 5°C also delayed deterioration and reduced browning but was ineffective when applied to mushrooms held at 22°C. The response of mushrooms to other chemical treatments has also varied according to the environmental conditions under which the fungi were growing (Hughes, 1958).

PROLONGING THE STORAGE LIFE OF CUT FLOWERS

Efforts to prolong the life of cut flowers have included experiments with several hundred chemicals, but most of them have not proven to be beneficial. Laurie (1936) found that if a piece of copper wire is placed in the glass container holding the cut flowers in water, the life of snapdragon, aster, stock, and marigold can be extended by 1.0 to 2.7 days. More recent reports have shown that materials ranging from commercial preservatives to growth regulators increase vase life by four to six days.

Snapdragon

The vase life of a cut snapdragon (*Antirrhinum majus*) that is maintained in tap water under favorable environmental conditions is usually less than one week. The commercial practice of cutting spikes when only a few florets are open often results in poor development of the spike and fading of color at the tip. Therefore, extension of vase life is extremely desirable. Chemicals have their greatest effect if

(1) they are used within a few hours after the flowers are cut and (2) the exposure is continuous. Several hundred chemicals have been tried, but relatively few have proved to be effective.

MH and certain chelating agents have been shown to extend the vase life of several varieties of snapdragon (Kelley and Hamner, 1958). Seven different varieties of snapdragon were cut and placed in solutions of seventeen different chelating agents. When MH at concentrations of 250 to 500 ppm was used alone, the vase life of some varieties was extended by two to four days. The most effective chelating agents were Cupferron (ammonium salt of N-nitroso-N-phenyl-hydroxylamine), diphenylamine, hexamethylenetetramine, and 1-nitro-2-naphthol-3,6-disulfonic acid, all compounds known to chelate iron and sometimes copper as well. Vase life was generally prolonged by four to six days depending on variety and concentration. For example, when Cupferron at a concentration of 200 ppm was applied to 'Rockwood's Summer Pink,' the average vase life was 10.0 days, whereas the control lasted only 4.7 days. No beneficial synergistic effect is achieved by combining MH and Cupferron. The investigators suggested that the chelating agents were effective because they reduced enzyme activity and that the MH acted by destroying the naturally occurring IAA.

Larsen and Scholes (1966) showed that the vase life of snapdragon can be increased 2.7 times by immersion of stems in a solution containing QC at a concentration of 300 ppm, SADH at concentrations of 10 to 50 ppm, and 1.5 percent sucrose. Use of SADH alone usually failed to increase vase life, although excellent results had been previously obtained by Halevy and Wittwer (1965c), as shown in Figure 9–9. The mixture of chemicals also promoted opening of a large number of undeveloped florets. The increase in number of opened florets was 3.3 times that of the control (maintained in water); the increase in spike length was 4.2 times that of the control. Natural floret color of spikes treated with these solutions was maintained, whereas control florets that opened after cutting evidenced little pigment development. Chemical treatment also resulted in growth of larger florets, stimulation of lateral shoot growth with florets, and elimination or reduction of growth of micro-organisms.

The primary function of the sucrose was to supply a source of energy for metabolic processes, and the primary purpose of the QC was probably to control growth of micro-organisms (Larsen and Scholes, 1965). The SADH may have served to reduce water requirement, slow metabolism, and, to a minor extent, control growth of micro-organisms (Larsen and Cromarty, 1966).

FIGURE 9-9

Snapdragon, cv. 'White Apollo,' 8 days after being harvested. Stem on right was immersed for 18 hours in SADH at a concentration of 10 ppm. Stem on left was immersed in water only. (After A. H. Halevy and S. H. Wittwer, 1965c.)

Larsen and Sholes (1966) showed that application of SADH alone at a concentration of 50 ppm drastically reduced spike length; some reduction also occurred at lower concentrations. This method can be advantageous because the combination of QC and sucrose sometimes results in an excessive spike length. The most desirable results were usually obtained with a combination of QC at a concentration of 300 ppm, 1.5 to 2.0 percent sucrose, and SADH at a concentration of 25 ppm. The response to treatment was similar for both greenhouse and outdoor cultivars. It has been shown that ^{14}C-labeled SADH spreads throughout the entire spike less than thirty minutes after stem immersion, indicating passive transport throughout the vascular stream (Larsen and Scholes, 1966).

Carnation

Halevy and Wittwer (1965c, 1966) tested the effects of three growth retardants and BA on five cultivars of greenhouse-grown carnation (*Dianthus caryophyllus*). The four cultivars of snapdragon that were also included in the experiment produced results similar to those obtained with carnation. Flowers of uniform size and color were selected; stems of carnation were trimmed to a length of 30 cm and those of snapdragon were trimmed to 40 cm. After the bases had been immersed in chemical solutions for sixteen to eighteen hours, they were placed in water and held at 22°C. The bases were then recut and the water was changed at two-day intervals. When other plants were sprayed with the chemicals, BA and Phosfon-D were found to be either ineffective or detrimental. CCC and SADH were most effective following overnight immersion of the bases of the cut stems. Spray treatments were less effective than immersion. Both CCC and SADH were found to prolong the life of most cultivars for two or three days, but the optimal chemical concentrations varied. Significant differences among the cultivars were observable, and some did not respond to any growth regulator. The most effective concentration of CCC in the summer, 500 ppm, was much higher than the optimal winter concentrations of 10 to 25 ppm.

The delay in senescence of carnation and chrysanthemum as a result of BA treatment was found to be accompanied by an inhibition of respiration rate (MacLean and Dedolph, 1962). The respiration rate of treated flower stalks of chrysanthemum was reduced by 22.2 percent compared with that of the controls.

In Washington State Larsen and Scholes (1965) demonstrated that a combination of SADH and QC retards senescence of the carnation cultivars 'Petersen New Pink' and 'Red Gayety.' Cuttings eighteen inches long were placed in jars containing only nutrient solution or nutrient solution plus SADH at a concentration of 500 ppm, 3 to 5 percent sucrose, and QC at concentrations of 300 to 500 ppm, and were stored at room temperature until petals lost their turgidity and began to curl upward. The additives more than doubled vase life and increased flower diameter and weight over those of the control, placed in water (Fig. 9–10). On the average, diameter of treated flowers was increased 2.5 times over that of the control. In one experiment with 'Red Gayety,' vase life of the control was 6.8 days; that of the treated flowers was 16.8 days. Increases in flower diameter were, in the same order, 4.6 mm and 19.0 mm. Increases in flower weight were 1.2 g and 3.0 g, respectively. In a subsequent experiment Larsen and Frolich (1969) demonstrated that the delay of flower senescence caused by these chemicals results at least in part from their delay of a day or two in the climacteric and their promotion of the water flow through the stem sections. According to these investigators, sucrose is an energy source that delays senescence, QC prevents blockage of the vascular

FIGURE 9-10

Effect of chemical treatment on longevity of cut carnation, cultivars 'Red Gayety' (*top*) and 'Petersen New Pink' (*bottom*), after 15 days at 22–23°C. *Left to right:* stems immersed in a combination of QC at a concentration of 400 ppm, SADH at a concentration of 500 ppm, and 5 percent sucrose; water; and a combination of QC at a concentration of 800 ppm, SADH at a concentration of 500 ppm, and 5 percent sucrose. High concentrations of QC weaken stems. (After F. E. Larsen and J. F. Scholes, 1965.)

tissue of the cut stems and thus delays senescence, and SADH postpones the climacteric by a day and promotes water flow so that it exceeds that of the controls.

In their studies of the opening of buds of six cultivars of carnation that were shipped by air from Colorado and California to Maryland, Hardenburg *et al.* (1970) tested three mixtures: Cornell solution (5 percent sucrose, 8-hydroxyquinoline sulfate at a concentration of 200 ppm, and silver acetate at a concentration of 50 ppm); Everbloom (product of the W. Atlee Burpee Company); and a solution containing 3 percent sucrose, QC at a concentration of 400 ppm, and SADH at a concentration of 300 ppm. The Cornell solution produced the largest blooms with the longest life, but Everbloom at a concentration of 2 percent and the third mixture were also satisfactory.

In Norway Heide and Øydvin (1969) found that the vase life and storage life of cut carnations were improved by postharvest treatment with BA. Flowers of 'Cardinal Sim,' 'William Sim,' and 'Crowley's Sim' were cut to a length of 40 cm and were then dipped for two minutes in BA at a concentration of 10^{-3} M. The vase life of the flowers was increased by three to five days, but immersion in the same solution for twelve hours was found to be detrimental. After treatment the flowers were placed in tap water or in 5 percent sucrose acidified to pH 3.5. Treated flowers had the same vase life after four weeks of storage at 0.5°C as did freshly cut flowers. When the flower alone was dipped in a solution of growth substance, deleterious color changes were observable in red flowers, although vase life of all flowers was increased to the same extent when stems also were treated. These investigators explain that the negative results obtained with BA by Halevy and Wittwer were caused by prolonged treatment.

Daffodil

In Canada Ballantyne (1963) submerged cut flowers of *Narcissus pseudonarcissus* 'King Alfred' for five seconds in BA at a concentration of 6.5×10^{-4} M or kinetin at a concentration of 5×10^{-4} M and then placed the cut ends in water at room temperature. When the flowers died the perianth segments lost their color and their edges became brown. Both dips extended flower life by approximately one day. BA was also effective in increasing the dry storage life of cut narcissus. Flowers were dipped, subjected to dry storage in a cardboard flower box for one day at room temperature, and then placed in water. Again storage life was increased by one day as a result of treatment. Since narcissus rarely keep longer than three days after removal from storage, extension of flower life by one day is very significant. Many of these flowers are shipped from western to eastern Canada at Easter, and an added day of bloom life means an increase of 50 to 100 percent in the length of time the flowers are presentable in the home of the consumer.

Further experimentation by Ballantyne (1965) demonstrated that freshly cut 'King Alfred' daffodils can be preserved by dipping them in a combination of BA at a concentration of 5×10^{-4} M plus 2,4-D at a concentration of 10^{-4} M, but not by dipping in either compound alone. The mixture of compounds was found to prevent the dehydration and wilting that accompany or cause flower senescence. Stems of treated flowers were placed in tap water at room temperature, and after six days the fresh weight of treated perianth segments was 3.93 g; that of control flowers was 1.56 g.

Stocks

Application of BA effectively extends the vase life of stock flowers. Uota and Harris (1964) sprayed plants of the cultivar 'Avalanche' in the field with the compound at concentrations of 25 and 50 ppm two hours before pulling. After the plants had been stored for five days at 10°C, the stalks were recut and placed in jars of water. After three days of storage at 21°C, leaves of the treated plants remained dark green and turgid, whereas leaves of untreated plants rapidly yellowed and became unsalable. Blooms on treated plants were also much superior to those on untreated plants. Postharvest delays in the application of BA decreased its effectiveness in retarding senescence.

PROLONGING THE STORAGE LIFE OF FLOWERS OF POT PLANTS

The use of growth regulators to prolong the life of flowers of pot plants has received much less scientific attention than has use of the compounds to prolong the life of cut flowers. Chemicals intended to control the height of flowers have sometimes extended flower longevity.

Buxton and Culbert (1967) demonstrated that spraying 'Yellow Delaware' chrysanthemum with SADH increases flower life up to five days. The longest flower life resulted from an application of the compound at a concentration of 2,500 ppm made three weeks after the start of short days, followed by a similar application made five weeks later.

Kelley and Schlamp (1964) demonstrated that GA_3 prolongs the keeping qualities of the uncut flowers of three varieties of Easter Lily (*Lilium longiflorum*). Best results were obtained when GA_3 at concentrations of 500 to 1,000 ppm was sprayed on the flowers six to fourteen days before full bloom. Daily observations were then made until 50 percent of the petal surface showed signs of necrosis or browning. The delay in the first signs of petal necrosis appearing after full bloom was greater in GA_3-treated flowers than in controls. The difference in delay was a maximum of 1.9 days for 'Croft,' 2.5 days for 'Ace,' and 2.2 days for 'Nellie White.' The delay evidenced by treated flowers, compared with that of the controls, was (in percent) 29.5, 35.2, and 24.7, respectively. Treatment with GA$_3$ sometimes delayed the time from full bloom to 50 percent necrosis by as much as 35 percent.

Supplementary readings on the general topic of senescence are listed in the Bibliography: Addicott, F. T., 1969; Leopold, A. C., 1967; Wangermann, E., 1965; and Woolhouse, H. W., ed., 1967.

Chapter 10

ABSCISSION

Abscission is the separation of a plant part, such as leaf, flower, fruit, or stem, from the parent plant. Examples of abscission are the autumnal coloration of leaves and their subsequent shedding and the dropping of some tree fruits following their maturation.

Many factors can initiate a train of events leading to the formation of an abscission zone and shedding of a plant part. Cold, heat, drought, chemicals, and various types of injury can produce abscission (Addicott, 1964). Growth regulators may be used to speed up or retard abscission processes. For example, such crops as cotton must be defoliated if mechanized picking is to be successful, and a loosening of the fruit is desirable for mechanical harvest of grape and of some tree fruits.

ANATOMICAL ASPECTS OF LEAF ABSCISSION

The leaves of most dicotyledonous plants have a structurally favored zone where abscission may occur (Webster, 1968). Some species, such as Coleus, have only one abscission zone, but others, such as citrus and bean, have two. The abscission zone is characterized by the formation of distinct, specialized cell layers that frequently are located at the base of the petiole (Fig. 10–1). The leaves of some species, such as tobacco, are not shed because these species have no abscission zone.

Abscission includes the functions of separation and protection. The process of leaf separation includes changes in the metabolism of the cell wall and in the chemistry of the pectins that form the middle lamella. In most species only the pectins in the middle lamella and some of the cellulose in the primary wall dissolve. In some plants the entire cell wall and the adjacent cell contents disappear (see Addicott,

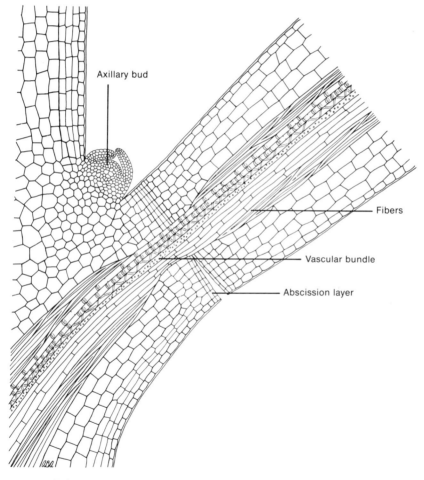

FIGURE 10-1
Typical abscission zone. (Diagram courtesy of F. T. Addicott.)

1970). The formation of the abscission layer is sometimes accompanied by a series of cell divisions proximal to it. Often these new cells differentiate into a periderm and form a protective layer across the scar left by abscission. However, separation and cell division are independent processes. In some plants cell division precedes separation, and in others it occurs after separation; in some species there is no cell division. Generally, the function of the cell division, when it occurs, is to form a protective layer; it is not to cause separation. The process of abscission is not merely a passive one, because metabolic energy must be expended for separation to occur.

HORMONAL CONTROL OF LEAF ABSCISSION

Auxins

Abscission is correlative because it is affected by the physiological status of the other parts of the plant. If a leaf blade is removed from some species, an abscission layer is promptly formed in the petiole and the petioles fall off in two or three days. However, if auxin is applied to the debladed stump, abscission is retarded. The naturally occurring auxin in the leaf blade is probably IAA (see Jacobs, 1968). As a leaf ages there is a decline in the inhibitory effect of auxin and abscission often commences.

There are three main theories that attempt to explain the role played by auxin in leaf abscission. The auxin gradient theory of Addicott et al. (1955) states that a gradient of auxin concentration located across the abscission zone controls abscission. According to this theory, abscission is enhanced when the proximal quantities of auxin (located on the stem side of the abscission zone) are equal to or larger than the distal quantities (located on the leaf side of the abscission zone). The theory also states that abscission is inhibited when large amounts of auxin are located distally from the abscission zone.

A second theory holds that the effect of auxin on abscission is determined by the concentration of the compound (Biggs and Leopold, 1957). Biggs and Leopold (1958) pointed out that since higher concentrations of auxin inhibit abscission but lower concentrations accelerate it, abscission has a two-phase response curve. However, Rubinstein and Leopold (1964) noted that the effect of auxin on abscission varies with time of application.

The third theory is called the two-phase theory. On the basis of the results they obtained with bean explants, Rubinstein and Leopold

(1964) concluded that the leaf abscission response to auxin can be divided into two phases if auxin treatments are applied at different intervals after the removal of the blade. The first stage (induction) is inhibited by auxin and the second stage is promoted by the same concentration of auxin. Concentrations of NAA that retard abscission when applied shortly after explant removal stimulate abscission when applied twelve hours after explant preparation.

These three theories imply that auxin directly initiates abscission. Another possibility is that auxin affects abscission only indirectly, by affecting plant growth.

Experiments by Osborne (1955) showed that diffusates from aging leaves contain materials that stimulate abscission and thus provided proof that substances other than auxins affect the process.

Accelerator Compounds

As a leaf grows older, its auxin level decreases and the inhibitory effect of the auxin on abscission wanes. At the same time abscission-accelerating compounds appear. Some of these compounds are always present and others appear only as the leaf ages.

Several accelerator substances, including ABA, GA_3, ethylene, and perhaps some unknown abscission-promoting factor or factors, are probably active in the abscission process. Some of these compounds appear in increasing amounts as the leaf ages.

GIBBERELLIN. This compound stimulates abscission of debladed petioles of some species (see Jacobs, 1968). However, the small accelerating effect of GA_3 is limited to younger tissues. The action is much more rapid when GA_3 is applied to the young petioles in combination with IAA. The abscission of older petioles is not affected by GA_3, whether IAA is added or not. Therefore, it is likely that gibberellin plays an important role in abscission in intact plants after they have passed the early stages of growth.

Some early work showed that spray application of gibberellin to deciduous woody plants delays development of autumnal foliage color, retards leaf abscission, and stimulates renewal of shoot growth (Brian et al., 1959a). The retardation of abscission by gibberellin is probably indirect because leaf drop is a symptom of aging and senescence, and in these species gibberellin induced a renewal of shoot growth, indicating that the delay of abscission was caused by the rejuvenation of the branches rather than any direct effect of gibberellin.

ABSCISIC ACID.   ABA apparently acts as a general senescence-inducing agent. For many years investigators have obtained from the leaves and fruits of many plant species diffusible substances that accelerate abscission; much of the activity of these diffusates results from the presence of ABA or related compounds. However, the theory that ABA is the normal agent that accelerates leaf abscission remains to be proven (Jacobs, 1968). This compound usually causes abscission in explants, but the leaves of many species of intact plants fail to abscise after application of ABA. To accelerate abscission effectively, high concentrations of the compound must be applied at frequent intervals (Smith et al., 1968).

ETHYLENE.   It has long been known that ethylene is a potent accelerant of abscission, and one theory holds that the auxin-ethylene balance controls abscission (see Carns, 1966). In this system auxin retards the initiation of the process while ethylene stimulates and initiates it. That many auxins stimulate ethylene production in the plant is now common knowledge. Ethylene apparently does not significantly speed abscission until the second stage of the abscission process is reached, indicating that ethylene does not initiate the process but is active in the later stages of abscission. It has been demonstrated that ethylene reduces both the synthesis of and the amount of auxin in leaves (Valdovinos et al., 1967).

OTHER SENESCENCE-INDUCING HORMONES.   This group includes compounds obtained from senescent petioles and plant parts that speed abscission. Some are probably related to ABA; some may also be amino acids that are released during protein breakdown in senescent leaves. Rubinstein and Leopold (1962) found that synthetic alanine and glutamic acid are effective in stimulating abscission in bean plants. These investigators extracted the amino acids from young and senescing bean leaves and applied them to bean explants; only the amino acids from the aging leaves stimulated abscission of the explants.

HORMONAL CONTROL OF FRUIT ABSCISSION

The hormones that are active in leaf abscission are probably also active in fruit abscission and probably have similar mechanisms of action. At first auxin was believed to control fruit abscission. Luckwill (1953b) demonstrated the existence of a positive correlation between the level of auxin in the seed and abscission of the fruit of 'Lane's Prince Albert' and 'Crawley Beauty.' Preharvest fruit drop in these cultivars was thought to be caused by a decline in the auxin level as the seed matured. Abbott (1959) removed the seeds from fruits of 'Cox's Orange Pippin' apple at weekly intervals and showed that fruit abscission is induced only until the end of the June drop. This finding makes Luckwill's theory untenable, because the June drop of fruits occurs six weeks after petal fall and may be caused by competition between fruits for assimilates.

The mechanism of fruit abscission, like that of leaf abscission, remains to be elucidated.

Some Effective Compounds

In their paper on control of abscission in agricultural crops, Cooper *et al.* (1968) list five types of chemicals that induce ethylene production in plants but that have no growth-promoting effect that would retard abscission. The first group includes GA_3 and ABA, which induce the leaves of some species to abscise but do not promote fruit abscission or induce production of ethylene by the fruit. The second group includes ascorbic acid, which, at relatively high concentrations, induces ethylene production and promotes abscission. However, leaves treated with ascorbic acid usually do not produce enough ethylene to cause much abscission. The third group includes cycloheximide (actidione) and other protein-synthesis inhibitors. Low concentrations of these compounds induce ethylene production, which temporarily overcomes any effect of the inhibitors. Group four includes ethephon and copper chelated with EDTA, which induce ethylene production in both fruit and leaves. Since more ethylene is produced in leaves than in fruit, considerable abscission results. Group five includes cotton defoliation chemicals, which produce ethylene by causing chemical injury to the leaves.

Work with bean leaf petiole explants had previously revealed that compounds that accelerate abscission also accelerate ethylene synthesis (Abeles, 1967).

NONHORMONAL CONTROL OF ABSCISSION

As previously stated, many factors can initiate the chain of events that lead to the abscission of plant parts by triggering changes in the level of endogenous hormones. Man has synthesized chemicals that cause plants to initiate the abscission process. Some defoliants or desiccants are effective because they mimic the effect of heat or cold. Many of these compounds are not hormones, but they induce abscission by injuring the leaf and thus cause changes in the levels of naturally occurring plant hormones.

CONTROL OF LEAF ABSCISSION IN AGRICULTURAL CROPS

Harvest-aid Practices Used in Cotton Production

In the past cotton bolls have been harvested by hand. However, in areas where the costs of harvesting by hand are high, the use of a machine harvester can reduce production costs. For efficient use of either the spindle picker (a harvest machine with rotating spindles that picks the cotton only from open bolls) or the stripper picker (a harvest machine that strips the entire plant except the main stems), much of the foliage must be removed before harvest. Harvest-aid practice usually consists of treating the cotton plant at the proper time to faciliate harvest by defoliation (Fig. 10–2). Chemical harvest aids are presently used on more than 75 percent of the cotton acreage in the United States (Walhood and Addicott, 1968). More than seven million acres of cotton are defoliated annually in the United States alone (Carns, 1966). In areas where cotton is spindle-picked, the use of chemicals in harvesting becomes increasingly essential as plant size increases. Either leaf moisture must be reduced or leaf fall must be induced before harvest to derive maximum benefit from the spindle picker.

The amount of foliage may be reduced by the use of either defoliants or desiccants. Defoliants induce leaf fall and must be applied seven to fourteen days before harvest so that the induced abscission process may be completed. Desiccants cause the foliage to lose water. The leaves, stems, and even branches of the plants are sometimes killed so rapidly by desiccants that an abscission layer has insufficient time to develop, and the drying leaves thus remain attached to the plant. Desiccants require one to three days to act before harvest can be started. The advantage of desiccants over defoliants is that desiccants can be applied at a later date than defoliants. Thus additional time is gained during which the leaves continue to function and to contribute to seed and fiber quality.

Of the several hundred chemicals that have been applied under

FIGURE 10-2
Cotton defoliated by sodium chlorate at a rate of 5 lb per acre. *Top,* defoliated cotton in center is surrounded by nondefoliated cotton. *Bottom,* close-up of defoliated cotton. (Photographs courtesy of V. T. Walhood.)

field conditions to defoliate and desiccate cotton, fewer than ten are now in general use (Table 10–1). Desiccants and defoliants are applied either by ground machine or by aircraft.

Paraquat is a relatively new compound that is now recommended

TABLE 10-1

Generalized Guide for Use of Chemicals in Harvesting of Cotton

Chemical	Principal Formulation (Percent)	Nature of Formulation	Suggested Rate of Application (per acre)
DEFOLIANTS			
Dusts			
Sodium chlorate (with fire suppressant)	12.0	dust	20–33 lb
	20.0	dust	15–25 lb
	23.5	dust	15–20 lb
Tributylphosphorotrithioate	5.0	dust	15–25 lb
	7.5	dust	15–25 lb
Tributylphosphorotrithioite	5.0	dust	20–40 lb
Sprays			
Ammonium nitrate	39.5	liquid	10–15 gal
Ammonia	100.0	anhydrous pressurized liquid	100 lb
	4.7	liquid	7–12 gal
	18.2	liquid	1.5–2.0 gal
	18.5	liquid	1–2 gal
Sodium chlorate (with fire suppressant)	19.0	liquid	1–2 gal
	22.0	liquid	1–2 gal
	28.0	liquid	1.0–1.3 gal
	40.0	wettable powder	7–10 lb
Tributylphosphorotrithioate	70.5	liquid	1.3–2.0 pt
Tributylphosphorotrithioite	75.0	emulsifiable concentrate	1–2 pt
DESICCANTS			
Sprays	39.5		
Ammonium nitrate	44.6	liquid	10–15 gal
Arsenic acid	75.0	liquid	1.0–1.5 qt
1,1'-Dimethyl-4,4'-bipyridinium (paraquat)	29.1	liquid	0.5 lb

SOURCE: Adapted from *Cotton Pest Control Guides*, National Cotton Council, 1968.

for use as either a defoliant or a desiccant. It acts very quickly under warm, sunny conditions. Discoloration of leaves occurs within a few hours after application and is followed, within the next twenty-four hours, by wilting and browning. When used as a desiccant, the compound is usually applied when 80 to 95 percent of the bolls are open and the remaining bolls to be harvested are mature. When used as a defoliant, paraquat is combined with another defoliant, such as phosphate or chlorate. Application is made when at least 60 to 70 percent of the bolls are open and the remaining bolls to be harvested are mature.

Defoliation of Bean Plants

The effectiveness of the mechanical harvesting of varieties of snap bean with dense foliage is greatly decreased. In Israel Palevitch (1970) demonstrated that a preharvest treatment with ethephon at a concentration of 1,000 ppm achieves nearly complete defoliation of snap bean, although some pod abscission and yellowing occur, lowering the marketable yield. Palevitch suggests that lower levels of ethephon, combined with a delay in the preharvest treatment to within four or five days of harvest, should limit the adverse effect of the compound on the pods.

Inducing Abscission in Woody Plants

Nursery plants grown in some areas of the United States can be shipped for fall planting if digging is done early. Early digging also decreases the danger from early fall freezes. In many locations nurserymen must dig the deciduous nursery stock before natural leaf abscission occurs because plants stored with leaves attached often develop leaf decay, which damages the bark and buds. Hand-stripping of leaves before digging is a common practice. A chemical defoliant that would induce leaf abscission without damaging bark or buds would be very useful.

Earlier researchers (Chadwick and Houston, 1948; Pridham, 1952) obtained promising results using Nacconol NR (an alkylarylsulfonate). Pridham used the compound at a concentration of 4.0 percent and also found endothall at a concentration of 0.3 percent to be an effective defoliating compound. However, the response of individual types of nursery stock to these agents varied considerably.

Potassium iodide has been shown to induce premature leaf abscission in intact leguminous plants (Herrett et al., 1962). Later, Larsen (1966) showed that potassium iodide at concentrations of 0.15 to 0.6 percent generally results in excellent defoliation of woody plants. The compound is also effective on 'Yellospur' apple, 'Hi-Early' apple, 'Italian Prune' plum, 'Bartlett' pear, 'Montmorency' cherry, 'Eva

FIGURE 10-3

Top left, 100 percent defoliation of *Spiraea billiardi* 3 weeks after spray application of 0.3 percent potassium iodide. *Top right,* control. *Bottom left,* 100 percent defoliation of 'Yellospur' apple 3 weeks after 2 consecutive spray applications (one week apart) of 0.3 percent potassium iodide plus 0.5 percent Nacconol NR. *Bottom right,* control. Not shown: application of potassium iodide and Nacconol NR alone resulted in 45 and 30 percent defoliation, respectively. (After F. E. Larsen, 1966.)

Rathke' weigela, *Spiraea billiardi* and *Prunus mahaleb* (Fig. 10-3). Poor results are obtained with 'Rome Beauty' apple and seedlings of French apple. A concentration of 0.15 or 0.3 percent is usually as effective, and occasionally more effective, than a concentration of 0.6 percent. Even lower concentrations may be satisfactory for some plants. The time required for complete defoliation varies from two to six weeks, depending on the variety of plant and where it is being grown. However, the leaves usually become loose enough in half this time to be removed using the normal digging and handling procedures, so it is not necessary in commercial practice to postpone digging until defoliation is complete. Potassium iodide at a concentration of 0.6 percent often kills buds and desiccates the bark, effects that become visible during the storage period. Part or all of the tops of some plants are also killed by this high concentration.

Chemical defoliation can be accomplished without inflicting damage if potassium iodide is repeatedly applied at low concentrations (Larsen, 1966). The initial application hastens leaf senescence and initiates the abscission process and is supplemented by subsequent applications. A single application may be sufficient with easily defoliated plants, but more resistant types require repeated applications. The addition of DEF at a concentration of 0.25 percent or Nacconol NR at a concentration of 0.5 percent to the potassium iodide frequently hastens defoliaton.

The effectiveness of potassium iodide in inducing abscission is increased by the addition of alanine at a concentration of 2.0 percent (Larsen, 1967). In a laboratory test with seedlings of French crab and *Pyrus communis* pear, application of potassium iodide alone at a concentration of 0.3 percent induced 25 and 60 percent defoliation of the two varieties, respectively. The addition of alanine at a concentration of 2.0 percent increased the defoliation of the two varieties to 60 and 90 percent, respectively. Alanine alone at a concentration of 2.0 percent resulted in little stimulation of leaf abscission in either variety.

Another iodine-containing compound, bromodine [40 percent alkyl (C_{12}) (ethylcycloimidinium) 3-hydroxy, 3-ethyl sodium alcoholate, 2-methyl sodium carboxylate-tridecylopolyoxyethylene ethanol-iodine complex and 20 percent triethanolamine sulfonate tridecylopolyoxyethylene ethanol-bromine complex] is an effective defoliant for the nursery stock of some species of woody plants. For such easily defoliated types as peach, pear, and cherry, one to three applications of a 0.25 percent solution at five- to seven-day intervals usually produce 70 to 100 percent leaf abscission (Larsen, 1970a). For more difficult types, such as apple, two to three applications of the compound at a concentration of 0.5 percent are effective. Thorough wetting of the leaves is essential for favorable results.

SADH is sometimes applied to nursery stock to control plant size. When subsequent application of defoliants is made, the amount of leaf abscission is usually not affected (Larsen, 1969a). However, SADH treatment sometimes retards abscission (as when Nacconol NR is applied to 'Law Red Rome' apple); it sometimes stimulates abscission when defoliating compounds are subsequently applied (as when potassium iodide is applied to 'Spartan' apple).

ABA is also an effective defoliant. Larsen (1969b) sprayed six tree fruit cultivars with the compound at concentrations of 0, 500, 1,000, and 2,000 ppm. He found that the higher the concentration of ABA, the more complete the defoliation.

Ethephon also causes leaf abscission in deciduous nursery stock. In the northwestern United States, three applications of ethephon at a concentration of 2,000 ppm, made at weekly intervals, were found to be effective on apple (Larsen, 1970b). 'Italian Prune' plum required two applications of the compound at a concentration of 2,000 ppm, whereas a single application of the compound at concentrations of 500 and 1,000 ppm produced satisfactory results on seedlings of 'Bartlett' pear and mazzard cherry (*Prunus avium*), respectively. In New York State, Cummins and Fiorino (1969) noted that ethephon causes some delay in growth during the spring following treatment.

Defoliants have been used in the war in Vietnam to eliminate or reduce the concealment afforded by jungle growth and other vegetation. Enough herbicide was used in Vietnam in 1967 to treat approximately 965,000 acres, but since many areas were retreated the total defoliated area was significantly less (Boffey, 1968). A 50:50 mixture of n-butyl esters of 2,4-D and 2,4,5-T was used for jungle defoliation. A combination of picloram and 2,4-D in a low-volatile amine formulation was used to control growth of woody plants and in areas in which accurate placement of spray was essential.

Inducing Abscission in Hydrangea

Flower bud initiation in the common florist hydrangea, *Hydrangea macrophylla*, occurs in late summer or early fall. A cold period of six to eight weeks is then required for flower bud development before plants can force normally. The cold requirement can be satisfied by placing the plants in dark storage at 1° to 4°C or by storing them outdoors in a cool location. However, the plants should be defoliated before storage to prevent growth of the fungus *Botrytis* sp. on dead or dying leaves. This fungus can cause infection of flower buds, resulting in distortion of flowers after forcing. Ethylene can be used to induce defoliation (Post, 1947). Some Pacific coast growers rely on the natural defoliation caused by frost, but often the frost may be too light and may cause incomplete defoliation.

Kofranek and Leiser (1958) found that Vapam (31 percent solution of sodium methyldithiocarbamate), used as a vapor treatment,

FIGURE 10-4

Effect of Vapam, Folex, and zinc chloride on defoliation of *Hydrangea* cv. 'Europa.' *Top row, left to right:* plants following 16-hour exposure to Vapam at concentrations of 2.50 ml/liter, 3.75 ml/liter, and 5.00 ml/liter, and the control. *Bottom row, left to right:* plants that received direct applications of Folex at concentrations of 1.0 percent, 1.5 percent, 2.0 percent, and 4.0 percent zinc chloride in water. Inflorescences of plants treated with highest concentrations of Vapam and Folex exhibit wilting and necrosis. Zinc chloride-treated plants show typical scorching of leaf margins and incomplete defoliation. Photographed 15 days after treatment. (After A. M. Kofranek and A. T. Leiser, 1958.)

and Folex (75 percent emulsion of tributylphosphorotrithioite) cause thorough defoliation. Vapam at concentrations of 2.5 ml/liter to 5.0 ml/liter, applied at the rate of 0.5 pint per square foot of soil and vaporized in 10 cubic feet of space, results in complete defoliation after the plants have been exposed for sixteen hours (Fig. 10–4). However, the highest concentration of Vapam causes root injury, which results in wilting of the plants at the time of forcing. A 1 to 2 percent Folex emulsion applied as a spray produces 91 to 100 percent defoliation, depending on concentration. The lowest concentration produces no injury to plants during treatment or storage or during the forcing period.

Delaying Abscission in Ornamentals

A method to delay abscission, especially of cut flowers and nursery stock, is often desirable. In 1936 La Rue reported that the fall of coleus leaves can be delayed by treatment with synthetic hormones. His results stimulated further work on the prevention of abscission of flowers as well as of other plant parts.

POINSETTIA. The leaf and bract drop of poinsettia plants that are exposed to the environmental conditions of the greenhouse and home during the Christmas holiday season often drastically shortens flower longevity. Spray applications of 2,4,5-trichlorophenoxyacetamide at a concentration of 6 ppm resulted in significant reductions in both leaf and bract loss (Carpenter, 1956). In one laboratory experiment treated plants retained approximately half their leaves and bracts after forty-nine days, whereas the untreated control plants retained only 13 percent. A slight epinasty of bracts and leaves developed within one day after plants were sprayed with the chemical. Under some conditions concentrations lower than 6 ppm must be used to prevent extreme epinasty.

HOLLY. Leaves of American holly (*Ilex opaca*) and English holly (*Ilex aquifolium*) often drop off during shipment, usually when branches are exposed to moist air. However, the losses cannot be prevented by storage in dry air because the leaves soon dry out and become discolored. When such an auxin as NAA is sprayed on the foliage, the leaves and berries remain attached much longer (Gardner and Marth, 1937). Leaves of treated English holly remain attached until they finally turn dark and must be discarded.

Holly grows very well in the northwestern United States, but before auxin treatment was introduced the plant had only a local market. The perishable product could not be moved until a few days before Christmas, when it was rushed to markets before the leaves and berries fell off. After auxin was utilized to prevent leaf drop, the market was rapidly expanded; at present the Northwest produces most of the holly in the United States.

FLOWERING CHERRY AND DOGWOOD. The blossom display of varieties of the Oriental or flowering cherry can be prolonged by spray application of NAA at a concentration of 10 ppm (Wester and Marth, 1950). Experiments performed on trees in the Tidal Basin in Washington, D.C., revealed that the most effective time for spraying is just as the trees come into full bloom. There is a varietal difference in response to the hormone spray. Treated 'Yoshino' and 'Akebono' trees were found to retain 35 to 80 percent of their blossoms three to seven days longer than the unsprayed trees; thirteen days after treatment, 'Kwanzan' trees still retained 23 percent of their flowers; untreated trees retained only 4 percent of their flowers.

Sprays of 4-CPA at a concentration of 10 ppm applied to white flowering dogwood (*Cornus florida*) caused the petal-like flower bracts to adhere four to six days longer than those on unsprayed plants. However, some foliage injury did occur.

CHEMICAL THINNING OF FLOWERS AND FRUITS

Thinning of grape and of the fruit of certain species and varieties of fruit trees is a necessary commercial practice. Thinning of fruit trees, especially most varieties of apple, eliminates or alleviates biennial bearing and also enhances fruit size, color, and quality. Thinning of grape results in less compact clusters that are less likely to rot. Hand-thinning is one of the greatest single costs in apple production, particularly in central Washington State (Batjer, 1965). It is very difficult to thin certain varieties of fruit adequately by hand to prevent the alternate-bearing tendency. Thinning of some fruit species is necessary to prevent excessive June drop, to balance the crop load of fruit in proportion to the vegetative development of the tree, and to increase the size of the individual fruits.

Apple

In the early 1930's Auchter and Roberts (1934) tested several chemicals in an attempt to find a compound that would prevent fruit set in some varieties of apple. They found that tar oil distillates kill the flower buds if applied at the cluster bud stage. A subsequent history of thinning sprays up to 1964 has been compiled by Batjer (1964, 1965). Much attention has been given more recently to the use of chemical sprays during bloom or during the early postbloom period to reduce fruit set and to eliminate, at least partially, the necessity of hand-thinning.

The most important chemicals used in thinning apples are listed in Table 10-2. Ethephon also shows promise as a thinning agent. Application of the compound at concentrations of 100 to 300 ppm either

TABLE 10-2

Chemicals Used for Thinning Apple

Symbol or Common Name	Chemical Name	Trade Name
DNOC*	Sodium 4,6-dinitro-*o*-cresylate	Elgetol
DNOC*	4,6-dinitro-*o*-cresol	Dinitro-dry
NAA	naphthaleneacetic acid	Fruitone-*N*, Fruit Fix, Fruit Set, Stafast, Kling-Tit
NAAm	naphthaleneacetamide	Amide-Thin and Anna-Amide
Carbaryl	1-naphthyl *N*-methyl carbamate	Sevin

* The same symbol is generally used for both forms of dinitro chemicals. One pint of Elgetol contains the same amount of chemical as 0.5 lb of Dinitro-dry.

SOURCE: Adapted from L. P. Batjer, *Fruit Thinning with Chemicals.* Agr. Information Bull. no. 289, U.S. Dept. Agr., Agr. Res. Serv.

before or during bloom or ten days after bloom was found to reduce set in several varieties of apple. However, size of treated fruits at harvest was smaller than that of the controls (Edgerton and Greenhalgh, 1969). Application of the compound at concentrations of 100 to 2,000 ppm caused abscission of almost all flowers and fruit with little or no phytoxicity. Attention to the proper timing and concentration of a specific compound is important if satisfactory thinning is to be obtained.

TIMING.   Applications of DNOC to apple at full bloom or at petal fall two or three days later result in approximately the same amount of thinning (Batjer, 1965). Therefore, timing is relatively not critical and the grower usually has three to four days during which to apply the compound under most conditions. DNOC is not useful as a postbloom thinning agent.

NAA, NAAm, and Sevin can be used as postbloom thinners (Southwick *et al.*, 1964; McKee and Forshey, 1966). These compounds can partially thin late-ripening varieties at bloom or at petal fall, but they are more effectively used as postbloom sprays. Postbloom thinning affords an opportunity to evaluate the degree of fruit set before applying sprays, and delayed application of the compound may mean that the danger of frost occurring before spraying can be avoided. NAA and NAAm are usually applied to late maturing varieties between ten and twenty-one days after full bloom. There is a wide latitude in time of application of these two compounds on late ripening varieties. Most varieties of apple are suitable for thinning during a thirteen-day period when fruits range from 5 to 16 mm in diameter (Batjer, 1968).

Timing is critical when the growth regulators NAA or NAAm are used on early ripening varieties. These compounds should be sprayed on early varieties at the petal fall stage since application of sprays later than ten days after bloom may result in fruit splitting and premature ripening.

There is also some latitude in timing of application of Sevin. Work performed in Australia and Washington State has demonstrated that application of Sevin sprays applied fifteen to twenty-seven days after full bloom thinned several varieties of apple (Batjer and Thompson, 1961). Sevin can probably be used on summer varieties at a later stage of fruit development than can NAA or NAAm.

CONCENTRATION. The proper concentration of thinning agent to use varies with weather conditions, tree vigor, variety, and other factors. Therefore, a safe rule is always to get specific recommendations from local authorities before spraying.

A good thinning agent is one that thins moderately but does not over- or underthin. Batjer (1965) found that Sevin is very effective when applied at a concentration of 1.25 pounds per 100 gallons of water to 'Delicious' and 'Winesap' and at a concentration of 1.5 pounds per 100 gallons of water to 'Golden Delicious.' Varieties that tend to set heavy crops can usually be thinned with greater safety and dependability than those that do not. The principal varieties of apple, grouped according to their response to thinning sprays, in order of their effectiveness, are listed in Table 10–3. DNOC is not included in the table because this material is not generally suitable for the more humid fruit areas of the United States, although it has been used extensively in the arid regions of the Northwest.

Application of NAA sometimes results in epinasty and dwarfing of foliage. To minimize these effects it is desirable to apply NAA as late as possible during the postbloom period. When late sprays are applied to early maturing varieties of apple, premature ripening and splitting of the fruit are likely to occur. For this reason, it is best to use NAA on fall and winter varieties, but the compound should be used for the earlier varieties only when Sevin or NAAm are not effective.

NAAm is less injurious to the foliage and is less likely to overthin than NAA when applied at the postbloom stage. NAAm should not

TABLE 10-3

Important Varieties of Apple Grouped
According to Their Response to Thinning
Sprays

Variety	*Chemicals**
HARD TO THIN	
Early McIntosh	A-N-S
Golden Delicious	N-S-A
Wealthy	N-A
Rome Beauty	S-A-N
Baldwin	N-A
INTERMEDIATE	
York Imperial	A-N-S
Jonathan	S-N-A
Yellow Newtown	A-S
Grimes Golden	A-S-N
Ben Davis	S-A-N
EASY TO THIN	
Delicious	S-N
McIntosh	S-A-N
Winesap	S-A-N
R. I. Greening	S-A-N
Stayman	A-N

* A = NAAm, N = NAA, S = Sevin. Chemicals
for each variety are listed left to right in order of
effectiveness. This listing is general in nature. Specific recommendations should be obtained from
local authorities.

SOURCE: Adapted from *Fruit Thinning with Chemicals*. Agr. Information Bull. no. 289, U.S. Dept. Agr.,
Agr. Res. Serv.

be used on the 'Delicious' variety after petal fall; if it is, young fruit
becomes stunted and does not abscise, and final fruit size is greatly
reduced.

It is often desirable to apply two sprays to hard-to-thin varieties,
and to other varieties only when the conditions are conducive to a
heavy set. Enough time must elapse between the two sprays for the
effects of the first spray to be evaluated before the second spray is

applied. A common practice in the northwestern United States is to apply a DNOC spray to such difficult-to-thin varieties as 'Golden Delicious' during the bloom period and to follow this application fourteen to twenty-one days later with a postbloom spray of either Sevin or NAAm. In the humid areas of the Midwest and East, NAAm should be applied at petal fall, followed two weeks later by an application of Sevin or NAA.

Thinning agents are usually applied as dilute sprays. However, in Maryland, Rogers and Thompson (1969) conducted a four-year experiment with 'Rome Beauty' apple and obtained excellent thinning by using concentrated sprays. Sevin, in the form of dilute sprays and also as sprays that were three to thirty-three times more concentrated than the dilute sprays, was applied to mature trees.

Sprays that were thirty-three times more concentrated than the dilute sprays were applied at the rate of 2.5 pints per 100 feet of orchard row using an Econ-O-Mist air blast sprayer. Approximately 1.25 pints was applied per tree. All sprays produced excellent thinning in three of the four years. In a one-year trial with 'Jonathan' apple, both Sevin and NAAm resulted in good thinning when applied as dilute sprays or as sprays that were either three or six times more concentrated. When NAA was applied to 'Golden Delicious,' the dilute spray was found to be more effective than more concentrated sprays.

EFFECTS. Chemical thinning results in an increase in fruit size at harvest that is roughly proportional to the amount of thinning done. The increase in size of fruit on thinned trees becomes apparent early in the growing season. However, applications of NAA or NAAm temporarily reduce the growth rate. Inhibition of growth of 'Golden Delicious' and 'Early McIntosh' apple as a result of application of these compounds was evident within five days or less and was measurable for ten days thereafter whether or not hand-thinning was employed simultaneously (Southwick et al., 1962). The eventual size at harvest of fruit that has been thinned with chemical sprays may sometimes be smaller than that of fruit that has been thinned by hand or by mechanical means.

Chemical thinning sprays also have a marked effect on alternate bearing. Apple growers have found that heavy and light crop years usually ensue when excess fruit is set during the bearing, or "on" year. During heavy overcropping, or when the quantity of fruit in relation to the amount of foliage is excessive, fruit bud formation is reduced or almost entirely prevented. As a result, in the season following the bearing year, reduced bloom results in a low crop yield and the consequent formation of too many fruit buds. Once an alternate fruiting habit is established, years of heavy bloom and heavy crop are followed by years of relatively light bloom and light crop.

Chemical thinning sprays reduce fruit set relatively early in the growing season and allow the tree to form more fruit buds for the next year's crop. Thompson (1957) suggested that the amount of blooming that occurs during the year following a NAA or NAAm spray may be a result of both the thinning action of the spray and the chemical's direct effect on fruit bud formation. Experiments conducted by Harley et al. (1958) supported this conclusion. The fruit-thinning variable was eliminated by partially defoliating trees of 'Golden Delicious' apple during the "off" year. When NAA was applied to "on-year" trees, formation of fruitful buds was increased even when the sprays had no thinning effect. It is probable that in some orchards the chemicals effect flower initiation only indirectly through their thinning action; in others they exert a direct effect on flower bud initiation.

MECHANISM OF ACTION. Elgetol has a caustic effect on the stem and other exposed parts of the flower. It also prevents pollen germination and inactivates the pollen tubes growing down the style of the pistil (Hildebrand, 1944).

Luckwill (1953a) advanced the theory that NAA causes thinning by effecting seed abortion. Later work showed that only 0.2 percent of the radioactive NAA applied to the leaves is recovered from the seed after five days, and that none of this compound is in the form of unmetabolized NAA (Luckwill and Lloyd-Jones, 1962). According to these investigators, this result supports the view that seed abortion and the consequent abscission of the seedless fruit are not caused by the direct action of NAA itself but rather by a breakdown product that, unlike NAA, has no auxin properties. However, the fact that in some orchards no relationship has been found between the number of viable seeds and the response to thinning casts some doubt on Luckwill's theory.

Williams and Batjer (1964) have concluded that Sevin or its metabolite in the vascular system probably interferes with the movement of vital chemicals to or from the fruit. Certain important fruit growth processes cease and abscission follows. These investigators consider seed abortion to be a secondary phenomenon and not a prerequisite for abscission. When they applied ^{14}C-labeled Sevin to either a leaf or fruit, most radioactivity appeared in the vascular tissues of the fruit. Williams and Batjer suggest that the apparent absence of activity in the seed in all the experiments they conducted using ^{14}C-labeled Sevin and the absence of correlation between the number of viable seeds in persisting fruit and the percent thinning indicate that seed abortion is not the primary cause of fruit abscission.

Stone Fruits

Thinning of most varieties and species of stone fruits is necessary to produce fruit of optimal size and quality. Thinning of stone fruits has been somewhat less successful than that of apple. Of the chemicals used on apple, DNOC has shown the most promise (Batjer, 1965). NAA has proved to be too erratic and NAAm has proved to be only slightly effective. Sevin has proved to be completely ineffective on stone fruits. Other materials that have been widely tested include CIPC, NPA, 3-CP, and N-(3,4-dichlorophenyl) methacrylamide.

PEACH. Most commercially important varieties of peach are self-fruitful, and if favorable weather exists at bloom many varieties set excessively heavy crops, thus necessitating thinning to obtain fruit of satisfactory size and quality. DNOC is probably the best available material for the chemical thinning of peach. The main objection to this compound is that it is a blossom spray, and when it is applied the danger from spring frost still exists in most locations.

DNOC should be sprayed when 60 to 90 percent of the blossoms are open (Batjer, 1965). Exact timing is required because if the sprays are delayed until full bloom or slightly thereafter, less thinning is obtained. Bloom sprays have been most effective in seasons when the blossoms are continually open for three to five days. In areas where there is insufficient winter chilling to break the rest period of the trees, bloom may continue over a relatively long period, and DNOC bloom sprays are then less effective.

DNOC is usually applied at a rate of one to two pints per hundred gallons. The higher rate is used on varieties that set heavily. Over- and underthinning with DNOC occurs more frequently with peach than with apple. In general, if 20 percent of the blossoms of peach trees set fruit, an adequate commercial crop of peaches can be expected (Batjer, 1965). Experiments that he conducted in Washington State for twelve years showed that approximately 60 percent of the necessary thinning was accomplished by spraying.

More recent experimental evidence indicates that sprays of ethephon may be useful for thinning peaches. Application of the compound at a concentration of 300 ppm to the variety 'Maygold' when 80 to 100 percent of the blossoms were open effectively thinned blossoms and young fruit without producing any phytotoxic effects (Buchanan and Biggs, 1969). Higher concentrations caused gumming in the leafy branches and dwarfing of leaves that developed within three weeks of treatment. Application of ethephon at a concentration of 30 ppm to 'Early Amber' peach when the endosperm was changing from the free nuclear to the completely cellular stage resulted in satisfactory thinning (Buchanan et al., 1970).

In other experiments with 'Redhaven' peach, application of ethephon at concentrations of 50 and 150 ppm from one to four weeks after full bloom caused abscission of some fruits (Edgerton and Greenhalgh (1969). The peaches grown on treated trees were larger at harvest than those grown on trees that were hand thinned. Edgerton and Greenhalgh (1969) have also reported a reduced fruit set in 'Earlyvee' and 'Reid's Seedling,' varieties of peach grown in Australia, following application of ethephon at a concentration of 200 ppm one month after bloom.

Much effort has been devoted to the development of a postbloom thinner, since there is always the danger that a killing frost will occur following the application of DNOC. Application of NPA at concentrations of 200 ppm to 'Elberta' peach and 300 ppm to such heavy-setting varieties as 'Redhaven' has proved to be as satisfactory in the state of Washington (Batjer, 1965). Sprays should be applied no later than five to seven days after full bloom; later sprays produce less thinning and may cause foliage injury.

NPA has not proved to be satisfactory in other areas of the United States in which stone fruits are grown. There is a need for a compound that will thin peaches effectively when applied from ten to twenty-eight days after full bloom. N-(3,4-dichlorophenyl) methacrylamide, CIPC, and 3-CP have been applied late in the postbloom period, but the results have been inconsistent and fruit and foliage have occasionally been injured.

NAA, one of the first growth regulators to be tested on peach for its thinning effect, has usually produced erratic results. Some successes were obtained when compound was applied thirty to forty-five days after bloom. Leuty and Bukovac (1968) showed that fruits that contain endosperm undergoing the various stages of cytokinesis, especially the early stages, are most likely to be sensitive to NAA. Further investigation might prove NAA to be a useful chemical for thinning peaches.

Brown et al. (1968) have suggested a possible new approach to the thinning of peach and other *Prunus* species by demonstrating that flower bud formation on these trees can be curtailed or completely prevented in the season following the use of GA₃, depending on the concentration used. The cling peach was used for these experimental trials because it generally flowers profusely and sets excessive numbers of fruits that must be reduced by hand thinning to obtain marketable fruit size at harvest.

Peaches grown in California's San Joaquin Valley begin to form flower buds for the next year in June or July. If cling peaches are sprayed with GA₃ in late July to coincide with the critical period of flower bud differentiation, much thinning results the following year. If spraying is done when the current crop is still on the trees, that fruit can be harvested at the same time and is the same size and color as fruit from unsprayed trees.

Application of gibberellin at a concentration of 50 ppm to the variety 'Paloro' reduced the crop enough that no supplemental hand-thinning was necessary. Application of this concentration to 'Peak' and 'Halford' resulted in moderate-to-good thinning, but there was little or no thinning of 'Fortuna' or 'Loadel,' showing that higher concentrations of the compound are necessary for these two varieties.

Spraying during fruit bud formation has the same disadvantage as blossom thinning: There is no chance to evaluate the amount of set before treatment, and fruit-set problems caused by frost and other factors can often be of paramount importance.

$$O-CH_2-\underset{\underset{\textstyle O}{\|}}{\overset{\overset{\textstyle CH_3}{|}}{C}}-OH$$

3-CP

$$O-CH-\underset{\underset{\textstyle O}{\|}}{\overset{\overset{\textstyle CH_3}{|}}{C}}-NH_2$$

3-CPA

3-chlorophenoxy-α-propionamide (3-CPA) has produced good thinning in several varieties of peach (Albrigo and Christ, 1968; Beutel, 1968). The compound should be applied at concentrations of 150 to 300 ppm when seed length is 7 to 10 mm. Since there is little translocation of the chemical within the tree, good spray coverage is essential. The chemical acts by increasing the June drop, so hand-thinning should be avoided until the June drop is completed. The hormone removes the smaller and medium-sized peaches, leaving the larger peaches on the tree. Application of 3-CPA at a concentration of 300 ppm to the varieties 'Halford,' 'Peak,' 'Paloro,' 'Carolyn,' 'Suncrest,' and 'July Elberta' in California and to 'Sunhaven' and 'Triogem' in New Jersey produced excellent thinning. In Florida, adequate thinning of 'Early Amber' peach was accomplished by applying 3-CPA at a concentration of 300 ppm during cytokinesis (Buchanan et al., 1970).

The best thinning effect is obtained from 3-CPA when both foliage and leaves are given a thorough spray coverage (Martin and Nelson, 1969). Studies with ^{14}C-labeled 3-CPA have revealed that there is very slow movement of the compound in the plant, and therefore poor spray coverage results in erratic thinning, especially if an effective concentration of 3-CPA fails to reach the fruit until it has grown beyond the responsive stage of development.

There are considerable varietal differences in the response of peach to 3-CPA, and timing of application is critical (Stembridge and Gambrell, 1969).

MH has also shown some promise as a thinning agent. Application of the compound at concentrations of 500 ppm or higher at various stages of bloom reduces fruit set in peach without concomitant injury to foliage (Langer, 1952). MH probably remains effective longer than do the dinitro compounds.

APRICOT. Large apricots are more valuable than small ones, and any practice that increases fruit size is desirable even though it may result in a moderate reduction in yield. DNOC is the most promising chemical (Batjer, 1965), and the spray is most effective in reducing set if applied when 60 to 75 percent of the blossoms are open. If the spray is delayed until all blossoms are open or are slightly past full bloom, considerably less thinning is obtained. The desirable range of concentration of DNOC is 0.67 to 1.33 pints per 100 gallons, depending on weather conditions and the setting tendency of the variety. If the weather during the bloom period and for several days following the spray application is cool and rainy, more thinning will be obtained than if the weather is warm and dry.

PLUM AND PRUNE. Some varieties of Japanese plum can be effectively thinned by spray application of DNOC (Batjer, 1965). The most effective concentration ranges from .33 to .67 pints per 100 gallons—the higher dosage should be used for varieties that tend to set heavily, such as 'Beauty' and 'Duarte.' If a large crop is expected, the sprays should be applied slightly before full bloom to obtain adequate thinning, but if unfavorable weather has caused a crop reduction, the sprays should be delayed until a day or two after full bloom. Results may vary from moderate thinning to overthinning, depending on spray, timing, and weather conditions.

DNOC has also proved to be the most effective chemical thinner for prune. The range of effective concentrations is similar for both prune and peach, and the results obtained with both fruits are similarly affected by timing and weather conditions.

There is some evidence that 3-CP shows promise as a postbloom thinner for plum and prune (Batjer, 1965).

SWEET CHERRY. DNOC at a concentration of .33 to .67 pints per hundred gallons will thin sweet cherry but foliage injury frequently results. In general, any benefit gained by thinning is usually accompanied by a substantial reduction in yield. Thinning of sweet cherry should be done only on an experimental basis.

PEAR. Chemical thinning of pear has generally been unsuccessful. DNOC seems to be the most effective compound, but if rain or high humidity follows application, severe leaf burning often occurs. No more than 0.67 to 1.00 pint per hundred gallons should be applied according to the schedule recommended for apples (see p. 336). DNOC has the weakness of being a blossom spray, and when fruit set is limited by lack of cross-pollination or unfavorable weather, thinning sprays may result in a greater set reduction than desirable (Batjer, 1965).

Such varieties of pear as 'Bosc,' 'Winter Nelis,' and 'Bartlett' set heavy crops that do not develop to good marketable size unless thinning is done. Other varieties require no thinning. For example, in California, the fruit set of 'Bartlett' pear is seldom heavy enough to require thinning.

Olive

Olive (*Olea europaea*) has a strong tendency toward alternate bearing. In some years the trees set an extremely heavy crop, probably as a result of favorable weather conditions. At least half the fruit should be removed from overcropped olive trees to gain the benefits of thinning. In California, because the industry is based on the production of large canning olives, the small fruit that results from a heavy set is considered unprofitable. Such fruit is left on the tree until winter, when it is harvested for oil extraction.

Heavy crops also mature late in the season. In California the fruit of overcropped trees does not reach the desired stage for canning until late November, when there is danger of frost injury (Hartmann, 1952). When an olive tree sets an extremely heavy crop, it nearly always fails to bloom the following year, and an alternate-bearing pattern is established; consequently, profitable crops are not produced either on the "on" or "off" years. Thinning is therefore desirable on the "on" years. However, hand-thinning of small fruits of large trees early in the season is a very laborious task when a large acreage is to be covered.

In California a reduction in the fruit set of olive has been accomplished by spraying NAA at concentrations of 40 to 50 ppm at full bloom. Concentrations of 75 to 100 ppm result in complete prevention of fruit set, which is desirable when olives are grown as ornamentals (Hartmann, 1952). However, spray thinning at the blossom stage is not recommended because the need for thinning is impossible to predict at that time.

Application of NAA at concentrations of 100 to 125 ppm results in satisfactory fruit thinning if applied fourteen to twenty-five days after full bloom, when the fruit is 3 to 5 mm in diameter (Hartmann, 1952). If the hormone is applied at concentrations of 150 ppm or higher, killing of terminal buds results. The chief benefits of application of NAA are increased fruit size, increased flesh-pit ratio, higher oil content, earlier fruit maturity, and a reduction in the tendency toward alternate bearing.

In the Mediterranean regions small fruit size and the alternate-bearing habit are the most serious problems of olive growers. In Israel Lavee and Spiegel (1958) performed spray thinning experiments with 'Suri' and 'Manzanillo,' that country's most widespread and important varieties. Their results generally agreed with those obtained by Hartmann (1952) in California. Sprays of NAA applied six to eighteen days after full bloom resulted in a 25 to 35 percent increase in fruit size over that of the controls, and sprayed fruits also evidenced a larger flesh-pit ratio. The 'Manzanillo' variety required an application of the compound at a concentration of 60 ppm six days after full bloom and an additional 10 ppm every subsequent day up to the eighteenth day. Smaller concentrations were found to be effective for the 'Suri' variety. The effect of thinning on alternate bearing of both varieties was not clear.

NAAm is also effective for thinning of olive (Hayto and Nobuhiro, 1963; Lavee and Spiegel-Roy, 1967). However, NAAm is not as effective as NAA; concentrations of NAAm that are 1.5 to 2.0 times those of NAA are required for equal activity. As time before bloom increases, a higher concentration of NAA is required for thinning. The thinning effect of NAAm applied at various times after bloom is continuous over a wide range of concentrations above a minimum level.

Grape

Thinning of grape with sprays presents certain difficulties not encountered in thinning tree fruits. The flower clusters of grape are very small until the shoots are three or four inches long. Hand-thinning cannot be properly done until the shoots are five or six feet long. Furthermore, in California and certain other areas, flowering does not commence until six weeks after shoot growth has begun. Therefore, by the time the clusters are large enough to be individually sprayed, they are obscured by foliage that is also subject to injury. The clusters of flowers and berries grow so closely together that it is very difficult to kill only a portion of either.

Since hand-thinning is a very expensive operation, chemical thinning is highly desirable. At present, only table grapes are hand-thinned, as the process is too expensive for wine grapes. Spray thinning might make feasible lighter pruning (leaving more and longer spurs or more canes), which would result in the production of more shoots and foliage to nourish a larger crop. Thinning sprays might be useful to kill or injure some flowers or berries growing in compact clusters, as they do in 'Zinfandel,' and to reduce the yield of over-cropped vines.

EARLY WORK. Many growth regulators, defoliants, and thinning agents have been tested to ascertain their effectiveness in thinning of grape. Early researchers (Weaver, 1954; Samish and Lavee, 1958) demonstrated that appropriate thinning of individual clusters at full bloom with chemical sprays is difficult because the flowers are borne in compact clusters. It is easy to overthin or underthin. In Israel application of water sprays at the time of bloom has successfully thinned the varieties 'Queen of Vineyards' and 'Shaslas Doré' (Samish and Lavee, 1958). Of course, from the horticultural point of view it is desirable to delay thinning until the amount of fruit set can be estimated. Samish and Lavee found that application of NAA at a concentration of 5 ppm at time of fruit set produced consistently good results, but overthinning occurred when a concentration of 10 ppm was applied. According to these investigators, the sprays had the advantage of increasing berry size and hastening berry maturation without increasing the number of "shot" berries (small berries that fail to develop) or reducing yield.

Cluster thinning of vines can be accomplished by killing groups of flower clusters with a spray of sodium monochloroacetate at a concentration of 0.5 percent when the shoots are approximately twenty inches long (Weaver, 1954). Another possible use of plant regulators is to remove clusters of second-crop fruits. These clusters develop on lateral shoots arising from the main shoots. By the time the second-crop flowers are blooming, the first crop has developed small berries. At this time concentrations of NAA that would arrest the growth of, or kill, the second-crop clusters have little or no effect on the first-crop clusters (Weaver, 1963a).

APPLICATION OF PREBLOOM SPRAYS OF GIBBERELLIN TO COMPACT CLUSTERED WINE VARIETIES. The incidence of rot in seeded wine grapes is excessive in varieties that produce compact clusters. Berries in such clusters are pushed off by other expanding berries, and the presence of juice from the fruit allows development of decay organisms. Tight clusters are slow to dry after rains. Prebloom sprays of gibberellin are used to loosen these clusters (Alleweldt, 1960; Julliard and Balthazard, 1965; Weaver and McCune, 1959; Rives and Pouget, 1959; Wilhelm, 1959). Vines are sprayed three weeks before bloom when the shoots are fourteen to sixteen inches long. Loosening of the cluster is a result of reduced set, cluster elongation, and/or production of shot berries (Fig. 10–5). In California gibberellin is applied at concentrations varying from 1 to 10 ppm, depending on variety. The sprays usually decrease bunch rot without decreasing yield (Weaver et al., 1962). Clusters with rot contribute little to the total yield because of their light weight. It is probable that any reduction in crop weight as a result of gibberellin application is more than compensated for by the elimination of or decrease in amount of rotting. Since the severeity of bunch rot varies from year to year, a prebloom application of gibberellin is considered to be a form of insurance. The decrease in the percentage of rot will be more pronounced in some years than in others.

Care must be taken not to exceed the recommended dosage of gibberellin since seeded vines may be injured by the hormone. If too high a concentration is applied to seeded varieties, bud growth the following spring is often sharply decreased. Interestingly, high concentrations of gibberellin can be applied to seedless varieties with no ill effects (Weaver, 1960). Gibberellin should not be applied for

FIGURE 10-5
'Zinfandel' grapes at harvest, taken from shoots that were sprayed at prebloom stage with increasing levels of gibberellin (*left to right*). Unsprayed control is on extreme left. (After R. J. Weaver and S. B. McCune, 1959.)

thinning purposes to seeded table grapes because the shot berries that develop as a result of the application detract from the appearance of the cluster.

APPLICATION OF BLOOM SPRAYS OF GIBBERELLIN TO 'THOMPSON SEEDLESS.' For several years the commercial practice in California was to spray gibberellin on seedless varieties of grape (especially 'Thompson Seedless') once at fruit-set stage to increase berry size. These vines were also girdled at fruit set, and the large-berried clusters that consequently developed had to be heavily berry-thinned (that is, some of the berries had to be removed from the clusters at fruit-set stage) to reduce compactness. Even then, the clusters were often quite compact, thus encouraging the development of summer bunch rot and making packing more difficult. Weaver and Pool (1965a) noted that application of gibberellin to 'Thompson Seedless' grape during bloom produces a very loose cluster. Application of the compound at concentrations of 2.5 to 20.0 ppm at 30 to 80 percent capfall was found to be most effective (Christodoulou *et al.*, 1968). The percent of berries set was reduced by approximately 50 percent. These loose clusters were less subject to summer bunch rot and were easier to pack. Application of gibberellin during the early stages of bloom increased berry size but not as much as when the compound was applied at later stages of bloom.

Application of gibberellin at the bloom stage results in striking longitudinal berry elongation; later treatments also cause radial expansion. Although all bloom sprays produce loose clusters, clusters treated with gibberellin at the later stages of bloom are loosened to a lesser degree. The present commercial practice in California is to spray twice, once at bloom for a loosening and sizing effect and again at fruit-set stage for an additional sizing effect.

Application of gibberellin reduces the set of 'Thompson Seedless' less in some areas than it does in California. In the Salt River Valley of Arizona, bloom sprays of gibberellin fail to reduce the number of berries set, an effect ascribed to differences in climate or vine growth pattern (Kuykendall et al., 1970). In India, higher concentrations of gibberellin are required for thinning of the variety 'Pusa Seedless' ('Thompson Seedless') than for other local varieties. Application of gibberellin at a concentration of 50 ppm at bloomtime reduced fruit set by 15.4 percent (Krishnamurthi et al., 1959).

APPLICATION OF BLOOM SPRAYS OF GIBBERELLIN TO SEEDED WINE AND TABLE VARIETIES.   This technique may be useful, but more experimentation is needed. 'Carignane' clusters that were treated at 50 percent capfall with gibberellin at a concentration of 10 ppm set less than half as many berries as the controls. In India clusters of the 'Anab-e-Shahi' table grape are dipped in low concentrations of gibberellin at bloom to reduce set and to increase berry size.

PREBLOOM CLUSTER AND/OR BERRY THINNING FOLLOWED BY BLOOM SPRAY OF GIBBERELLIN.   The effect of gibberellin on loosening clusters of seedless varieties when applied at bloom makes feasible cluster thinning (removal of whole clusters) or berry thinning by hand at the prebloom stage (Christodoulou et al., 1967). Cluster and berry thinning of 'Thompson Seedless' (also known as 'Sultanina') grape normally is done at the fruit-set stage, after the fall of impotent flowers and berries (Winkler, 1931). According to Winkler, more berries are obtained if grapes are thinned before the flowering stage. At maturity such clusters frequently are too compact. Thinning at the shatter stage is laborious and expensive, since the clusters often are concealed by the abundant foliage. Thinning before the flowering stage, when clusters are smaller, more tender, and better exposed to view, is much easier and less expensive.

Bloom sprays of gibberellin produce loose clusters on vines that were cluster-thinned at the prebloom stage (Christodoulou *et al.*, 1967). Bloom sprays also produce moderate loosening of clusters if both cluster thinning (in which whole clusters are removed) and berry thinning (in which the apical halves of retained clusters are removed) are done before bloom. Prebloom cluster thinning of 'Thompson Seedless' in conjunction with application of gibberellin at bloom is usually a good commercial practice, but more experimentation is necessary to determine the value of prebloom cluster and berry thinning.

THINNING 'PERLETTE' GRAPE WITH GIBBERELLIN. 'Perlette' is a popular early ripening variety of grape that is widely grown in the hot desert table grape regions of California's Coachella Valley and in northern India. Because the clusters are overly compact, extensive removal of berries by hand-thinning is required following fruit set. In many vineyards, application of gibberellin at concentrations of 10 to 15 ppm during the bloom period has resulted in satisfactory thinning (Kasimatis *et al.*, 1971). However, in other vineyards satisfactory thinning of clusters was not attained. Application of the compound at concentrations of 40 to 80 ppm increases berry size.

OTHER COMPOUNDS USED FOR THINNING. The morphactins, a class of compounds that produce morphological changes and a striking suppression of growth in many plant species (Schneider, 1964), have been shown to be very effective in causing abscission in *Vitis vinifera* grape (Weaver and Pool, 1968). Two morphactins, IT 3233 and IT 3456, have been found to be very effective on the 'Muscat of Alexandria,' 'Thompson Seedless,' and 'Black Corinth' varieties.

Weaver and Pool (1968) found IT 3456 to be much more effective than IT 3233 in causing abscission in all varieties tested. Within four days of application of IT 3456, abscission zones develop at the base of the pedicel or cap stem and the berries drop. If the compound at a concentration of 0.5 ppm is applied to 'Thompson Seedless' after fruit set, set is reduced by more than 25 percent; a concentration of 2.0 ppm applied at the fruit-set stage reduces set by approximately 70 percent. The same concentration reduces fruit set in 'Black Corinth' by approximately 80 percent. A concentration of 10 ppm reduces set at fruit-set stage in 'Muscat of Alexandria' by 50 percent.

Applied auxin counteracts the abscission-promoting effect of morphactin, thus indicating that morphactin may influence auxin metabolism in grape (Weaver and Pool, 1968). Their experiments showed that when morphactins were applied to 'Zinfandel' and 'Carignane' both at the end of and following late fruit-set stage, little or no stimulation of abscission resulted, possibly because these seeded varieties of wine grape produce so much auxin that abscission is prevented.

Ethephon is also an effective thinning agent. In California, tests in which ethephon was applied to four varieties of Vitis vinifera grape ('Thompson Seedless,' 'Perlette,' 'Carignane,' and 'Muscat of Alexandria') showed that ethephon is only slightly less effective than morphactin (Weaver and Pool, 1969). Stimulation of abscission by both ethephon and ABA decreased as flowers and berries became older. ABA was found to be much less effective in inducing abscission than either ethephon or morphactin.

Date

Thinning of Phoenix dactylifera, cultivar 'Deglet Noor,' was successful when NAA was applied as an aqueous spray after pollination (Nixon and Gardner, 1939). The compound was applied at a concentration of 100 ppm ten days after pollination and again at a concentration of 200 ppm six days later. Fruit set was reduced by 50 percent. Most of the pollinated dates that developed after treatment with NAA were similar to untreated pollinated dates.

PREVENTING FORMATION OF FRUIT ON LANDSCAPE TREES

There are numerous reasons for desiring trees without fruit. A fruit crop may make a tree unattractive and retard its growth and development, stain cars and patios, produce unpleasant odors, attract unwanted birds and insects, clog drains, and make mowing and garden maintenance difficult. In general, the hormones and caustic agents that are used for thinning of fruit remove all fruit when used at high concentrations. Sprays are usually applied either at full bloom or to young fruit. Care must be taken not to use too high a concentration because phytotoxic effects may result. Hormone-type thinners can usually be used at concentrations of 20 to 100 ppm, depending on the variety and species. Concentrations of 1 to 10 ppm usually suffice for such caustic agents as Elgetol.

Another approach for prevention of fruiting is to cover the stigma and other flower parts with a thin, impervious film that might act as a physical barrier to pollination and fertilization. Some success has been achieved with this technique but the results have been erratic (Lumis and Davidson, 1967).

PREVENTING PREHARVEST FRUIT DROP

Frequently the mature fruits of apple, pear, and other varieties drop just before harvest. Fallen fruit is often injured and is thus less valuable. Growers used to spread leaves or straw under the trees of certain varieties to lessen such injury. Preharvest fruit drop also means that harvesting is sometimes started earlier than desirable, and the resultant fruit is poor and inferior in quality.

It was reported by Gardner *et al.* (1939) that application of certain growth regulators to trees of apple just before harvest reduces fruit drop; the fruit sometimes remained on the trees almost indefinitely. This discovery was put to commercial use immediately.

Apple

Early researchers found NAA and 2,4-D to be the most effective chemicals in preventing fruit drop of apple (Fig. 10–6). In eastern Washington State, both NAA and 2,4,5-T have been used for this purpose (Anonymous, 1968b). When applied to such varieties as 'Delicious' and 'Winesap,' NAA remains effective for three to four weeks. Separate sprays are necessary in order to obtain the greatest protection for each variety, since they mature at different times. To be most effective, NAA must be applied at a concentration of 20 ppm before the apples start to drop; it becomes effective two or three days after application. When ground application is impossible, a concentrated spray should be applied by aircraft at a rate of 48 g per acre. No more than two applications should be made and applications should not be made more than two days before harvest (Anonymous, 1968b).

The period of effectiveness of 2,4,5-TP (five to seven weeks) is longer than that of NAA when application is made with ground spray equipment. This advantage of 2,4,5-TP makes it possible to spray 'Delicious' and the later-maturing 'Winesap' simultaneously with the chemical and obtain a high degree of control over the fruit drop of both. If both varieties are sprayed before 'Delicious' matures, fruit drop in that variety is controlled and the chemical persists long

FIGURE 10-6

A plant scientist tests the strength of the stem of an apple growing on a tree sprayed with NAA, which makes the fruit remain attached to the tree until harvest time. The large pile of apples on the right fell from the nearby unsprayed tree. (After J. W. Mitchell and P. C. Marth, 1962.)

enough to prevent preharvest drop in 'Winesap' when it matures. The effectiveness of 2,4,5-TP begins seven to twelve days after application. It should be used at a concentration of 20 ppm, but not as a concentrate because when it is applied as such by aircraft the compound frequently causes injury to some of the lateral buds on the terminal growth. This injury is noticeable the following year when terminals at the tops of trees are naked for a distance of six to eight inches.

Stimulation of color and maturity of 'McIntosh' apples growing in orchards in Washington State does not always result unless a spray of 2,4,5-TP at a relatively high concentration is applied considerably in advance of harvest. Even when such sprays were used in excessive concentrations the maturity of 'Early McIntosh,' 'Delicious,' and 'Winesap' apple was not advanced. Accelerated maturation of 'McIntosh' was obtained only with an early spray at a concentration of 50 ppm (Batjer *et al.*, 1954). In orchards in the eastern United States, stimulation of color and maturity of 'McIntosh' resulted from application of a spray of 2,4,5-TP at a concentration of 20 ppm twenty to twenty-five days before harvest (Southwick *et al.*, 1953).

More recent results have shown that SADH at concentrations of 1,000 to 2,000 ppm is very effective in reducing harvest drop (Batjer and Williams, 1966; Edgerton *et al.*, 1967). In the northwestern United States, SADH is just as effective or more effective than NAA or 2,4,5-TP. It also delays the development of water core in 'Delicious' and 'Winesap' apple. Water core is a change in the tissues surrounding the locules of the fruit that results in a watery or translucent appearance. Later, during storage, the tissues break down and become brown. Water core frequently accompanies advanced maturity. SADH-treated fruit has a lower level of soluble solids and is smaller and firmer than untreated fruit, indicating a delay in maturation and therefore a delay in the development of water core of treated fruit. Sprays of SADH also reduce storage scald. (Scald is a condition in which circular areas of the fruit skin take on a bronze cast.) An early SADH treatment has been proven more effective than a later one in the Northwest (Williams *et al.*, 1964).

In New York State application of SADH at concentrations of 500 to 2,000 ppm as a spray to 'McIntosh' apple from full bloom until a few days before harvest inhibited fruit drop (Edgerton and Hoffman, 1966). At normal harvest time fruit drop of control trees was as high as 56 percent in some of the tests, whereas trees sprayed with SADH at a concentration of 500 ppm dropped less than 6 percent of their crop. An application made in late August, one month before harvest, showed that the retardant was most effective at lower concentrations. Fruit maturation was delayed, but this was shown by Edgerton and Hoffman not to be the sole cause of the delay in abscission. Enhanced formation of anthocyanin and firmer fruit also resulted from the use of SADH.

These investigators also found that applications of SADH made just before harvest were ineffective in controlling drop. In contrast, sprays of such auxins as NAA have been shown to retard abscission and control drop within a few days after treatment (Gardner *et al.*, 1939).

Pear

NAA at a concentration of 10 ppm is frequently used to prevent drop of 'Bartlett' pear. These trees should not be sprayed with 2,4,5-TP because it causes injury to foliage and fruit.

SADH has shown much promise for preventing preharvest drop of 'Bartlett' pear and is just as effective for this purpose as NAA (Martin and Griggs, 1970). SADH allows considerable flexibility because it can be applied throughout the year. Martin and Griggs found that a single spray of SADH at a concentration of 1,000 ppm applied late in the season was just as effective as a double spray applied early in the season or a double spray applied late in the season.

Prune

Experiments in the northwestern United States have shown that to prevent or reduce preharvest drop of prune, 2,4,5-TP at concentrations of 15 to 20 ppm should be applied two weeks after the beginning of the pit-hardening stage, about mid-May (Anonymous, 1968b). Earlier applications may result in an excessive fruit load. The compound should not be applied in combination with other chemicals.

Citrus

2,4-D is very effective in reducing or preventing preharvest drop of citrus, mainly because the compound delays development in the abscission zone of the fruit stem, thus allowing it to remain alive and functional longer. Since the compound is a potent weed killer, it must be handled with care to avoid plant damage. In California only a high-volatile ester is recommended, because other types may cause damage to the trees (Hield et al., 1964). Sprays of 2,4-D should never be applied to citrus fruit less than seven days before harvest and should be timed to avoid periods of immature leaf growth, because such sprays cause leaf curling and distortion of young growth even at recommended concentrations. Applications made after a flush or before the start of new growth minimize or eliminate leaf distortion. The growth regulator should not be used on trees that are less than six years old. The 2,4-D treatments become effective approximately one week after application and may be expected to reduce drop by 60 percent. The maximum period of effectiveness of a drop spray is three or four months. Since sprays applied while the fruit has a partially green rind may delay color development for approximately ten days, the compound should not be applied to orchards when an early harvest is desired.

'WASHINGTON NAVEL' ORANGE. In California, the trees of 'Washington Navel' orange flower in the spring and their fruit matures in the winter. In southern California the fruit may be harvested as late as

May 15 (Stewart *et al.*, 1951). A large percentage of the 'Washington Navel' orange orchards in southern California are treated with 2,4-D to reduce fruit drop. The recommended treatment is either a 16 ppm water spray in October or November or an 8 ppm water spray in December or January.

'VALENCIA' ORANGE. In California, trees of 'Valencia' orange flower in the spring and the fruit is harvested the following year, usually between May 15 and November 15. Toward the end of the season a severe fruit drop may occur as a result of maturity and such unfavorable environmental conditions as high temperatures and wind (Stewart *et al.*, 1952). An application of 2,4-D at a concentration of 8 ppm made when fruit averages at least .5 inch in diameter reduces the drop of mature fruit but does not increase the size of fruit in the next year's crop. A reduction in the drop of mature fruit and an increase in fruit size in the following year's crop may be effected by a later spray.

GRAPEFRUIT. The dosage and timing necessary to reduce fruit drop of grapefruit are the same as for 'Valencia' orange. A spray of 2,4-D at a concentration of 60 ppm applied in June or July will reduce fruit drop during the summer and early fall (Hield *et al.*, 1964). If final size is to be unaffected, the young fruit should average at least .75 inch in diameter. If fruit-sizing sprays are applied before June or July, the drop of mature fruit is reduced but not as much as when drop sprays are applied later in the season. To prevent the serious winter fruit drops that sometimes occur in California's San Joaquin Valley, a spray of 2,4-D at a concentration of 24 ppm should be applied in an October whitewash; alternatively, a 16 ppm water spray can be applied in October or November.

LEMON. In inland areas of southern California, a fall oil-spray application of 2,4-D at a concentration of 4 ppm reduces the fruit drop that occurs in the early spring (Hield *et al.*, 1964). These hormone sprays also have the advantage of delaying the yellowing of the fruit's rind. Similar results can be obtained by applying 2,4-D at concentrations of 8 to 12 ppm in a water spray during November and December.

Preventing Seed Shedding in *Phalaris*

In Australia seed shedding in *Phalaris tuberosa* causes serious harvesting problems and reduces seed production. A means to reduce natural shedding would make harvesting with a combine possible. To avoid severe seed loss the crop is usually cut with a reaper and binder and is then shocked.

Sprays of NAA at a concentration of 20 ppm applied as soon as the panicle has emerged from the uppermost leaf sheath and/or the full flowering stage greatly decrease the number of seeds of *Phalaris* that abscise (Mullet, 1966). The greatest increase in seed production was found to result from one early spray and one late spray of NAA. The bloom spray probably enhanced seed set. The weights of seeds retained by sixteen plants that were sprayed twice with NAA and that of the seeds retained by an equal number of control plants were 78.35 g and 14.30 g, respectively. The weights per 100 seeds in the same order were 0.181 g and 0.147 g; thus NAA also increased individual seed size.

INDUCTION OF FRUIT ABSCISSION

Research and field trials have resulted in procedures that largely eliminate the problem of premature loosening and dropping of fruit. However, the fruits of some plants are held so tightly that removal is extremely difficult, and efficient mechanical harvesting is virtually impossible. The shaking of trees, vines, and other plants on which fruits are held too tenaciously can cause bruising and crushing of fruit, excessive breakage of limbs, spurs, and other plant parts, and even a reduction in yield because much fruit is left on the plant.

Apple

It is well known that ethylene gas can stimulate abscission processes in plants, including fruit abscission. However, ethylene is difficult to apply to large fields of crops for commercial purposes. Edgerton (1968) demonstrated that ethephon, like ethylene, promotes loosening and separation of fruit abscission sites. When ethephon was applied to such early blooming varieties as 'Melba,' 'Gravenstein,' 'Milton,' and 'Early McIntosh' two to three weeks before harvest, loosening of the fruit was stimulated and was accompanied by a hastening of maturation, as measured by softening of the flesh and accelerated color development, and a higher soluble solids level. A detectable response often occurred within a week after application. Confirming results were obtained by Edgerton (1968). An orchard of young trees of 'Rome Beauty' apple was sprayed with ethephon at a concentration of 1,000 ppm on October 10. By October 27 apples on the sprayed trees were appreciably looser. The fruit drop of sprayed trees was 20 percent, compared with 3 percent for the controls.

These studies were subsequently extended to 'Ben Davis,' 'Twenty Ounce,' 'Rome Beauty,' and 'Cortland,' apple varieties in which the abscission process occurs very slowly (Edgerton and Hatch, 1969). These varieties are machine harvested for processing, but their slender, drooping lateral branches and medium-to-long fruit stems are not conducive to effective transmission of the shaking force to the fruit. Five to seven days after application, ethephon reduced the amount of force required to remove the fruit of all varieties, and ten days to two weeks after application, fruits of most varieties were loosened sufficiently to allow easier removal. Ascorbic acid at a concentration of 0.25 percent reduced the fruit removal force required by 'Twenty Ounce,' 'Ben Davis,' and 'Rome Beauty' but was less effective than ethephon.

Interesting experiments on 'McIntosh' apple have demonstrated that application of ethephon one to two weeks before harvest can overcome the delay in abscission that is induced by SADH (Edgerton and Blanpied, 1970). Ethephon did not decrease the firmness produced by SADH, but it increased the soluble solids level, and development of red color was accelerated. The addition of NAA counteracted the effect of ethephon on abscission, but not on maturation.

Cherry

In Michigan it was demonstrated that application of foliar sprays of ethephon to trees of the sour cherry Prunus cerasus 'Montmorency' three to fifteen days before harvest reduces the required fruit removal force (Bukovac et al., 1969). Application of ethephon at a concentration of 500 ppm reduced this force to 224 g after six days; a force of 503 g was required for the controls (Fig. 10–7). Fruit removal was facilitated further because there was less variation in required fruit removal force (treated compared with controls). The investigators view this reduction in variability as a reflection of an increased uniformity in maturity, which may be related to an acceleration of ripening resulting from ethephon treatment. Higher concentrations of ethephon, such as 2,000 and 4,000 ppm, may cause defoliation and excessive gummosis (Fig. 10–8). However, ethephon at concentrations of 500 and 1,000 ppm proved to be only slightly phytotoxic and therefore commercially acceptable.

Bukovac et al. (1969) also tested iodoacetic acid at a concentration

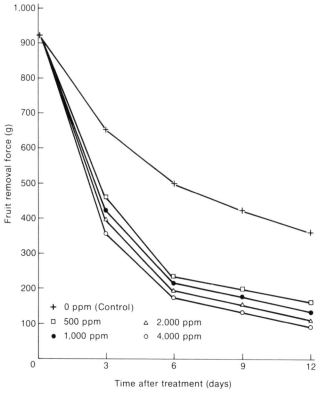

FIGURE 10-7

Effect of various concentrations of ethephon on force required to remove fruit of sour cherry 3, 6, 9, and 12 days after foliar application. (After M. J. Bukovac, F. Zucconi, R. P. Larsen, and C. D. Kesner, Chemical promotion of fruit abscission in cherries and plums with special reference to 2-chloroethyl-phosphonic acid. *Jour. Amer. Soc. Hort. Sci.* 94:226–230, 1969.)

of 300 ppm and salicylic acid at a concentration of 500 ppm, but these compounds either were not effective or were less effective than ethephon.

Studies made in Utah on 'Montmorency' cherry have confirmed that preharvest ethephon sprays reduce the required fruit removal force and that treated fruit has a higher soluble solids level and darker red color than the controls (Anderson, 1969). In these tests, ethephon at concentrations of 1,000 and 2,000 ppm caused heavy leaf drop and other phytotoxic effects, but a concentration of 500 ppm produced satisfactory results.

FIGURE 10-8

Effect of ethephon on trees of sour cherry. *Top,* dieback; *middle,* gummosis on current season's wood; *bottom,* gummosis on older wood. (After M. J. Bukovac, F. Zucconi, R. P. Larsen, and C. D. Kesner, 1969.)

362

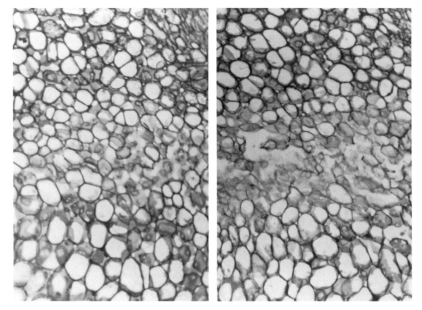

FIGURE 10-9

Longitudinal cross section of abscission layer of detached fruits of sour cherry showing effect of ABA treatment on abscission layer development. *Right*, fruit treated with ABA at a concentration of 5 × 10⁻⁴ M; left, control. Photographed 3 days after treatment, magnified approximately 230 times. (After F. Zucconi, R. Stösser, and M. J. Bukovac, 1969.)

Research conducted in New York State has also clearly demonstrated that ethephon promotes abscission in 'Montmorency' cherry and in sweet cherry, cultivars 'Emperor Francis' and 'Windsor' (Edgerton and Hatch, 1969). Application of the compound at a concentration of 500 ppm to 'Montmorency' cherry often reduced the required fruit removal force by one-third within six days, and the time required to harvest the treated trees with a shaker was less than half that required for untreated controls.

Repeated applications of ABA at concentrations of 0.5×10^{-6} M and 5×10^{-5} M to branches of trees of sour cherry ten days before maturity stimulate abscission layer development (Zucconi *et al.*, 1969)—see Figure 10–9. Within forty-eight hours after the first treatment, there was a 25 percent reduction in the required fruit removal force of the ABA-treated fruit compared with that of the control. Application of the compound at a concentration of 5×10^{-5} M caused leaf abscission of 10 to 20 percent.

Citrus

In Florida most oranges are harvested before abscission layers are adequately developed to facilitate separation of the fruit from the stem. W. C. Cooper and his colleagues have done much detailed work on this problem over a period of years. Sprays of 1 percent ascorbic acid applied to trees of mature 'Valencia' orange in one grove during July and August effectively loosened the fruit. In September 0.05 percent ascorbic acid was required to loosen the fruit; in October 0.01 percent was sufficient (Cooper and Henry, 1967). However, higher concentrations (2 to 5 percent) are generally required to obtain satisfactory results. The dextro isomer of ascorbic acid, called erythorbic acid, was as effective as ascorbic acid in inducing abscission.

There is much varietal variation in sensitivity to ascorbic acid. Application of the substance at a concentration of 5 percent to a tree bearing both mature and ripe 'Lisbon' lemon produced no effect, but when the compound at a concentration of 1 percent was applied to a tree of 'Dream' navel orange that bore full-sized but green, immature fruit, a substantial amount of fruit was loosened. The same tree dropped all its fruit and leaves when the concentration was increased to 3 percent.

When trees of 'Robinson' tangerine were covered with polyethylene tents and exposed to ethylene at a concentration of 200 ppm for a period of twelve hours, both fruit coloration and abscission were hastened. However, these trees experienced abscission of more leaves and less fruit than trees treated with ascorbic acid (Cooper and Henry, 1967). In tests on trees of 'Valencia' orange and 'Robinson' tangerine, GA_3 failed to cause fruit abscission, although extensive leaf abscission occurred.

In subsequent Florida studies with calamondin (*Citrus reticulata* var. *austera?* × *Fortunella* sp.?, widely known as *Citrus mitis*), Rasmussen and Cooper (1968) determined that all chemicals and combinations tested that induced abscission of citrus fruits also increased ethylene production. However, probably because of some uncontrollable factors, all chemicals that induced ethylene formation in calamondin did not induce abscission in citrus fruit. A new finding by these investigators was that addition of iron chelated with EDTA to ascorbic acid causes more of a decrease in the firmness of attachment of mature fruit than does ascorbic acid alone. Citric acid at a concentration of 1.0 percent also causes a great deal of abscission when applied to trees of 'Hamlin' orange; its use as an additive to ascorbic or isoascorbic acid is promising (Cooper and Henry, 1968a).

Field trials in Florida with 'Valencia' orange revealed that cycloheximide at concentrations of 2 to 25 ppm effectively induces abscission (Cooper and Henry, 1968a). It was further shown that a mist application of the compound at a concentration of 1,000 ppm causes more abscission than does a high-volume spray of the compound at concentrations of 2 to 25 ppm, even though both applications contain the same amount of chemical. In these experiments, seven days after treatment with the mist, 100 percent of the fruit had dropped, whereas only 26 percent of the fruit dropped as a result of the higher-volume spray of the compound at a concentration of 25 ppm. Other studies showed that the concentration of cycloheximide has a positive correlation with ethylene production in the fruit and an inverse correlation with the force required to remove the fruit (Cooper et al., 1969).

Ethephon was found to be less effective than cycloheximide in inducing abscission in 'Valencia' orange and was also shown to cause excessive leaf damage (Cooper and Henry, 1968a). In their research with five varieties of citrus, Cooper and Henry (1968b) noted that high concentrations of ABA applied during the summer resulted in defoliation; winter applications caused no defoliation but did cause fruit coloration.

In California a search is also being made to find effective abscission-inducing compounds for use on citrus. Hield et al. (1968) found that potassium iodide at a concentration of 1 percent decreases fruit attachment force. In one experiment with trees of navel orange, fruit drop of trees thus treated was 47.2 percent twelve days after treatment; fruit drop of controls was only 7.3 percent. When the trees were shaken, 2.5 percent of the fruit on control trees fell, but 22.5 percent of the fruit of treated trees fell.

Ascorbic acid and erythorbic acid effectively loosen fruits of 'Valencia' orange three to fourteen days after treatment, but these chemicals also cause severe rind damage. These compounds also were effective in removing 'Valencia' fruit from the trees after it had been frozen.

Experiments in which some of the fruit was enclosed in plastic bags to protect it from the spray showed that no induction of abscission occurs unless the fruit is directly treated with the ascorbic acid spray because there is insufficient absorption by or translocation from treated leaves. Ascorbic acid sometimes fails to cause abscission under the very dry climatic conditions that prevail in California. The problem of absorption of ascorbic acid and certain other compounds is critical; effective wetting agents are needed.

In California fruit and leaf drop of orange was increased by application of ethephon at concentrations of 100 to 1,000 ppm (Hield *et al.*, 1969b). However, unless a means is found to minimize citrus leaf drop while maintaining the fruit-loosening ability of ethephon, its commercial outlook as an aid to mechanical harvesting is not promising. There is a marked effect of temperature on amount of abscission of ethephon-treated leaves and fruit. In one experiment in which large seedlings of sweet orange were sprayed with ethephon at a concentration of 1,000 ppm, 90 percent leaf drop occurred within ten days under warm greenhouse conditions, but in cooler outside temperatures leaf drop was slow to begin and proceeded at a slower rate: 85 percent defoliation occurred in twenty-six days (Hield *et al.*, 1969b).

Fruit of citrus does not drop only as a result of formation of an abscission layer; it may fall when the stem is pulled from the button (leaving a stem-size cavity), or when the stem is broken above the button, or when a plug is removed from the rind. The first, or stem-slip type of separation, in which the woody stem slips from the bark, may result from increased cambial activity in the region directly above the button. California tests have demonstrated that a combination of NAA at a concentration of 1,000 ppm or 2,4-D at a concentration of 10 ppm increases the percentage of fruit that separates from the tree in this manner, although the compounds have no effect on abscission (Hield *et al.*, 1969a).

Olive

Mechanical harvesting of olive with tree shakers has been only partially satisfactory because the energy output of such shakers is usually insufficient to remove most of the fruit. Therefore, a chemical treatment that would reduce the fruit-stem attachment would allow tree shakers to drop the olives more easily on catching frames. Shaking treatments are preferable to hand-picking, which is a very laborious operation, but such treatments must not cause any appreciable abscission of the leaves.

In one of the first successful attempts to induce fruit abscission, foliar sprays of MH at a concentration of 1 percent were applied to trees of olive (*Olea europaeae*) during California's winter months under conditions of high humidity (Hartmann, 1955). In one experiment, when branches were shaken ten days after treatment, 95 percent of the sprayed fruit dropped, but only 15 percent of the unsprayed fruit dropped. However, when the humidity was low, little or no abscission occurred as a result of treatment, probably because of poor absorption of the chemical.

ABA at concentrations of 1,000 and 2,000 ppm caused complete leaf abscission in 'Manzanillo' sixteen days after application but had no effect on fruit abscission (Hartmann et al., 1968). An appreciable leaf drop in the fall is very injurious to olive trees because defoliation at this time interferes with subsequent flower formation.

Ethephon was also found to cause fruit loosening. Application of the compound at concentrations of 1,000 or 1,500 ppm on October 2 caused heavy leaf and fruit drop in 'Manzanillo' (Hartmann et al., 1968). The average force required to remove fruit from trees sprayed with ethephon at a concentration of 1,000 ppm was 428 g on October 6; by October 18 this force had been reduced to 0 g. There was also 100 percent leaf abscission. The addition of urea at a concentration of 1.35 percent often accentuated the loosening effect but also caused severe defoliation. A varietal difference was observed in the response of different varieties of olive to ethephon. Much loosening of fruit occurred in 'Manzanillo,' moderate loosening occurred in 'Mission' and 'Ascolano,' and only slight loosening occurred in 'Barouni.' Leaf drop was heavy in 'Manzanillo' but was only moderate in the other varieties.

Ethephon effectively induces abscission in cool, rainy weather but not in hot, dry weather. Hartmann and his colleagues demonstrated that the addition of NAA at a concentration of 150 ppm to an ethephon-urea combination completely blocked the abscission-inducing effect of the mixture on both leaves and fruit. However, if NAA was applied two or three days after ethephon, the former compound reduced leaf abscission but did not affect fruit abscission (Hartmann et al., 1970).

Cycloheximide also reduced the required fruit removal force and increased the amount of leaf abscission but to a lesser degree than did ethephon (Hartmann et al., 1970).

Ascorbic acid at a concentration of 3 percent and iodoacetic acid at a concentration of 150 ppm, when used with appropriate wetting agents, weaken the attachment of olive fruits to the tree (Hartmann et al., 1967). However, these compounds are effective only when applied under very humid conditions. There is no practical way to increase artifically the humidity of the environment surrounding the trees during or following spray applications. Conditions of low humidity prevent surfactants and penetrants from enabling ascorbic acid and iodoacetic acid to loosen olive fruits. Dry conditions prevail in California during the harvest of table olives.

Trials in California and Australia showed that the sodium, potassium, and ammonium salts of salicylic acid at a concentration of 0.5 percent significantly reduce the force required to remove olives from the trees (Hartmann et al., 1968). Again the loosening effect of the compounds was correlated with high humidity. Application of salicylic acid to 'Manzanillo' olive at the onset of the harvest season in California was very effective in cool, rainy weather, but in hot, dry weather there was no reduction in fruit attachment force.

Trials extending over a nine-year period in Israel revealed that glycerin and related alcohols effectively loosen the connection between olive fruit and branch (Lavee, 1965). Branches or whole trees of the 'Manzanillo' and 'Souri' varieties were usually sprayed when the fruit was mature (black). Seven days after treatment, trees were shaken with a small boom shaker or branches were manually shaken. Application of ethylene glycol and low concentrations of triethanolamine to 'Manzanillo' revealed that these chemicals were more effective than glycerin, which caused leaf damage. Glycerin effectively caused abscission of fruit of 'Souri' with less damage to foliage than the other two chemicals. Glycerin at a concentration of 2 percent increased fruit drop of 'Souri' by approximately 30 percent. Ethylene glycol was less effective on 'Souri' than on 'Manzanillo.'

Pear

Mechanical harvesting of pear has the disadvantage that this fruit is highly susceptible to impact bruising during shaking and fruit fall. The formation of an abscission layer would thus be very advantageous. Griggs et al. (1970) reported that application of ethephon at a concentration of 1,000 ppm to 'Bartlett' pear nine days before harvest reduced the required fruit removal force. However, tissues of ethephon-treated fruit deteriorated during storage at 0°C more rapidly than did the controls.

In California almost all 'Bartlett' orchards are sprayed with NAA at a concentration of 10 ppm seven to fourteen days before harvest to prevent preharvest drop. Griggs et al. (1970) noted that ethephon failed to reduce the required fruit removal force when NAA was also used.

Plum

Ethephon is effective in reducing the required fruit removal force of trees of 'Stanley' and 'Damson' plum growing in Michigan (Bukovac et al., 1969). Within ten days following preharvest treatment with ethephon at concentrations of 500 and 1,000 ppm, there was a reduction in the force necessary to remove fruit of 18 and 27 percent, respectively; no severe phytotoxic effects were observed. Most of the treated fruit that was removed did not have the stems attached, as did many of the control fruits.

At harvest, the souble solids level of fruit that received ethephon at a concentration of 1,000 ppm was 12.7 percent; the comparable figure for controls was 10.8 percent. Severe phytotoxic effects resulted at higher concentrations (2,000 and 4,000 ppm), especially in the 'Stanley' variety. Leaf abscission, gummosis, and necrosis of terminal parts of shoots were the main deleterious side effects of high concentrations of ethephon.

Grape

When grapes are harvested mechanically, the canes or cordons of vines must be shaken to remove the fruit. Less injury is inflicted to the vines if the fruit is not attached too firmly. Berries are attached so tightly in some varieties that the shaker type of mechanical harvester is unsatisfactory; a chemical means to loosen berries is therefore desirable.

Experiments have shown that morphactins, ethephon, and ABA are effective in stimulating formation of abscission layers and thus loosening the berries (Weaver and Pool, 1969). The loosening is greater at early stages of fruit development than at later stages.

The berries of the *Vitis labrusca* grape, a species grown extensively in the eastern United States, readily form abscission layers, and preharvest application of ethephon has been shown to reduce the fruit removal force and facilitate mechanical harvesting. In Washington State, Clore and Fay (1970) found that ethephon at concentrations of 250 ppm and higher induced berry abscission in six days or less. Concentrations higher than 250 ppm, however, induced leaf yellowing and accelerated leaf senescence.

Walnut

The edible walnut kernel is generally mature and of highest market quality three to four weeks before harvest. In the interim between maturity and harvest that is necessary for the dehiscence of the hull from the shell (hull loosening) to occur, kernel quality decreases because of heat, mold, and insect damage. In addition, the nuts of such cultivars as 'Eureka' and 'Gustine' are difficult to remove from the tree even when they are hullable. For these reasons, a chemical means of facilitating mechanical harvest would be of value.

Martin (1971) sprayed several of the walnut varieties grown in the six major walnut-producing areas of California with ethephon at concentrations of 500 or 1,000 ppm ten to twenty-seven days before normal harvest. At each location more than 95 percent of the ethephon-treated varieties were removed by a mechanical shaker ten days before normal harvest. During the same period, 5 to 90 percent of the control nuts at each location were harvestable. Moreover, 95 percent of all ethephon-treated walnuts were hullable ten days before harvest, whereas only 5 to 80 percent of the nuts on untreated trees were hullable. These results indicate that preharvest treatment of walnut trees with ethephon makes possible removal of most of the crop with a single mechanical shaking operation, thereby eliminating the high cost of multiple harvest.

A matter of concern to growers was leaf drop, especially drop of leaves from weak trees, following ethephon application. A determination must be made of the percentage of leaf loss that can be sustained by a walnut tree without a reduction in the yield and quality of nuts in subsequent years.

POSTHARVEST ABSCISSION EFFECTS

Prevention of Berry Drop in Grape

A serious problem encountered in grape handling is that after the grapes are packed in boxes and stored or shipped, many berries fall from the clusters. Therefore, any treatment that keeps the berries more tightly attached is of value. In India Narasimham et al. (1967) sprayed 'Bangalore Blue' grape with six different growth-regulating chemicals seven days before harvest. NAA checked the amount of berry drop during transit and also during storage both at room temperature and at cold temperatures. A concentration of 100 ppm was found to be optimal, although a concentration as high as 250 ppm had no deleterious effects on the vines or bunches.

Rao (1970), also working in India, demonstrated that a preharvest spray of BA at a concentration of 500 ppm reduced postharvest berry drop of 'Anab-e-Shahi' grape but that least drop occurred as a result of treatment with BA at a concentration of 100 ppm plus NAA at a concentration of 100 ppm.

In Israel Lavee (1959) had also noted that berry drop of 'Muscat of Hambourg' could be decreased by spraying fruit with NAA or 4-CPA at concentrations of 10 to 20 ppm four days before picking, thus retarding formation of an abscission layer. The growth substances are effective in preventing berry drop after picking if the grapes are not placed in cold storage until thirty-six hours or more after they are picked. In one experiment in which cold storage was delayed for forty-eight hours after picking, each of the growth substances reduced berry drop by approximately 40 percent. However, Pentzer (1941), working in California, was unable to demonstrate that application of NAA to several varieties just before harvest strengthens berry attachments.

Supplementary readings on the general topic of abscission are listed in the Bibliography: Addicott, F. T., 1970; Carns, H. R., 1966; Symposia on ethylene and fruit abscission, Miami Beach, Florida, 1971; and Symposium on leaf abscission, Fort Detrick, Frederick, Maryland, 1968.

Chapter 11

SIZE CONTROL AND RELATED PHENOMENA

Compounds are now available that can increase or scale down plant size. For example, gibberellins make most plants taller and make some larger. On the other hand, several growth retardants are especially useful when there is a need to slow down growth to obtain small or compact plants. The use of retardants to stimulate flower initiation is discussed in Chapter 7. Other interesting effects that have been obtained experimentally by application of growth retardants include increased salt tolerance and increased resistance to drought, heat, low temperature, and disease.

INCREASING PLANT SIZE AND/OR YIELD

The most frequent result of application of gibberellin to plants is the stimulation of shoot growth (Fig. 11–1). It is therefore not surprising that as soon as the compound was made available for experimentation, there were high hopes that it could be used to increase the size and yield of many kinds of plants to commercial advantage. Although these early hopes have not yet been fully realized, a number of interesting and valuable uses of gibberellin have been developed.

FIGURE 11-1
Eight consecutive weekly applications of 100 μg of gibberellin to growing tips of cabbage plants on right resulted in elongation and earlier flowering. The 2 untreated plants on left were vegetative and formed heads. All plants were grown at 10–13°C. (Photograph courtesy of S. H. Wittwer.)

Effect of Gibberellin on Grass

Gibberellin stimulates growth of grass grown at temperatures lower than those normally required for growth (Imperial Chemicals Industries Ltd., 1955; Morgan and Mees, 1956; Wittwer and Bukovac,

1957c). Yields of several hundred pounds per acre may be obtained by treating grass with gibberellin in early spring before growth would otherwise occur. Two or three times more forage than is usually obtainable from untreated grass is also possible during the early part of the season when growth of grass is usually slow (Fig. 11-2).

In Indiana, Leben and Barton (1957) sprayed 'Kentucky' bluegrass (*Poa pratensis*) with GA₃ at rates of 0, 28, 56, or 112 g per acre. Each amount of GA₃ was dissolved in one hundred gallons of water. Spraying was done in the fall, an unfavorable time of year for growth,

FIGURE 11-2

Application of gibberellin promotes growth of grass (predominantly *Poa pratensis*, cv. 'Kentucky' bluegrass) in early spring. Rectangular plots, each 15 square feet, were sprayed March 5, 1957, with aqueous solutions containing gibberellin at concentrations of (*from front to rear*) 1,000, 0, 100, 1,000, and 10 ppm. Plots at extreme rear, which show marked stimulation of growth, received the compound at concentrations of 100 or 1,000 ppm. Photographed April 1, 1957. (After S. H. Wittwer and M. J. Bukovac, 1957c.)

using fertilizer (10 percent N–10 percent P–10 percent K) at levels
of 0, 215, and 645 pounds per acre. Within four days after treatment
the slow-growing grass began to grow rapidly again as indicated
by intensification of the green color and development of new shoots.
Fifteen days after treatment plants were harvested by clipping them
4 cm above the ground. Gibberellin brought about a significant
increase in both the fresh and dry weights of plants. The largest increase occurred when fertilizers were used in conjunction with the
growth regulator.

In Michigan, Wittwer *et al.* (1957) studied the effect of gibberellin
on the nutrient composition of 'Kentucky' bluegrass. Plots were well
fertilized (12 percent N–6 percent K at 450 pounds per acre) in the
fall. Gibberellin at a rate of two ounces per acre was applied early
the following spring. Twelve days after treatment the grass was harvested and analyzed for ten nutrient elements, dry matter, and total
sugars. Although growth was greatly accelerated, the percentages of
dry matter and mineral composition were not altered. There was,
however, a decrease in amount of total sugars.

Poa pratensis and Bermuda grass (*Cynodon dactylon*) were most
responsive to gibberellin, followed by *Agrostis, Festuca,* and *Lolium,*
in that order. *Zoysia* did not respond (Wittwer and Bukovac, 1957c).

In an experiment in Australia, Arnold *et al.* (1967) studied the effect of winter application of GA$_3$ to pastures planted with a mixture
of *Phalaris,* annual grasses, and subterranean clover. The response
to gibberellin was greater in pastures that were producing poor
growth. A sixfold increase in growth rate was obtained as a result of
application of GA$_3$ at a rate of 4 g per acre when the normal growth
rate was two pounds of dry matter per acre per day. The length of
time during which cumulative growth on gibberellin-treated pastures
exceeded that of the unsprayed pastures varied proportionately with
the growth rate of the latter. These investigators also showed that
NAA or kinetin, applied at a rate of 4 g per acre, combined with GA$_3$
at a rate of 4 g per acre, increased the response to equal that obtained
by application of GA$_3$ alone at a rate of 20 g per acre. *Phalaris* was
more responsive to GA$_3$ than either annual grasses or subterranean
clover. In an extended field trial the cumulative growth produced by
a gibberellin-treated pasture was still double that of the control
ninety-five days after application.

Several investigators have demonstrated that although gibberellin-
treated grasses grow more rapidly and initially produce an increase
in the fresh weight yield, the size of the total harvest after the first
cutting is reduced proportionately with that increase. The elongated

leaves produced by gibberellin are often easily bruised and susceptible to frost damage, and they contain approximately the same amount
of dry matter as the shorter, untreated leaves. Thus the main benefit
is an early increase in growth; best results are obtained when slow-
growing species are grown at low temperatures. In areas where stock
maintenance is based on marginal production of grassland, such an
early increase could be of economic value even though the size of
the final yield is not increased.

Effect of Simazine on Grass

Application of simazine has dramatically increased the yield and protein content of both ryegrass forage and pea and bean seed. Ries
et al. (1968), working both in the subtropical conditions of Costa
Rica and in the temperate conditions of Michigan, applied simazine
at low rates (.06 to .50 pounds per acre), rates that are much lower
than those used when the compound is applied for weed control. In
Michigan, ryegrass (Lolium perenne) that was sprayed with simazine
at a rate of .06 pounds per acre yielded 12.9 tons per acre on a fresh
weight basis, compared with the 8.7 tons yielded by the untreated
controls. When simazine was applied at a rate of .25 pounds per
acre, the resultant dry weight of total crude protein was 190 mg per
gram, an increase of 52 percent over the protein content of unsprayed
plants. Another interesting finding in both locations was that the
highest level of simazine resulted in a more than eightfold increase
in the nitrate content of the dry forage. This result may be related
to the fact that the nitrate reductase activity in plants treated with
simazine is increased (Ries et al., 1967).

Application of simazine to flooded soil at flowering time has been
found to increase the percentage of protein in the rice grain (Vergara
et al., 1970). However, the increase is accompanied by a decrease
in grain yield that can be attributed to increased sterility. The decrease in grain yield results in a decrease of total grain protein per
unit area of crop.

Use of Gibberellin to Increase Sugar
Production in Sugarcane

Several cultural techniques have been used to increase the sucrose
content of sugarcane without resort to chemicals. One is to top the
canes above the mature internodes that are to be included at harvest.
Another is to fertilize in such a manner that the nitrogen is depleted

FIGURE 11-3

Sections from stalks of Badila sugar cane. *Top left,* short winter joints; *bottom left,* normal-size joints; *right,* enlongated joints resulting from treatment with GA₃. (After T. Tanimoto and L. G. Nickell. In 1967 *Reports, Hawaiian Sugar Technologists,* Jour. Series Exptl. Sta. HSPA Paper no. 199, p. 140. Used by permission.)

by the time of maturation. On irrigated plantations water can be withheld to "dry off" the crop. These three methods have been only partially successful.

Hundreds of chemicals have been tested in Hawaii to control sugar production of sugarcane (Nickell and Tanimoto, 1967; Tanimoto and Nickell, 1967). The most promising chemical in these field trials was GA_3. This compound increased length of the stalk (Fig. 11–3) and tonnage of cane and sugar at harvest if the plant was allowed sufficient time to mature after application. The nodal tissue of sugarcane contains less sugar than the internodal tissue. Therefore, a stalk of a given length of cane treated with GA_3 contains more sugar than a similar length of stalk of control, which has more nodes per linear foot.

In Australia application of gibberellin at concentrations of 100 to 200 ppm to both young and older sugarcane plants grown under greenhouse conditions increased both concentration of total sugars and amount of millable stalk (Bull, 1964). In field trials sugar yields were increased up to 25 percent by repeated sprays of gibberellin, the first applied when the cane was 6.5 feet high and the second applied three or four weeks later. Best results were obtained when the harvest was postponed for at least three months after the second spray to allow proper ripening.

In Puerto Rico Alexander (1968) applied a spray of gibberellin at a concentration of 10 ppm to sugarcane plants growing in crocks and obtained internode elongation while retaining stockiness of the cane. Significant increases in the sucrose content of storage tissue occurred only when the nitrate supply was low. Sucrose losses due to a high nitrate level could not be offset by gibberellin-induced sucrose increases. These results suggest that application of gibberellin should be delayed after a heavy nitrogen fertilization. Alexander et al. (1970) also found that three applications of GA_3 made at ten-day intervals resulted in a larger increase in green weight, internode length, and stalk length than did one application.

Use of Auxins to Increase Plant Yield

Application of 2,4-D and related compounds at low concentrations has been shown to increase the size and yield of several plants if made at the proper stage of plant growth. The range of concentrations of 2,4-D that will safely stimulate growth is restricted because of the compound's toxic effect on sensitive plants, but this range can be widened by the inclusion of micronutrients.

In experiments conducted in California, Huffaker et al. (1967) showed that spray application of the isopropyl ester of 2,4-D increases the yields of both barley and wheat when these plants are treated at the five-to-seven-leaf stage. Highest yields were obtained when the compound was used in combination with either $FeSO_4$ or FeEDDA. An untreated plot of barley yielded 2,440 kg per hectare; a plot treated with 2,4-D at a concentration of 2,000 ppm plus FeEDDA containing 500 ppm of Fe yielded 3,020 kg per hectare. The figures for untreated and treated plots of wheat were 595 and 719 kg per hectare, respectively. The protein content of wheat was increased by the treatment. Positive results were also obtained when this treatment was applied to 'Sutter Pink' bean.

In Canada, Wort (1966) demonstrated that application of sublethal concentrations of 2,4-D to the foliage of potato, pea, bean, sugar beet, maize, and other crops resulted in an increase in dry matter and in yield of tubers, pods, and seeds. Both sprays and dusts were effective, but greatest increases were obtained when the compounds were fortified with such minor elements as Fe, Cu, Mn, and B. Wort also found that the quality of the crops improved as a result of treatment. The red skins of potato tubers treated with 2,4-D plus a mineral formulation became more intense in color than untreated tubers; they also lost less weight during storage. Pods of similarly treated bean plants were found to contain more vitamin C; treatment retarded the degeneration of the vitamin during storage.

Although the use of 2,4-D to increase plant yield shows promise, it has not yet been utilized commercially for this purpose on a wide scale.

Use of Naphthenic and Cycloalkanecarboxylic Acids to Increase Plant Yield

These compounds have also been used to stimulate plant growth and yield. Foliar or soil application of potassium naphthenate to young bush bean plants increases the yield of bean pods per plant by approximately 25 percent (Wort and Patel, 1970a). Foliar applications of potassium naphthenate to fourteen-day-old plants of maize, spring wheat, sugar beet, and radish resulted in an increase in foliage of 8 to 21 percent (Wort and Patel, 1970b). The fresh weight of the radish taproot was increased by 13.5 percent over that of the control. Application of the potassium salts of cyclohexanecarboxylate at a concentration of 1×10^{-2} M increased the yield of green pods of bush bean plants by 35 percent.

Effect of Gibberellin on Flowers

Experiments to test the effect of gibberellin on flowering plants used for decoration were initiated as soon as experimental quantities of the compound became available in the mid-1950's (see Cathey, 1959c; Lindstrom et al., 1957; Wittwer and Bukovac, 1957b, 1958; Marth et al., 1956). Some trials were conducted in Michigan during the fall and winter under low-sunlight conditions (Lindstrom et al., 1957). When GA_3 was applied to stem apices at a concentration of 20 μg or as a foliar spray at concentrations of 10 to 100 ppm, flowering was hastened by ten days to four weeks in petunia, stock, larkspur, English daisy, China aster, and gerbera. No excessive stem elongation or deleterious effect on market quality was observed in these experiments. A cold-requiring biennial, foxglove, required a concentration of 600 μg to induce flowering, but again no excessive stem elongation occurred. Stock treated with GA_3 matured earlier and produced longer and more uniform flower spikes than untreated plants. Treated forget-me-not and pansy flowered while controls remained vegetative. Hastening the maturation of decorative flowers decreases production costs; these investigators suggest that, by using gibberellin, greenhouse flower growers in the North might be able to compete successfully with those in more southerly regions.

Gibberellin treatment often increases the size of vegetative growth and flowers, as well as the size of peduncle, pedicel, and petal. However, to date gibberellin has been used commercially only to a limited degree to increase flower size and hasten overall development.

GERANIUM. Sprays of gibberellin at concentrations of 1 to 10 ppm have been shown to increase the size of inflorescences of the 'Spartan White' and 'Brick Red Irene' varieties (Lindstrom and Wittwer, 1957). These experiments indicated that 'Spartan White' should be sprayed when a few florets of each inflorescence are beginning to open and show color; treatment at earlier stages resulted in excessive elongation of the peduncles. Inflorescences of the treated 'Spartan White' variety remained in a marketable condition two weeks longer

FIGURE 11-4

Effect of gibberellin on flower development and longevity of 'Spartan White' geranium. *Left,* untreated control; *right,* flower of plant that received a foliar spray of gibberellin at a concentration of 10 ppm when florets were beginning to open and to show color. (After R. S. Lindstrom and S. H. Wittwer, 1957.)

than untreated ones. Application of gibberellin at a concentration of 5 ppm to 'Brick Red Irene' increased the diameter of the inflorescence but did not excessively elongate the peduncles. Gibberellin increased the flower size in both varieties (Fig. 11–4) and also produced larger pedicels and longer peduncles (Fig. 1).

ROSE. The longer the stem, the higher the value of a cut rose. The size, form, and color of the flower and the strength of stem and condition of foliage are other less important standards for grading. One application of gibberellin at concentrations of 10 to 100 ppm to 'Better Times' roses increased stem length and fresh weight of cut flowers (Mastalerz, 1965). Three-year-old 'Better Times' roses were cut back to the second five-leaflet leaf, and gibberellin was applied when new shoots were .75 to 1.00 inch long. When flowering commenced, stem lengths of controls and plants treated with gibberellin at a concentration of 100 ppm were 27.7 and 31.8 cm, respectively. Fresh stem weights in the same order were 30.0 and 34.7 g. On the other hand, more than one application of gibberellin decreased quality by increasing the number of outer petals that developed an elongated shape. These outer petals also developed sepal-like features, such as crinkling, pronounced veins, and interveinal areas that turned green.

CAMELLIA. Applications of gibberellin at concentrations of 8,000 to 12,000 ppm have been shown to induce earlier, larger, longer-lasting camellia blooms (Gill, 1966). Two techniques have proved successful. In the first method, the leaf bud adjacent to a flower bud is removed and the hormone is applied to the wound "cup" so formed. The second, more effective, method is to inject gibberellin into the lower portion of a flower bud. An injection of gibberellin at a concentration of 11,444 ppm into the base of the flower bud of 'Ville de Nantes' on October 9 reduced days to flowering from 91 to 56, increased flower diameter from 8.9 to 12.5 cm, and increased the average time that flowers were retained on plants from 6.2 to 8.3 days (Gill, 1966). Painting of the flower buds themselves also produces bigger and earlier blooms, but defoliation and bud drop may occur if more than a limited number of buds are treated.

Although gibberellin produces excellent results in warm climates, less is known about the response of flowers growing in colder areas, where the interval between bud maturity and the onset of cold weather is short. Maryott (1969) tested gibberellin on eighty-eight varieties growing at the National Arboretum near Washington, D.C. He classified the varieties into three categories: (1) good response,

(2) some response, and (3) no good blooms obtained. Seventeen varieties were classified in group one; the remainder were divided about equally between groups two and three.

SNAPDRAGON. Stimulation of stem elongation is expected as a result of gibberellin application, but this response is rarely induced by a chemical that ordinarily acts as a growth retardant. However, spraying F_1 hybrids of snapdragon cultivars 'Rocket White' and 'Snowman' with CCC at concentrations of 50 to 2,000 ppm markedly increases stem height as well as dry weight of the leaves and stems (Halevy and Wittwer, 1965b). A concentration of 500 ppm is optimal; concentrations of 3,000 ppm and higher are toxic. Spray application of SADH and soil application of CCC or Phosfon-D do not stimulate growth. No explanation can be offered as to why foliar sprays of CCC induce a response in snapdragon similar to that induced by gibberellin, but such response is a reminder that there are sometimes exceptions to the generally accepted responses of plants to a given growth regulator.

Effect of Gibberellin on Cherry Trees

Gibberellin applications can hasten the growth of certain young trees. Trees of 'Montmorency' cherry that had produced terminal buds initiated a second flush as a result of application of gibberellin at concentrations of 100 to 1,000 ppm (Hull and Lewis, 1959). All treatments produced an increase in trunk diameter. Fresh and dry weights of the aerial portion of the trees increased proportionately with the concentration of gibberellin applied, but weight of roots was not affected. The growth resulting from one spray application was stocky, well proportioned, and commercially desirable. A repeated spray of gibberellin at concentrations of 500 or 1,000 ppm applied approximately two months after the first spray was not beneficial because it resulted in weak growth that wilted and died several weeks later.

Effect of Ethylene on Tree Trunks

Container-grown nursery trees are often tall and spindly as a result of crowded growth conditions. Such trees require staking, which can interfere with normal strengthening of the trunk (rapid trunk growth is desirable), and thus the subsequent growth of these trees may be modified in an adverse manner.

Neel (1969) has shown that a three-week application of ethylene gas at concentrations of 10 to 20 ppm to the lower halves of young tree trunks enclosed in plexiglass tubes causes extreme thickening of the trunks (Fig. 11-5). He found that the thickening caused by ethylene in *Eucalyptus fasiculosa* is mainly a result of the chemical's stimulation of phloem development. Neel believes that trunk flexing affects ethylene production in the trunk and that this increase in ethylene production may be one reason why trunk diameters of trees that are free to move are larger than those of staked trees.

INCREASING LATEX YIELD OF RUBBER TREES.   Stimulation of latex yield by such chemicals as 2,4,5-T has been important for sufficient production from most clones of rubber trees (*Hevea brasiliensis*) at some stage during their average life-span of thirty years. According to de Wilde (1971), ethephon was first used commercially to stimulate latex production. If ethephon at a concentration of 10 percent in palm oil is brushed on a strip of bark measuring 1.5 to 2.5 inches located directly below the tapping cut, latex flow and dry rubber yields of commercially important clones are increased 100 percent or more (see de Wilde, 1971). The greatest increase in dry rubber content is achieved when ethephon is applied to a conventional half-spiral cut that is tapped every two days. New techniques can be used with

FIGURE 11-5

Trunk thickening resulting from ethylene application. Portions of the trunks between T_1 and T_2 were enclosed in plexiglass tubes. Air flowed through tube in tree on right at rate of 1 liter per hour. Ethylene at concentrations of 6–15 ppm flowed through tube in tree on left at the same rate. (Photograph courtesy of P. L. Neel.)

ethephon that were not formerly possible with commercial stimulants. For example, increased yields can be obtained by using short (.25- or .33-inch) tapping cuts and by tapping at more frequent intervals (three to six days). The use of ethephon conserves bark and thereby prolongs the economic life of the tree; it also increases labor productivity while maintaining or increasing latex yields.

Use of Gibberellin to Induce Elongation of Celery Petioles

Gibberellin sprays sometimes increase the length of celery petioles and thereby induce higher yields. Figure 11–6 shows the results of some Michigan experiments. Tests made with varieties of celery grown in four counties of California also showed that gibberellin treatment sometimes has beneficial effects (Takatori et al., 1959). Most tests were conducted on short-shanked varieties, which are inherently better adapted for cool-season production but which often produce stalks so short that the product is not of top commercial quality. Single spray applications of GA_3 at concentrations of 25 to 100 ppm made four weeks before the intended harvest date increased the length of petioles of 'Utah 52–70' by approximately one to two inches; petioles of 'Utah 16–11' and 'Utah 10–B' were increased by two to three inches. The tests showed that earlier applications stimulate seedstock formation and thus should be avoided.

An increase in total plant height resulted from elongation of the heart and adjacent whorl of leaves. There was no elongation of the outer whorl of leaves. The weight per field-trimmed plant was almost always increased by the gibberellin sprays when they were applied at optimal concentrations. Most response occurred when environmental conditions were the least favorable. For example, in California in 1957, when the treated plants were growing actively, response to the compound was limited, but pronounced responses to sprays occurred in 1958 when plant growth was very slow.

In Taiwan, the Western variety of celery 'Utah 52–70' was found to be more responsive with gibberellin than the local variety (Huang, 1961). Plants sprayed with gibberellin at the late stages of growth grew faster than those sprayed at the early stages. In the northern part of Taiwan this variety should be sprayed with gibberellin at a concentration of 100 ppm approximately one month before harvest. Research done on the 'Giant Pascal' variety showed that increases in plant height are caused mainly by an increase in cell size rather than by an increase in cell number (Huang, 1963).

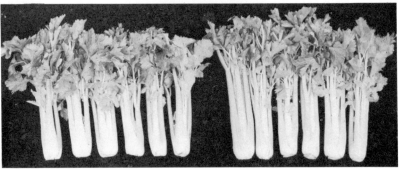

FIGURE 11-6

Effect of gibberellin on growth of 'Early Fortune' (golden plume type) celery 19 days after treatment. Top photograph shows: *left,* untreated control; *right* (*row in background*), effect of applying a foliar spray of gibberellin at a concentration of 100 ppm—the equivalent of 15 g per acre. In the bottom photograph the average petiole length of the controls (*left*) was 8 inches, compared with an average length of 10 inches for the treated celery (*right*). Treated stalks were 40 percent heavier and were also better blanched. Photographed at celery farm of Thomas Bosgraaf and Sons, Hudsonville, Michigan. (After S. H. Wittwer and M. J. Bukovac, 1957a.)

Use of Ethephon to Initiate Bulbing in Onion

Bulbing of onion (*Allium cepa*) is induced by photoperiodism, and at normal temperatures bulb formation requires a minimum of twelve to sixteen hours of light per day. At the initiation of bulbing the leaf bases swell, cell enlargement occurs, and assimilates translocate to these tissues. It has been shown that application of ethephon stimulates bulb production under noninductive short-day conditions (Levy and Kedar, 1970). In a field trial ethephon was applied once or several times to five cultivars of onion beginning at the stage at which seedlings had four to five true leaves. Except for one cultivar, 'Aichi

Shiro,' which developed bulbing under natural day-length conditions, treatments resulted in early bulb initiation, accelerated maturation, and increased bulbing. Concentrations of 5,000 and 10,000 ppm were more effective than lower concentrations, but they also retarded leaf growth and final bulb size (Levy and Kedar, 1970).

WIDENING THE CROTCH ANGLES OF FRUIT TREES

Primary branches that form narrow angles with the trunks of fruit trees produce structurally weak crotches that often break under a normal crop load and thus reduce the tree's productivity. Such crotches are also subject to winter injury and susceptible to disease. Methods to increase the crotch angles of young trees include tying the branches down, placing disks or other devices between the main stem and the lateral branches, attaching weighted objects to the laterals, and heading young trees at a low level, but such methods are not totally effective.

Sprays of TIBA at concentrations of 30 or 50 ppm applied to one-year-old trees of 'Delicious' apple slightly retarded terminal growth and produced trees that were more open and spreading than the untreated trees (Bukovac, 1963). The mean primary branch angles for trees sprayed with TIBA at concentrations of 0, 30, and 50 ppm were 47, 57, and 61 percent, respectively.

Applications of cytokinins alone or in combination with gibberellin also increase crotch angles of trees of 'Delicious' apple. Williams and Billingsley (1970) applied lanolin paste containing BA at a concentration of 0.02 percent, PBA at a concentration of 0.02 percent, or PBA at a concentration of 0.02 percent plus GA_{4+7} at a concentration of 0.01 percent to dormant buds of young nursery trees several weeks before planting. The crotch angles of shoots that developed after planting were as follows: controls, 34°; BA, 54°; PBA, 66°; PBA plus GA_{4+7}, 59°. Satisfactory permanent primary scaffold limbs could be selected from the treated branches (Fig. 11–7). Another interesting effect was that total shoot growth on the trees treated with BA plus GA_{4+7} or PBA plus GA_{4+7} was more than double that of the controls. This was the result of an increased number of actively growing shoots on a tree with a larger leaf surface available for photosynthetic activity.

Another advantage of an open-spreading branch structure is that it renders some trees more suitable for mechanical harvest. In citrus, for example, a low, spreading canopy that is open in the center allows

FIGURE 11-7
Shoot of 'Red Delicious' apple on right, treated with BA and
GA_{4+7} in lanolin, developed a larger number of primary
branches with wider crotch angles than the control (*left*).
(After M. W. Williams and H. D. Billingsley, 1970.)

more fruit to reach the catching frame without hitting interfering
branches. Boswell and McCarty (1970) topped one-year-old seedlings
of four varieties of citrus. Soon after the new shoots began to grow,
trees were sprayed with solutions of TIBA. The compound at a concentration of 500 ppm resulted in an outward bending of branches,
providing a good selection of branches from which to form an open
scaffold.

CHEMICAL PRUNING OF ORNAMENTAL PLANTS

Pinching, or the removal of the apical portion of plant shoots, and
pruning are laborious and costly operations. Pinching and pruning of

ornamental plants to control growth has four main purposes: (1) to increase the number of shoots, (2) to control plant shape, (3) to increase the number of flowers, and (4) to control time of flowering.

It has been reported that the fatty acid methyl esters and alcohols with chain lengths of C_8 to C_{12} are highly active in killing or inhibiting axillary bud growth of tobacco plants (Cathey *et al.*, 1966a; Tso, 1964). The effectiveness of the chemicals varies with the length of the carbon chain; best results are usually obtained with the C_{10} chain. These chemicals are referred to as chemical pruning agents or tobacco sucker control compounds. The compounds destroy only the meristematic cells, and the toxic effect is localized in the area of application; there is no apparent translocation of chemical to other plant parts.

Cathey *et al.* (1966a) reported that the lower alkyl esters of fatty acids effectively and selectively destroy the terminal buds of many

FIGURE 11-8

Number of flowers of *Chrysanthemum morifolium*, cv. 'Fred Shoesmith,' can be controlled by chemical means. Plants were propagated on July 15, 1965, potted on August 5; growing point was removed on August 19; 8-hour days at 17°C commenced on September 1. Plants were photographed on November 17. *Left*, plant sprayed with 1 percent HAN and 0.1 percent Triton X-100 (a wetting agent) on September 15. *Center*, plant sprayed with 0.1 percent Triton X-100 on September 15. *Right*, plant whose lateral flower buds were removed by hand on October 6. Two flower buds developed subsequently. (After H. M. Cathey, A. H. Yoeman, and F. F. Smith, 1966b.)

plants; in other words, the compounds effectively "pinch" the shoot. The effectiveness of the concentration varies with the woodiness of the tissue. Methyl nanoate and methyl decanoate pinch shoots of such herbaceous plants as coleus, cotton, ageratum, peanut, marigold, soybean, snapdragon, and tomato, as well as such semiwoody plants as carnation, chrysanthemum, forsythia, geranium, hydrangea, and poinsettia, and such woody plants as apple, azalea, chamaecyparis, juniper, elm, kolkwitzia, ligustrum, lonicera, maple, paper birch, pyracantha, taxus, weigela, pear, and euonymus. The ranges of concentrations used for chemical pruning of herbaceous, semi-woody, and woody plants are 0.025 to 0.05 M, 0.05 to 0.16 M, and 0.16 to 0.27 M, respectively. Cathey et al. (1966a) have also reported that axillary buds are killed when too high a concentration of compound is applied, but nonactive, resting terminal buds are not killed (Fig. 11-8). DMSO at a concentration of 0.26 M increases the toxicity of the emulsions of fatty acid esters and alcohols applied to woody plants (Cathey et al., 1966a).

Results obtained from application of fatty acid derivatives to seven different plant species showed that compounds with chain lengths shorter than C_9 or longer than C_{11} were generally less effective than those with lengths of C_9, C_{10}, or C_{11} (Cathey and Steffens, 1968). However, the fatty acid methyl esters with chain lengths of C_8, C_{10}, and C_{12} produced a satisfactory pinching effect on 'Improved In-

FIGURE 11-9
Axillary bud growth of plants of 'Improved Indianapolis Yellow' chrysanthemum after being sprayed with emulsions containing 0.16 M of varying chain lengths. *Left to right:* plants treated with C_6, C_8, C_{10}, C_{12}, C_{14}, and C_{16} fatty acids. Photographed 21 days after treatment. (Photograph courtesy of H. M. Cathey.)

dianapolis Yellow' chrysanthemum (Fig. 11–9); C_{10} produced the greatest weight of lateral growth.

The selective action of the fatty acid derivatives as chemical pruning agents depends on the type and amount of surfactant used to emulsify them (Steffens and Cathey, 1969). The emulsions formed by the blending of the two compounds penetrate only the youngest meristematic tissues and kill them. The emulsions of fatty acid derivatives and surfactant probably remain on or near the surfaces of more mature tissues without injuring them. In the absence of a suitable surfactant, nonselective tissue injury often occurs.

Azalea

Chemical pruning has been best adapted for use on the azalea plant. Killing of the terminal meristem stimulates the growth of lateral shoots, which greatly increases the number of flowers per plant. Growth of shoots developing from chemically pinched stems may be initially delayed by a few days, but usually within a few weeks growth is equal to, or perhaps even greater than, that of unsprayed plants.

Such chemicals as methyl caproate, methyl caprylate, methyl caprate, methyl laurate, methyl myristate, methyl palmitate, and methyl stearate were found to produce the pinching effect when sprayed on azalea (Furuta, 1967). Methyl caprylate is most effective in killing the terminal bud of the variety 'Pride of Dorking.' The specific components of the spray are important. Since the chemicals are not soluble in water, a surfactant must be added to form a stable emulsion. The surfactant also facilitates uniform application to the foliage and aids penetration of the active material. Manufacturers incorporate the appropriate amount of surfactant in the pinching compounds. The proper selection of the exact ester or combination of esters and use of the proper kind and concentration of surfactant are essential for the preparation of an effective formulation. Much research has been conducted on prepared formulations. A new and different formulation requires experimentation to determine its suitability and effectiveness.

The spray should be prepared in such a way as to ensure the production of small droplets of the pinching chemical. The method recommended for azalea by Furuta (1967) is as follows:

1. To the proper amount of concentrate, add a small amount of warm water—about one-fourth that of the concentrate. Stir until uniformly blended.

2. Add and blend a second and equal portion of water.
3. After the first, and possibly after the second portion of water, the mixture will thicken into a gel. Further additions of water produce a cloudy, thin mixture.
4. Continue to add small amounts of water and blend until the mixture becomes thin and cloudy.
5. While continuing to stir, bring the mixture to a final volume with cool water. Agitate and shake the solution thoroughly. The emulsion will be cloudy or opalescent and will be stable.

Since the chemicals are not translocated, the terminal buds of the plant must receive the proper amount of the pinching agent. A proper concentration wets the plant, deposits an effective amount of chemical on the terminal buds, and does not cause foliage injury. If shoots are protected with leaves or flowers or if nonuniform applications of spray are made on plants with many shoots, some of the shoots will not be properly pinched. A fine mist, applied through a solid-cone nozzle at 40 to 60 pounds of pressure, is very effective (Furuta, 1967). Plants should be sprayed until some excess liquid drips off in order to produce even pinching and adequate kill of the terminal buds. A concentration that is too weak may injure only the young, partially expanded foliage. As a result, some lateral shoots begin to grow because terminal growth is inhibited for only a short time. Growth of lateral shoots is often undesirable from a horticultural point of view.

An overdose of compound may cause excessive injury to foliage and inhibit growth of lateral buds for a considerable distance from the terminal portion of the stem (Furuta, 1967). Overdose may result either from the application of too strong a concentration of the chemical or from too frequent an application of a normal dose. Injury often occurs when large droplets of spray collect on the leaves and overlapping leaves hold drops of spray between them. The most effective pinching chemicals usually cause the greatest injury to foliage. The use of unstable emulsions results in foliage injury, and with these there is little or no control of the growing plant.

The effectiveness of the concentration depends somewhat on temperature. Optimal temperatures range from 21° to 29°C. At temperatures below 21°C, a slightly stronger solution (up to 0.5 percent) should be used; slightly weaker concentrations are preferable at temperatures above 29°C. Plant temperatures are pertinent only at the time of application and for thirty to sixty minutes afterward.

The more succulent the foliage, the more likely it is to be injured by the pinching compounds. Plants grown under controlled conditions in a greenhouse are often severely damaged by concentrations that cause no injury when applied to plants grown outdoors. Immature leaves are more sensitive to injury than are mature ones. The injury becomes apparent within a few minutes after spraying, when the meristems blacken. The small amount of foliage injury caused by use of the proper concentration and rate of application for pinching does not decrease the potential sale value of azalea.

Foliage injury may be lessened by washing off the chemicals with water ten to twenty minutes after application (Furuta, 1967). Washing too soon after application reduces the effectiveness of the chemical. Washing is practiced only when severe injury is expected either because of the physiological condition of the plant or because of environmental conditions. Under usual conditions, washing can be omitted because foliage damage is not excessive.

If plants have been sprayed with pesticide within one week of treatment, they should be washed before the chemical pinching agents are applied to remove the possible influence of residual surfactants, emulsifiers, or pesticides on the action of the pinching agents. The plants should be thoroughly watered before treatment, and foliage should be dry when the pinching agents are applied.

Varieties of azalea differ in sensitivity to the chemicals. In a comprehensive varietal study of azalea in which a formulation containing chemicals having chain lengths of C_8 and C_{10} was used, a 5 percent spray proved to be an overdose for the cultivar 'L. J. Bobbink' but not for any of the other twenty-four varieties treated (Furuta, 1967). In this study, a 3 percent spray did not kill the terminal bud in most cultivars. If there is any doubt as to the correct concentration, only a few plants should be treated with the selected strength. Necessary adjustments of concentrations can be determined within a few hours. Application of pinching chemicals must be more frequent than manual pinching to prevent development of straggly plants.

The varietal difference in effectiveness of pinching agents may be explained at least in part by differences in bud morphology. The vegetative buds of 'White Gish,' a variety that responds poorly to methyl decanoate, are enclosed by an abnormally long sheath of leaves that is not present in 'Coral Bells,' a variety that responds readily to the chemical (Sill and Nelson, 1970). The path of downward movement of pinching agent emulsion is 59 percent longer in 'White Gish' than in 'Coral Bells,' and this longer path causes a reduction in movement of compounds to the meristematic tissue. The pinching agent is less effective on the young reproductive buds of 'Red Wing' than on the vegetative buds, partly because the reproductive buds are surrounded by a greater number of immature leaves that act as barriers. Penetration of the compound into older reproductive buds is even more difficult because of the overlapping bud scales that obstruct the small aperture that was present when the bud was still in the vegetative condition (Sill and Nelson, 1970).

Chrysanthemum

Chrysanthemums with large flowers are usually disbudded manually. For each flower bud that is allowed to develop, growers must remove ten to twenty-five lateral flower buds, a tedious and time-consuming operation. Alkylnaphthalenes, which are used as solvents for chlorinated insecticides and herbicides, can cause the death of apical meristems of chrysanthemums (Cathey et al., 1966b). HAN (a product of the Humble Oil and Refining Company of Houston, Texas), a selected petroleum fraction of high aromatic content (approximately 87 percent by weight), is very effective. Approximately one-half of the aromatics in HAN are alkylnaphthalenes, mostly C_{10} to C_{13}. The action of HAN may be similar to that of the methyl esters of the fatty acids that kill all tissues with high rates of cell division.

Cathey et al. (1966b) used two methods of application. One method was to place plants in large drums and expose them for one to two hours to oil aerosols atomized with air pressure of 10 psig. The petroleum fractions caused some distortion of the leaves and, depending on the state of development of the plant, death of the apical meristem and abortion of the lateral flower buds. The heavy aromatic naphthas were particularly active in the tests.

In the second method, spray application tests were performed on *Chrysanthemum morifolium* cv. 'Fred Shoesmith.' The aromatic products were emulsified in water and were applied to plants when the apical flower bud had formed but before initiation of the lateral flower buds. HAN caused the abortion of partially initiated flower buds. The terminal flower was not affected by the chemical and developed at the same rate as terminal flowers of plants from which the side flower buds had been removed manually. If applied too early, while the plants were rapidly growing, HAN also killed the growing point and no flowers were produced. If applied too late, after lateral flower buds had completely formed, the treatment was also ineffective.

The time of treatment with HAN, after plants are placed under eight-hour, short-day conditions for flower induction, is of utmost importance. In this experiment, Cathey et al. (1966b) found that treatment with HAN on the ninth and twelfth short days killed the apical meristem, but the lateral flower buds seven or eight nodes down the stem developed flowers. Treatment on the thirteenth, fourteenth, and fifteenth short days caused abortion of the lateral flower buds, but the terminal flower bud was not affected and flowered at the same time as terminal flower buds of plants from which the lateral flower buds had been removed by hand. Treatment on the sixteenth short day or later caused the abortion of only a few of the lateral flower buds.

Cathey and his colleagues also found that HAN was most effective on 'Fred Shoesmith' when applied as a 1 percent foliar spray; a 2 percent foliar spray was most effective during the summer. The optimal time of treatment, as measured by the number of days of exposure to short days, was the fourteenth or fifteenth short day. Treatments made on sunny days were less effective than those made on cloudy days.

The response of various cultivars of chrysanthemum to treatment with HAN varies greatly. In a September test conducted by Cathey et al. (1966b), 'Yellow Delaware' and 'Princess Anne,' when treated with 3 percent and 5 percent emulsions of HAN, respectively, aborted only a portion of their lateral flower buds. 'Shasta' and 'Improved Indianapolis Yellow' developed yellow leaves with black margins in response to treatment with a 5 percent emulsion of HAN, and the lateral flower buds were unaffected by all dosages tested.

Shanks (1969) demonstrated that the number of branches formed by a plant increase when ethephon is applied in conjunction with chemical pinching agents. Ethephon at concentrations of 1,000 or 2,000 ppm was applied to chrysanthemum cv. 'Golden Yellow Princess Anne' following treatment with different concentrations of a pinching compound. Ethephon increased the number of branches produced at all concentrations of the pinching compound; the greatest increase resulted when application of ethephon followed application of a suboptimal level of the pinching compound.

CHEMICAL PRUNING OF ROOTS

Kinked, twisted, or circled root patterns often form when nursery plant seedlings are not pruned and carefully transplanted (Nussbaum, 1969). Such plants often have a short life-span or are not vigorous. If a species has tap roots, more lateral branches are obtained by pinching the root tip. The resulting well-branched root system usually provides better support and anchorage as the plant develops.

Nussbaum studied techniques by which the roots would be pinched as they reached the bottom or chemically treated layers in the seedbed, thus eliminating the need for time-consuming manual root pruning. In one experiment germinating acorns of cork oak (*Quercus suber*) were planted in flats three inches above layers of Osmocote (a controlled-release fertilizer), Treflan (a preemergence herbicide), or Perlite soaked in copper naphthenate (a wood preservative). The tap roots were unable to penetrate any of the chemical treatments, and upon transplanting, a more fibrous, compact root system was formed in each seedling as a result of the chemical pinching of the primary root. Nussbaum (1969) found that a better and more convenient method of chemically pinching roots is to paint copper naphthenate on the inside bottom of the flat. This method reduced the length of the primary root of Jeffrey pine (*Pinus jeffreyi*), roundleaf eucalyptus (*Eucalyptus polyanthemos*), and mesquite (*Prosopis tamarugo*) compared with that of the untreated controls. The average lengths of primary roots of Jeffrey pine, measured thirty days after planting, were 23.8 cm for controls and 6.2 cm for plants grown in flats painted with copper naphthenate. The seedlings of all species growing in the flat painted with copper naphthenate had 50 percent more secondary roots per unit length of primary root than did the controls and also had more secondary roots on the upper 6 cm of the primary root.

RETARDING PLANT GROWTH

It is often desirable to minimize the elongation of plants. Although a general retardation of plant growth may often be desirable, maintenance of a certain level of growth may sometimes be necessary. Overly elongated plants are spindly and unattractive (an important consideration with ornamentals) and develop poorly after transplanting. Plant growth retardants frequently can control the amount of vegetative elongation (Fig. 11–10). Other inhibitors, such as MH, TIBA, and morphactins, are sometimes also effective.

Flowers

Losses are suffered by the nursery industry each year as a result of container-grown plants that grow too large. Another important effect of growth-retardant sprays is the promotion of precocious and heavy flowering in many woody plants. This effect was thoroughly discussed in Chapter 7.

FIGURE 11-10

Effect of growth retardants on 'Princess Anne' chrysanthemum (*top*) and 'Blue Blazer' ageratum (*bottom*). *Left*, plant that received Phosfon-D soil drench; *center*, control; *right*, plant that received SADH foliar spray. (Photographs courtesy of H. M. Cathey.)

CHRYSANTHEMUM.   Plant size can be reduced by applying Phosfon-D as a soil drench to recently potted chrysanthemum plants. Cathey (1967) applied 200 to 250 ml of a dilute Phosfon-D solution (one part of 10 percent powder or liquid per 160 to 800 parts of water, depending on the cultivar) per 15-cm pan. SADH is also effective and should be applied as a foliar spray at concentrations of 2,500 to 5,000 ppm. Sprays of SADH should be applied to chrysanthemum two weeks after the start of short days to reduce the height of potted plants and should also be applied at time of disbudding to retard further elongation of the pedicel, to improve flower form, and to increase flower size.

RHODODENDRON.   To promote compact growth and early flower bud initiation of actively growing hybrid rhododendron plants, 0.1 to 0.2 g of Phosfon-D should be applied to the soil in which the plants are grown, in a 15-cm container (Cathey, 1967). Rhododendron plants should be grown on long days (natural days plus incandescent light of 20 foot-candles from 10 P.M. to 2 A.M.) at 18° C. Flower buds normally appear on treated plants three to four months after treatment with the chemical. The plants are then placed on eight-hour days to stimulate flower bud development. When the buds have formed, the plants are transferred to temperatures below 10° to 13°C for a minimum of eight to ten weeks to terminate the dormancy of the flower buds. Plants are then returned to the greenhouse, where they flower in four to eight weeks, depending on cultivar.

POINSETTIA.   To retard growth of poinsettia, CCC should be applied as a soil drench in a ratio of one part of an 11.8 percent solution to forty parts water (Cathey, 1967). The solution (60 ml per three-inch pot, 200 to 250 ml per six-inch pot) should be applied two weeks after potting or when the root system is well established.

In tests with potted poinsettias, Besemer (1967) compared the effectiveness of CCC used as a pot drench or foliage spray and SADH applied as a foliage spray. The shortest plants were of the varieties 'Elizabeth Ecke' and 'Paul Mikkelson,' grown in six-inch pots and given a single drench of CCC at a rate of 3 ounces per gallon. This treatment caused more retardation when applied ten days after planting than when applied the day following planting. Sprays of SADH at concentrations of 0.5 percent and 1.0 percent were much less effective. Stem diameters were not markedly reduced by any of the treatments. Lindstrom (1967) also noted that CCC was much more effective than SADH on 'Paul Mikkelson.'

The response to growth retardants is influenced by timing, environmental factors, and variety. The 'Paul Mikkelson' variety evidenced less response than did 'Elizabeth Ecke.' The United States Department of Agriculture poinsettia introduction 'Stoplight' showed little response to either SADH or CCC at the usual concentrations.

AZALEA.   Growth retardants are effective in producing more compact growth. Both CCC and SADH are applied for this purpose. The primary commercial benefit of the use of plant growth retardants on azalea is to induce more profuse flowering. The details of this application were discussed in Chapter 7.

KALANCHOE.   The development of a number of hybrids has improved the usefulness of this plant for indoor landscaping. The hybrids have varied and attractive foliage and a wide range of flower color. However, the plants often grow tall and have spreading inflorescences, and are thus top-heavy and difficult to handle during production and shipping. Frequent pinching must be performed by hand to maintain a desirable compactness.

Applications of SADH can produce a compact plant and replace the pinching operation (Nightingale, 1970). The hybrid variety 'Mace' was sprayed with SADH at concentrations of 0, 1,000, 5,000, and 10,000 ppm at the time that the apical bud was normally removed by

FIGURE 11-11

Effect of applications of SADH on growth of *Kalanchoe* hybrid 'Mace.' *Left to right:* control; plants that received SADH at concentrations of 1,000 ppm, 5,000 ppm, and 10,000 ppm. (After A. E. Nightingale, 1970.)

hand. Plants were sprayed again fourteen and twenty-eight days later with each concentration. Plant height and span of the inflorescences were reduced in proportion to the concentration of compound used (Fig. 11–11). Although there was a reduction in the number of flowers as a result of SADH treatment, the overall result was a well-proportioned, desirable plant. The compound at a concentration of 5,000 ppm delayed full bloom by approximately one week, compared with the control; a concentration of 10,000 ppm caused a delay of ten days.

Woody Ornamentals

The growth rate of inherently tall or rapidly growing species may be sharply reduced to the desired level by application of the appropriate chemical. MH, SADH, Phosfon-D, CCC, TIBA, and morphactins are among the useful and effective growth inhibitors.

MH is one of the oldest growth inhibitors on the market and has herbicidal effects when used at high concentrations. However, because it is an inhibitor of meristematic activity in general, it prevents leaf and flower initiation and fruit set and enlargement and also retards stem elongation. MH usually kills the terminal bud; some of the lateral shoots then commence vigorous growth and produce a ragged shrub or tree. Even though MH is rapidly absorbed by the foliage and translocated from shoots to roots, it does not accumulate at high concentrations in all axillary buds. A major difficulty in using MH is the selection of a dosage high enough to stop growth of the terminal bud without killing it (Sachs *et al.*, 1967). In some species inhibition of leaf initiation and expansion of newly formed leaves results in an unsightly shrub.

When application of MH is timed to follow pruning or bud break in the spring, concentrations of 0.1 to 0.5 percent but no higher than 1.0 percent (of a commercial preparation that contains 58 percent diethanolamine salt) are satisfactory for *Rhamnus alaternus, Cotoneaster pannosa, Ligustrum lucidum, Ulmus parvifolia, Pittosporum undulatum, Viburnum japonicum,* and several other species (Sachs and Maire, 1967).

Figures 11–12 and 11–13 illustrate the growth control afforded by MH. Higher dosages are required to inhibit growth when applications are made late in the season after shoots are further developed. The species *Pinus* and *Juniperus* may tolerate higher dosages of MH with little of the foliar damage or tip dieback that frequently occurs in many treated broad-leaved plants.

High humidity increases the penetration of MH. Smith *et al.* (1950)

FIGURE 11-12

Portions of hedges of *Rhamnus alaternus* (*left*) and *Cotoneaster pannosa* (*right*) in foreground were sprayed with MH in October after having been pruned in September. Both photographs were taken the following April. (After R. M. Sachs, W. P. Hackett, R. G. Maire, T. M. Kretchun, and J. de Bie, 1970.)

noted that after forty-eight hours, twice as much MH is absorbed at 100 percent relative humidity as at 75 percent. Sachs and Maire (1967) made field plot studies on newly rooted cuttings of woody ornamental plants at Davis, Westwood, and Tustin, California. Davis, located in the Sacramento Valley, has low humidity from late spring to early fall, with occasional exceptions. The two coastal cities in the Los Angeles area are within six to nine miles of the ocean and are thus cooler and more humid than Davis. MH was considerably more effective in Westwood and Tustin than in Davis. A concentration of 1,000 ppm caused a 42 percent reduction in growth in Westwood, whereas concentrations of up to 2,500 ppm had no significant effect in Davis. Similar results were obtained with SADH. In Westwood and Tustin, SADH caused reduction in growth of *Cotoneaster pannosa*, but the compound had no effect in Davis even at a concentration of 10,000 ppm. April applications of MH made in Davis reduced vegetative growth of *Cotoneaster, Pyracantha,* and *Oleander* more effectively than applications made in July.

FIGURE 11-13

Hedge of *Ligustrum lucidum* in foreground was sprayed with MH in October after having been pruned in September. The original trim line was maintained for several weeks during the following spring, as shown by this photograph taken in April. (After R. M. Sachs, W. P. Hackett, R. G. Maire, T. M. Kretchun, and J. de Bie, 1970.)

Greenhouse experiments with seedlings and rooted cuttings treated with ^{14}C-labeled MH and SADH revealed large daily variation in foliar penetration, suggesting that relatively short-term changes in climatic conditions play a paramount role in controlling movement of these substances into the leaves (Sachs and Maire, 1967). Since the daily relative humidity and evaporation rates vary as much as 50 percent or more, from a high before sunrise to a low in midafternoon, the time of day that application is made may play an important role in determining response to MH and SADH. Multiple applications of low dosages of MH and SADH may be preferable to single high dosages from the standpoint of improved penetration. Multiple applications of MH result in more control of the growth of axillary branches than does a single application of the compound. However, increased axillary branching is desirable during the early stages of propagation and in certain landscape situations for rejuvenation of old, heavily pruned shrubs.

The activity of MH and SADH is usually greatly enhanced by the addition of wetting agents at concentrations of 0.1 to 0.5 percent (Sachs and Maire, 1967). Penetration rates of compounds vary considerably among species, and this variability may at least partly explain differences in susceptibility of species to the compounds.

Regrowth after pruning is not affected by chemical treatment, which indicates that there is apparently very little translocation of MH or SADH into new branches that arise from previously dormant basal axillary buds. Another explanation is that because MH and SADH are inactivated within two months after application, there can be some carryover when these compounds are applied late in the fall (Sachs and Maire, 1967).

Long-term MH treatments (lasting twelve years or more) have no deleterious effects on the longevity of healthy shrubs (Sachs et al., 1970). Professor J. E. Knott at the University of California at Davis has sprayed a Pyracantha hedge with MH at a concentration of 0.5 percent yearly since 1945 except for 1950, 1965, and 1966. He has found no evidence of injury to the hedge (personal communication).

A concentration of 1 to 2 percent of SADH is the most effective for control of the vegetative growth of shrubs and trees, although concentrations of 0.25 to 0.5 percent are optimal for some species, such as Ligustrum japonicum. As with MH, applications must be made within one or two weeks after bud break in the spring or after pruning to ensure success for most species. Sachs and Maire (1967) recommend addition of a wetting agent at a concentration of 0.1 to 0.5 percent to improve the effectiveness of the treatment. Alcohol (mehyl, ethyl, or isopropyl) at 30 percent greatly increases the initial SADH-induced inhibition of elongation and prolongs the effect of a single application, but alcohols may not be practical adjuvants for commercial formulations.

Applications of SADH have varied results if the time of application is not carefully selected. Sachs et al. (1967) found that the 1 percent solution caused as much as a 90 percent reduction in stem elongation under optimal conditions but that inhibition was as low as 10 percent when applications were made several weeks after bud break or between flushes. SADH, or more active analogues, offer great promise for landscape and nursery application because of the high tolerance of most plants to these compounds, which cause few phytotoxic side effects. Treatment usually improves the plant's appearance because it results in darker green foliage, shorter internodes, and, in some species, heavier flowering.

Phosfon-D and CCC are best used as soil drenches, a method of application that is too expensive to be used commercially to solve most field landscape problems. TIBA inhibits the growth of soybeans and, when applied under experimental conditions, has increased yield up to 20 percent.

Morphactins are relatively new experimental compounds having fluorene nucleii (Schneider *et al.*, 1965a). Several derivatives are available, of which IT 3456 is one of the most active. Morphactins have universally demonstrated extensive herbicidal growth-regulating properties. At the most effective concentrations (ranging from 0.001 percent to 0.1 percent as a spray application), morphactins have caused foliar distortion, inhibition of stem elongation, and considerable axillary bud break.

The effect of morphactins on axillary bud development may be of value because of the need to improve branch development during the early stages of propagation of seedlings and rooted cuttings. Encouraging results have been obtained with *Callistemon citrinus, Juniperus tamariscifolia,* and *Euonymus japonicus* when morphological aberrations were temporary or not too severe (Sachs *et al.*, 1967). MH and the fatty acid esters are also useful for increasing branching but they simply act as a pinching agent, killing the terminal bud. The morphactins, however, appear to cause a release in inhibition of axillary buds that is systemic (all buds begin growing), and their mode of action is doubtlessly very different from that of MH and the fatty acid esters.

Nursery Trees

Because nursery stock is often larger than the size desirable for salable trees and because larger trees require higher storage and shipping costs, nurserymen spend large amounts of money each year pruning plants before shipping them.

Since digging is often done in the fall, leaves must be removed from the trees by hand in advance. Terminal buds should be formed and hardened to some degree before leaves are stripped. In the western part of the United States, withholding water generally is effective for hardening of terminal buds, although some trees become too dry even under ideal conditions. In addition, enough moisture may remain in some parts of the nursery to allow tree growth and thus delay hardening. A spray that would induce terminal bud formation and prevent further growth would require less rigid water control and result in smaller trees that were more suitable for storage and subsequent planting.

Sprays of SADH applied to trees of apple, pear, cherry, and plum reduce growth and induce terminal bud formation (Stahly and Williams, 1967). In one experiment, the following combinations of variety and rootstock were used: 'Golden Delicious' apple/'EM VII' apple, 'Bartlett' pear/'Bartlett' pear, 'Early Italian' plum/'Lovell' peach, 'Stanley' plum/'Lovell' peach, 'Van' cherry/mazzard cherry, 'Chinook' cherry/mazzard cherry, and 'Bing' cherry/mazzard cherry. Actively growing trees 5.0 to 5.5 feet high were sprayed in late July with SADH at a concentration of 2,000 ppm. Growth of all varieties was reduced and terminal bud formation was induced. There was no effect on trunk diameter, and tree growth during the year following treatment was normal. Application of SADH prevents excessive growth of nursery trees and results in more trees of salable size.

Marcelle (1966), working in Belgium, demonstrated that the growth of young trees of 'Tydeman's Early Worcester' apple is effectively checked if they are sprayed in the nursery with either CCC or MH. Application of CCC at a concentration of 2,000 ppm reduced internode length, and treated shoots developed fewer nodes than control shoots. The number of nodes was probably decreased because CCC caused the terminal bud to become relatively dormant. MH at a concentration of 750 ppm reduced growth of the main stem, but stronger dosages often killed the terminal buds. This compound had the advantage of stimulating the growth of lateral branches that had wide crotch angles, but such angles were formed only when the terminal buds were not killed by the chemical. If terminal buds were killed by MH or if pinching was done by hand, lateral branches would emerge but with narrow angles.

In his investigations of several varieties of one-year-old apple trees, Barden (1968) confirmed that SADH treatment retards terminal growth and also obtained a reduction in weight of new stems and fewer nodes. However, the decrease in the weight of new stems was offset by an equal increase in the weight of the old stems. The weight of roots was slightly increased as a result of SADH treatment. The fact that SADH can cause desirable inhibitory effects to the portion of the apple tree growing above ground without suppressing root growth adds to its potential for successful commercial use.

Spraying the terminal twelve inches of cherry trees is just as effective for controlling growth as is spraying whole trees (Stahly and Williams, 1967) but is advantageous because rapid application of the spray can be made either just before or at the time that the trees reach the desired size. Effects of the spray last long enough to prevent late flushes. Two sprays may sometimes be required to control growth adequately.

The growth regulator EDNA may be of value when used in nursery production and in landscaping of trees of red cedar (*Juniperus virginiana*). Sucoff (1968) sprayed young potted red cedar trees with EDNA at concentrations of 1,000 and 4,000 ppm and found that after 110 days the heights of treated plants were decreased to 91 percent and 36 percent of the controls, respectively. Similar results were obtained with arborvitae, but there was less of a decrease. These results are especially interesting since a previous report revealed that low concentrations of EDNA stimulated the growth of bean, cabbage, and a rye-grass-clover mixture (Holmsen, 1966).

Fruit Trees

Controlling the size of fruit trees has held the interest of researchers and fruit growers alike for several decades. Dwarfing of rootstocks has been the chief means of producing smaller trees, but chemicals now offer a second method of controlling tree size. Figure 11–14 illustrates the retarding effect of SADH on the growth of peach shoots.

SADH shows considerable promise for controlling the growth of apple, cherry, and pear trees (Batjer *et al.*, 1963, 1964). Three spray applications of SADH at concentrations of 500 to 2,000 ppm were made ten days apart, fifteen to seventeen days after full bloom. Most treatments greatly reduced shoot growth; shoot growth of trees treated with a concentration of 2,000 ppm was reduced by one-fourth to one-half, compared with that of unsprayed trees. A single spray of 2,000 ppm applied to 'Starking' was not as effective as three sprays of the same concentration. The latter treatment caused earlier formation of terminal buds and formation of shorter internodes, both of which resulted in a reduction in shoot growth. Generally, internodes of sprayed trees were only one-half to two-thirds as long as those of untreated trees. Sprayed trees, particularly apple, had larger terminal shoot leaves. The larger size, combined with the greater number of leaves per unit length, resulted in much denser foliage.

SADH reduced the size of apple and pear fruit both during the season the sprays were applied and the year following. However, the size of 'Lambert' cherry was not affected. Generally, the increase in trunk circumference of treated apple trees was greater than that of unsprayed trees. The reverse was true with cherry trees, but the treatment had no apparent effect on pear trees.

The 'Golden Delicious' variety showed a tendency toward shorter

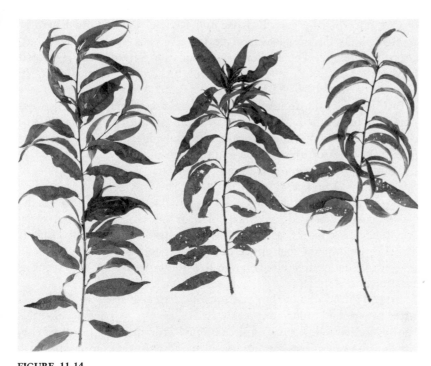

FIGURE 11-14

Effect of SADH on growth of 'Redhaven' peach shoots. Typical shoot from control tree (*left*), compared with shoot treated with SADH at concentrations of 2,000 ppm (*center*) and 5,000 ppm (*right*) on July 23, 1964. Photographed in September. (After L. J. Edgerton, 1966.)

stems the year following treatment, but no such tendency was observed in the 'Delicious' variety or in any variety of cherry or pear.

Other results of application of SADH to apple trees were an increase in leaf area and the production of thicker leaves (Halfacre *et al.*, 1968). In their studies of young trees of 'Golden Delicious' and 'York Imperial' apple, Halfacre and Barden (1968) found that SADH treatment caused development of longer palisade cells and a looser arrangement of the spongy mesophyll cells, both of which caused thickening of the leaves.

Leaves of apple trees also often develop a darker green color as a result of SADH application (Edgerton and Hoffman, 1965; Halfacre *et al.*, 1968). Halfacre *et al.* (1968) found that there was more chlorophyll per unit area of fresh tissue as a result of SADH treatment. However, the net assimilation rate of leaves, as determined by the amount of CO_2 absorbed by each leaf per unit of time, was less for leaves of treated plants than for leaves of untreated plants except when senescence had begun. The higher net assimilation rate of SADH-treated leaves after onset of senescence might be a result of the delay in senescence caused by SADH.

In California, the trend is to plant pear trees in narrow hedgerows that can be pruned and harvested from mobile platforms. Heavy pruning is required to maintain the trees in hedgerows. The vegetative growth of trees of 'Bartlett' pear trained in hedgerows was found to be suppressed by repeated sprays of SADH at a concentration of 3,000 ppm (Griggs and Iwakiri, 1968). The first spray was applied when 85 percent of the flowers had shed their petals and when the oldest leaves averaged two-thirds of their full size; the second spray was made two weeks later. When elongation ceased, treated shoots growing six to eight feet above the ground were 25.4 percent as long as control shoots; treated shoots in the tree tops were 47.6 percent as long as control shoots.

In general, to obtain the greatest retardation of growth, the sprays must be applied sufficiently early in the season to become effective during the period of most rapid terminal growth. An appreciable reduction in growth becomes apparent one to two weeks after spraying. The amount of growth inhibition of a given variety is determined for the most part by the concentration and the number of applications.

Cutting the tops of lemon trees has been a commercial practice in California for several years. MH and ethyl hydrogen 1-propylphosphonate have been found to retard the regrowth of the cut shoots (Burns et al., 1970).

Shade Trees and Other Large Trees

The maintenance problems presented by the need to restrict the size of trees growing along city streets and in parks and cemeteries will, of course, become more difficult and expensive with the increase in population and the corresponding increase in the demand for public services. For example, utility companies have spent millions of dollars in some years to have pruned by hand the trees that interfered with transmission lines. However, chemical control of vegetative growth makes possible a considerable reduction in these labor and maintenance costs.

MH is a valuable aid in suppressing the growth of shade trees (Hamner and Rai, 1958; Sachs et al., 1970). Control of the growth of sycamore trees by MH is shown in Figure 11–15. Improper timing of

FIGURE 11-15

Sycamore trees on this street were pruned in April. Trees on the left were sprayed with MH in July; photograph was taken the following April. New growth appeared on both rows of trees but branch elongation of treated trees was inhibited. (Photograph courtesy of Sohner Tree Service, San Anselmo, California.)

application of this compound can result in unnecessary damage or can reduce effectiveness of the treatment. Although winter application does not affect plant growth, foliar application in early spring results in striking inhibition. Hamner and Rai (1958) obtained satisfactory results from spring application of MH at concentrations of 0.2 and 0.8 percent to trees whose leaves were fully expanded or three-fourths expanded. The compound suppressed apical dominance; the degree of inhibition was correlated with species, stage of growth of the tree, and amount of chemical used.

Experiments conducted in California have shown that the proper time for spraying is determined by the number of fully expanded leaves on newly formed shoots. Excessive damage can occur to deciduous species if MH is applied before the new leaves have expanded in the spring (Sachs *et al.*, 1970). Since MH prevents or decreases the rate of meristematic activity in plant tissues, stem elongation and growth of young leaves are prevented. If MH is applied after leaf expansion but before extensive stem elongation, the growth of the terminal bud is retarded, and further stem elongation is prevented without damage to the expanded leaves. If MH application is delayed until the stem has almost ceased elongating and leaf formation has almost ceased, the retarding effect is only slight.

In order to attain maximum inhibition of growth of evergreen trees, they should be treated with MH either after pruning or in the spring when axillary buds commence growth. This timing is optimal because newly developing evergreen leaves absorb MH more readily than do fully expanded leaves. The compound induces inhibition in terminal buds that are close to the site of absorption of newly developing leaves, and application during bud break prevents excessive elongation of new shoots. A dosage of 0.1 to 0.25 percent (of a commercial formulation containing 30 percent MH), when applied properly, inhibits growth of pine and juniper species for at least four months. Some plant species require a higher concentration of 1.33 percent of solution (Sachs et al., 1970).

Field tests performed on *Ulmus parvifolia* demonstrated that 0.1 to 0.25 percent of a 30 percent solution effectively controls plant growth, causing only slight foliar damage and no delay in leaf initiation. Nursery trials with *Eucalyptus viminalis* showed that 0.1 percent of the solution inhibits growth for sixty days or longer (Sachs et al., 1970).

Preventing Growth of Suckers on Tobacco Plants

Thinning tobacco suckers by hand is a very laborious and costly process. Most tobacco growers in the United States and in some of the other tobacco-growing countries have accepted as standard practice the use of MH to prevent the growth of suckers on flue-cured and burley tobacco (Fig. 11–16), thereby increasing yield and reducing brown spot fungus (Zukel, 1963). However, MH application does result in the production of fewer cigarettes per pound of treated leaves.

MH inhibits the growth of suckers but does not kill them. One pint of a 30 percent solution per 1,000 plants or 20 to 50 gallons per acre are the correct rates. Since MH is absorbed and translocated internally, spraying the entire plant is not necessary; best results are obtained by spraying the upper one-third or one-half of the plant. It is important to apply equal amounts of MH to each plant at the time of topping.

Satisfactory sucker control depends on correct timing of the spray application. The plants should be topped as soon as the plants are in full flower (the stage when 90 percent of the plants have begun to shed the first flowers). Within the next twenty-four hours, all suckers

FIGURE 11-16
Growth of suckers in tobacco can be controlled by chemical means. Plant on left was topped and chemically suckered. Plant on right was neither topped nor suckered. (Photograph courtesy of W. K. Collins.)

should be pulled from early maturing plants and MH should be sprayed uniformly over the entire field (Zukel, 1963). If there is a wide range in time of flowering, early and late flowering plants should be treated separately. Apical leaves that are to be harvested should be at least six inches long at time of spraying.

For best results, the spray should be applied early in the morning on sunny days or at any time of day, with the exception of late afternoon or night, on cloudy days. Spraying should not be performed in the middle of hot, dry days when leaves are severely wilted.

Closely hand-suckered or MH-treated tobacco will not ripen as rapidly as poorly suckered tobacco. Under some conditions, shoot tips that have been treated with MH have a tendency to turn yellow prematurely. However, this condition is not harmful and does not necessarily mean that the tobacco is ripe. Ripeness of MH-treated tobacco is indicated by leaf texture and intensity of color, and by fading of color from the midrib. If tobacco plants are grown under drought conditions, absorption of MH will be slow and the amount of suckering may not be as satisfactory as it would be if the plants had been treated under normal conditions.

In North Carolina Collins *et al.* (1970a) found that the highest yields and values per acre of flue-cured tobacco were obtained when a contact-type chemical was applied in the early flower stage, followed seven to ten days later by an application of MH. This dual treatment also resulted in a higher degree of sucker control than that obtained when MH was used alone. Off-Shoot-T (Procter and Gamble Co.), Penar (Pennwalt), and Conoco T-504 (Continental Oil Co.) were among the effective contact-type chemicals used (Collins *et al.*, 1970b).

Prevention of Lodging

Severe losses to cereal crops often occur as a result of lodging, especially when such crops are grown in extremely fertile soil. Such plants may be elongated and weak and thus easily flattened to the ground by rain or wind (Fig. 11–17).

WHEAT.   Wheat that has lodged is difficult or impossible to harvest. Application of CCC reduces culm length by retarding internode elongation and thus renders plants resistant to lodging.

Increases in wheat yield have resulted from the use of CCC in continental Europe, England, and Israel. The increases have been attributed mainly to the prevention of lodging of the plants, but CCC has sometimes increased yields even when all grain from the unsprayed lodged plants has been recovered. In Israel application of CCC at a rate of 10 kg per hectare resulted in a 30 percent increase in grain yield by causing the production of a greater number of kernels per spike (Pinthus and Halevy, 1965). The varieties tested were 'Florence Aurore 8193,' which is the most commonly grown wheat variety in Israel, and 'M-745,' a variety that lodges. CCC (the active ingredient) at a rate of 10 kg per hectare was applied to the soil twice—one-third on January 7 immediately after seedling emergence and two-thirds on February 25 at the onset of internode elongation. Nitrogen was applied at a rate of 16 kg per acre or at the extremely high fertilization rate of 48 kg per acre. Severe lodging occurred in all the untreated plots, but practically no lodging occurred in the CCC-treated plots of 'Florence Aurore 8193' and only slight lodging occurred in the treated plots of 'M-745.' CCC had reduced culm length by retarding internode elongation.

In England, Humphries *et al.* (1965) found that even when control

FIGURE 11-17

Control of lodging of winter wheat, cv. 'Record.' *Left,* control; *right,* wheat treated with 4 kg of CCC and fertilized with 120 kg of nitrogen per hectare. (Photograph courtesy of H. H. Mayr, Stickstoff Linz, Austria.)

plants did not lodge so that all grain could be recovered and weighed, CCC treatments (2.5 or 5.0 pounds per acre applied at the six-leaf stage) nevertheless increased mean grain yields by 5 percent by increasing the number of ears and the number of grains per ear.

In Austria, Primost (1967a) found that applications of CCC made at the proper time to lessen or prevent lodging increased the yield of 'Tassilo' (a winter wheat) by as much as 60 percent or more over that of the control plots. However, when there was no lodging of the controls, application of CCC resulted in only a slight increase or no increase in yield. In those varieties in which there was a yield increase following treatment, such increase was caused by an increase in the number of kernels per ear, or in the yield from a single ear. The yield increase of one variety resulted from an increase in number of heads per plot. When CCC produced a decrease in kernel weight, an increase in the number of kernels per ear usually compensated for this loss. The degree of shortening of plants and the increase in breadth of leaves as a result of application of CCC varied according to the variety of wheat (Primost, 1967b). Primost (1970b) also found that the best time to apply CCC is between tillering and rapid shoot growth.

Numerous field experiments conducted in Austria showed that in all varieties under study, the shortening of the three lowest internodes that resulted from CCC treatment was more marked than that of the apical two internodes (Primost and Rittmeyer, 1968a). Increasing rates of nitrogen elongated all internodes, but the elongation of the lower internodes was most pronounced. According to these investigators, application of CCC increased the plant's resistance to lodging by inhibiting the elongation of the lower internodes and increasing the thickness of the culm wall.

Long-term field experiments conducted in Austria have shown that a high rate of fertilization, especially with nitrogen, is necessary to obtain the highest yield and best quality of wheat (Primost, 1968). Another advantage of increased nitrogen fertilization is an improvement in baking quality (Steiger et al., 1969).

RYE. Field experiments conducted in Austria have shown that application of CCC to rye increases resistance to lodging (Primost and Rittmeyer, 1968b). Lodging was delayed in some trials and completely eliminated in others. CCC treatment did not affect the baking quality of the grain. In some localities, the compound increased grain yield. The combination of CCC and nitrogen fertilizers was found to be most effective on some soils (Primost, 1970a). Varieties of rye showed greater differences in their response to CCC than did varieties of wheat (Primost and Rittmeyer, 1970).

BARLEY. Barley is another cereal crop that is very susceptible to lodging. In Canada, Larter et al. (1965) made detailed studies of the effects of CCC on the barley varieties 'Parkland' and 'Hannchen.' Under controlled environmental conditions, seedling elongation was retarded by a soil drench of CCC and also by a foliar application of the compound, but the soil drench was much more effective because it increased both number of tillers and grain yield per plant. When CCC at concentrations of 0, 10^{-4}, 10^{-3}, 10^{-2}, and 10^{-1} M was applied as a soil drench to 'Parkland' barley, the grain yields per plant were 100.0, 117.2, 127.6, 148.3, and 172.4 percent, respectively. However, under conditions of restricted soil moisture, neither plant weights nor grain yields were significantly altered by the CCC treatments. Foliar sprays were ineffective in increasing grain yields even under an optimal moisture regime.

Field trials demonstrated that CCC does have a dwarfing effect on the 'Parkland' and 'Hannchen' varieties and thus is of value for reducing losses caused by lodging (Larter, 1967). However, because barley is relatively insensitive to CCC, high concentrations or repeated applications of the compound throughout the growing period are required. Application of CCC at a concentration of 10^{-1} M at three developmental stages (three-leaf, five-leaf, and flag-leaf) reduced the height of sprayed plants to approximately 75 percent of that of the controls, a reduction sufficient to prevent lodging. The inhibition caused by CCC was more marked under moist soil conditions. The maturation of treated plants was delayed, on the average, 2.5 days for 'Hannchen' and 4 days for 'Parkland.' Grain yield was not increased by CCC treatment.

Inhibition of Grass Growth

Control of grass growth is desirable along highways and golf courses and around military installations and industrial locations. Many compounds have been tested for their ability to retard growth of turf grasses, but at present a satisfactory one has yet to be found. MH has inhibited the growth of pasture grasses in Florida (Ruelke, 1961), and morphactins have inhibited the growth of Bermuda grass, 'Kentucky' bluegrass, and 'Penncross' creeping bent grass in California (Madison *et al.*, 1969), but neither has produced satisfactory control without causing undesirable yellowing or phytotoxic effects.

INCREASING DROUGHT RESISTANCE

Growth retardants have been found to increase the drought resistance of some plants (Halevy, 1967). In some arid regions, after the emergence and initial growth of the crop following early rains, a prolonged drought occurs and causes severe wilting that is not always ameliorated by subsequent rainfall. A possible conclusion is that plants treated with growth retardants stay alive longer under dry conditions and thus are able to utilize the later rainfall.

Some promising possibilities for the use of growth retardants to increase the drought resistance of some crops are suggested in the following paragraphs.

Mode of Action

The mechanism by which plant growth retardants increase the drought tolerance of plants is not known. However, the effect of CCC and SADH in increasing the ability of the plant to withstand drought is thought to be related to the ability of these chemicals to delay the senescence of detached leaves. The biochemical changes that accompany the senescence of leaves, either attached or detached (Varner, 1961), are comparable to those that occur in plants deprived of water (Vaadia et al., 1961)—a decrease in the amount of protein, nucleic acids, and chlorophyll. Even brief periods of water deficit can hasten senescence (Gates, 1957). By slowing down the breakdown of protein that results from water deprivation, growth retardants may enable the tissue to stay alive longer and respond to rewatering after a prolonged wilting period.

The reduction in the number of stomata per unit area caused by soil applications of CCC to Brussels sprouts could decrease the rate of water loss from the leaf and contribute to the plant's drought tolerance (van Emden and Cockshull, 1967). Increases in leaf thickness as a result of application of CCC could also contribute to drought resistance.

Wheat

A greenhouse experiment conducted by Plaut and Halevy (1966b) demonstrated the effect of SADH and CCC on the ability of wheat plants to recover from prolonged wilting. Drought conditions were approximated by withholding water from the pots containing wheat plants of the variety 'Florence Aurore 8193' for intervals ranging from two to eighteen days. The first drought cycle started three weeks after plant emergence; the second began forty days later, just before earing. Three sprays of CCC at concentrations of 500, 1,000, and 2,000 ppm or SADH at concentrations of 1,000, 2,000, and 4,000 ppm were applied to the plants: twice at one-week intervals before the first drought cycle and once before the second drought cycle. Both growth retardants caused a significant increase in dry weight and grain yield of plants exposed to two drought cycles of five to six days each. The chemical treatments also greatly increased the ability of wilted plants to regenerate new shoots upon being rewatered. Longer drought periods of ten to twelve days caused complete desiccation of both treated and untreated plants. The growth retardants had no effect on root growth in any of the treatments. The increase in yield resulting from application of the compounds was caused exclusively by the increase in the production of late shoots and ears.

Barley

Greenhouse experiments with 'Blanco' barley showed that application of CCC produces shorter stems and wider leaves, increases seed production, and results in a favorable ratio of seed yield to moisture used (Goodin *et al.*, 1966). CCC in the form of a soil drench was added to pots of young barley at rates of 0, 3, and 6 ounces per 6-inch pot (equivalent to 222.9 and 445.8 mg of active ingredient per pot). The ratios of water used to seed yield for CCC at rates of 0, 3, and 6 ounces were 1,275, 1,051, and 630, respectively. Since the barley plants receiving 6 ounces of CCC used less than half the amount of water used by the nontreated plants to produce the same amount of seed, Goodin *et al.* suggest that CCC may be useful for increasing grain production in dry regions. Larter *et al.* (1965) also found that 6 ounces of CCC increased the efficiency of water utilization in barley.

Goodin and his coworkers (1966) also found that the higher concentration of CCC increased seed yield by approximately 40 percent over the controls. Greater seed production resulted from more seeds per head rather than from more seed heads or an increased seed weight.

Grape

Another interesting greenhouse study on the effect of SADH on the water utilization of grape was made by Bukovac *et al.* (1964). Rooted cuttings of *Vitis labrusca* cv. 'Concord' were grown in quartz sand, and after four to six nodes had formed, shoots were sprayed with SADH at concentrations of 0, 500, 1,000, 2,000, 4,000, and 6,000 ppm. After the shoots had developed thirteen nodes, watering with nutrient solutions was withheld. When approximately 80 percent of the control plants showed pronounced wilting, all treatments were rated from 1 to 10 according to degree of wilting. Shoots of plants treated with SADH were approximately 30 percent shorter than those of nontreated plants. However, in general, the same number of nodes were produced per shoot. As a result of SADH treatment, there was a dramatic reduction in water consumption and a greater tolerance to drought. The degree of wilting under the stress of being deprived of water was less in plants treated with SADH. There was an inverse relationship between the degree of wilting and/or percentage of plants that wilted, and the concentration of SADH.

Bean

Some experiments conducted in Israel by Halevy and Kessler (1963) demonstrated that when water was withheld from potted bean plants (*Phaseolus vulgaris* cv. 'Brittle Wax'), untreated plants wilted sooner and more severely than those treated with growth retardants. When the first pair of leaves on each plant was fully expanded, aqueous solutions containing 50 mg of Phosfon-D or 500 mg of CCC were added to the cans containing the plants. After the third leaf had expanded, irrigation of the soil was stopped. Within five days of the last watering, the leaves of the control plants began to wilt and plant growth ceased. Thirty days after the last irrigation most of the control plants were completely desiccated, but the treated plants remained turgid (Fig. 11–18).

Plaut *et al.* (1964) have suggested that growth-retarding substances might increase resistance to drought because they sometimes increase the root weight and decrease the top-to-root ratio (ratio of weight of that part of the plant growing above the ground to that part growing below). Drought resistance would thus be increased even if the retardants did not affect it directly. These investigators also used potted bean plants. After the first trifoliate leaf had expanded, irrigation treatments were begun at moisture levels of 7.0, 9.5, 13.0, and 18.0 percent, which corresponded to soil moisture tensions of 15.0, 4.0, 1.5, and 0.3 atmospheres, respectively. (Soil moisture tension is the force with which water is held in the soil. One atmosphere equals 14.7 pounds per square inch.) Each of the chemicals used, Phosfon-D (applied as a soil drench), Phosfon-S (foliar spray), CCC (soil drench or foliar spray), and SADH (foliar spray) was applied twice: first at full expansion of the primary leaves, and again nine or ten days later.

Application of any of the four substances, within a range of optimal concentrations, increased the dry weight of plant roots. CCC,

FIGURE 11-18

Effect of Phosfon-D (*left*) and CCC (*right*) on bean plants. Plant in center is control. Photographed 30 days after last irrigation. (After A. H. Halevy and B. Kessler, 1963.)

Phosfon-D, and Phosfon-S usually decreased the dry weight of tops, but SADH had no effect. All compounds and application methods tested resulted in a decrease in the top-to-root ratios of all plants.

Both deprivation of water and application of CCC retarded longitudinal plant growth. The effect of CCC was most apparent in soils having higher moisture contents (13 and 18 percent), whether the chemical was applied as a soil drench or as a foliar spray. When CCC was applied at a concentration of 250 ppm as a soil drench, the dry weights of roots and tops were increased in the soil having the highest moisture content (18 percent). Under all moisture conditions except the driest, foliar sprays of CCC at concentrations of 500 to 1,000 ppm caused a significant increase in the dry weight of pods. Application of CCC as a soil drench decreased the transpiration rate at all moisture levels. Foliar sprays of the compound had much less effect and reduced transpiration mainly in the dry and very dry soils.

Apple

Martin and Lopushinsky (1966) made an interesting study of the effect of SADH on the drought tolerance of apple. Samples of fruit and leaves were taken from trees of 'Red Delicious' apple that had been treated six weeks earlier with two sprays of SADH at a concentration of 2,000 ppm. The water deficit of intact leaves and fruit was determined by obtaining the fresh weight, the saturated weight after water had been absorbed through the petioles for twenty-four hours at 100 percent relative humidity, and the dry weight after being placed in an oven at 70°C. Both leaves and fruit from treated trees had a smaller water deficit than the controls. Treated leaves also lost water at a slightly lower rate than did the controls.

Sunflower

In another experiment conducted by Martin and Lopushinsky (1966), plants of sunflower (*Helianthus annuus*) at the primary-leaf stage were sprayed three times at five-day intervals with SADH at a concentration of 2,000 ppm. When a difference in size between treated and nontreated plants was observable, watering was discontinued and the plants were allowed gradually to dry out the soil in the pots. The plants were watered only after they had become severely wilted. Treated plants recovered much more rapidly than did the controls. These investigators concluded that application of SADH to sunflower does not have a consistent effect on the magnitude of water deficit developed by plants subjected to drought, nor does it delay the onset of wilting; however, the compound does seem to enhance the ability of the plants to recover from severe drought conditions after being rewatered.

Gladiolus

In Israel, Halevy (1962) found that CCC and Carvadan, an isomer of Amo-1618, increase the drought tolerance of corms of the gladiolus variety 'Sans Souci.' Corms were planted in pots and after the leaves had fully emerged, aqueous solutions containing 50 mg of Carvadan or 500 mg of CCC were added to the soil. Application of CCC was repeated one month later. Irrigation was stopped when the first leaf was fully expanded and the second leaf had just emerged. Thereafter the pots were divided into three groups for irrigation treatments: One group (moist) was irrigated every morning; a second group (dry) was irrigated when 65 percent of the available water had been removed; and a third group (very dry) was irrigated when 95 percent of the available water had been removed. At harvest there was an increase in the corm weight of treated plants as compared with that of the controls in the two drier groups: In the dry regime, corm weights for control plants and for plants given the CCC and Cardavan treatments were 5.48, 7.22, and 7.36 g, respectively. Corm yield of plants in the moist group was not affected. The increased drought tolerance of gladiolus was not accompanied by a significant retardation of shoot growth.

INCREASING SALT TOLERANCE

Wheat

Plant growth retardants increase the tolerance of some plants to high concentrations of salt. Miyamoto (1962) soaked seeds of winter wheat cv. 'Carstens VIII' in an aqueous 0.5 percent solution of CCC for fourteen hours at room temperature. The seeds were then dried and planted in a neutral sandy soil. Eleven days after sowing, ammonium nitrate was added to the soil at the rate of 7 g per 200 g of soil. The water content of the soil was allowed to fall until it reached approximately 50 percent of field capacity (the amount of water in the soil after all the excess water has drained off) and was then held at this level. Within seventy-two hours after the ammonium nitrate was added 100 percent of the control plants were wilted, compared with 40.5 percent of those treated with CCC.

A greenhouse experiment showed that CCC increases the resistance of 'Opal' spring wheat to salinity (El Damaty *et al.*, 1964). Wheat kernels were soaked in CCC at a concentration of 500 ppm for fourteen hours at room temperature and were then planted in coarse white sand having a moisture content of approximately 50 percent of field capacity. Ten days after planting, the seedlings were irrigated with seven different applications of saline water containing the following concentrations of added salts (sodium chloride, calcium chloride, and magnesium chloride in a ratio of 1.0:0.85:0.15): 0, 1,250, 2,500, 5,000, 12,500, 25,000, and 50,000 ppm. Concentrations of salt higher than 5,000 ppm resulted in much more wilting and damage to the untreated plants than to those treated with CCC. Even at salinity levels below 5,000 ppm, plants treated wih CCC were healthier, more turgid, and more erect than untreated ones. The osmotic pressure of the treated plants seven days after planting was 7.521 atmospheres, compared with 6.682 atmospheres for the untreated plants. Water attraction in plants treated with CCC might be greater than that in the untreated ones; this could explain why plants treated with CCC can withstand drought.

Soybean and Spinach

Application of Amo-1618, Phosfon-D, or CCC to soybean plants increases the plants' tolerance to toxic levels of salt placed around their roots (Marth and Frank, 1961). Young, chemically retarded soybean plants were able to withstand application of commercial 5–10–5 fertilizer to the surface of the soil in amounts that killed comparable plants not so treated. (The numbers 5, 10, 5 refer to the percentages of N, P_2O_5, and K_2O, respectively, in the fertilizer.) Miyamoto (1963) found that treatment of spinach seeds with CCC also increases the resistance of seedlings to herbicides.

INCREASING RESISTANCE TO LOW TEMPERATURES

Inhibitors, especially plant growth retardants, have been used experimentally to increase the frost resistance of some plant species.

This result is not surprising because cold hardening (physiological changes that make plants resistant to frost) is generally accompanied by a decrease in rate of growth.

Cabbage

Frost damage to cabbage is noticeably reduced by application of growth retardants before the plants are exposed to low temperatures (Marth, 1965). In Maryland young plants of the early maturing variety 'Early Jersey Wakefield' and of the late maturing variety 'Late Flat Dutch' were grown in pots in the greenhouse. Early in October, when they had developed three secondary leaves, the plants were sprayed with CCC or SADH at concentrations of 2,500 or 5,000 ppm. They were then exposed to outdoor winter temperatures, the critical range being from −1 to −18°C. All treated plants survived and 90 percent of them quickly resumed growth when brought into a warm greenhouse the following spring (Fig. 11–19), but 40 to 60 percent of the control plants were killed by frost. The surviving control plants grew very slowly and their vegetative growth was greatly reduced.

Tomato

Soaking tomato seeds in CCC at a concentration of 500 ppm before planting was found to result in a small but significant increase in the percentage of seeds that germinated when grown at 12°C, the minimal temperature for germination of tomato seeds (Michniewicz and Kentzer, 1965). However, application of CCC had no effect on the resistance of the tomato seedlings to low temperatures. In another experiment young tomato seedlings were grown in a nutrient solution containing CCC at a concentration of 200 ppm until one or two pairs of leaves had developed. They were then placed in cold temperatures ranging from −2 to −10°C for intervals of from twenty minutes to twelve hours. Plants treated with CCC developed short, thick stems, and the percentage of treated plants that were killed by frost was much reduced. In one experiment 92.5 percent of the controls were killed but all of the treated plants survived.

Woody Plants

Development of cold hardening in woody plants has been shown to be dependent on photoperiod (Irving and Lanphear, 1968). Short

FIGURE 11-19

Effect of growth retardants on survival and growth rate of cabbage plants when applied after plants were exposed to frost throughout the winter and then maintained for 2 weeks in a warm greenhouse. *Top row,* 'Early Jersey Wakefield' variety treated with CCC. *Second row from top,* the same variety, untreated. *Second row from bottom,* 'Late Flat Dutch' variety treated with SADH. *Bottom row,* the same variety, untreated. (After P. C. Marth, 1965.)

days induce hardening and long days are inhibitory. In their experiments with box elder (*Acer negundo*), Irving and Lanphear (1968) found that applications of gibberellin made either during or after a short photoperiod decreased the amount of hardiness obtained during a hardening treatment. GA_3 stimulated vegetative development of the seedlings, even when they were grown under short-day conditions, and negated the beneficial effect of short days on hardening.

Irving (1969), carrying this research further, demonstrated that weekly applications of the plant retardants CCC at a concentration of 1,000 ppm, SADH at a concentration of 3,000 ppm, or Amo-1618 at a concentration of 1,000 ppm to box elder seedlings growing under short days, followed by a hardening period of three weeks in darkness at 4°C, caused greater hardening than did short days alone. In one experiment the gains in hardiness from treatment with CCC, SADH, and Amo-1618 were 3°, 3°, and 4°C, respectively. Such results indicate that application of growth retardants to plants growing under short days may increase their hardening. On the other hand, the treatment of hardened, nondormant plants with CCC, SADH, or Amo-1618 failed to retard the loss of hardiness after a dehardening treatment of five days of long-day conditions at 21°C.

Apple

Sprays of ABA have been shown to increase the cold hardening of 'Rome Beauty' apple both before and during the dormant period (Holubowicz and Boe, 1969). Seedlings that were sprayed with ABA at a concentration of 20 ppm and subjected to a hardening treatment consisting of decreasing temperature and day length (21°C under sixteen hours of light, 7°C under ten hours of light, and 0°C under eight hours of light for seven-day periods) showed more resistance to cold than did the controls. For example, the lowest temperature that could be endured by seedlings treated before dormancy was −26°C, compared with −20°C for the controls. Comparable figures for plants treated after onset of the dormant stage were −30°C and −28°C, respectively. It is well known that concentrations of growth inhibitors in the growing points increase as the plants become dormant, but it is not known whether exogenous ABA acts in the same way as the naturally occurring hormone.

Pear

At England's East Malling Research Station, Modlibowska (1965) demonstrated that CCC increases frost resistance in pear. One-year-old pear trees of the variety 'Williams' Bon Chrétien' were sprayed in the spring with CCC at a concentration of 10,000 ppm. The following year, when both treated and untreated trees were in full bloom, two trees of each group were exposed to a freezing temperature of −4°C for fifteen minutes. Blossoms on the trees that had received a CCC spray the preceding year had a greatly increased resistance to frost: Frost killed 81 percent of the blossoms of control trees but only 50 percent of the blossoms of sprayed trees.

Wheat

Frost hardiness of the winter wheat variety 'Starke' is increased by CCC and decreased by gibberellin following hardening treatment (Wünsche, 1966). In one experiment performed in Sweden, CCC at a concentration of 4×10^{-3} M or gibberellin at a concentration of 4×10^{-4} M were applied to the soil seventeen days after emergence of the seedlings. Seven days later the plants were maintained at 20°C for two days and were then placed in growth cabinets at 1°, 3°, or 5°C. After 0, 2, and 7 days of this hardening treatment, the cut plants were frozen in glass tubes at four different temperatures and were then transferred to a growth cabinet at 10°C for four days to recover. The estimate of the degree of frost damage was based on the percentage of killed tissue. The increase in frost hardiness as a result of CCC treatment was generally less than 1°C, but the effectiveness of the chemical was clearly visible. There often was a noticeable difference between the type of frost damage inflicted on CCC-treated and that suffered by untreated and gibberellin-treated plants. The tops of the two oldest leaves of CCC-treated plants were sometimes damaged, whereas the base of the youngest leaf of untreated and gibberellin-treated plants was usually damaged.

Citrus

After the discovery that MH retards the growth of orange trees and sometimes induces dormancy, studies were made of the compound's effect on the trees' frost tolerance. Dormant trees can survive lower temperatures than can trees in a growing condition. In 1960 Stewart and Leonard, working in Florida, found that application of a spray of MH at a concentration of 1,000 ppm to trees of 'Ruby Red' grapefruit that had been previously defoliated by cold reduced subsequent frost injury. Also, young orange trees that were rendered dormant by sprays of MH that were applied in November lost very few leaves as a result of frosts that occurred the subsequent January. Unsprayed trees either died or were defoliated.

Experimental field applications of MH at concentrations of 1,000 and 2,000 ppm that were made early in November of 1959 and 1960 (Hendershott, 1962) rendered the trees resistant to temperatures several degrees below those found to be critical for untreated trees. A similar graft application of MH to one-year-old trees of 'Valencia' orange on rough lemon rootstock resulted in an increase in cold resistance of approximately −15°C. Wood of actively growing untreated trees was generally killed when held at −3°C for four hours. No defoliation occurred but wilting of foliage did occur shortly after the temperature was increased to 26°C. Trees treated with MH required temperatures of −6°C or lower before wood damage occurred; however, defoliation often occurred as a result of subjection to this temperature.

In Florida, MH should be applied to young, nonbearing citrus trees between November 1 and 15 or at least two weeks before the onset of freezing weather (Hendershott and Matsolf, 1962). Actively growing trees may not be protected by the chemical unless sufficient time elapses between time of treatment and onset of freezing weather for the tender wood to "harden" or become dormant. It is well known that dormant citrus trees can withstand lower temperatures than can actively growing trees. MH should be applied to mature trees recently defoliated by a freeze as soon after defoliation as possible and before new growth begins.

Hendershott and Matsolf recommend an application of two quarts of a 30 percent solution of MH per one hundred gallons of water for both young, nonbearing citrus trees and mature citrus trees recently defoliated by a freeze. Repeat applications should consist of one quart of solution per one hundred gallons made at thirty-day intervals or two quarts per one hundred gallons made at sixty-day intervals, depending on the duration of effectiveness desired.

Results similar to those of Hendershott and Matsolf were obtained in South Africa by Edwards (1961) and in California by Burns et al. (1962). However, the effectiveness of MH in preventing low-temperature injury to citrus has varied, and more effective compounds are needed. Some antitranspirants, including film-forming, stomata-closing, and reflecting types, as well as some growth retardants, show promise (Burns, 1970).

Other Florida experiments have demonstrated that a quaternary ammonium derivative of (+)-limonene, 1-p-menthanol-2-dimethyl-amino heptyl bromide consistently retards growth of grapefruit and lemon trees (Pieringer and Newhall, 1970). Two applications of the compound at a concentration of 6,000 ppm retarded the growth of field-grown lemon trees for a period of twelve weeks, indicating that it might be used for reduction of tree size. In addition, the prevention of new growth during warm periods in winter would protect trees from cold injury. These investigators also found SADH to be effective for retarding growth of shoots of grapefruit.

Raspberry

In England, Modlibowska and Ruxton (1954) applied a spray of MH at a concentration of 750 ppm to red raspberry plants of the variety 'Malling Exploit' at bud burst. The plants were subjected to low temperatures when green buds appeared and again when flowers had opened. Some degree of frost resistance was attained in shoots whose growth was only slightly inhibited. When there was strong inhibition of shoot growth, susceptibility to frost was increased. The MH treatment did afford some frost protection by causing a delay in flowering, but it also reduced total crop yield.

In Quebec, raspberry yields are frequently reduced as a result of the severe dieback of canes during the winter. Granger and Hogue (1968) sprayed young raspberry plants of the varieties 'Latham' and 'Trent' with SADH at concentrations of 0, 1,000, and 2,000 ppm on June 21, July 14, and August 19, 1966. Observations made the following spring when new leaves were just expanding showed that SADH reduced cane dieback. The first two applications were more effective than the last in preventing winter injury, especially that of 'Latham.' The higher concentration of SADH applied in June and July was more effective than the lower one. A reduction in cane length was observed following the June application. Canes treated in July were moderately decreased in length, but otherwise SADH seemed to have little effect. Crop yields of canes treated in both June and July were estimated to be greater than those of untreated canes.

Grape and Mulberry

Application of a 0.05 percent concentration of a 30 percent formulation to MH to the leaves of growing grapevines decreased the cold damage to young shoots (Akabane et al., 1957). Sakai (1957) found that MH inhibits shoot growth of mulberry and increases its resistance to frost. When more than 0.1 percent of the compound was sprayed on the vines at the end of August, twig growth and bud formation were inhibited.

Strawberry

In regions where winters are characterized by periods of relatively low temperatures interspersed with periods of mild weather, strawberries often fail to become and remain dormant. Working in the coastal region of British Columbia, Freeman and Carne (1970) found evidence to suggest that winter injury to 'Northwest' strawberry plants can be reduced by SADH. Each of the plots that had received an application of the compound at a concentration of 5,000 ppm on October 3 produced 238 marketable berries, compared with the 139 marketable berries produced by the unsprayed plots. The investigators suggest that the main beneficial effect afforded by the treatment was its flower bud protection, since there was little difference in plant growth between the treated and untreated plants.

Azalea

In their studies on the effects of CCC and SADH on eight cultivars of evergreen azaleas growing in ground beds of peat moss, Rennix *et al.* (1964) observed that treated plants were much more tolerant to storage in darkness at $2°-3°C$ than were untreated ones. SADH at a concentration of 1,515 ppm was applied once on August 15 and once on August 22. CCC at a concentration of 2,260 ppm was applied similarly: on August 15, August 22, and August 29. Late in September the plants were dug, potted, and stored in darkness at $2°-3°C$ to break dormancy. After approximately two months the plants were exposed to environmental conditions favorable for growth. Observations made subsequent to storage showed that treated plants, whether sprayed two or three times, usually evidenced better leaf retention and a higher survival rate than did control plants. Treated plants of the cultivars 'Red Wing,' 'Sweetheart,' 'Alaska,' and 'Variegated Triomphe' were in excellent condition after the cold temperature treatment, whereas the control plants had dropped much foliage.

Loading the Plant System

There has been much speculation concerning the possibility of loading the plant system with a chemical that may serve as an "antifreeze." Polyethylene glycol (which increases resistance to freezing) at a concentration of 20 M has been shown to move through plants from roots to leaves without adversely affecting growth (Zelitch, 1964). A relatively new family of compounds have been shown to increase the resistance of plants to frost damage (Kuiper, 1964). In these experiments one of the compounds, n-decenyl-succinic acid, appeared to be particularly promising in increasing the frost resistance of flowering peach, apple, pear, and young bean plants. However, some researchers who have used other plants have failed to obtain positive results with these compounds.

INCREASING RESISTANCE TO SMOG

Pinto bean plants and petunias that received a soil drench of Phosfon-D before being exposed to irradiated automobile exhaust were protected from damage, but untreated plants were severely damaged (Seidman and Ibrahim, 1962).

INCREASING RESISTANCE TO DISEASE

Growth regulators can be used to increase the resistance of plants to some fungal and virus diseases and to some physiological disorders. Examples of such application are discussed in this section.

Verticillium alboatrum

Sinha and Wood (1964) observed that spraying tomato plants with CCC greatly decreases their susceptibility to Verticillium alboatrum. In one experiment 50 ml of a 200 ppm solution of CCC was applied to the soil around the bases of seven-week-old tomato plants of the susceptible 'Gem' variety, growing in pots five inches in diameter. Application was made both four days before and four days after roots had been dipped in a suspension of spores and hyphae of the parasite. Four weeks after such inoculation, the mean disease index for leaves of plants treated with CCC was 0.10, compared with 0.95 for untreated plants. (The disease index scale ranged from 0 for a healthy leaf to 4 for a leaf that was completely wilted or desiccated.) The compound induced formation of a large number of tyloses in vessels of both uninoculated and inoculated plants. This observation is significant because formation of a relatively large number of tyloses is a characteristic response of plants resistant to this disease.

Rhizopus nigricans

CCC decreases growth of the fungus *Rhizopus nigricans*. To demonstrate this effect, El-Fouly and Jung (1966) treated cotton seed with CCC both as a solution and in powder form. Similar results were observed when fungus cultures were grown on cotton seed pulp. These investigators suggest that the compound acted on the fungus either directly, by influencing the synthesis of one or more growth substances, or indirectly by causing a metabolic change in the host. On the other hand, Beck (1968) found that CCC produces no fungicidal and only small fungistatic effects on plants and on fungal cultures.

Southern Stem Rot, Wheat Stem Rust, Powdery Mildew, and Tobacco Mosaic

Tahori *et al.* (1965c) tested the effect of plant growth retardants on three fungal diseases, southern stem rot (*Sclerotium rolfsii*), wheat stem rust (*Puccinia graminis*), and powdery mildew (*Erysiphe graminis*), and on one viral disease (tobacco mosaic virus). The effect of retardants on the southern stem rot fungus was analyzed by applying Phosfon-D or CCC to young bean seedlings, either by soil application or by foliar spray. Two days later a homogenate containing the stem rot fungus was added to the soil. Both compounds reduced the infection. In earlier laboratory experiments Lavee (1959) had demonstrated that coumarin, MH, and triethanolamine inhibit germination of sclerotia and reduce the number of sclerotia per culture.

Tahori *et al.* (1965c) treated thirty young plants of 'Etit 38,' a rust-susceptible variety of wheat, by applying three growth retardants to the soil. Five days later each group of ten plants was inoculated with wheat stem rust. Eleven days after inoculation plants treated with SADH at a rate of 600 mg per 3 liters of soil had an average of 2.07 rust pustules; untreated controls had an average of 3.47 pustules. Phosfon-D and CCC had no effect.

Neither CCC, SADH, nor Phosfon-D increased the resistance of young barley plants to powdery mildew. In another experiment Tahori *et al.* (1965c) sprayed *Nicotiana glutinosa* plants with solutions of SADH, CCC, or Phosfon-D. Two days later the plants were infected with tobacco mosaic virus. Interestingly, the lesions on plants treated with Phosfon-D were stronger and redder than those on untreated plants, and plants treated with almost all compounds used were observed to have more lesions than untreated plants.

Virus Yellows Infection in Sour Cherry

Trees of sour cherry affected by virus yellows infection produce an excess number of blossom buds in terminal and lateral shoots. The result is a temporary increase in fruit production and production of more fruit on the individual twig. However, a few years after onset of the disease there is a drastic reduction in crop yield because no lateral vegetative buds are developed to produce fruiting spurs (Parker *et al.*, 1964).

Treatment of yellows-infected trees with GA_3 at a concentration of 10 to 15 ppm, depending on the severity of the disease as indicated by the presence of long, barren twigs, reduces the percentage of blossom buds and increases the number of vegetative buds on terminal and lateral shoots. The year following treatment most of the vegetative buds that develop produce spurs that bear fruit the next year. Treatment should be made any time from shuck fall, when five to six well-developed leaves are present, until approximately two weeks after shuckfall, before flower primordia are formed (Fliegel *et al.*, 1966). Cool temperatures and a low relative humidity are required at time of spraying to avoid poor vegetative bud development.

There are several other advantages of treatment. Leaf color is enhanced, especially early in the growing season. Wind damage to fruit is decreased because the twigs are stronger and more rigid. Increased twig growth results in an increased number of leaves, which also decreases wind damage to fruit (Parker *et al.*, 1964).

Yield may be reduced at first by the replacement of some flower buds with vegetative buds, but this problem is soon overcome by the fruit that forms on the additional spurs produced. The level of soluble solids of the fruit of the current crop may be reduced by the treatment, but it is increased in subsequent years as a result of the heavier leaf growth.

Chocolate Spot in Potato

In many potato-producing areas of the world, tubers frequently are seriously damaged by a physiological disease known as internal brown spot or chocolate spot. In Lebanon this disease was responsible for losses estimated at $2 million during 1969 (see de Wilde, 1971). The prevalence of chocolate spot in tubers of the cultivar 'Arran Banner' is sometimes as high as 60 percent. Kamal and Marroush (1971) found that in the Bekaa plain of Lebanon the disease in 'Arran Banner' could be completely controlled by foliar sprays of ethephon at concentrations of 200 to 600 ppm. A single application at concentrations of 200 to 600 ppm resulted in 98 to 99 percent control, and complete control was obtained by applying ethephon at a concentration of 200 ppm as a foliar spray five weeks after planting and repeating this treatment four times at two-week intervals (see de Wilde, 1971). However, all these treatments reduced the percentage of large tubers and increased the percentage of medium- and small-sized tubers in comparison with the controls.

INCREASING RESISTANCE TO INSECTS

Aphids are insects that feed on the host plant and that are closely adapted to its physiological condition. Thus it is not surprising that growth regulator treatment of plants affects the aphids that feed on them. Van Emden (1964) demonstrated that the size of an aphid population of the species *Brevicoryne brassicae* feeding on 'Wroxton' Brussels sprouts plants treated with CCC was smaller than that feeding on untreated controls. Young plants received two soil drench applications of CCC at a rate of 10 g per cubic foot of soil. On the day the second treatment was made, one alate (winged) aphid was transferred to a mature leaf of each plant that had been placed in a small cage. Approximately half as many aphids were produced on leaves of plants treated with CCC as on leaves of untreated plants (Fig. 11–20). Van Emden concluded that aphids living on plants treated with growth retardants are not as able to reproduce as aphids living on untreated plants and that feeding on retardant-treated plants causes many aphids to die before reaching maturity. He found that adulthood in those that attained it was very short and was accompanied by failure to reproduce.

Other interesting results were obtained with *Nerium oleander* by Tahori *et al.* (1965a). Stem bases of oleander plants with branches having two or three leaves were placed in solutions of Phosfon-D, CCC, or SADH. Then, by means of a camel-hair brush, young apterae (wingless forms) of the oleander aphid *Aphis nerii* were placed on both treated and untreated leaves, or foliage containing aphids was placed in contact with the treated leaves for two or three days so that the aphids could infest the plants. Leaves on branches of plants in the solution of Phosfon-D had less aphid infestation than did the untreated leaves or the leaves on branches of plants growing in solutions of CCC or SADH. In one experiment fifty young apterae were deposited on the leaves of shoots placed in water or in Phosfon-D at molar concentrations of 10^{-6}, 10^{-4}, 10^{-3}, or 3×10^{-3}. After ten days

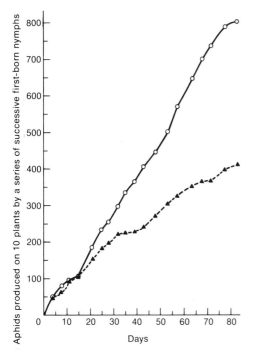

FIGURE 11-20
Number of aphids produced on CCC-treated plants (▲) and untreated plants (○). (After H. F. van Emden, 1964.)

the aphids on the leaves numbered 0, 135, 63, 61, and 6, respectively. Thus, the aphid population on shoots in water increased greatly but increased only slightly on shoots placed in Phosfon-D at concentrations of 10^{-6} M or 10^{-4} M. The compound at a concentration of 10^{-3} M or higher completely prevented aphid reproduction.

Cotton Leafworm

Antifeeding compounds protect plants from insect attacks not by killing the pests directly but by starving them to death because they are unable to feed. Carbamate derivatives, an acetanilide and an organotin compound, have antifeeding qualities (see Tahori *et al.*, 1965b).

In antifeeding trials with the cotton leafworm *Prodenia litura*, Tahori *et al.* (1965b) found that Phosfon-D and other plant growth retardants have a pronounced antifeeding effect. In one experiment, Phosfon-D, SADH, CCC, or Cardavan were applied to two-week-old plants of *Phaseolus vulgaris* cv. 'Brittle Wax' either as a soil application or as a foliar spray. One day later twenty-five young cotton leafworm larvae were deposited on the plants. Phosfon-D was the most

a   b   c   d

FIGURE 11-21

Effect of Phosfon-D as an antifeeding agent. Stems of chrysanthemum flowers were placed in solutions of Phosfon-D at the following concentrations: b, 10^{-6} M; c, 5×10^{-4} M; d, 10^{-3} M. (a is control, in water.) Plants were then infested with 10-day-old cotton leafworm larvae. (After A. S. Tahori, Z. Zeidler, and A. H. Halevy, 1965b.)

effective of the compounds tested, and foliar application was more effective than soil treatment.

In other experiments cotton leaves were dipped in Phosfon-D for three minutes; alternatively, the petiole bases were maintained in a solution of the compound. The leaves were placed in close proximity to each other so that the larvae could move freely among all of them, feeding on those they preferred. The percentages of leaf area consumed by the larvae on leaves dipped in water and in Phosfon-D at molar concentrations of 10^{-4}, 10^{-3}, 3×10^{-3}, 7×10^{-3}, and 10^{-2} were 55.8, 47.5, 36.7, 24.8, 21.2, and 6.0, respectively. Phosfon-D at a concentration of 10^{-2} M protected the cotton leaves almost completely.

Phosfon-D also afforded protection to the flowers of chrysanthemum plants growing in solutions of the compound at concentrations of 5×10^{-4} M and 10^{-3} M. The flowers of untreated plants were destroyed (Fig. 11-21). In a field trial, fewer cotton leafworm larvae were found on groundnut plants sprayed with Phosfon-D at concentrations of 10^{-3} or 10^{-2} M than on unsprayed plants. Young pepper plants sprayed with the compound at a concentation of 3×10^{-2} M also evidenced less damage from larvae than did untreated plants.

STIMULATION OF ENZYME ACTIVITY IN MALTING BARLEY

Soon after GA_3 became available for experimentation it was found to stimulate the germination of barley, wheat, and rice seeds and to increase the amount of amylase in the germinated barley and wheat grains (Hayashi, 1940). Workers in the brewing industry were quick to recognize the possible value of gibberellin as a promoter of enzyme activity in malting barley (Dahlstrom and Sfat, 1961). The addition of 1 to 3 mg of gibberellin per kilogram of barley during the early stages of germination causes an increase in enzymes and in enzyme diastatic power. An increase in the latter accelerates the malting process by approximately 1.5 days. One disadvantage is that amino acid production is also increased by gibberellin. Nevertheless, the compound is commonly used in the production of distiller's malt for the manufacture of neutral grain spirits.

Chapter 12

WEED CONTROL

Weed control is the most widespread practical use of synthetic plant regulators. Although many types of herbicides are used, the auxin-type compounds are most important and may be grouped into three general classes: (1) phenoxy-type compounds, (2) benzoic acids, and (3) heterogeneous compounds that do not fit in either the phenoxy or the benzoic acid classifications. Table 12–1 lists the more important auxins presently used as herbicides. For maximum effectiveness, the phenoxy compounds should be applied as a foliar spray because they are not readily translocated from the roots following soil application and because they are relatively unstable in soil. Some compounds other than the phenoxy type are effective if applied either to the foliage or to the soil.

IMPORTANCE OF HERBICIDES

Weeds are very costly because they rob crop plants of water, nutrients, and light. They also harbor disease and the insect organisms that attack crops. The result is a sharp reduction in both crop yield and the quality of crop and livestock products. The annual cost of weeds to agriculture is approximately $7.5 billion (Ashton and Harvey, 1971). Losses, sometimes exceeding $150 per acre, often spell the difference between success and failure for the farmer.

Approximately 75 percent of the total weed control in the United States is achieved by application of three chemicals and their analogues and homologues (van Overbeek, 1964). The most widely used herbicide is 2,4-D, followed by simazine and diuron. The auxin-type herbicides constitute only about one-fourth of the total number of commonly used herbicides, but their value in weed control is proportionally much greater. Much of the total increase in the use of herbicides is a result of increased use of 2,4-D and related compounds. It is estimated that 50,000 tons of 2,4-D are manufactured annually in the United States and that approximately 30 million acres of land are treated with this herbicide alone (Cherry, 1970).

The use of herbicides in the United States seems to be increasing exponentially. In 1965 herbicides were applied to nearly 120 million acres of land, compared with 70 million acres in 1962 and 53 million acres in 1959. In the three-year period 1962–1965 there was a 70 percent increase in the use of herbicides; in the preceding three-year period (1959–1962) the increase was only 34 percent (Anonymous, 1968a).

USE OF HERBICIDES ON PLANTS

Herbicides have been used very successfully in the culture of small grains including wheat, oats, maize, and rice. Favorable results have also been obtained by applying herbicides to cotton, peanuts, safflower, sorghum, soybeans, sugar beets, sugarcane, tobacco, alfalfa, grasses, trefoil, flax, and barley. The recommended technique often consists of two or more treatments. General recommendations for treating many field crops as well as other crops can be found in tables included in the *Suggested Guide for Weed Control 1969* (Agricultural Handbook no. 332), prepared by the Crops Research Division of the United States Department of Agriculture.

Herbicides fill an important role in the production of the more than fifty vegetable crops that are produced commercially in the United States. Herbicides are also valuable when applied to deciduous tree fruits and nuts, citrus fruits and subtropical fruits and nuts, and ornamental plants. Herbicides are also used profitably to control the weeds that plague forage crops as well as those found in nurseries

TABLE 12-1

Designation, Common and Chemical Names, and Structural Formulas of Important Auxin-Type Herbicides

Designation and Common Name	Chemical Name	Structural Formula
	CHLOROPHENOXY COMPOUNDS	
2,4-D	2,4-dichlorophenoxyacetic acid	
2,4-DB	4-(2,4-dichlorophenoxy)butyric acid	
2,4-DEB	2-(2,4-dichlorophenoxy)ethyl benzoate	
2,4-DEP	tris[2-(2,4-dichlorophenoxy)ethyl]phosphite	
2,4-DP (dichlorprop)	2-(2,4-dichlorophenoxy)propionic acid	

MCPA [(4-chloro-o-tolyl)oxy]acetic acid

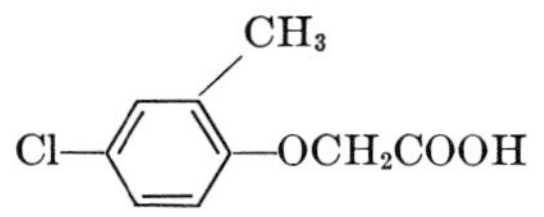

CH_3, $-OCH_2COOH$, Cl

MCPB 4-[(4-chloro-o-tolyl)oxy]butyric acid

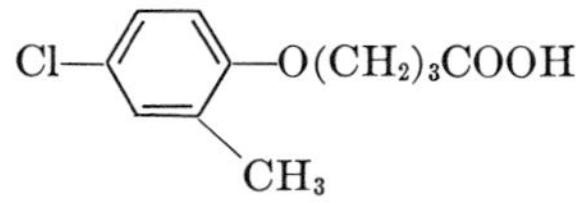

$-O(CH_2)_3COOH$, CH_3, Cl

MCPES 2-[(4-chloro-o-tolyl)oxy]ethyl sodium sulfate

$-O(CH_2)_2OSO_3Na$, CH_3, Cl

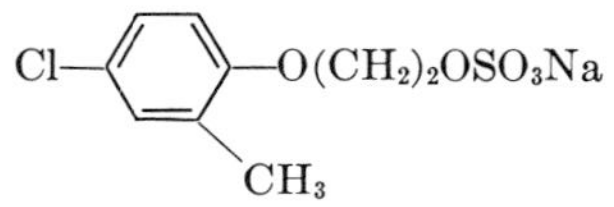

MCPP (mecoprop) 2-[(4-chloro-o-tolyl)oxy]propionic acid

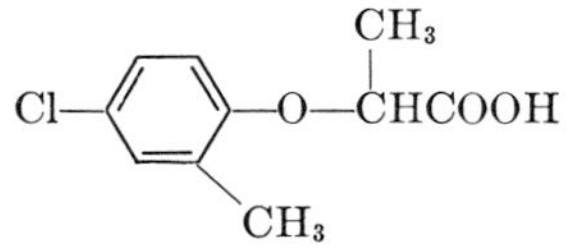

CH_3, $-O-CHCOOH$, CH_3, Cl

sesone (2,4-DES) 2-(2,4-dichlorophenoxy)ethyl sodium sulfate

$-OCH_2CH_2OSO_3Na$, Cl, Cl

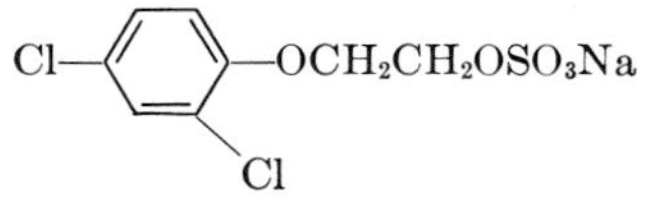

silvex (2,4,5-TP, fenoprop) 2-(2,4,5-trichlorophenoxy) propionic acid

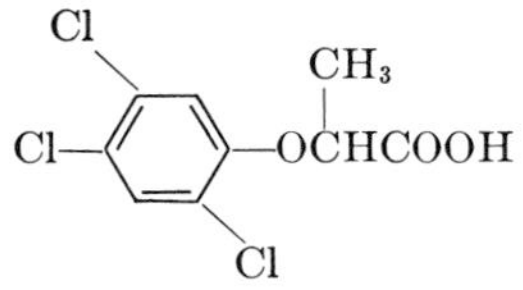

CH_3, $-OCHCOOH$, Cl, Cl, Cl

2,4,5-T 2,4,5-trichlorophenoxyacetic acid

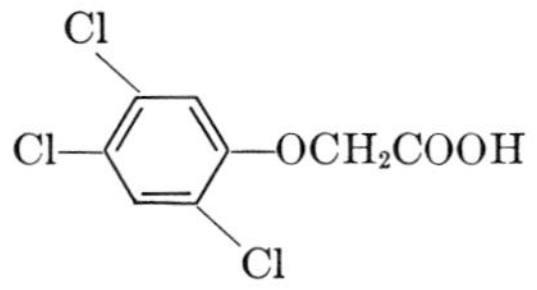

$-OCH_2COOH$, Cl, Cl, Cl

(*Continued*)

TABLE 12-1 (Continued)

Designation and Common Name	Chemical Name	Structural Formula
2,4,5-TES	sodium 2-(2,4,5-trichlorophenoxy)ethyl sulfate	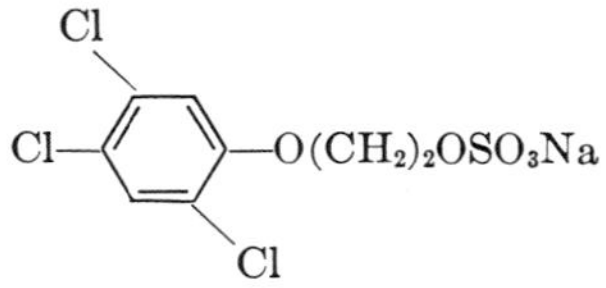

CHLOROBENZOIC ACIDS

2,3,6-TBA	2,3,6-trichlorobenzoic acid	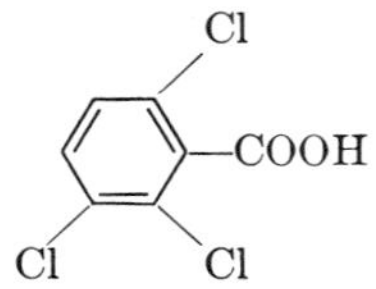
2,3,5,6-TBA	2,3,5,6-tetrachlorobenzoic acid	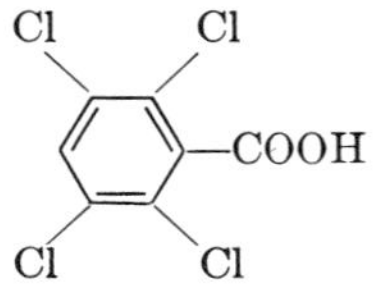
dicamba (Banvel D)	3,6-dichloro-o-anisic acid	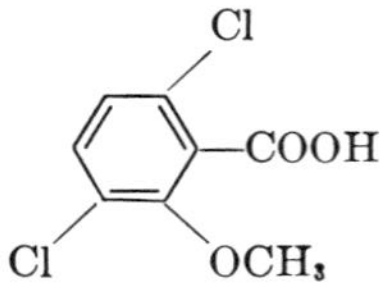

amiben (chloramben)

3-amino-2,5-dichlorobenzoic acid

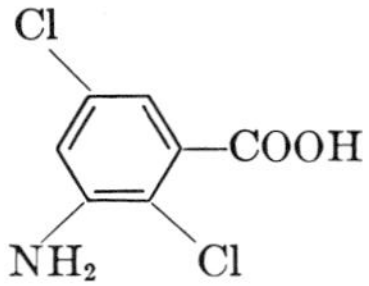

TYPES OTHER THAN PHENOXY OR BENZOIC ACIDS

picloram (Tordon)

4-amino-3,5,6-trichloropicolinic acid

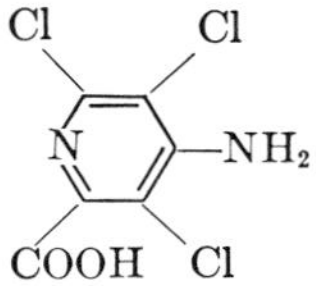

NPA (naptalam)

N-1-naphthylphthalamic acid

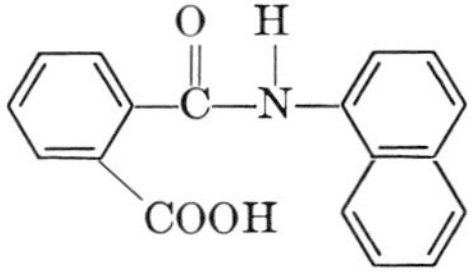

and greenhouses, pastures and rangelands, lawns and other turf areas, and on land not used for crops (ditch banks, fencerows, road and utility rights-of-way, and floodplains).

Table 12–2 lists the recommended dosages and spray volumes of the important hormone-type herbicides used in weed control. All materials used to treat food and forage crops should be cleared by local authorities before application.

HISTORICAL BACKGROUND OF HERBICIDES

Chlorophenoxy Compounds

Research conducted in the early 1940's revealed that such auxins as 2,4-D and MCPA can be used to kill crop plants. However, this discovery was obscured by wartime security policies. Some of the background of the discovery and development of 2,4-D and MCPA as herbicides was discussed in Chapter 1. MCPA is used mostly in England and its Dominions; 2,4-D is widely used in the United States and other parts of the world. 2,4,5-T is very effective on mesquite and other woody plants. 2,4,5-TP is very effective when used to eradicate oak, elm, beech, and maple and is frequently superior to 2,4,5-T. 2,4-DB is valuable because of its selectivity between certain crops and weeds (see p. 479). The phenoxy compounds listed in Table 12–1 were introduced as screening programs progressed, but they are of less value and are less widely used than the five chlorophenoxy compounds just discussed.

Chlorobenzoic Acids

The growth-regulating activity of the substituted benzoic acids was described in 1942 by Zimmerman and Hitchcock, and in 1950 Bentley noted the high physiological activity of 2,3,6-TBA. There are many chlorine-substituted benzoic acids, but 2,3,6-TBA and 2,3,5,6-TBA are two of the most phytotoxic. Dicamba, which was first field-tested in 1961 (Crafts and Robbins, 1962), is also very toxic and is widely used.

Other Types

NPA was discovered in the mid-1950's and has proved to be the outstanding herbicide of a new group of compounds, the N-aryl phthalamic acids described by Hoffman and Smith (1949). It exerts little or no toxic effect on foliage but it readily enters the plants through the roots, and its auxin-type action is evidenced by the subsequent epinasty and interruption of the normal geotropic response of roots. Roots treated with NPA often grow upward out of the soil surface.

Picloram was discovered by Hamaker *et al.* in 1963. It is of value in controlling the growth of woody plants and most perennial broad-leaved plants.

GENERAL NATURE OF HORMONE-LIKE HERBICIDES

The toxicity of hormone herbicides is closely correlated with the growth activity of the plant to which the compound is applied. Maximum effectiveness usually occurs when the herbicide is applied during the most rapid growth of the plant. However, treatment of weeds during their period of maximum growth is sometimes impossible. Young plants undergoing rapid growth are often easily killed by hormone herbicides, but conditions that tend to reduce this rapid growth, such as low temperatures, insufficient amounts of nitrogen, potassium, phosphorus, or other nutrients, and insufficient moisture, may also decrease sensitivity to the herbicide. Plants become progressively more difficult to kill because stress caused by lack of moisture is increased.

To kill plants, there must be basipetal translocation of assimilates to the roots at the time of spraying, especially the roots of perennial dicotyledons. Unfortunately, at the time of maximum growth rate, basipetal translocation of the applied herbicide is insufficient to effect a good kill. Therefore, in determining the optimal spraying period a compromise must be made between decreasing tissue sensitivity to the herbicide, thereby rendering the plant resistant to it, and increasing basipetal translocation. (Both of these processes occur as the plant grows older.) If perennial plants are under stress caused by lack of moisture, insufficient herbicide is usually translocated to the underground parts to kill the plant. However, this problem is not so important in most annuals, only the shoots of which must be killed.

It is sometimes possible to apply herbicidal concentrations to the underground tissues at the time when the weeds are most sensitive. If some species of evergreen are sprayed when in the dormant condition, sufficient herbicide is translocated basipetally to be toxic, although growth of meristematic tissue does not become maximal until

TABLE 12-2

Recommendations for Use of Hormone-Type Herbicides in Weed Control. (Adapted from *Weed Control Recommendations*, compiled by F. M. Ashton *et al.*, distributed by the Calif. Agr. Exptl. Sta. Ext. Serv., Univ. of California, 1969, and from *Suggested Guide for Weed Control, 1969*, Agr. Handbook no. 332, Agr. Res. Serv., U.S. Dept. Agr.)

Crop	Herbicide	Rate/Acre	Spray (Vol/Acre) or Type Solid Material	Weeds Controlled	Preferred Time of Application	Remarks
			2,4-D AND DERIVATIVES			
Alfalfa	2,4-D amine	1 lb	20–60 gal water	Morning glory (field bindweed)	When weed is in early bloom, before crop starts blooming.	Apply shielded spray to row crop grown for seed.
Apple	2,4-D sodium salt	1–2 lb	40–60 gal water	Broad-leaved perennials	Repeat applications throughout the growing season. Late fall applications after harvest and before frost are preferred.	Apply to well-established trees 1 year old or older. Do not use in desert valleys or on shallow or sandy soils. Avoid spraying fruit or foliage. Allow maximum time after application and before next irrigation.
Asparagus (in cutting beds)	2,4-D sodium salt	1–2 lb	40–60 gal water	Broad-leaved weeds	During harvest. Make no more than 2 applications spaced 1 month apart.	

Crop	2,4-D form	Rate	Water	Weeds controlled	When to apply	Remarks
Barley, wheat	2,4-D amine	0.50–0.75 lb	10–50 gal water	Annual broad-leaved weeds	When crop is well established and tillered, but not after crop is in boot stage.	Do not forage or graze livestock on treated fields for 2 weeks after treatment or use treated straw for stock feed.
Barley, wheat	2,4-D low-volatile esters	0.50–0.75 lb	10–50 gal water	As above	As above	As above
Oats	2,4-D amine	0.50–0.75 lb	10–50 gal water	As above	As above	
Corn, field	2,4-D amine	0.50–0.75 lb	20–40 gal water	Broad-leaved weeds	From emergence to tasseling.	Use as a directed spray with drop nozzles when corn is more than 10 in high.
Flax	2,4-D amine	0.50–0.75 lb (0.25 lb in San Mateo County, California)	20–70 gal water by ground; 10 gal by air	Annual broad-leaved weeds	When crop is 4–8 in high, before bud stage.	
Grape	2,4-D acid or amine	1.5 lb	60–80 gal water	Morning glory	When weed is in bloom stage and growing vigorously. Safest after shatter following bloom but before shoots reach the ground.	Safest as a spot application. Apply carefully to avoid drift; avoid spraying vine shoots. Use hooded boom and low-pressure flooding nozzles to deliver coarse droplets.
Ladino clover (for seed only)	2,4-D amine	0.75 lb	20–60 gal water	Broad-leaved weeds	In spring when soil is moist, when crop is growing vigorously, and when there is enough new crop growth to cover stolons.	Do not irrigate immediately after spraying or graze livestock on treated areas within 7 days after treatment.

(Continued)

TABLE 12-2 (Continued)

Crop	Herbicide	Rate/Acre	Spray (Vol/Acre) or Type Solid Material	Weeds Controlled	Preferred Time of Application	Remarks
Pasture grass, clover mixtures	2,4-D amine	0.50–0.75 lb	20–60 gal water	Broad-leaved weeds, especially annuals	When soil is moist, pasture is well established, and crop and weeds are growing vigorously (preferably when weeds are small).	Do not irrigate immediately after spraying. Do not graze livestock on treated areas within 7 days after treatment.
Milo (grain sorghum)	2,4-D amine	0.5–1.0 lb	20–60 gal water by ground; 10 gal water by air	Broad-leaved weeds	After crop is 6 in high but before boot stage.	Directed spray using drop nozzles is preferred. If overall spray is used, do not graze dairy or forage dairy animals or animals being readied for slaughter.
Pasture grass, clover mixtures	2,4-D acid, amine, or ester	2 lb	40–60 gal water	Broad-leaved weeds	For spot treatment of perennial broad-leaved weeds. Spray foliage with 2 lb per 100 gal of water. Avoid runoff. This treatment will damage most legumes in the pasture mixture.	
Pear	2,4-D			Broad-leaved weeds	Repeat applications throughout the growing season. Late fall applications after harvest and before frost are preferred.	Apply to well-established trees 1 year old or older. Do not use in desert valleys or on shallow or sandy soils. Avoid spraying fruit or foliage. Allow maximum time after application and before next irrigation.

Rangeland plants	2,4-D	Maximum rate: 3 lb for broad-leaved weeds, 6 lb for woody plants		Follow directions for individual weed or brush species in O. A. Leonard and W. A. Harvey, *Chemical control of woody plants.* Calif. Agr. Exptl. Sta. Bull. 812, 1965.		
Red clover	2,4-D amine	1 lb	20–60 gal water	Morning glory	When weed is in early bloom, before crop starts blooming.	Apply shielded spray. Use only on crop grown for seed.
Strawberry	2,4-D amine	1 lb	20–60 gal water	Broad-leaved weeds, morning glory	Before blossom, after picking season.	Apply shielded spray to prevent leaf bending.
Sundan grass	2,4-D amine	1 lb	10–50 gal water	Broad-leaved weeds	After crop is 6 in high and before panicle has emerged from uppermost leaf sheath.	Do not spray seedling grass; do not spray between boot and milk stages if crop is grown for seed.
Trefoil	2,4-D	0.75 lb	20–60 gal water	Broad-leaved weeds	In the spring, when soil is moist and after new growth begins.	Do not irrigate immediately after spraying. Do not graze dairy animals on treated area within 7 days after treatment.
Plants growing on noncultivable land	2,4-D amine + dalapon + wetting agent	2 lb (acid equiv) + 6 lb, 16 fl oz/ 100 gal water	Sufficient spray to wet all leaves (100 gal water). Volume of spray/acre depends on height and density of growth.	Mixture of annual grasses and broad-leaved weeds	During the growing season as a foliar spray, when weeds are 3–4 in high. Repeat application when additional weeds appear.	

(Continued)

TABLE 12-2 (*Continued*)

Crop	Herbicide	Rate/Acre	Spray (Vol/Acre) or Type Solid Material	Weeds Controlled	Preferred Time of Application	Remarks
Plants growing on noncultivable land	Amitrole + 2,4-D amine + wetting agent	1 lb + 1 lb, 16 fl oz/100 gal water	Sufficient spray to wet all leaves.	Mixture of annual grasses and broad-leaved weeds	During the growing season as a foliar spray.	
Plants growing on noncultivable land	2,4-D amine + wetting agent	2–4 lb (acid equiv) + 16 fl oz/100 gal water	Sufficient spray to wet all leaves (80–100 gal water).	Perennial broad-leaved weeds (morning glory)	During the growing season as a foliar spray, when stems are at least 18 in long (normally at early bloom stage). Repeat application as new growth demands. When warranted by regrowth, October applications are very effective.	
Plants growing on noncultivable land	2,4-D acid (emulsifiable)	2 lb (acid equiv)				
Plants growing on noncultivable land	2,4-D amine + wetting agent	6 lb (acid equiv) + 16 fl oz/100 gal water	200 gal water	Russian knapweed	In October at full bloom stage or late bloom stage, as a foliar spray. Repeat application the following year commencing in the spring.	Prevent seedling growth with cultivations or 2,4-D applications when new plants are small. Thorough cultivation in June followed by October spraying appears to increase kill.
		3 lb (acid equiv)/100 gal water	Sufficient spray to wet all leaves and stems.	Russian knapweed		
Plants growing on noncultivable land	2,4-D acid (emulsifiable)	6 lb (acid equiv)/100 gal water	200 gal water	Russian knapweed	As above	As above
		3 lb (acid equiv)/100 gal water	Sufficient spray to wet all leaves and stems.	Russian knapweed	As above	As above

Site	Herbicide	Rate	Spray/Water	Weed	Remarks
Plants growing on noncultivable land	2,4-D low-volatile ester	4 lb (acid equiv) 2 lb/100 gal water	200 gal water Sufficient spray to wet all leaves and stems.	Russian knapweed	As above
Plants growing on noncultivable land	2,4-D low-volatile ester / 2,4-D acid (emulsifiable)	4 lb actual or 2 lb/100 gal (16 fl oz/100 gal water)	Sufficient spray to wet all leaves (200 gal water).	Hoary cress (whitetop and perennial pepperweed, perennial pepper grass)	Between bud stage and early bloom stage. Re-treatments are necessary for complete eradication.
	2,4-D amine + wetting agent	As above	As above		
Plants growing on noncultivable land	2,4-D amine + wetting agent (16 fl oz/100 gal water)	2 lb (acid equiv) or 1 lb/100 gal water	Sufficient spray to wet all leaves (200 gal water).	Canada thistle	Make first application between early bud and full bloom stages; respray new growth at early bud stage. Several re-treatments may be necessary for complete eradication.
	2,4-D acid (emulsifiable)	As above	As above	Canada thistle	
Plants growing on noncultivable land	2,4-D low-volatile ester	1.5 lb (acid equiv)/100 gal water	Sufficient spray to wet all leaves (50–100 gal water).	Curled dock (*Rumex crispus*)	At early bloom stage. Re-treatments are necessary for complete eradication.
Plants growing on noncultivable land	2,4-D amine + wetting agent	2 lb (acid equiv) + 16 fl oz/100 gal water	100 gal water	White horse nettle	During the growing season as a foliar spray, between early bloom and full bloom stages. Several re-treatments are necessary for complete eradication.
Plants growing on noncultivable land	2,4-D ester	2 lb (acid equiv)		White horse nettle	
Plants growing on noncultivable land	2,4-D acid (emulsifiable)	2 lb (acid equiv)	100 gal water	White horse nettle	Addition of a wetting agent at 1 qt/100 gal may increase effectiveness of spray, particularly with 2,4-D amine and amitrole.

(Continued)

TABLE 12-2 (*Continued*)

Crop	Herbicide	Rate/Acre	Spray (Vol/Acre) or Type Solid Material	Weeds Controlled	Preferred Time of Application	Remarks
Plants growing on noncultivable land	2,4-D volatile ester	2 lb (acid equiv)	Sufficient spray to wet all leaves (100 gal water).	Nut grass (nutsedge)		When weed is 8 in high. Re-treat when regrowth reaches height of 4-6 in. Many re-treatments are necessary for complete eradication.
Plants growing on noncultivable land	2,4-D acid (emulsifiable)	2 lb (acid equiv)				
Plants growing on noncultivable land	2,4-D amine + wetting agent	1.5 lb (acid equiv) + 16 fl oz/100 gal water	Sufficient spray to wet all leaves.	Bank or shoreline weeds (arrowhead and water plantain)	Apply early before weed becomes a problem. Repeat as necessary.	
Plants growing on noncultivable land	2,4-D low-volatile ester	1.5 lb (acid equiv)	Sufficient spray to wet all leaves (100 gal water).	Bank or shoreline weeds (umbrella sedge)	Apply early before weed becomes a problem. Repeat as necessary.	
Plants growing on noncultivable land	2,4-D low-volatile ester	1 lb (acid equiv)/100 gal water	Sufficient spray to wet all leaves.	Bank or shoreline weeds (lady's thumb, pale smartweed, and other annual smartweeds)	Apply early before weed becomes a problem.	

Plants growing on noncultivable land	2,4-D low-volatile ester + diesel oil	2 lb (acid equiv) + 1 gal/100 gal water	Immersed aquatic weeds (cattails)	Sufficient spray to wet all leaves (500–1,000 gal water).	In the spring when flower heads are emerging and until fully emerged. Respray when regrowth is 3–5 ft high.	Mix 2,4-D and oil before adding to agitating water. Do not apply to drinking water.
Plants growing on noncultivable land	2,4-D low-volatile ester + diesel oil	As above	Immersed aquatic weeds (hardstem bulrush)	As above	As above	As above
Plants growing on noncultivable land	2,4-D low-volatile ester	1.5 lb (acid equiv)/100 gal water	Floating aquatic weeds (water primrose and water hyacinth)	Sufficient spray to wet all leaves (100 gal water).	During the growing season as a foliar spray.	When extensive weed beds are to be sprayed, care should be taken to avoid contaminating water. Do not apply to drinking water.
			2,4-DB AND DERIVATIVES			
Alfalfa	2,4-DB amine	1 lb	Seedling or established broad-leaved weeds	10 gal water by air or 20–60 gal by ground spray	Apply to seedlings or established crops when weeds are at 1- to 3-leaf stage.	Do not forage or graze livestock on treated fields within 30 days of application.
Alfalfa	2,4-DB ester	0.50–0.75 lb	Seedling or established broad-leaved weeds	10 gal water by air or 20–60 gal by ground spray	Apply to seedlings or established crops when weeds are at 1- to 3-leaf stage.	Do not forage or graze livestock on treated fields within 30 days of application.
Ladino clover	2,4-DB amine	1 lb	Seedling or established broad-leaved weeds	20–60 gal water by ground spray or 10 gal by air	When weeds are at 1- to 3-leaf stage, before flowering of clover.	Do not forage or graze livestock on treated fields within 30 days of application. *(Continued)*

TABLE 12-2 (*Continued*)

Crop	Herbicide	Rate/Acre	Spray (Vol/Acre) or Type Solid Material	Weeds Controlled	Preferred Time of Application	Remarks
Ladino clover Pasture grass, clover mixtures	2,4-DB ester 2,4-DB amine	0.75 lb 1 lb	20–60 gal water by ground spray or 10 gal by air	Seedling or established broad-leaved weeds.	When weeds are at 1- to 3-leaf stage, before flowering of clover.	Do not forage or graze livestock on treated fields within 30 days of application.
Pasture grass, clover mixtures	2,4-DB ester	0.75 lb		As above	At crop seedling stage, when weeds are at 1- to 3-leaf stage.	Do not graze or forage livestock on treated fields.
Peas	2,4-DB amine	1 lb	20–60 gal water	As above		
Red clover	2,4-DB amine	1 lb	10 gal water by air or 20–60 gal by ground spray	As above	When weeds are at 1- to 3-leaf stage, before flowering.	As above
Trefoil	2,4-DB amine	1 lb	As above	As above	As above	Do not graze or forage livestock on, or cut hay from, treated fields within 30 days of application.
			2,4-DEP			
Corn, sweet	2,4-DEP	6 lb		Many germinating annual broad-leaved weeds and weed grasses.	Immediately after planting.	

Crop	Herbicide	Rate	Weeds controlled	Time of application	Remarks
Peanut	2,4-DEP	3 lb	Most small-seeded annual weeds.	After planting but before emergence.	Herbicide may cause crop injury (usually temporary), especially in cold, rainy periods during crop emergence. Do not use seed of poor vitality.
Potato	2,4-DEP	4 lb	Many germinating annual broad-leaved weeds and weed grasses.	After last cultivation.	
Strawberry	2,4-DEP	4 lb	Many germinating annual broad-leaved weeds and weed grasses.	Before planting or after transplanting during nonharvest year; after harvest in fields in commercial production.	
SESONE					
Asparagus	Sesone	2.0–5.5 lb	Germinating broad-leaved weeds and weed grasses.	Before and after harvest.	
Peanut	Sesone	2.7 lb	Most small-seeded annual weeds.	After planting but before emergence.	Herbicide may cause crop injury (usually temporary), especially in cold, rainy periods during crop emergence. Do not use seed of poor vitality. *(Continued)*

TABLE 12-2 (*Continued*)

Crop	Herbicide	Rate/Acre	Spray (Vol/Acre) or Type Solid Material	Weeds Controlled	Preferred Time of Application	Remarks
Potato	Sesone	3.6 lb		Germinating broad-leaved weeds and weed grasses.	Before and after emergence, after last cultivation.	
Strawberry	Sesone	3–4 lb		As above	Soil treatment should be applied after transplanting and after harvest, but not during flowering and fruiting periods.	
			MCPA AND DERIVATIVES			
Oats	MCPA amine	0.50–0.75 lb	10–50 gal water	Annual broad-leaved weeds	When crop is well established and tillered but not after crop is in boot stage.	
Flax	MCPA amine	0.6 lb	20–70 gal water by ground; 10 gal by air	Annual broad-leaved weeds	When crop is 4–8 in high but before bud stage.	
Milo (grain sorghum)	MCPA amine	0.75 lb	20–60 gal water by ground; 10 gal by air	Broad-leaved weeds	When crop is 6 in high but before boot stage.	Directed spray using drop nozzle is preferred. If overall spray is used, do not graze or forage dairy animals or animals being readied for slaughter.

Crop	Chemical	Rate	Spray volume	Weeds controlled	Time of application	Remarks
Peas	MCPA amine or sodium salt	0.75 lb		Annual broad-leaved and perennial weeds, including Canada thistle	After emergence, before bloom.	Safer to use than 2,4-D. Leaf burn may occur if spray is applied in hot weather.
Rice	MCPA amine or sodium salt	0.75–1.25 lb	10–30 gal water	Broad-leaved weeds, sedges	35–65 days after planting.	
Plants growing on noncultivable land	MCPA amine + wetting agent (16 fl oz/100 gal water)	2 lb (acid equiv) or 1 lb/100 gal water	Sufficient spray to wet all leaves (200 gal water).	Canada thistle	Make first application between early bud and full bloom stages; respray new growth at early bud stage. Several re-treatments may be necessary for complete eradication.	
MCPB						
Peas	MCPB	0.25–0.75 lb		Growing broad-leaved annual weeds and perennial weeds, including Canada thistle	Soon after emergence.	
2,4,5-T						
Rangeland plants	2,4,5-T	4 lb for range-land clearance				Follow directions for individual weed or brush species in O. A. Leonard and W. A. Harvey, *Chemical control of woody plants*. Calif. Agr. Exptl. Sta. Bull. 812, 1965.

(Continued)

TABLE 12-2 (*Continued*)

Crop	Herbicide	Rate/Acre	Spray (Vol/Acre) or Type Solid Material	Weeds Controlled	Preferred Time of Application	Remarks
Rice	2,4,5-T	0.5–1.5 lb		Broad-leaved weeds	When longest internodes are 0.125–0.500 in. long.	Severe crop injury may occur if rice is treated in early tillering, late jointing, booting, or early heading stages.
Plants growing on noncultivable land	2,4,5-T	2–4 lb		Broad-leaved herbaceous weeds	Between period of rapid vegetative growth and early bloom in the spring or summer; at time of vigorous rosette growth in the fall.	Use on weeds on which 2,4-D is ineffective; e.g., perennial white horse nettle, nightshade.
Plants growing on noncultivable land	2,4,5-T	2–4 lb		Undesirable woody plants	At full-leaf stage during rapid growth.	Especially effective on brambles, mesquite, oak, osage orange.
Plants growing on noncultivable land	2,4,5-T	2–4 lb	200 gal water/acre	Immersed and marginal aquatic weeds (arrowhead, smartweed, water hyacinth, water primrose)	When weeds are actively growing; repeat as necessary.	Reduce volume but not rate when foliage is wet.

Pasture grass	Silvex	0.50–0.75 lb		Chickweed, henbit, knotweed	After grass seedlings have reached 2- to 4-leaf stage.	Do not exceed .5 lb/acre until grasses become well established.
Rangeland plants	Silvex	4 lb for rangeland clearance				Follow directions for individual weed or brush species in O. A. Leonard and W. A. Harvey, *Chemical control of woody plants*. Calif. Agr. Exptl. Sta. Bull. 812, 1965.
Rice	Silvex	0.5–1.5 lb		Broad-leaved weeds	When longest internodes are 0.125–0.500 in. long.	Severe crop injury may occur if rice is treated in early tillering, late jointing, booting, or early heading stages.
Plants growing on noncultivable land	Silvex	2–4 lb		Undesirable woody plants	At full-leaf stage, during rapid growth.	Especially effective on maple, mulberry, palmetto, redbud, salt cedar, trumpet vine.
Plants growing on noncultivable land	Silvex ester	8 lb	150–200 gal water	Floating aquatic weeds (alligator weed, floating mats)	At first bloom and again when regrowth is 2–4 in above water.	2–4 applications are required for adequate control; do not use for irrigation or for domestic purposes.
Plants growing on noncultivable land	Silvex	2–4 lb	200 gal water	Immersed and marginal aquatic weeds (arrowhead, smartweed, water hyacinth, water primrose)	When weeds are actively growing; repeat as necessary.	Reduce volume but not rate when foliage is wet.

(Continued)

TABLE 12-2 (*Continued*)

Crop	Herbicide	Rate/Acre	Spray (Vol/Acre) or Type Solid Material	Weeds Controlled	Preferred Time of Application	Remarks
2,3,6-TBA						
Plants growing on noncultivable land	2,3,6-TBA	10–20 lb		Undesirable woody plants	When conditions are suitable for active growth.	Apply spray to both foliage and soil. Especially effective on woody vines.
Plants growing on noncultivable land	2,3,6-TBA	20 lb (acid equiv)	Apply in concentrated form or in water (50–100 gal water).	Perennial broad-leaved weeds (morning glory) and Russian knapweed	Apply to soil in fall or winter, when precipitation will leach herbicide into root zone.	Approximately 4- to 8-in precipitation following application ensures proper leaching. Supplemental irrigation may be necessary in areas of low rainfall. Avoid drift onto crops and ornamentals.
Plants growing on noncultivable land	Borate-2,3,6-TBA mixtures	20 lb active 2,3,6-TBA	Apply dry (granular)	Perennial broad-leaved weeds (morning glory) and Russian knapweed	Apply to soil in fall or winter, when precipitation will leach herbicide into root zone.	In areas where precipitation is less than 8 in, use supplemental irrigation to leach herbicide to deeper roots.
Plants growing on noncultivable land	Borate-2,3,6-TBA mixtures	20 lb active 2,3,6-TBA	Apply dry (granular)	White horse nettle	As above	As above

DICAMBA

Plants growing on noncultiva-ble land	2,3,6-TBA	20 lb active 2,3,6-TBA	50 gal water	White horse nettle	As above	As above
Barley, oats, wheat (fall-seeded)	Dicamba	0.25 lb		Corn cockle, cow cockle, dog fennel, smartweed, wild buckwheat, and most weeds susceptible to 2,4-D	After emergence in the spring, before joint stage.	Mixtures of 0.12 lb each of dicamba and 2,4-D per acre control mustards more effectively than dicamba alone.
Oats, wheat (spring-seeded)	Dicamba	0.12 lb (oats) 0.12–0.25 lb (wheat)		As above	After emergence, during 2- to 5-leaf stage.	As above
Corn	Dicamba	0.25 lb		Most annual broad-leaved weeds; controls or stunts broad-leaved perennial weeds	When corn is 4–18 in tall and weeds are shorter than corn. Avoid application when temperatures are high and corn is growing rapidly.	Apply as foliar spray to corn and weeds until corn is 12 in high; thereafter use in basally directed spray to avoid spraying into corn whorl.
Pasture grass	Dicamba	0.25–0.75 lb		Most winter and spring broad-leaved weeds	At single-leaf stage.	Grass may be temporarily injured; use on seed crop only.
Pasture grass	Dicamba + wetting agent	4 lb + 16 fl oz/100 gal water	80 gal water	Perennial broad-leaved weeds (morning glory), Russian knapweed, and Canada thistle	During the growing season as a foliar spray. Compound may persist in soil for one year or longer. Re-treatment is required.	Avoid drift to neighboring ornamentals or crops.

(Continued)

TABLE 12-2 (*Continued*)

Crop	Herbicide	Rate/Acre	Spray (Vol/Acre) or Type Solid Material	Weeds Controlled	Preferred Time of Application	Remarks
Pasture grass	Dicamba + wetting agent	0.75 lb (acid equiv)/ 100 gal + 16 fl oz/100 gal water	Sufficient spray to wet all leaves (50–100 gal water).	Curled dock (*Rumex crispus*)	At early bloom stage; re-treatment is required.	Do not contaminate water used for domestic, livestock, or irrigation purposes. Dicamba is active in the soil; therefore, do not apply it to or near crops or ornamentals.
				AMIBEN		
Asparagus	Amiben	3 lb		Germinating barnyard grass, crabgrass, curled dock, lamb's-quarters, pigweed, rag-weed, smart-weed	Before emergence.	
Pumpkin	Amiben	3–4 lb		As above	Before emergence.	
Squash, summer and winter	Amiben	3–4 lb		As above	Before emergence.	

Crop	Herbicide	Rate	Carrier	Weeds	Time of application	Remarks
Sweetpotato	Amiben	3 lb		As above	At transplanting.	
Tomato	Amiben	3–5 lb		As above	After emergence, after clean cultivation.	

PICLORAM

Crop	Herbicide	Rate	Carrier	Weeds	Time of application	Remarks
Plants growing on noncultivable land.	Picloram + wetting agent	1–2 lb + 16 fl oz/100 gal water	80 gal water (Higher rate required for eradication except on light, sandy soils.)	Perennial broad-leaved weeds (morning glory), Russian knapweed, Canada thistle, and white horse nettle	During the growing season as a foliar spray. Picloram may persist in soil for one year or longer. Repeat as necessary.	Avoid drift to neighboring ornamentals or crops.
Plants growing on noncultivable land	Picloram	2–4 lb	Sufficient spray to wet all leaves (80–100 gal water).	Bank or shoreline weeds (swamp smartweed)	At bloom stage or later. Several retreatments are necessary when regrowth occurs.	Complete coverage is essential. Avoid contamination of water used for irrigation.
Plants growing on noncultivable land	Picloram	6.0–8.5 lb		Undesirable woody plants	Broadcast granules just before or during period of adequate precipitation. Picloram persists in some soils for 3 years.	Highest concentration is necessary for ash, blackgum, and oak. For spot treatments under individual trees or clumps of brush, broadcast 1–2 tablespoonfuls per 30 sq ft of soil surface. *(Continued)*

TABLE 12-2 (Continued)

Crop	Herbicide	Rate/Acre	Spray (Vol/Acre) or Type Solid Material	Weeds Controlled	Preferred Time of Application	Remarks
			NPA AND DERIVATIVES			
Cantaloupe	NPA (sodium salt)	4 lb		Germinating broad-leaved weeds and weed grasses	Before emergence.	Herbicide may cause crop injury (usually temporary), especially in cold, rainy periods during crop emergence. Do not use seed of poor vitality.
Cucumber	NPA (sodium salt)	2–6 lb		As above	Before emergence, after transplanting.	
Peanut	NPA	2–4 lb		Most small-seeded annual weeds	After planting but before emergence.	
Pumpkin	NPA (sodium salt)	4 lb		Germinating broad-leaved weeds and weed grasses	Before emergence.	
Soybean	NPA	3–4 lb		Most small-seeded annual grasses, some broad-leaved weeds	After planting but before emergence.	Herbicide is likely to cause injury to crops grown in very coarse-textured soils.

a later date. However, absorption by the dormant foliage of other species may be insufficient for successful treatment; then the herbicide can be applied to cuts made in the stem. Application of herbicides to trunks, either through cuts or by basal sprays, should be made just before or at the time of maximum cambial activity in the spring for best results. Effective control of *Equisetum* spp. has been obtained by subsurface applications of MCPB to cut rhizomes (Holz, 1963).

An excellent method of applying herbicides to the underground parts of plants is to apply a readily absorbed compound such as picloram to the soil in the winter or early spring. The compound is then leached to the roots by rain before emergence of the weeds, ensuring that sufficient herbicide will be in the plant when its growing rate is at the maximum.

SELECTIVITY

Auxin-type herbicides can be used to control weeds without injuring the crop. In other words, the chemical, because of its unique toxic properties, can "select" and kill the weed. Some selective herbicides are applied to the foliage and others are applied to the soil; some may be applied in either way.

When foliage is sprayed, the herbicide moves either from the foliage to the young shoot tips or from the older leaves to the roots; movement generally follows the translocation of the photosynthate in the phloem. The selectivity of the compound depends primarily on inherent physiological differences between the crop and the weed.

Herbicides can be applied directly to the soil surface (but must then be washed into the soil by rainfall or irrigation to be effective), or, if they are volatile, they can be incorporated into the soil by cultivation or by injection below the soil surface. The stage of development of the crop and weeds at time of treatment is very important for best results. Herbicides may be applied at the preplanting, preemergence, or postemergence stage of the crop.

Variables that Affect Selectivity

A plant's selectivity is influenced by many variables, which include leaf structure, position of growing points, and extent to which the herbicide is absorbed and translocated; biochemical processes within the plant (enzyme inactivation, herbicide activation and inactivation); and biophysical processes (adsorption). An example of plant selectivity is that many broad-leaved weeds are killed by low concentrations of auxin-type herbicides, whereas other types are resistant even to high amounts. Herbicides, too, are selective: 2,4-D kills most of the broad-leaved weeds that plague grain crops without injuring those crops (Fig. 12–1).

LEAF STRUCTURE AND POSITION OF GROWING POINTS.   Cereals, onions, and certain other crops have narrow, upright leaves. Leaves may have corrugated surfaces consisting of many small ridges or waxy or hairy surfaces. Water droplets hitting waxy and hairy surfaces usually bounce off and thus only a small area of the leaf surface is wetted. On the other hand, most broad-leaved plants have smooth leaf surfaces to which much of the spray adheres.

The growing points of broad-leaved plants are located at the tips of the shoots and in the leaf axils where they are easily wetted by the spray (Fig. 12–2). The growing points of cereal crops are located at

FIGURE 12-1

Growth of weeds on right in this field of small grain was controlled with 2,4-D. (After D. L. Klingman and W. C. Shaw, 1967.)

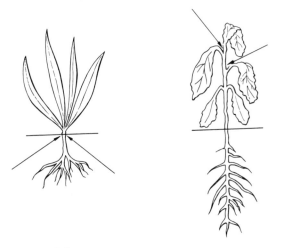

FIGURE 12-2
Selectivity of herbicide is dependent on location of growing points. *Left,* growing points (*arrows*) are protected from the spray and crop remains healthy. *Right,* growing points (*arrows*) are exposed and weed is killed. (After F. M. Ashton, W. A. Harvey, and C. L. Foy, 1961.)

the base of the plant and are protected from the spray by the surrounding leaves. These structural differences are important to consider when applying contact herbicides, which are not as readily translocated as auxin-type herbicides.

ABSORPTION OF HERBICIDE. Compounds are usually absorbed by the roots or leaves. The amount of absorption depends on the stage of plant development. The cuticle or waxy surface is thicker on old leaves than on young ones, and plants growing in the shade have a thinner cuticle than those growing in the sun. Cuticle thickness also varies among species. Since the cuticle is a major barrier to absorption of compounds, plants with a thin cuticle readily absorb the compound and may be killed, whereas those with a thick cuticle may absorb little and survive.

The number and size of stomata sometimes have an important effect on amount of absorption. In some plants, stomatal penetration is important only when oils or wetting agents are used (Fig. 12–3). Plant species differ in the number of stomata per unit leaf area and in location of stomata: In some species stomata are located only on the lower leaf surface; in others they are found both on upper and lower surfaces.

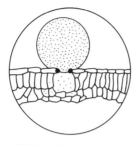 

FIGURE 12-3

Selectivity of herbicide is dependent on absorption. *Left,* in absence of wetting agent, stomatal absorption remains low. *Right,* wetting agent favors good stomatal absorption. (After F. M. Ashton, W. A. Harvey, and C. L. Foy, 1961.)

TRANSLOCATION OF HERBICIDE. After an auxin-type herbicide is absorbed by the roots, leaves, or stems, it usually is then translocated throughout the plant. Translocation can occur upward from the roots to the aerial portion of the plant or downward from the tops to the

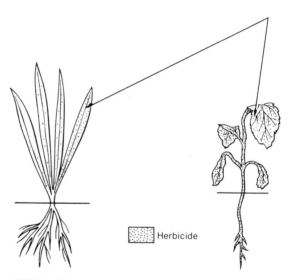

Herbicide

FIGURE 12-4

Selectivity of herbicide is dependent on translocation. *Left,* poor translocation of 2,4-D in grasses makes them resistant to weed killer. *Right,* good translocation of 2,4-D in broad-leaved weeds makes them susceptible to weed killer. Arrows indicate points of application. (After F. M. Ashton, W. A. Harvey, and C. L. Foy, 1961.)

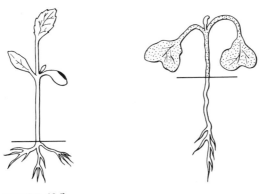

FIGURE 12-5

Selectivity of herbicide is dependent on activation of herbicide. *Left,* in a resistant crop, such as alfalfa, harmless 2,4-DB is not converted to plant killer 2,4-D. *Right,* in a susceptible weed, 2,4-DB is converted to 2,4-D. (After F. M. Ashton, W. A. Harvey, and C. L. Foy, 1961.)

roots. There is also much movement within shoots and stems. The amount of herbicide translocated depends on type of herbicide, plant species, and environmental conditions. In general the rate and amount of translocation of 2,4-D within susceptible species is greater than that within resistant species (Ashton *et al.,* 1961)—see Figure 12–4.

PHYSICAL AND CHEMICAL VARIABLES. Herbicides are sometimes so tightly bound to certain plant constituents that they are either unable to translocate from the site of application (Fig. 12–5) or ineffective even if they are translocated.

Sometimes an herbicide interferes with the plant's normal metabolic processes. Thus certain plants can be killed while others remain unharmed. Sometimes a relatively nontoxic herbicide is changed into a highly toxic one within the plant and kills it (Fig. 12–5). For example, in some sensitive plants 2,4-DB is changed into 2,4-D, but no such change occurs in resistant plants like alfalfa.

UPTAKE OF HERBICIDES IN PLANTS

The effectiveness of herbicides is determined mainly by the extent of their penetration into the plant and their subsequent translocation within the plant. An understanding of these processes, as well as of plant anatomy and physiology, is important for successful application of herbicides.

Plant Structure and Physiology

A study of the uptake and distribution of chemicals in plants should include a discussion of the terms "apoplast" and "symplast." The *apoplast* is a system of interconnecting cell walls and intercellular spaces, including the water-filled (or air-filled) xylem elements. The *symplast,* which constitutes the remaining part of the plant, is a system of interconnected protoplasm that is connected from cell to cell by means of the plasmodesmata, excluding the vacuoles. The apoplast protects the symplast from unfavorable environmental conditions that may cause desiccation and abrasion. Of course, any chemical substances that penetrate the symplast must first penetrate the apoplast. Herbicides injure or kill plants by their action within the symplast.

The protoplasm circulates within the cells by a streaming action, and the solutes, including those absorbed from outside, move within the plant body. The translocation rate is relatively slow (from a few millimeters to a few centimeters per hour) in parenchyma but can be very rapid (over one hundred centimeters per hour) in tracheary tissues of the xylem or in the sieve tube system of the phloem (Crafts, 1964), "a highly specialized system of conduits that serve as a functional part of the symplast continuum" (Crafts, 1961a).

The cuticle, which is part of the apoplast, and the plasma membrane, which covers all surfaces of the symplast including the plasmodesmata, are the two main barriers that a chemical must penetrate to enter the symplast. The cuticle is a noncellular layer of cutin that covers the outer epidermal cell walls and is composed of polymerized long-chain fatty acids and alcohols. It is separated from the underlying cell walls by pectic substances, but the demarcation area is frequently indistinct. Waxes in varying amounts are embedded in the cutin layer and are deposited on the outer surface. It should be noted that plant species vary considerably within this generalized structure.

The basic structure of the upper cuticle of a pear leaf is schematically illustrated in Figure 12-6. The cutin matrix is separated from the underlying cell wall by a zone of pectic substances that extends into the cutin matrix but not to the outer surface of the cuticle as in apple, as was demonstrated by Roberts *et al.* (1948). Chemicals that have penetrated the outer cuticle travel to the plant cells assisted by the pectic substances, which serve as pathways. Epicuticular waxes having a platelet-like structure are deposited in zones on the outer

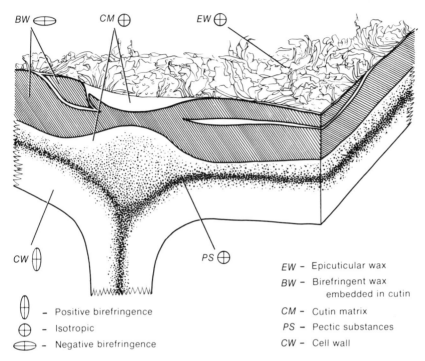

BW ⊖   CM ⊕   EW ⊕

CW ⊖   PS ⊕

⊖ - Positive birefringence
⊕ - Isotropic
⊖ - Negative birefringence

EW - Epicuticular wax
BW - Birefringent wax
        embedded in cutin
CM - Cutin matrix
PS - Pectic substances
CW - Cell wall

FIGURE 12-6

Structure of pear leaf, upper cuticle (not drawn to scale). Lower cuticle may differ from upper in certain morphological characteristics and in quantity of oriented wax molecules. (After R. F. Norris and M. J. Bukovac, 1968.)

surface of the cutin matrix, which contains areas of embedded waxes that are birefringent in polarized light. The negatively birefringent waxes are separated by layers of isotropic materials that extend from the cuticular surface to the region of embedded pectic substances. Norris and Bukovac (1968) suggest that such layers of isotropic materials act as pathways of penetration through an otherwise relatively impermeable wax layer.

Compounds that penetrate the stomata must traverse the walls of the substomatal chambers. Although these surfaces are cutinized, this cuticle is very water-permeable (*see* Crafts, 1964).

The cuticle covers the outer epidermal cell walls, which consist chiefly of cellulose and pectins. Cellulose, a carbohydrate, is hydrophilic (that is, has a strong affinity for water) and extremely elastic. Pectins are composed largely of polyuronides and are also hydrophilic. They have an important effect on the water-holding capacity of the cell.

Penetration of Foliage by Herbicides

Successful use of an herbicide depends on its ability to penetrate the cuticle and the plasma membrane. Fat-soluble compounds and lipoidal compounds readily penetrate the cuticle, and water and polar solutes readily penetrate the pectin and cellulose portions. Crafts (1956) has presented evidence that the fat-soluble acid molecules of the chlorophenoxy compounds and esters of dalapon can enter leaves through the epidermis even if no stomata are present. The cuticle is also somewhat permeable to polar molecules, such as ions of the chlorophenoxy compounds, dalapon, and MH.

Penetration of the leaf by polar molecules is greatly assisted by high humidity, which indicates that entry is through an aqueous medium (Clor *et al.*, 1962). A high water content in the leaf as a result of high humidity enables the water system of the leaf cells to become almost contiguous with the outer surface of the cuticle and provides an aqueous pathway from the cuticle to the symplast. Compounds diffusing by the lipoid pathway are taken up more rapidly than those diffusing by the aqueous pathway because at the leaf surface, even under conditions of high humidity, there are only a few aqueous components. However, the aqueous pathway is also operational, as evidenced by the successful use of MH and dalapon as herbicides.

The uptake of hormone-type herbicides is probably a two-step process (van Overbeek, 1956). The first step is probably simple physical adsorption and is very rapid. The second step is slow and steady and probably requires metabolic energy. The adsorption process explains why 2,4-D and related compounds are so rapidly bound to plants.

Penetration of herbicides can occur through the stomata if they are open at the time of treatment (Dybing and Currier, 1959). Because plain aqueous solutions usually do not penetrate stomata, a wetting agent is required to lower surface tension. Wetting agents improve spreading of the solution by decreasing the surface tension so that air films between solution and cuticle that prevent contact of the solution with the cuticle surface are destroyed. The solution can then spread along the intercellular spaces and wet the walls of the mesophyll and the bundle sheath cells; in addition, penetration between the wax particles of the cuticle is increased when it is stretched by the turgid cells below. If cracks in the cuticle exist, wetting agents facilitate penetration.

Wetting agents also facilitate good cuticular absorption of many systemic and contact herbicides, including the heavy ester and emulsifiable acid formulations of 2,4-D, 2,4,5-T, and other chlorophenoxy compounds, the chlorobenzoic acids, and dalapon.

Penetration of Stems by Herbicides

Application of herbicides to the green or succulent stems of plants can frequently be just as effective or even more effective than foliar application. One reason is that the stems are closer to the roots than the leaves. Thus if there is good penetration into the stems more herbicide may reach the roots than if the chemical were applied to the foliage. A second reason is that chemicals are distributed differently depending on whether they are applied to leaves or stems. When an herbicide is applied to the leaves, it moves out of them through the phloem and continues to move in the stream of phloem assimilates. But when stems are sprayed, large amounts of chemical can move through the phloem into the xylem, where rapid upward movement can occur in the transpiration stream. This is the reason why many leaves on plants are killed when the basal portions of the stems are sprayed with 2,4-D or 2,4,5-T.

Penetration of Bark by Herbicides

The bark of woody stems usually presents a strong barrier against herbicidal penetration. There is a wide variation in type of bark. Some types have splits or cracks that expose considerable amounts of cortical tissue that are not suberized; in other types the bark has no splits or cracks and may or may not be penetrated by some corky lenticels. Because aqueous sprays are usually inadequate to attain good penetration of the latter type, an oil-soluble formulation is required. An ester of 2,4-D or related compounds is usually dissolved in an aromatic weed oil or diesel oil. Basal sprays of 2,4-D, 2,4,5-T, or other phenoxy compounds at a concentration of approximately 5 percent sometimes kill trees and brush. Large volumes of solution should be used so that the liquid runs down the crown of the plant and kills any buds present to prevent regrowth of new shoots.

It is sometimes desirable or necessary to apply the herbicide to a cut surface around the trunk. An aqueous solution or an oil carrier of the herbicide is applied to the cuts and moves rapidly into the tracheary elements of the tree or shrub. Solutions of both 2,4-D and 2,4,5-T are extremely effective when applied in this manner.

Absorption of Herbicides by Roots and Subsequent Upward Movement

The most important avenue of entry of many herbicides into the plant is through the roots. Since many hormone-type compounds are applied as preemergence or postemergence soil-borne herbicides, it is logical that much research has been done on the absorption of solutes and water and its effect on weed control.

The description of the mechanism of absorption by Crafts and Broyer (1938), which is still accepted by most plant physiologists, is as follows. Most of the water and salts are taken up by root hairs and cortex cells in the primary region behind the root tip. Ions from the ambient medium diffuse across the apoplasts of the root hair and of the other cortex cells until they contact the outer layer of the symplast. They are then absorbed and accumulate in the symplast in concentrations higher than those in the ambient solution, a process requiring metabolic energy resulting from respiration. The ions move from the epidermis across the cortex and endodermis and into the stele. In the stele solutes leak from the symplast to the apoplast, where water is gained through osmosis. Excess hydrostatic pressure develops and the solution follows the walls to the xylem, enters the vessels, and moves as a mass acropetally into the stem and leaves. This sap (which is the sap that exudes or guttates from the trunk or stem of a plant when the top is cut off) carries mineral nutrients, as well as any herbicides present, to the stems and foliar portions of the plant.

Herbicides differ in the rate at which they are taken up by the roots. Entry of 2,4-D is rapid, but that of MH and dalapon is slow (Crafts, 1964). However, once these compounds are inside the roots, their rate of movement upward to the tops, as well as their rate of absorption, often differs. For example, 2,4-D enters the roots much more rapidly than MH or dalapon but is the last to reach the leaves. Roots evidently have the capacity to absorb different organic molecules selectively in the same way that they do different inorganic ions. Some organic molecules are also bound in the roots in some manner so that some or all of them fail to reach the stele and move to the upper plant parts. Other compounds, such as 2,4-D, apparently are released by the roots but are bound in the stems and fail to reach the foliage.

TRANSLOCATION OF HERBICIDES FROM FOLIAGE

Once an herbicide has penetrated the plant through the foliage, stems, or roots, its effectiveness depends largely on its pattern of translocation. The two main transport systems are the phloem and the xylem. The herbicides, along with the assimilates, move in a stream from the leaves mainly through the phloem and continue to travel in a source-to-sink direction: The main sources are expanded leaves; the sinks are shoot tips, roots, fruits, and other plant parts undergoing expansion or development.

The translocation pattern followed by a substance from a leaf into the plant can be readily determined by applying a tracer such as ^{14}C to the single leaf. Hale and Weaver (1962) used autoradiographic techniques to determine changes in the direction of translocation of ^{14}C following the assimilation of $^{14}CO_2$ by single leaves or shoot tips of 'Muscat of Alexandria' grape as the shoots matured (Fig. 12–7). The grape leaves began to export photosynthate when they had expanded to approximately one-half their full size. The first assimilate exported by the treated leaf moved to the shoot tip. When the treated leaf was separated from the oldest importing leaf on the shoot tip by two or three other exporting leaves, the assimilate from the leaf was translocated both to the shoot tip and basipetally. With further shoot growth basipetal movement increased until translocation from a leaf below the shoot tip was only basipetal. When fruit development began, translocation from the leaves below the cluster was partially reversed. These leaves exported assimilates both to the fruit cluster and to the parent vine. When the rate of shoot elongation sharply decreased, assimilates from the shoot tip moved in a basipetal direction. This general pattern of food movement has also been demonstrated in many other plants.

A strong sink is necessary to achieve good export of herbicide from the leaves. The roots often serve this purpose, and should be actively growing at time of treatment. If soil moisture has been depleted by the time the leaves mature, a poor kill often results even though penetration of the foliage may be satisfactory.

A chemical such as 2,4-D, which is strongly bound in living cells, requires both an active source and an active sink for extensive and rapid translocation. Herbicides that are not tightly bound are not so adversely affected by a slow movement of assimilates.

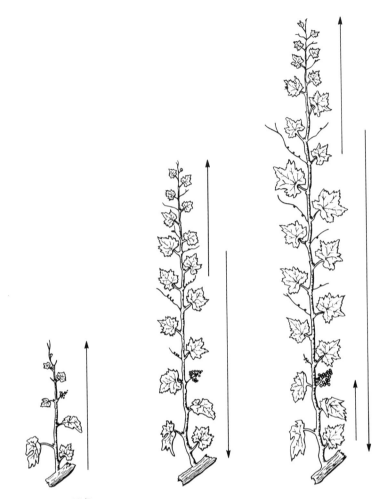

FIGURE 12-7

Main direction of movement of photosynthate at 3 different developmental stages of a rapidly growing grape shoot. *Left,* movement in a very young shoot or shoot tip is apical. *Center,* at the prebloom or bloom stage, export is bidirectional from 2 or 3 leaves below the shoot tip. Below this region movement is basal. *Right,* after the set of fruit, photosynthate also moves apically into the cluster from leaves below the cluster. After the rate of shoot growth decreases several weeks later, photosynthate moves basally from the tip. (After C. R. Hale and R. J. Weaver, 1962.)

In order to hasten export of herbicides from the leaves, foliar treatments should be made when the leaves are mature and exporting assimilates. Herbicidal sprays should be applied to such deciduous woody plants as maple, elm, and oak only when the foliage is mature and when export of assimilates is proceeding. Because flowers and young, rapidly growing shoots (including those of certain evergreen shrubs) accumulate a high concentration of food material and thus do not readily export, 2,4-D should be applied early in the spring before the seasonal flush occurs (Leonard and Crafts, 1956).

Hormonal herbicides usually accumulate in the symplast and move almost exclusively in the phloem. However, such hormonal herbicides as MH penetrate the cuticle, but instead of entering the phloem, they move with the transpiration stream within the apoplast. The resultant acropetal movement in the fine veins of the leaf emerges as a wedge-shaped pattern on an autoradiograph (Fig. 12–8). On rare occasions 2,4-D exhibits apoplastic movement. Some chemicals, including MH, exhibit both symplastic and apoplastic movement.

When some hormone-type herbicides are applied to foliage they move basipetally into the phloem and then move upward in the xylem into the stems and leaves. These compounds, including dalapon and 2,3,6-TBA, often accumulate in large quantities in the roots, and then, because of their molecular properties, apparently move from the phloem into the xylem. In the leaves the compounds can reenter the symplast and can again be carried to the root system (Crafts, 1964).

Injury to leaf tissue reduces or blocks export of herbicide from the leaf. If extremely high concentrations of compounds are used, export is usually reduced even if no leaf injury is visible for a day or so after application.

Alteration of Translocation Patterns by Herbicides

It is often difficult or impossible to kill woody plants by foliar application of herbicides even when the foliage is exporting the assimilates and the root system is actively growing and hence serves as a strong sink. Frequently the aerial portions are killed, but sprouting and regrowth emerge from the base of the plant. Experiments conducted with grape suggest that hormonal herbicides tend to defeat their own transport capability by quickly creating sinks in the sprayed stems and foliage.

Leonard et al. (1967) demonstrated that high concentrations of 2,4-D and picloram interfere with the downward movement of ^{14}C

FIGURE 12-8

Translocation within *Phaseolus vulgaris* of ^{32}P (*left*), ^{65}Zn (*center*), and ^{45}Ca (*right*) as shown in autoradiographs (*top*). Mounted plants are shown at bottom. ^{32}P is rapidly absorbed and translocated to the shoot tip and roots via the phloem. Translocation of ^{65}Zn is restricted because of slower absorption. ^{45}Ca is localized within the apoplast because this element does not enter and move within the phloem. (Photographs courtesy of A. S. Crafts and S. Yamaguchi.)

assimilates in field-grown *Vitis vinifera* grape cv. 'Ribier.' Shoots were covered with a polyethylene bag and 50 μc of $^{14}CO_2$ gas was injected into the bag with a syringe. After one hour the bags were opened and the foliage of the shoots was uniformly treated with aqueous drops of 2,4-D or picloram (20 ml total) at concentrations of 10,000 to 50,-000 ppm. Picloram at concentrations of 2,000 or 20,000 ppm was more effective than 2,4-D in preventing downward movement of ^{14}C out of the shoot and into the parent vine. Translocation within the treated shoots was not prevented by treatment with either 2,4-D or picloram; transport of ^{14}C from the vegetative parts of the shoots into the clusters continued regardless of treatment.

These results also explain why application of 2,4-D to the herbaceous shoots of a plant does not cause symptoms to appear on another part of the plant. It also explains why application of 2,4-D can kill the aerial portion of one-half of a cordon-trained grapevine without inflicting injury to the other half. Alteration of transport is probably caused by the development of strong sinks as a result of the treatment.

BINDING AND PERSISTENCE OF HERBICIDES IN PLANTS AND SOIL

Binding and Persistence of Herbicides in Plants

After uptake, the herbicides must translocate to active sites in the plants and produce injury. Poor translocation sometimes occurs despite strong phloem movement because the compounds either have accumulated in and are bound to proteins in the protoplasm, or perhaps have accumulated in the vacuoles, so that only relatively small amounts reach the active meristematic sites. More binding occurs if the food is conducted slowly in the phloem than if it is conducted rapidly. This explains why the effectiveness of 2,4-D is frequently dependent on an active phloem movement, as only with such movement does sufficient compound reach the root and shoot meristems and other sinks before being bound along the translocation route.

Compounds vary greatly in their ease of translocation. 2,4,5-T seems to be much less mobile than 2,4-D and thus is probably more readily bound in the plant (Crafts, 1964).

Hormone-type herbicides can persist in plants for several months or even occasionally for several years. The persistence of 2,4-D in foliage, as measured by development of formative effects on new leaves, may last throughout the growth of only four or five nodes, after which the leaves are normal in such plants as bean and cotton, but the compound may persist during the growth of ten to fifteen nodes in grape. The best evidence of long persistence of herbicides is the presence of auxin-produced effects (epinasty and abnormal venation) in seedlings grown from seeds of plants previously sprayed with herbicides. Bradley and Crane (1957) noted that apricot seedlings that developed seventeen months after application of chlorophenoxy compounds showed marked formative effects. Similar results have been obtained by applying 2,4-D to cotton (McIlrath et al., 1951) and by applying 2,4-D and dalapon to cotton and wheat (Foy, 1961).

Binding and Persistence of Herbicides in Soil

Herbicides are frequently applied to the soil before or after plant emergence. Foliar application can also result in dripping of excess solution from the foliage to the soil surface. The duration of weed control depends on the length of time that an herbicide persists in the soil. If the herbicide persists too long, it may injure succeeding crops and may adversely affect soil fertility by upsetting the components of the populations of micro-organisms that inhabit the soil. Herbicides that have accumulated in the soil can be leached out and may flow into rivers, canals, and reservoirs, thus constituting a danger to man and animals. Other considerations that influence persistence of herbicides in soil are microbial and chemical decomposition, volatility, and adsorption on soil colloids.

Herbicides differ greatly in their ability to persist in soil. Some may last for only a week or so, whereas others may persist in toxic concentrations for over two years. Soil temperature, moisture content, fertility, and microbial population all affect the persistence of a given herbicide, so its duration under average conditions is difficult to estimate. 2,4-D usually lasts for approximately two to four weeks; MH lasts for four to ten weeks, and MCPA lasts for eight to twelve weeks. 2,3,6-TBA is very persistent and lasts from ten to thirty weeks (Audus, 1964b).

The persistence of an herbicide in the soil is often determined by the ease with which it is decomposed by micro-organisms. 2,4-D is easily decomposed by micro-organisms and is one of the least persistent chemicals, whereas the more persistent 2,3,6-TBA is more resistant to microbial attack. Early work revealed that such herbicides as 2,4-D rapidly disappear in warm, moist, organic soils that are favorable to bacterial growth and demonstrated a correlation between the number of bacteria in the soil and the rate of disappearance of an herbicide. When soil was sterilized in an autoclave to kill the micro-organisms, persistence of herbicides was found to be greatly prolonged (see Audus, 1964b). Similar results have been obtained when chemical treatments are made to block microbial metabolism before the herbicide is applied. There are many bacterial and fungal organisms that can break down herbicides; a list of some of them is given by Audus (1964b).

Herbicides frequently can be removed from the soil by leaching. Water-soluble compounds are usually most easily leached, but these compounds sometimes react with other chemicals to form insoluble complexes that resist leaching. The soil particles may adsorb chemicals and thus prevent their downward movement with water. Salts of 2,4-D are adsorbed by soils with high organic content, but in light, sandy soils these water-soluble salts readily leach downward. Adsorption is probably the main factor that prevents many herbicides from being removed from the soil by leaching. Another important effect of the adsorption of an herbicide on soil particles is to reduce the concentration of herbicide freely available in the soil solution, which decreases the rate of entry of compound into the roots. The degree of plant response to an herbicide is probably determined by a concentration of chemical that is intermediate between free (that portion not adsorbed by soil particles) and the total herbicide (Hartley, 1964). Because of adsorption, soils high in organic matter and clay content must be treated with relatively high amounts of herbicide for preemergence weed control. These soils also tend to retain herbicides much longer than do light, sandy soils.

Chemical decomposition, which includes such processes as oxidation, reduction, hydrolysis, and hydration, destroys some herbicides in the soil. For example, dalapon undergoes slow hydrolysis in the presence of water, which renders the compound inactive. When chemicals are in the adsorbed state they are especially likely to undergo rapid oxidation. Some compounds are nontoxic until they undergo chemical decomposition. The herbicide sesone is nontoxic, but when it is applied to moist soil it is rapidly hydrolyzed to 2,4-dichlorophenoxyethanol, which, in turn, is oxidized to 2,4-D. This is a situation in which partial decomposition is required to produce an active compound.

Such volatile compounds as the ester forms of 2,4-D may be lost from the soil or soil surface by evaporation. These gases may drift onto and injure susceptible crops (cotton, grape, tomato). The ester forms of 2,4-D are lost to the atmosphere in large quantities on a day when the soil surface is hot. When these compounds are washed into the soil and adsorbed, evaporation is decreased.

DETOXIFICATION OF HERBICIDES IN PLANTS

After entering plants most herbicides are metabolized or broken down into nontoxic substances. If one plant species metabolizes an herbicide more rapidly than another species, there is a basis for herbicidal selectivity. Many auxin-type herbicides are detoxified in plants, and detoxification occurs to different extents in different species (Wain, 1964). Sometimes the herbicide condenses with a cell constituent to form an inactive substance. IAA is detoxified in plants in part by its conversion to indoleacetyl DL-aspartic acid (Good *et al.*, 1956). The relative toxicity of herbicides is often based on the ability of the plant to detoxify the less active herbicides more rapidly. 2,4,5-T is more toxic to cucumber than 2,4-D because cucumber can detoxify 2,4-D more rapidly than 2,4,5-T (Slife *et al.*, 1962). Simazine is one of the best examples of a compound that is selectively detoxified by plants. Corn can hydroxylate simazine to form a nontoxic derivative and thus is protected from the lethal action of the herbicide, but susceptible plants are unable to alter the compound and it accumulates to toxic levels (Gysin and Knüsli, 1960)—see Figure 12–9. As a result of these metabolic differences between weeds and corn, simazine is useful for the latter.

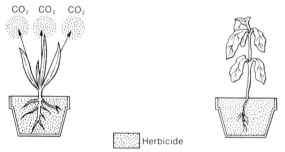

CO_2 CO_2 CO_2

Herbicide

FIGURE 12-9

Selectivity of herbicide is dependent on decomposition of herbicide. *Left,* herbicide simazine is decomposed by corn, liberating CO_2 gas and eliminating most of the plant killer. *Right,* simazine is absorbed by and remains in weed. (After F. M. Ashton, W. A. Harvey, and C. L. Foy, 1961.)

CONVERSION OF INACTIVE COMPOUNDS TO ACTIVE COMPOUNDS IN PLANTS

Some chemicals that have little or no biological activity can be activated by certain plant enzymes. 2,4-DB, which is of low toxicity, is converted in some plants to 2,4-D. Crafts (1960) demonstrated that esters of 2,4-D are hydrolyzed by plant enzymes. Fawcett *et al.* (1960), using chromatographic techniques and sensitive methods of bioassay on a wide range of phenoxy and indole derivatives, found that whenever growth was induced, a physiologically active acid compound was formed in the tissues; other derivatives were inactive. Some compounds are converted in one species of plant but not in another; for example, IAN is rapidly hydrolyzed to IAA in wheat but not in pea (*see* Wain, 1964). Since species differ in their rate of conversion of inactive derivatives to active derivatives, there is opportunity for herbicidal selectivity.

The γ-phenoxybutyric acids are perhaps the best example of herbicides whose selectivity depends on the capacity of the susceptible plant to convert an inactive compound to an active one. This type of conversion is effected by the β-oxidation process, which occurs in some plant tissues and degrades certain of these growth substances to toxic compounds. (The term "β-oxidation" refers to the breaking off of two carbon fragments of chains that occur in fatty acid metabolism.) In 1947 Synerholm and Zimmerman, using the tomato leaf epinasty test, bioassayed seven (2,4-dichlorophenoxy) alkane carboxylic acids and found that compounds with an even number of carbon atoms in the side chain (acetic, butyric, caproic, and octanoic derivatives) were highly toxic but that those with an odd number of carbon atoms (propionic, valeric, and heptanoic) were not. They explained their results by suggesting that the tomato degrades the even-numbered side chains down to a stable two-carbon side chain but that odd-numbered chains are degraded to an unstable one-carbon side chain that is broken down to a phenol. For example:

Even-numbered chain

2,4-DB
(slight activity)

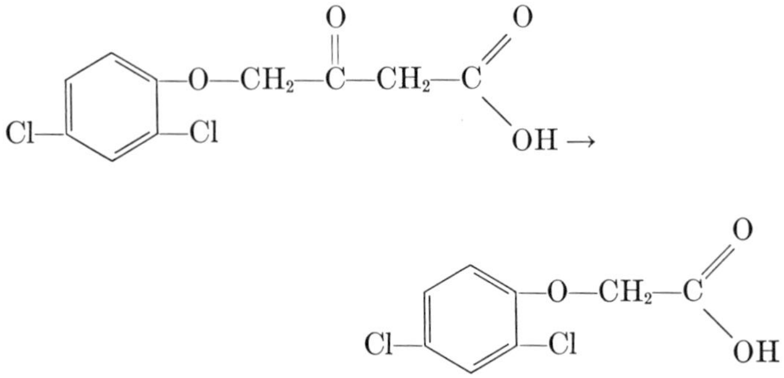

2,4-D
(stable; active)

Odd-numbered chain

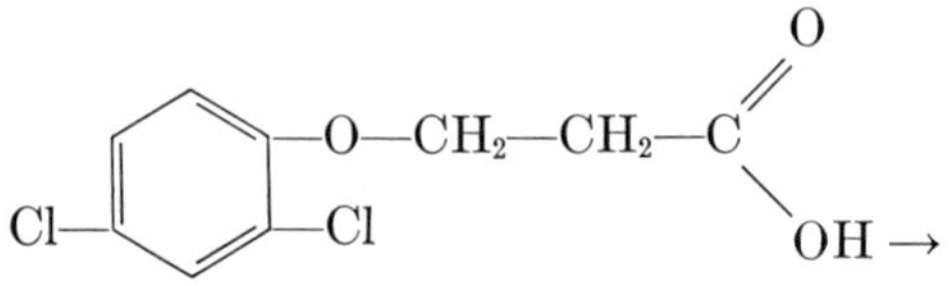

2,4-DP

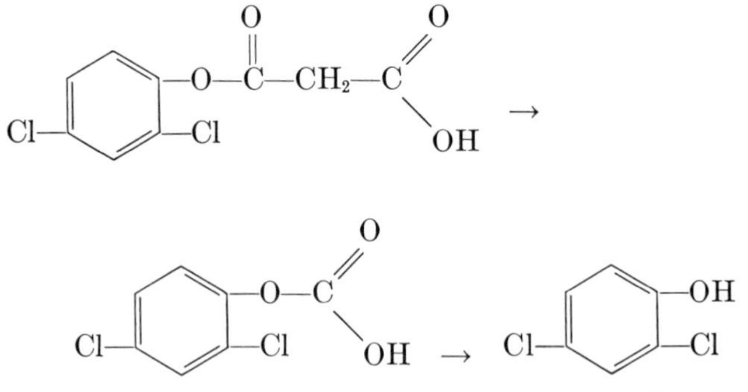

2,4-dichlorophenol
(unstable; not active)

Further work performed at Wye College in England (see Fawcett *et al.*, 1956) showed that some plants can degrade even-numbered carbon chains to acetic acid whereas others cannot. Since some leguminous crops lack this mechanism, a compound such as 2,4-DB, to which many weeds are susceptible, will kill weeds growing among these crops that do possess the β-oxidation mechanism without affecting the crops (Fig. 12–10).

a b c d e f

FIGURE 12-10

Effect of spraying 0.2 percent solutions of 2,4-D (*a*), 2,4-dichlorophenoxypro-
pionic acid (*b*), 2,4-DB (*c*), 2,4-dichlorophenoxyvaleric acid (*d*), 2,4-dichloro-
phenoxycaproic acid (*e*), and 2,4-dichlorophenoxyheptanoic acid (*f*), on charlock
(*top row*) and clover (*bottom row*). Control plants are on extreme right.
Photographed 2 weeks after spraying. (After R. L. Wain, 1964.)

CHLOROPHENOXY HERBICIDES

The chemical 2,4-D became commercially important soon after the
first trials were conducted. One advantage of the compound is its
selectivity: Broad-leaved weeds growing in grain fields can be killed
without injuring adjacent cereal plants. Mechanical removal of weeds
growing in grain fields is difficult or impossible. Since cereals are a
major field crop, the chemical was immediately in great demand.
Another advantage of the 2,4-D is its low cost per acre: The compound
is very potent and often controls weeds at rates of one pound per
acre or less. When used at recommended rates it is not hazardous to
man, domestic animals, or game; it does not accumulate in the soil or
harm soil organisms. Spray equipment is not corroded by 2,4-D.

The four other major phenoxy herbicides, 2,4,5-T, MCPA, silvex,
and 2,4-DB, have the same advantages as 2,4-D (Klingman and
Shaw, 1967).

Response of Plants to Chlorophenoxy Herbicides

In weed control an effort is made to spray the tops of the weeds,
including the leaves, twigs, green stems, flowers, and fruits. Roots

can also absorb these herbicides when they are sprayed on the soil but larger doses are required. After absorption the chemical is rapidly translocated throughout a plant, often causing death. Annual plants are most sensitive when they are young; perennial weeds are most easily killed when they are seedlings.

MORPHOGENETIC EFFECTS. The correlative relationships in plants are upset when they are sprayed with 2,4-D or other chlorophenoxy herbicides. Different growth responses occur throughout the plant because the sensitivities of the plant parts to the herbicide vary. When plants are severely injured but not killed, new organs may be initiated. For example, if the tip of the shoot is killed, some lateral buds can begin growth because they are released from the apical dominance of the shoot tip.

Auxin-type herbicides often stop elongation and increase the width of such plant parts as stems, leaf petioles, axes of flowers and inflorescences, rhizomes, tubers, and roots. At the same time cell division is stimulated, so that many lateral root primordia are produced. Sometimes the lateral primordia are so close together that when they break through the external tissues of the root they form a single fused sheet or sheets that run along the root. Auxins have an irregular effect on the cell-division processes of many root crops, such as carrot, radish, and sugar beet, thus inducing localized root-thickening (Fig. 12–11). In grasses, cereals, and other monocotyledonous plants, auxins may cause premature death of the primary root system and an increase in the number of adventitious roots (Gorter and van der Zweep, 1964). High concentrations of auxin may also cause abnormal swellings at the tips of root hairs and thus hinder their capacity to absorb.

Within a few hours after treatment, weed stems and petioles may exhibit a characteristic bending or twisting. Downward bending (epinasty) occurs more frequently than upward bending (hyponasty). Often the stem gradually reorients itself, at least partially. Sometimes the curvature is fixed by anatomical changes in the stem. Lateral swellings often occur as a result of enhanced lateral cell extension, which is usually accompanied by decreased cell elongation. (Treated seedlings are often short, thick structures.) Auxins often stimulate cell divisions that can lead to production of galls and tumors on the stem. Disorganization of tissues may follow and allow entry for various pathogenic organisms.

The amount of mesophyll tissue in leaves that develop after plants are treated with auxin-type herbicides is sometimes decreased. Slightly

FIGURE 12-11

Hour-glass-shaped radishes produced by soaking seed for 20 hours in NAA at a concentration of 200 ppm before sowing. (Photograph courtesy of L. C. Luckwill.)

affected leaves show only small loss of mesophyll or abnormal venation, but in severe cases leaves may be so narrow that they have almost no mesophyll (Fig. 12–12). Leaves sometimes become crinkled as a result of differential growth between veins and the mesophyll. The degree of injury to leaves usually decreases as the leaves mature. Leaves that are mature at the time of treatment often show no visible effects of the herbicide.

Many treated plants slowly die within two to three weeks of treatment. Plants that are not killed may recover and grow to almost full size.

BIOCHEMICAL EFFECTS. The effects of chlorophenoxy and other herbicides on the composition of a plant are very complex. The type of herbicide and type of plant and the environmental conditions largely determine the changes in the plant's metabolic and compositional status. The chlorophenoxy herbicides can effect changes in the composition of carbohydrates, lipids, nitrogen, organic acids, vitamins, aromatics, alkaloids, steroids, ethylene, auxins, water, minerals, nu-

FIGURE 12-12

Typical leaves from vines of 'Black Corinth' grape growth in Davis, California. *Top,* controls; *middle,* leaves sprayed with 4-CPA at a concentration of 40 ppm; *bottom,* leaves sprayed with BOA at a concentration of 100 ppm. Spraying with 4-CPA resulted in stunted growth and abnormal venation resembling that caused by spraying with 2,4-D. Foliage sprayed with BOA formed many cup-shaped leaves. Similar leaves sometimes develop when the regulators are applied at concentrations of 10 or 20 ppm. Chlorine-substituted benzoic acids produce symptoms similar to those produced by BOA.

cleic acids, and enzymes. They can also cause changes in photosynthesis and respiration and in the metabolism of nitrogen and nucleic acid. These effects have been reviewed by Wort (1964a), Penner and Ashton (1966), and Moreland (1967).

Compounds and Formulations

Acids, salts, esters, and amines are the usual formulations for phenoxy herbicides. The salt and ester formulations are most frequently available as liquid concentrates (Klingman and Shaw, 1967) and are mixed before use. Salt concentrates readily form solutions with water, and ester concentrates form milky-white emulsions when mixed with water. In oil, ester concentrates form solutions.

Esters can be classified into high-volatile and low-volatile forms. Above 32°C both types release vapors, but at lower temperatures high-volatile esters are far more volatile than low-volatile esters. Therefore, crops growing adjacent to sprayed areas are usually much less likely to be injured from the low- than from the high-volatile compounds. Salt formulations are safe because of their low volatility. However, high-volatile esters are cheaper and are recommended if no susceptible crops are growing nearby.

Ester formulations are usually more toxic than salts because the former more effectively penetrate leaves and other plant surfaces. Therefore, esters can usually be applied at lower rates. Esters have a more toxic effect than salts on weeds that are growing slowly because of adverse environmental conditions, such as low temperatures or a shortage of moisture. Because esters are formulated in oils, an ester solution spreads and remains in moist contact with the leaves far longer than does a salt solution, which is water soluble. Rains that follow spray application are much less likely to wash off the oily ester formulation than the aqueous salt formulation.

Amine salts of 2,4-D are highly soluble in water and hence are valuable for use in low-gallonage spray application. The amines usually used are diethanolamine, triethanolamine, dimethylamine, and trimethylamine, or mixtures of them. There are also oil-soluble amine products whose oil solubility depends on the lipophilic characteristics of the amine portion. As the length of the carbon chain increases, the lipophilic attributes of the compound also increase so that oil solubility is increased and water solubility is decreased.

Dust forms are very dangerous from a drift standpoint, and such applications are generally prohibited by law in the United States. Solid forms, both granular and pellet, are available and are often useful for preemergence weeding and aquatic weed control.

Application

Phenoxy herbicides must be applied with care to avoid crop damage. Attention should be paid to the formulation used because these compounds differ in solubility, volatility, specific gravity, and toxicity. When susceptible crops are growing nearby, high-volatile esters should be avoided and spraying should not be done on a windy day. To minimize damage from spray drift, nozzles should be used that provide a course spray that is applied at a low pressure: a maximum of thirty-five pounds per square inch for boom sprayers and one hundred pounds per square inch for spray guns. Spraying should be done when the wind, if any, blows from an area where susceptible crops are grown toward the area being sprayed. The spray equipment should be cleaned after use so that residues will not subsequently be reapplied.

CULTIVABLE LAND.   Herbicides can be applied to the soil as preemergence sprays, at planting time or shortly thereafter. The chemical is applied at a concentration that kills the weeds as they germinate but leaves the crop undamaged. Phenoxy compounds are usually used as postemergence sprays after the weeds have begun to grow. Modern spray equipment utilizes volumes of from five to twenty gallons per acre. If the crops are not susceptible to the herbicide, a broadcast spray can be made. If the crops are sensitive, then a directed spray must be used so that the weeds are sprayed but the crops are not. Spraying by airplane is frequently done on such nonrow crops as grains and rice.

NONCULTIVABLE LAND.   In large areas or areas where the terrain is rough or that have too many obstructions for use of ground equipment, low-volume application is frequently made by airplane. A ground sprayer equipped with a boom for low-volume broadcast spray can be used to control weeds growing along irrigation canals or on grazing land. Brush and trees growing along roads and utility lines and aquatic weeds and brush growing along irrigation and drainage canals can be controlled most effectively by a high-volume (over one hundred gallons per acre) directed spray that thoroughly wets the leaves and leaf stems.

Trees with a trunk diameter of less than four inches can be killed by spraying the bark on the basal six to twelve inches of the trunk with ester formulations of phenoxy compounds in oil. However, such a spray frequently cannot penetrate the bark of a tree whose trunk is more than four inches in diameter. The base of the tree should then be ringed with ax cuts that go through the bark and into the sapwood; the solutions (water-soluble formulations or oil carriers) are then applied into the cuts—see Figure 12–13 (Leonard and Harvey, 1965).

FIGURE 12-13

Method used to apply herbicides to hatchet cuts made in tree trunks. (Photograph courtesy of O. A. Leonard.)

Mechanism of Action

Van Overbeek (1964) suggests that the mechanism of action of 2,4-D is its effect on the concentrations of naturally occurring auxins in the plant. Considerable fluctuations in the amount of indole auxins occur during plant growth and development. Since 2,4-D is also an auxin (although it is far more potent and persistent than the naturally occurring ones), a logical assumption is that the entrance of 2,4-D into the cells saturates them with auxin and upsets the auxin fluctuations that are required for normal growth and differentiation. Many plant functions, such as phloem transport, absorption, and photosynthesis, are upset by the morphological and biochemical changes that occur when the orderly course of development is disrupted by application of 2,4-D. As a result of such treatment, cell division in the meristems may be disrupted, cell elongation is decreased but the cells continue to grow in width, and young leaves do not expand properly. Root elongation is also inhibited; cells then swell and divide in the functional part of the roots and stems, producing callus growth and splitting of stems and root primordia. The end result of these effects is usually death of the plant.

Application of 2,4-D can have many other effects on the cell (Penner and Ashton, 1966). The compound prevents immature cytoplasm from changing to mature cytoplasm (although it does not prevent transformation of mature cytoplasm to immature cytoplasm). Studies made at the biochemical level have shown that 2,4-D increases the level of RNA, and this process can lead to the extra and abnormal growth just described. Moreover, cytokinins are important in cell division, and the imbalance in the auxin-to-cytokinin ratio caused by 2,4-D treatment might lead to abnormal growth and destruction of the plant (see van Overbeek, 1964).

Chlorophenoxy compounds cause plant tissues to revert to a meristematic state, a condition often noted in stems of treated plants. Excessive meristematic activity in stems, roots, or petioles, for example, creates metabolic sinks that can weaken plant parts that have no metabolic sinks. Cell proliferation may sometimes be sufficient to crush the phloem, a result that damages the plant by blocking translocation.

The fact that many of a plant's physiological activities are affected by the chlorophenoxy compounds indicates either that the mode of action of these compounds is very nonspecific or that they affect a specific process basic to all the observed responses (Penner and Ashton, 1966). Research on the influence of these herbicides on the physical and chemical structure of DNA and the consequent alteration in RNA metabolism might shed light on the mode of action (see Key, 1969).

A hypothesis advanced by Cherry (1970) is that the reaction of the auxin herbicides 2,4-D or 2,4,5-T with a receptor in the cytoplasm is the first step in herbicidal action. The receptor is more readily available for auxin interaction in plants that are susceptible to 2,4-D and 2,4,5-T than in plants that are not. The auxin receptor complex or a modified receptor is translocated through the nuclear membrane into the nucleus. The receptor then either reacts with RNA polymerase or changes its specificity so that the enzyme can transcribe different genomes.

Other Phenoxy Herbicides

There are several important phenoxy-type herbicides in addition to 2,4-D, 2,4,5-T, and MCPA. These compounds, esters of the chlorophenoxy alcohols, include 2,4-DES, MCPES, 2,4-DEB, 2,4,5-TES, and 2,4-DEP. In the soil these compounds are converted by microbial action to their corresponding acetic acids, which are effective in preemergence weed control. These esters are nontoxic until hydrolyzed and thus can be safely applied to foliage.

CHLOROBENZOIC ACIDS

Several derivatives of chlorinated benzoic acids are used extensively for perennial weed control in uncultivated areas or areas that are not suitable for cultivation. These compounds are effective on a wide range of plants and are especially valuable for weeds that are resistant to the chlorophenoxy acids. Canada thistle, honeysuckle, sassafras, persimmon, sumac, leafy spurge, wild onion, cattail, garlic, Russian knapweed, pine, locust, trumpet vine, bur ragweed, and skeleton weed are among the weeds that are susceptible to the chlorobenzoates but are, to varying extents, difficult to control with chlorophenoxy herbicides. Rates for soil application that range from twenty to forty pounds per acre of acid equivalent and approximately eight pounds per acre are used for foliage sprays of chlorobenzoates. These rates are considerably higher than those used for the chlorophenoxy compounds. Since the chlorobenzoic acids are relatively persistent and can persist in moist soil for several years, residual herbicides in soil to be cropped later can be a problem. Nevertheless, the substituted benzoic acids are used more frequently than chlorophenoxy compounds when soil application is desirable. Chlorobenzoic acids are active against both grasses and broad-leaved plants. Preemergence applications to cotton, corn, and many other species have proved to be effective.

2,3,6-TBA and 2,3,5,6-TBA are the most toxic of the chlorine-substituted benzoic acids. However, their effect is mild compared with that of 2,4-D. Formative effects, such as abnormal venation, may develop in treated plants, but this type of damage is easily distinguished from that of 2,4-D (Fig. 12–12).

Mechanism of Action

The substituted benzoic acids probably act fundamentally in the same manner as 2,4-D (van Overbeek, 1964). Differences in toxicity between these acids and 2,4-D are caused by variations in chemical persistence and mobility within the plant.

AUXIN-TYPE HERBICIDES OTHER THAN CHLOROPHENOXY AND CHLOROBENZOIC ACIDS

This section discusses the auxin-type herbicides that do not fit into either of the other two major classifications. The most important herbicides in this group are NPA and picloram (Table 12–1).

The growth-regulating properties of the N-aryl phthalamic acids were described by Hoffman and Smith in 1949, and the compounds were introduced as herbicides a year later. Preemergence applications at a rate of two to ten pounds per acre kill both monocotyledenous and dicotyledonous weeds. NPA and its derivatives are effective on seedlings and enter the plants through the roots. Such compounds thus are usually applied to germinating weed seeds before crop emergence. When applied in nonlethal amounts, NPA causes epinastic effects and abnormal venation, indications of its hormone-like properties. The geotropic response of roots is also upset and they often grow out of the soil surface. This group of chemicals is particularly promising as a selective herbicide for cucurbit crops. A preemergence application to such cucurbit crops as pumpkin, cantaloupe, cucumber, and watermelon affords excellent weed control for several weeks.

Picloram (Tordon)

Picloram, which was introduced in the early 1960's, induces many of the same physiological responses as are induced by the substituted benzoic acids (Hamaker et al., 1963). This herbicide is useful to control most perennial broad-leaved plants and woody plants. Growth of broad-leaved weeds among grass crops can be controlled because most grasses are resistant to picloram. The compound is rapidly absorbed by foliage and roots and translocates both up and down in plants, much of it accumulating in developing tissues.

Soil applications of the chemical in granular form are most effective when made in spring and early summer. Foliar sprays can be made up to three weeks before frost if other growing conditions are favorable.

HERBICIDES AS POLLUTANTS

The rapid advances that have been made in the introduction of new herbicides and pesticides have revolutionized many phases of crop production, including forestry and range management. However, many new and perplexing problems have arisen (volatilization, drift, mis-application, soil residues) to plague people applying the new chemicals (Crafts, 1970) and worry ecologists and biologists, among others. Solutions to these problems are difficult because without pesticides and herbicides, crop yields are limited and starvation is the ultimate result. However, permanent damage to the atmosphere and soil must be avoided.

Unexpected damage is sometimes inflicted on nontreated plants as a result of movement of 2,4-D or related compounds, which is affected by formulation of the compound, technique of application, and environmental conditions. Horticultural plants, including trees and ornamental plants, some fruits, and vegetables, vary widely in their sensitivity to 2,4-D (Sherwood *et al.*, 1970). Many trees and woody plants growing in a large area of St. Louis, Missouri, for example, have also been injured to various degrees by industrial emissions of herbicides (Lanphear and Soule, 1970).

Supplementary readings on the general topic of weed control are listed in the Bibliography: Anonymous, 1970; Audus, L. J., 1964a; Crafts, A. S., 1961a and 1961b; Crafts, A. S., and Crisp, C. E., 1971; Crafts, A. S., and Robbins, W. W., 1962; King, L. J., 1966; and Klingman, G. C., 1961.

BIBLIOGRAPHY

Abbott, D. L. 1959. Growth substances as fruit thinning agents for apples: progress report. The effects of seed removal on the growth of apple fruitlets. In *Ann. Rept. Agr. Hort. Res. Sta.*, Long Ashton, Bristol, England, 1958, pp. 52–56.

———. 1960. The bourse shoot as a factor in the growth of apple fruits. *Ann. Applied Biol.* 48:434–438.

Abdel-Kader, A. S., Morris, L. L., and Maxie, E. C. 1966. Effects of growth-regulating substances on the ripening and shelf-life of tomatoes. *Hort-Science* 1:90–91.

Abeles, F. B. 1967. Mechanism of action of abscission accelerators. *Physiol. Plantarum* 20:442–454.

Acker, R. M. 1949. Growth of three varieties of lilium from bulbs stored in vapors of methyl ester of naphthaleneacetic acid. *Bot. Gaz.* 111:21–35.

Addicott, F. T. 1964. Physiology of abscission. In Ruhland 1965b, pp. 1094–1126.

———. 1969. Ageing, senescence, and abscission in plants: Phytogerontology. *HortScience* 4:114–116.

———. 1970. Plant hormones in the control of abscission. *Biol. Rev.* 45:485–524.

Addicott, F. T., Carns, H. R., Cornforth, J. W., Lyon, J. L., Milborrow, B. V., Ohkuma, K., Ryback, G., Smith, O. E., Thiessen, W. E., and

Wareing, P. R. 1968. Abscisic acid: a proposal for the redesignation of abscisin II (dormin). In Wightman and Setterfield 1968, pp. 1527–1529.

Addicott, F. T., Carns, H. R., Lyon, J. L., Smith, O. E., and McMeans, J. L. 1964. On the physiology of abscisins. In Nitsch 1964, pp. 687–703.

Addicott, F. T., Lynch, R. S., and Carns, H. R. 1955. Auxin gradient theory of abscission regulation. *Science* 121:644–645.

Addicott, F. T., and Lyon, J. L. 1969. Physiology of abscisic acid and related substances. *Ann. Rev. Plant Physiol.* 20:139–164.

Akabane, N., *et al.* 1957. Prevention of cold and frost damage of fruit trees, III: Influence of maleic hydrazide on the properties of the grapevine. *Hokkaidoritsu Nogyo Shikenjo Shuho* 1:1–9. In *Chem. Abstr.* 52:5725 (Pref. Agri. Exptl. Sta., Hokkaido, Sapporo, Japan).

Albrigo, L. G., and Christ, E. G. 1968. Fruitone CPA thinning of peaches in New Jersey in 1967. *Hort. News,* vol. 48.

Alexander, A. G. 1968. Interrelationships of gibberellic acid and nitrate in sugar production and enzyme activity in sugarcane. *Jour. Agr. Univ. Puerto Rico.* 52:19–28.

Alexander, A. G., Montalvo-Zapata, R., and Kumar, A. 1970. Gibberellic acid activity in sugarcane as a function of the number and frequency of applications. *Jour. Agr. Univ. Puerto Rico.* 54:477–503.

Alleweldt, G. 1960. Die Wirkung der Gibberellinsaüre auf Reben. *Kali-Briefe* 5:1–8. Bern: Internationales Kali-Institute.

Amen, R. D. 1968. A model of seed dormancy. *Bot. Rev.* 34:1–31.

Anderson, I. C., Greer, H. A. L., and Tanner, J. W. 1965. Response of soybeans to triiodobenzoic acid. In F. A. Greer and T. J. Army, eds., *Genes to genus,* pp. 103–115. Skokie, Ill.: International Minerals and Chemical Corp.

Anderson, J. L. 1969. The effect of ethrel on the ripening of Montmorency sour cherries. *HortScience* 4:92–93.

Anonymous. 1955. *Natl. Acad. Sci. Natl. Res. Counc. Publ. 384.* Chemical-Biological Coordination Center Positive Data Series no. 2, 45 pp.

―――. 1968a. *Extent and cost of weed control with herbicides and an evaluation of important weeds, 1965.* U.S. Dept. Agr., Agr. Res. Serv., pp. 34–102.

―――. 1968b. *Spray recommendations for tree fruits in eastern Washington.* Wash. State Univ. Extension Bull. 419 (see pp. 28–29).

―――. 1969. *Suggested guide for weed control.* Agric. Handbook no. 332. Crops Research Division, U.S. Dept. Agr.

―――. 1970. *Herbicide handbook of the weed society of America.* 2d ed. Geneva, N.Y.: W. F. Humphrey.

Antcliff, A. J. 1967. A field trial with growth regulators on the Zante currant (*Vitis vinifera* var.). *Vitis* 6:14–20.

Anthony, A., and Street, H. E. 1970. A colorimetric method for the estimation of certain indoles. *New Phytol.* 69:47–50.

Arnold, G. W., Bennett, D., and Williams, C. N. 1967. The promotion of winter growth in pastures through growth substances and photoperiod. *Australian Jour. Agr.* 18:245–257.

Ashiru, G. A. 1969. Effect of kinetin, thiourea and thiourea dioxide, light and heat on seed germination and seedling growth of kola (*Cola nitida* (Ventenant) Schott and Endlicher). *Jour. Amer. Soc. Hort. Sci.* 94:429–432.

Ashton, F. M., and Harvey, W. A. 1971. *Selective chemical weed control.* Calif. Agr. Exptl. Sta. Ext. Serv. Circ. 558.

Ashton, F. M., Harvey, W. A., and Foy, C. L. 1961. *Principles of selective weed control.* Calif. Agr. Exptl. Sta. Ext. Serv. Circ. 505.

Ashton, F. M., et al. 1969. *Weed control recommendations.* Calif. Agr. Exptl. Sta. Ext. Serv.

Auchter, E. C., and Roberts, J. W. 1934. Experiments in spraying apples for the prevention of fruit set. *Proc. Amer. Soc. Hort. Sci.* 30:22–25.

Audus, L. J. 1947. Effects of certain organic metabolic products on plant nutrition and growth. In *Int. Cong. Pure Appl. Chem. Rpt.* XI.

Audus, L. J. 1959. *Plant growth substances.* 2d ed. London: Leonard Hill.

————, ed. 1964a. *The physiology and biochemistry of herbicides.* London and New York: Academic Press.

Audus, L. J. 1964b. Herbicide behaviour in the soil, II: Interactions with soil micro-organisms. In Audus 1964a, pp. 163–206.

————. 1968. Plant growth substances—past, present and future. *Advancement of Science* 25:1–12.

Avery, G. S., Jr., Burkholder, P. R., and Creighton, H. B. 1937. Production and distribution of growth hormone in shoots of *Aesculus* and *Malus*, and its probable role in stimulating cambial activity. *Amer. Jour. Bot.* 24:51–58.

Avery, G. S., and Johnson, E. B. 1947. *Hormones and horticulture.* New York: McGraw-Hill.

Bachelard, E. P., and Stowe, B. B. 1963. Rooting of cuttings of *Acer rubrum* L. and *Eucalyptus camaldulensis* Dehn. *Australian Jour. Biol. Sci.* 16:751–767.

Badizadegan, M., and Carlson, R. F. 1967. Effect of N^6 benzyladenine on seed germination and seedling growth in apples (*Malus sylvestris* Mill.). *Proc. Amer. Soc. Hort. Sci.* 91:1–8.

Badr, S. A., Bradley, M. V., and Hartmann, H. T. 1970. Effects of gibberellic acid and indoleacetic acid on shoot growth and xylem differentiation and development in the olive, *Olea europaea* L. *Jour. Amer. Soc. Hort. Sci.* 4:431–434.

Bailey, W. K., Toole, E. H., Toole, V. K., and Drowne, M. E. 1958. Influence of temperature on the after-ripening of freshly harvested Virginia bunch peanut seeds. *Proc. Amer. Soc. Hort. Sci.* 71:422–424.

Baldev, B., Lang, A., and Agatep, A. O. 1965. Gibberellin production in pea seeds developing in excised pods: Effect of growth retardant AMO-1618. *Science* 147:155–157.

Ballantyne, D. J. 1963. Note on the effect of growth substances on the bloom life of narcissus cut flowers. *Canadian Jour. Plant Sci.* 43:225–226.

————. 1965. Senescence of daffodil (*Narcissus pseudonarcissus*) cut flowers treated with benzyladenine and auxin. *Nature* 205:819.

————. 1966. The influence of low temperature and gibberellin on development and respiration of flower buds of the redwing azalea (*Rhododendron* cv.). *Proc. Amer. Soc. Hort. Sci.* 88:595–599.

Barden, J. A. 1968. Effects of Alar on the growth and distribution of the growth increment in one-year-old apple trees. *Proc. Amer. Soc. Hort. Sci.* 93:33–39.

Barker, W. G., and Collins, W. B. 1965. Parthenocarpic fruit set in the lowbush blueberry. *Proc. Amer. Soc. Hort. Sci.* 87:229–233.

Barritt, B. H. 1970. Fruit set in seedless grapes treated with growth regulators Alar, CCC and gibberellin. *Jour. Amer. Soc. Hort. Sci.* 95:58–61.

Barton, L. V. 1956. Growth response of physiological dwarfs of *Malus Arnoldiana* Sarg. to gibberellic acid. *Contrib. Boyce Thompson Inst.* 18:311–318.

Batjer, L. P. 1964. *Apple thinning with chemical sprays.* Washington Agr. Exptl. Sta. Bull. 651.

————. 1965. *Fruit thinning with chemicals.* Agr. Information Bull. no. 289. U.S. Dept. Agr., Agr. Res. Serv.

————. 1967. Chemical control of tree size. *Proc. 17th Int. Hort. Cong.* 2:71–75.

————. 1968. Effectiveness of thinning sprays as related to fruit size at time of spray application. *Proc. Amer. Soc. Hort. Sci.* 92:50–54.

Batjer, L. P., Siegelman, H. W., Rogers, B. L., and Gerhardt, F. 1954. Results of four years' tests of the effect of 2,4,5-trichlorophenoxypropionic acid on maturity and fruit drop of apples in the northwest. *Proc. Amer. Soc. Hort. Sci.* 64:215–221.

Batjer, L. P., and Thompson, B. J. 1961. Effect of 1-naphthyl N-methylcarbamate (Sevin) on thinning apples. *Proc. Amer. Soc. Hort. Sci.* 77:1–8.

Batjer, L. P., and Williams, M. W. 1966. Effect of N-dimethyl amino succinamic acid (Alar) on watercore and harvest drop of apples. *Proc. Amer. Soc. Hort. Sci.* 88:76–79.

Batjer, L. P., Williams, M. W., and Martin, G. C. 1963. Chemicals to control tree size. *Proc. Wash. State Hort. Assoc.* 59:107.

Batjer, L. P., Williams, M. W., and Martin, G. C. 1964. Effects of N-dimethyl amino succinamic acid (B-nine) on vegetative and fruit characteristics of apples, pears, and sweet cherries. *Proc. Amer. Soc. Hort. Sci.* 85:11–16.

Bayliss, W. M., and Starling, E. 1904. The chemical regulation of the secretory process. *Proc. Royal Soc.* (series B) 73:310–322.

Bayzer, H. 1964. Dünnschichtchromatographische Trennung quaternärer Ammoniumverbindung auf Celluloseschichten. *Experientia* 20:233.

———. 1966. Dünnschichtelektrophoretische Trennung quaternärer Ammoniumverbindungen. *Jour. Chromatog.* 24:372–375.

Beck, W. 1968. Die bisher bekannten Wirkungen von CCC auf die entwicklung pflanzenparasitärer Pilze. *Pflanzenschutz-Berichte* 27:10/11: 145–160.

Benes, J., Veres, K., Chvojka, L., and Friedrich, A. 1965. New types of kinins and their action on fruit tree species. *Nature* 206:830–831.

Bennet-Clark, T. A., and Kefford, N. P., 1953. Chromatography of the growth substances in plant extracts. *Nature* 171:645–647.

Bentley, J. A. 1950. Growth-regulating effect of certain organic compounds. *Nature* 165:449.

———. 1958. The naturally occurring auxins and inhibitors. *Ann. Rev. Plant Physiol.* 9:47–80.

———. 1961a. Chemistry of the native auxins. In Ruhland 1961, pp. 485–500.

———. 1961b. Extraction and purification of auxins. In Ruhland 1961, pp. 501–512.

———. 1961c. Chemical determination of auxins. In Ruhland 1961, pp. 513–520.

Besemer, S. T. 1967. Growth retardant tests on potted poinsettias. *Calif. Agr.* 21:6–7.

Bessey, P. M. 1960. *Effects of a new senescence inhibitor on lettuce storage.* Univ. Ariz. Exptl. Sta. Rept. 189, 5–8.

Beutel, J. 1968. Chemical peach thinning. *The Blue Anchor* 45(2):31–32.

Beyerinck, M. W. 1888. Über das *Cecidium* von *Nematus capreae* auf *Salix amygdalina. Zeitschr. Bot.* 46:1–11, 17–28.

Biale, J. B. 1954. The ripening of fruit. In *Scientific American* (May), p. 40. Reprinted in Janick *et al., Plant agriculture.* San Francisco: W. H. Freeman, 1970, pp. 70–74.

Biggs, R. H., and Leopold, A. C. 1957. Factors influencing abscission. *Plant Physiol.* 32:626–632.

———. 1958. The two-phase action of auxin on abscission. *Amer. Jour. Bot.* 45:547–551.

Blackman, G. E. 1945. Plant growth substances as selective weed killers:

A comparison of certain plant-growth substances and other selective herbicides. *Nature* 155:500–501.

Blumenthal-Goldschmidt, S., and Rappaport, L. 1965. Regulation of bud rest in tubers of potato *Solanum tuberosum* L., II: Inhibition of sprouting by inhibitor β complex and reversal by gibberellin A₃. *Plant Cell Physiol.* 6:601–608.

Boffey, P. M. 1968. Defense issues summary of defoliation study. *Science* 159:613.

Bonner, J. 1944. The role of toxic substances in the interactions of higher plants. *Bot. Rev.* 16:51–65.

————. 1965. *The molecular biology of development.* Oxford: Clarendon Press.

Bonner, J., and English, J. 1938. A chemical and physiological study of traumatin, a plant wound hormone. *Plant Physiol.* 13:331–348.

Bonner, J., and Galston, A. W. 1944. Toxic substances from the culture media of guayule which may inhibit growth. *Bot. Gaz.* 106:185–198.

————. 1952. *Principles of plant physiology.* San Francisco: W. H. Freeman.

Boodley, J. W., and Mastalerz, J. M. 1959. The use of gibberellic acid to force azaleas without a cold temperature treatment. *Proc. Amer. Soc. Hort. Sci.* 74:681–685.

Borrow, A., Brian, P. W., Chester, V. E., Curtis, P. J., Hemming, H. G., Henehan, C., Jeffreys, E. G., Lloyd, P. B., Nixon, I. S., Norris, G. L. F., and Radley, M. 1955. Gibberellic acid, a metabolic product of the fungus *Gibberella fujikuroi:* Some observations on its production and isolation. *Jour. Sci. Food Agr.* 6:340–348.

Borthwick, H. A., Hendricks, S. B., Parker, M. W., Toole, E. H., and Toole, V. K. 1952. A reversible photoreaction controlling seed germination. *Proc. Natl. Acad. Sci. U.S.* 38:662–666.

Borthwick, H. A., and Robbins, W. W. 1928. Lettuce seed and its germination. *Hilgardia* 3:275–304.

Boswell, S. B., and McCarty, C. D. 1970. Effects of TIBA growth regulator on open branching of citrus for mechanical shaking. *Calif. Agr.* 24:6–7.

Bottelier, H. P. 1956. The significance of the age and the presence of the lamina for the petiole test. *Ann. Bogorienses* 2:183–192.

Bouillenne, R., and Went, F. W. 1933. Recherches expérimentales sur la néo-formation des racines dans les plantules et les boutures des plantes supérieures. *Ann. Jard. Bot. Buitenzorg* 43:25–202.

Boysen-Jensen, P. 1913. Über die Leitung des phototropischen Reizes in der *Avena*-koleoptile. *Ber. Deut. Bot. Ges.* 31:559–566.

Boysen-Jensen, P. 1935. *Die Wuchsstofftheorie, und ihre Bedeutung für die Analyse des Wachstums und der Wachstumsbewegungen der Pflanzen.* Jena: Verlag Gustav Fischer.

Boysen-Jensen, P. 1936. *Growth hormones in plants.* Trans. G. S. Avery, Jr., and P. R. Burkholder. New York: McGraw-Hill.

Bradley, M. V., and Crane, J. C. 1957. Effects of auxins on development of apricot seeds and seedlings. *Amer. Jour. Bot.* 44:164–175.

————. 1960. Gibberellin-induced inhibition of bud development in some species of Prunus. *Science* 131:825–826.

Brantley, B. B., and Warren, G. F. 1960. Sex expression and growth in muskmelon. *Plant Physiol.* 35:741–745.

Brian, P. W., and Hemming, H. G. 1955. The effect of gibberellic acid on shoot growth of pea seedlings. *Physiol. Plantarum* 8:669–681.

————. 1961. Promotion of cucumber hypocotyl growth by two new gibberellins. *Nature* 189:74.

Brian, P. W., Hemming, H. G., and Lowe, D. 1960. Inhibition of rooting of cuttings by gibberellic acid. *Ann. Bot. N.S.* 24:407–419.

Brian, P. W., Hemming, H. G., and Radley, M. 1955. A physiological comparison of gibberellic acid with some auxins. *Physiol. Plantarum* 8:899–912.

Brian, P. W., Petty, J. H. P., and Richmond, P. T. 1959a. Effect of gibberellic acid on development of autumn colour and leaf fall of deciduous woody plants. *Nature* 183:58–59.

————. 1959b. Extended dormancy of deciduous woody plants treated in autumn with gibberellic acid. *Nature* 184:69.

Bringhurst, R. S., Voth, V., and Crane, J. C. 1956. Blackberry yields increased. *Calif. Agr.* 10:5.

Brown, D. S., Griggs, W. H., and Iwakiri, B. T. 1960. The influence of gibberellin on resting pear buds. *Proc. Amer. Soc. Hort.* 76:52–58.

Brown, L. C., Crane, J. C., and Beutel, J. A. 1968. Gibberellic acid reduces cling peach flower buds. *Calif. Agr.* 22:7–8.

Browning, G., and Cannell, M. G. R. 1970. Use of 2-chloroethane phosphonic acid to promote the abscission and ripening of fruit of *Coffea arabica* L. *Jour. Hort. Sci.* 45:223–232.

Buchanan, D. W., and Biggs, R. H. 1969. Peach fruit abscission and pollen germination as influenced by ethylene and 2-chloroethane phosphonic acid. *Jour. Amer. Soc. Hort. Sci.* 94:327–329.

Buchanan, D. W., Biggs, R. H., Blake, J. A., and Sherman, W. B. 1970. Peach thinning with 3CPA and Ethrel during cytokinesis. *Jour. Amer. Soc. Hort. Sci.* 95:781–784.

Buchanan, D. W., Hall, C. B., Biggs, R. H., and Knapp, F. W. 1969. Influence of Alar, Ethrel and gibberellic acid on browning of peaches. *HortScience* 4:302–303.

Bukovac, M. J. 1963. Wide angle crotches are essential for structural strength in apple trees. *Ann. Rept. Mich. State Hort. Soc.* 93:63–67.

Bukovac, M. J., and Honma, S. 1967. Gibberellin-induced heterostyly in the tomato and its implications for hybrid-seed production. *Proc. Amer. Soc. Hort. Sci.* 91:514–520.

Bukovac, M. J., Larsen, R. P., and Bell, H. K. 1960. Effect of gibberellin on berry set and development of Concord grapes. *Quar. Bull. Mich. Agr. Exptl. Sta.* 42:503–510.

Bukovac, M. J., Larsen, R. P., and Robb, W. R. 1964. Effect of *N,N*-dimethylamino-succinamic acid on shoot elongation and nutrient composition of *Vitis Labrusca* L. cv. Concord. *Quar. Bull. Mich. Agr. Exptl. Sta.* 46:488–494.

Bukovac, M. J., and Nakagawa, S. 1967. Comparative potency of gibberellins in inducing parthenocarpic fruit growth in *Malus sylvestris* Mill. *Experientia* 23:865.

———. 1968. Gibberellin-induced asymmetric growth of apple fruits. *HortScience* 3:172–174.

Bukovac, M. J., and Wittwer, S. H. 1957. Gibberellin and higher plants, II: Induction of flowering in biennials. *Quar. Bull. Mich. Agr. Exptl. Sta.* 39:650–660.

———. 1961. Gibberellin modification of flower sex expression in *Cucumis sativus* L. *Advances Chem. Series* 28:80–88.

Bukovac, M. J., Zucconi, F., Larsen, R. P., and Kesner, C. D. 1969. Chemical promotion of fruit abscission in cherries and plums with special reference to 2-chloroethylphosphonic acid. *Jour. Amer. Soc. Hort. Sci.* 94(3):226–230.

Bull, T. A. 1964. The effects of temperature, variety, and age on the response of *Saccharum* spp. to applied gibberellic acid. *Australian Jour. Agr. Res.* 15:77–84.

Burg, S. P. 1962. The physiology of ethylene formation. *Ann. Rev. Plant Physiol.* 13:265–302.

———. 1968. Ethylene, plant senescence, and abscission. *Plant Physiol.* 43:1503–1511.

Burg, S. P., and Burg, E. A. 1966. The interaction between auxins and ethylene and its role in plant growth. *Proc. Natl. Acad. Sci. U.S.* 55:262–269.

———. 1968. Ethylene formation in pea seedlings; its relation to the inhibition of bud growth caused by indole-3-acetic acid. *Plant Physiol.* 43:1069–1074.

Burns, R. M. 1970. Effects of foliar sprays for frost protection in tests with young citrus trees. *Calif. Agr.* 24:10–11.

Burns, R. M., Boswell, S. B., and Hield, H. Z. 1970. Chemical inhibition of lemon top regrowth. *Calif. Citrograph* 55:251–252.

Burns, R. M., and Coggins, C. W., Jr. 1969. Sweet orange germination and growth aided by water and gibberellin seed soak. *Calif. Agr.* 23:18–19.

Burns, R. M., Garber, M. J., and Hield, H. Z. 1962. *Calif. Citrograph* 47:384, 392–395.

Burns, R. M., Hench, K. W., and Hield, H. Z. 1964. MH growth inhibition effects on young navel trees. *Calif. Citrograph* 49:396–400.

Burns, R. M., Mircetich, S. M., Coggins, C. W., Jr., and Zentmeyer, G. A. 1966. Gibberellin increases growth of Duke avocado seedlings. *Calif. Agr.* 20:6–7.

Burns, R., Rosedale, D. O., Pehrson, J. E., Jr., and Coggins, C. W., Jr. 1964. Gibberellin sprays delay lime maturity. *Calif. Agr.* 18:14–15.

Buxton, J. W., and Culbert, J. R. 1967. Effects of N-dimethyl amino succinamic acid (B-nine) on flower longevity and vegetative growth of pot chrysanthemums (*Chrysanthemum morifolium* Ramat). *Proc. Amer. Soc. Hort. Sci.* 91:645–652.

Byers, R. E., Dostal, H. C., and Emerson, F. H. 1969a. Regulation of fruit growth with 2-chloroethanephosphonic acid. *BioScience* 19:903–904.

Byers, R. E., and Emerson, F. H. 1969b. Effects of succinamic acids 2,2-dimethyl hydrazide (Alar) on peach fruit maturation and tree growth. *Jour. Amer. Soc. Hort. Sci.* 94:641–645.

Carlson, R. F., and Badizadegan, M. 1967. Peach seed germination and seedling growth as influenced by several factors. *Quar. Bull. Mich. Agr. Exptl. Sta.* 49:276–282.

Carns, H. R. 1966. Abscission and its control. *Ann. Rev. Plant. Physiol.* 17:295–314.

Carpenter, W. J. 1956. The influence of plant hormones on the abscission of poinsettia leaves and bracts. *Proc. Amer. Soc. Hort. Sci.* 67:539–544.

Casilli, O. 1969. Earlier artichokes following treatment with gibberellin. *L'Informatore Agrario* 25:511–512.

Cathey, H. M. 1959a. Effects of gibberellin and Amo-1618 on growth and flowering of *Chrysanthemum morifolium* on short photoperiods. In R. B. Withrow, ed., *Photoperiodism and related phenomena in plants and animals*, pp. 365–371. Washington, D.C.: Amer. Assoc. Advance. Sci.

————. 1959b. Poinsettia study: Changing growth and flowering by use of carvacrol form of Amo-1618. *Florists' Rev.* 124:19, 20, 83, 84.

————. 1959c. Gibberellins in horticulture: A preliminary review. *Nat. Hort. Mag.* 38:215–231.

Cathey, H. M. 1964. Physiology of growth retarding chemicals. *Ann. Rev. Plant Physiol.* 15:271–302.

————. 1965. Initiation and flowering of rhododendron following regulation by light and growth-retardants. *Proc. Amer. Soc. Hort. Sci.* 86:753–760.

————. 1967. Labor- and time-saving chemicals and techniques. *Florist and Nursery Exchange.* February 18.

————. 1968a. Response of *Dianthus caryophyllus* L. (carnation) to synthetic abscisic acid. *Proc. Amer. Soc. Hort. Sci.* 93:560–568.

————. 1968b. Response of some ornamental plants to synthetic abscisic acid. *Proc. Amer. Soc. Hort. Sci.* 93:693–698.

————. 1969. Plant selectivity in response to variation in the structure of succinic acid 2,2-dimethylhydrazide (B995). *Phyton* 26:77–85.

Cathey, H. M., and Steffens, G. L. 1968. Relation of the structure of fatty acid derivatives to their action as chemical pruning agents. In *Plant growth regulators*. Monograph no. 31, pp. 224–235. London: Society of Chemical Industry.

Cathey, H. M., Steffens, G. L., Stuart, N. W., and Zimmerman, R. H. 1966a. Chemical pruning of plants. *Science* 153:1382–1383.

Cathey, H. M., and Stuart, N. W. 1961. Comparative plant growth-retarding activity of Amo-1618, phosphon, and CCC. *Bot. Gaz.* 123:51–57.

Cathey, H. M., Yoeman, A. H., and Smith, F. F. 1966b. Abortion of flower buds in chrysanthemum after application of a selected petroleum fraction of high aromatic content. *HortScience* 1:61–62.

Cavell, B. D., MacMillan, J., Pryce, R. J., and Sheppard, A. C. 1967. Thin-layer and gas-liquid chromatography of the gibberellins; direct identification of the gibberellins in a crude plant extract by gas-liquid chromatography. *Phytochem.* 6:867–874.

Chacko, E. K., and Singh, R. N. 1969. Induction of parthenocarpy in mango (*Mangifera indica* L.) using plant growth regulators. *HortScience* 4:121–123.

Chadwick, L. C., and Houston, R. 1948. A preliminary report on the prestorage defoliation of some trees and shrubs. *Proc. Amer. Soc. Hort. Sci.* 51:659–667.

Chailakhian, M. K. 1968. Flowering hormones of plants. In Wightman and Setterfield 1968, pp. 1317–1340.

Challenger, S., Lacey, H. J., and Howard, B. H. 1965. The demonstration of root-promoting substances in apple and plum rootstocks. In *Ann. Rpt. East Malling Res. Sta. for 1964*, pp. 124–128.

Chan, B. G., and Cain, J. C. 1967. The effect of seed formation on subsequent flowering in apple. *Proc. Amer. Soc. Hort. Sci.* 91:63–68.

Chaplin, M. H., and Kenworthy, A. L. 1968. The influence of N-dimethyl-amino succinamic acid (Alar) on growth of the sweet cherry, *Prunus avium*. In *Abstract 65th Annual Meeting Amer. Soc. Hort. Sci., Davis, Calif.*, p. 93.

————. 1970. The influence of succinamic acid 2,2-dimethyl hydrazide on fruit ripening of the 'Windsor' sweet cherry. *Jour. Amer. Soc. Hort. Sci.* 95:532–536.

Cherry, J. H. 1970. Effect of auxin-herbicides on plant and animal biochemistry. *HortScience* 5:205–207.

Chrispeels, M. J., and Varner, J. E. 1967. Hormonal control of enzyme synthesis: On the mode of action of gibberellic acid and abscisin in aleurone layers of barley. *Plant Physiol.* 42:1008–1016.

Christodoulou, A. J., Pool, R. M., and Weaver, R. J. 1966. Prebloom thinning of Thompson Seedless grapes is feasible when followed by bloom spraying with gibberellin. *Calif. Agr.* 20:8–10.

Christodoulou, A., Weaver, R. J., and Pool, R. M. 1967. Response of Thompson Seedless grapes to prebloom thinning. *Vitis* 6:303–308.

———. 1968. Relation of gibberellin treatment to fruit-set, berry development, and cluster compactness in *Vitis vinifera* grapes. *Proc. Amer. Soc. Hort. Sci.* 92:301–310.

Chrominski, A. 1967. Effect of (2-chloroethyl) trimethylammonium chloride on protein content, protein yield, and some qualitative indexes of winter wheat grain. *Jour. Agr. Food Chem.* 15:109–112.

Chvojka, L., Veres, K., and Kozel, J. 1961. The effect of kinins on the growth of apple-tree buds and on incorporation of radioactive phosphate. *Biol. Plant.* 3:140–147.

Clark, H. E., and Kerns, K. R. 1942. Control of flowering with phytohormones. *Science* 95:536–537.

———. 1943. Effects of growth-regulating substances on a parthenocarpic fruit. *Bot. Gaz.* 104:639–644.

Clark, R. K., and Kenney, D. S. 1969. Comparison of staminate flower production on gynoecious strains of cucumbers, *Cucumis sativus* L., by pure gibberellins (A_3, A_4, A_7, A_{13}) and mixtures. *Jour. Amer. Soc. Hort. Sci.* 94:131–132.

Clor, M. A., Crafts, A. S., and Yamaguchi, S. 1962. Effects of high humidity on translocation of foliar-applied labeled compounds in plants, part I. *Plant Physiol.* 37:609–617.

Clore, W. J. 1965. Responses of Delaware grapes to gibberellin. *Proc. Amer. Soc. Hort. Sci.* 87:259–263.

Clore, W. J., and Fay, R. D. 1970. The effect of pre-harvest applications of ethrel on Concord grapes. *HortScience* 5:21–23.

Coggins, C. W., Jr. 1969. Gibberellin research on citrus rind aging problems. In H. D. Chapman, ed., *Proceedings First Intern. Citrus Symposium*, III, 1177–1185. Riverside: Univ. of Calif.

Coggins, C. W., Jr., Eaks, I. L., Hield, H. Z., and Jones, W. W. 1963. Navel orange rind staining reduced by gibberellin A_3. *Proc. Amer. Soc. Hort. Sci.* 82:154–157.

Coggins, C. W., Jr., and Hield, H. Z. 1968. Plant-growth regulators. In W. Reuter, L. D. Batchelor, and H. J. Webber, eds., *The citrus industry*, II, 371–389. Univ. of Calif., Div. Agr. Sci.

Coggins, C. W., Jr., Hield, H. Z., and Burns, R. M. 1962. The influence of potassium gibberellate on grapefruit trees and fruit. *Proc. Amer. Soc. Hort. Sci.* 81:223–226.

Coggins, C. W., Jr., Hield, H. Z., Burns, R. M., Eaks, I. L., and Lewis, L. N. 1966. Gibberellin research with citrus. *Calif. Agr.* 20:12–13.

Coggins, C. W., Jr., Hield, H. Z., Eaks, I. L., Lewis, L. N., and Burns, R. M. 1965. Gibberellin Research on Citrus. *Calif. Citrograph* 50:457, 466–468.

Coggins, C. W., Jr., Hield, H. Z., and Garber, M. J. 1960. The influence of potassium gibberellate on Valencia orange trees and fruit. *Proc. Amer. Soc. Hort. Sci.* 76:193–198.

Coggins, C. W., Jr., and Lesley, J. W. 1968. Attempts to improve flower bud retention and development in tomatoes with grafts, nutrition and growth regulators. *HortScience* 3:237–238.

Coggins, C. W., Jr., and Lewis, L. N. 1962. Regreening of Valencia orange as influenced by potassium gibberellate. *Plant Physiol.* 37:625–627.

———. 1965. Some physical properties of the navel orange rind as related to ripening and gibberellic-acid treatments. *Proc. Amer. Soc. Hort. Sci.* 86:272–279.

Coggins, C. W., Jr., Platt, R. G., and Opitz, K. W. 1969. Changes in recommendations for gibberellic acid. *Calif. Citrograph* 54:532.

Coggins, C. W., Jr., Scora, R. W., Lewis, L. N., and Knapp, J. C. F. 1969. Gibberellin-delayed senescence and essential oil changes in the navel orange rind. *Jour. Agr. Food Chem.* 17:807–809.

Cohen, D., Robinson, J. B., and Paleg, L. G. 1966. Decapitated peas and diffusible gibberellins. *Australian Jour. Biol. Sci.* 19:535–543.

Collier, H. O. J. August, 1962. Kinins. *Scientific American* 207(1):111–118.

Collins, W. B., Irving, K. H., and Barker, W. G. 1966. Growth substances in the flower bud and developing fruit of *Vaccinium angustifolium* Ait. *Proc. Amer. Soc. Hort. Sci.* 89:243–247.

Collins, W. K., Hawks, S. N., Jr., and Kittrell, B. U. 1970a. Effect of contact and systemic sucker control agents on yield and value of flue-cured tobacco. *Tobacco Sci.* 14:65–68.

———. 1970b. Effects of systemics alone and contacts followed by a systemic sucker control agent on some agronomic-economic characteristics of flue-cured tobacco. *Tobacco Sci.* 14:86–88.

Cooke, A. R., and Randall, D. I. 1968. 2-haloethanephosphonic acids as ethylene releasing agents for the induction of flowering in pineapples. *Nature* 218:974–975.

Coombe, B. G. 1960. Relationships of growth and development to changes in sugars, auxins, and gibberellins in fruits of seeded and seedless varieties of *Vitis vinifera*. *Plant Physiol.* 35:241–250.

———. 1965. Increase in fruit set of *Vitis vinifera* by treatment with growth retardants. *Nature* 205:305–306.

Cooper, W. C. 1938. Hormones and root formation. *Bot. Gaz.* 99:599–614.

Cooper, W. C., and Henry, W. H. 1967. The acceleration of abscission and coloring of citrus fruit. *Proc. Fla. State Hort. Soc.* 80:7–14.

———. 1968a. Field trials with potential abscission chemicals as an aid to mechanical harvesting of citrus in Florida. *Proc. Fla. State Hort. Soc.* 81:62–68.

———. 1968b. Effect of growth regulators on the coloring and abscission of citrus fruit. *Israel Jour. Agr. Res.* 18:161–174.

Cooper, W. C., Rasmussen, G. K., and Hutchison, D. J. 1969. Promotion of abscission of orange fruits by cycloheximide as related to site of treatment. *BioScience* 19:443–444.

Cooper, W. C., Rasmussen, G. K., Rogers, B. J., Reece, P. C., and Henry, W. H. 1968a. Control of abscission in agricultural crops and its physiological basis. *Plant Physiol.* 43:1560–1576.

Cooper, W. C., Rasmussen, G. K., and Smoot, J. J. 1968b. Induction of degreening of tangerines by preharvest applications of ascorbic acid, other ethylene-releasing agents. *Citrus Industry* 49:25–27.

Cooper, W. C., Young, R. H., and Henry, W. H. 1969. Effect of growth regulators on bud growth and dormancy in citrus as influenced by season of year and climate. *Proc. First Int. Citrus Symp.* 1:301–314.

Corcoran, M. R., West, C. A., and Phinney, B. O. 1961. Natural inhibitors of gibberellin-induced growth. *Advances Chem. Series* 28:152–158.

Cornforth, J. W., Milborrow, B. V., and Ryback, G. 1966. Identification and estimation of (+)-abscisin II ("dormin") in plant extracts by spectropolarimetry. *Nature* 210:627–628.

Cornforth, J. W., Millborrow, B. V., Ryback, G., and Wareing, P. F. 1965. Identity of Sycamore "dormin" with abscisin II. *Nature* 205:1269–1270.

Cotton pest control guides. 1968. Memphis, Tenn.: National Cotton Council.

Coyne, D. P. 1969. Effect of growth regulators on time of flowering of a photoperiodic sensitive field bean (*Phaseolus vulgaris* L.). *HortScience* 4:100–117.

———. 1970. Effect of 2-chloroethylphosphonic acid on sex expression and yield in butternut squash and its usefulness in producing hybrid squash. *HortScience* 5:227–228.

Crafts, A. S. 1956. The mechanism of translocation, I: Methods of study with C^{14}-labelled 2,4-D. *Hilgardia* 26:287–334.

———. 1960. Evidence for hydrolysis of esters of 2,4-D during absorption by plants. *Weeds* 8:19–25.

———. 1961a. *The chemistry and mode of action of herbicides.* New York: Interscience.

———. 1961b. *Translocation in plants.* New York: Holt, Rinehart and Winston.

———. 1964. Herbicide behavior in the plant. In Audus 1964a, pp. 75–110.

————. 1970. Pest control: A new industrial revolution. *Pure Appl. Chem.* 21:295–308.

Crafts, A. S., and Broyer, T. C. 1938. Migration of salts and water into xylem of the roots of higher plants. *Amer. Jour. Bot.* 25:529–535.

Crafts, A. S., and Crisp, C. E. 1971. *Phloem transport in plants.* San Francisco: W. H. Freeman.

Crafts, A. S., and Robbins, W. W. 1962. *Weed control.* New York: McGraw-Hill.

Crane, J. C. 1949. Controlled growth of fig fruits by synthetic hormone applications. *Proc. Amer. Soc. Hort. Sci.* 54:102–108.

————. 1952. Ovary wall development as influenced by growth regulators inducing parthenocarpy in the Calimyrna fig. *Bot. Gaz.* 114:102–107.

————. 1954. Frost resistance and reduction in drop of injured apricot fruits effected by 2,4,5-trichlorophenoxyacetic acid. *Proc. Amer. Soc. Hort. Sci.* 64:225–231.

————. 1955. Preharvest drop, size and maturity of apricots as affected by 2,4,5-trichlorophenoxyacetic acid. *Proc. Amer. Soc. Hort. Sci.* 65:75–84.

————. 1964. Growth substances in fruit setting and development. *Ann. Rev. Plant Physiol.* 15:303–326.

————. 1965. The chemical induction of parthenocarpy in the *Calimyrna* fig and its physiological significance. *Plant Physiol.* 40:606–610.

————. 1969. The role of hormones in fruit set and development. *Hort-Science* 4:108–111.

Crane, J. C., and Blondeau, R. 1949. Controlled growth of fig fruits by synthetic hormone application. *Proc. Amer. Soc. Hort. Sci.* 54:102–108.

Crane, J. C., and Brooks, R. M. 1952. Growth of apricot fruits as influenced by 2,4,5-trichlorophenoxyacetic acid application. *Proc. Amer. Soc. Hort. Sci.* 59:218–224.

Crane, J. C., and Hicks, J. R. 1968. Further studies on growth-regulator-induced parthenocarpy in the 'Bing' cherry. *Proc. Amer. Soc. Hort. Sci.* 92:113–118.

Crane, J. C., Marei, N., and Nelson, H. M. 1970. Growth and maturation of fig fruits stimulated by 2-chloroethylphosphonic acid. *Jour. Amer. Soc. Hort. Sci.* 95:367–370.

Crane, J. C., Primer, P. E., and Campbell, R. C. 1960. Gibberellin-induced parthenocarpy in *Prunus. Proc. Amer. Soc. Hort. Sci.* 75:129–137.

Crane, J. C., Rebeiz, C. A., and Campbell, R. C. 1961. Gibberellin-induced parthenocarpy in the J. H. Hale peach and the probable cause of "Button" production. *Proc. Amer. Soc. Hort. Sci.* 78:111–118.

Criley, R. A. 1969. Effect of short photoperiods, cycocel, and gibberellic acid upon flower bud initiation and development in azalea 'Hexe.' *Jour. Amer. Soc. Hort. Sci.* 94:392–395.

Crocker, W. 1948. *Growth of plants: Twenty years' research at Boyce Thompson Institute.* New York: Reinhold.

Crocker, W., and Knight, L. I. 1908. Effect of illuminating gas and ethylene upon flowering of carnations. *Bot. Gaz.* 46:259–276.

Cross, B. E. 1954. Gibberellic acid, part I. *Jour. Chem. Soc. London,* pp. 4670–4676.

Cummins, J. N., and Fiorino, P. 1969. Preharvest defoliation of apple nursery stock using Ethrel. *HortScience* 4:339–341.

Curtis, O. F. 1918. Stimulation of root growth in cuttings by treatment with chemical compounds. *Cornell Univ. Agr. Exptl. Sta. Mem.* 14:75–138.

Dahlstrom, R. V., and Sfat, M. R. 1961. Relationship of gibberellic acid to enzyme development. *Advances Chem. Series* 28:59–70.

Dalgliesh, C. E. 1955. Isolation and examination of urirary metabolites containing an aromatic system. *Jour. Clin. Path.* 8:73–78.

Daris, B. T. 1966. Effect of gibberellin on the variety Perlette (translated title). *Bul. Inst. de la Vigne,* Lycovrissi, Athens.

Darrow, G. M. 1956. Use of napththalene acetamide in blueberry breeding. *Proc. Amer. Soc. Hort. Sci.* 67:341–343.

Darwin, C. R. 1880. *The power of movement in plants.*

Dass, H. C., and Randhawa, G. S. 1968a. Effect of gibberellin on seeded *Vitis vinifera* with special reference to induction of seedlessness. *Vitis* 7:10–21.

———. 1968b. Response of certain seeded *Vitis vinifera* varieties to gibberellin application at postbloom stage. *Amer. Jour. Enol. Vitic.* 19:56–62.

Davis, L. A. 1968. Gas chromatographic identification and measurement of abscisic acid and other plant hormones in the developing cotton fruit. Ph.D. thesis, Univ. of Calif. Davis.

Davis, L. A., Heinz, D. E., and Addicott, F. T. 1968. Gas liquid chromatography of trimethylsilyl derivatives of abscisic acid and other plant hormones. *Plant Physiol.* 43:1389–1394.

Davis, E. F. 1928. The toxic principle of *Juglans nigra* as identified with synthetic juglone and its toxic effects on tomato and alfalfa plants. *Amer. Jour. Bot.* 15:620.

Davis, R. M., Jr. 1961. Ingrown sprouts in potato tubers; factors accompanying their origin in Ohio. *Amer. Potato Jour.* 38:411–413.

Dayawon, M. M., and Shutak, V. G. 1967. Influence of N^6-benzyladenine on the postharvest rate of respiration of strawberries. *HortScience* 2:12.

Dedolph, R. R. 1962. Effect of benzothiazole-2-oxyacetate on flowering and fruiting of papaya. *Bot. Gaz.* 124:75–78.

Dedolph, R. R., Wittwer, S. H., and Larzelere, H. E. 1963. Consumer verification of quality maintenance induced by N^6-benzyladenine in the storage of celery (*Apium graveolens*) and broccoli (*Brassica oleracea* var. *Italica*). *Food Tech.* 17:111–112.

Dedolph, R. R., Wittwer, S. H., Tuli, V., and Gilbart, D. 1962. Effect of N^6-benzylaminopurine on respiration and storage behavior of broccoli (*Brassica oleracea* var. *Italica*). *Plant Physiol.* 37:509–512.

Denffer, D. V., Behrens, M., and Fischer, A. 1952. Papierelecktrophoretische Trennung von Indolderivaten Aus Pflanzenextrakten. *Naturwiss.* 39:258–259.

Dennis, F. G., Jr. 1967. Apple fruit-set: Evidence for a specific role of seeds. *Science* 156:71–73.

————. 1968. Growth and flowering responses of apple and pear seedlings to growth retardants and scoring. *Proc. Amer. Soc. Hort. Sci.* 93: 53–61.

————. 1970. Effects of gibberellins and naphthaleneacetic acid on fruit development in seedless apple clones. *Jour. Amer. Soc. Hort. Sci.* 95: 125–128.

Dennis, F. G., Jr., and Bennett, H. O. 1969. Effect of gibberellic acid and deflowering upon runner and inflorescence development in an evergreen strawberry. *Jour. Amer. Soc. Hort. Sci.* 94:534–537.

Dennis, F. G., Jr., and Edgerton, L. J. 1961. The relationship between an inhibitor and rest in peach flower buds. *Proc. Amer. Soc. Hort. Sci.* 77: 107–116.

————. 1966. Effect of gibberellins and ringing upon apple fruit development and flower bud formation. *Proc. Amer. Soc. Hort. Sci.* 88:14–24.

Dennis, F. G., Jr., and Nitsch, J. P. 1966. Identification of gibberellin A_4 and A_7 in immature apple seeds. *Nature* 211:781–782.

Dennis, F. G., Jr., Wilczynski, H., de la Guardia, M., and Robinson, R. W. 1970. Ethylene levels in tomato fruits following treatment with Ethrel. *HortScience* 5:168–170.

Denny, F. E., Guthrie, J. D., and Thornton, N. C. 1942. Effect of the vapor of alpha-naphthalene acetic acid on the sprouting and sugar content of potato tubers. *Contrib. Boyce Thompson Inst.* 12:253–268.

De Wilde, R. C. 1971. Practical applications of (2-chloroethyl)phosphonic acid in agricultural production. *HortScience* 6(4):364–370.

Dilley, D. R. 1969. Hormonal control of fruit ripening. *HortScience* 4:111–114.

Domanski, R., and Kozlowski, T. T. 1967. Variations in kinetin-like activity in buds of *Betula* and *Populus* during release from dormancy. *Canadian Jour. Bot.* 46:397–403.

Donoho, C. W., and Walker, D. R. 1957. Effect of gibberellic acid on breaking of rest period in Elberta peach. *Science* 126:1178–1179.

Doran, W. L. 1953. Effects of treating cuttings of woody plants with both a root-inducing substance and a fungicide. *Proc. Amer. Soc. Hort. Sci.* 60:487–491.

Dostal, H. C., and Leopold, A. C. 1967. Gibberellin delays ripening of tomatoes. *Science* 158:1579–1580.

Doughty, C. C. 1962. The effects of certain growth regulators on the fruiting of cranberries, *Vaccinium macrocarpon*. *Proc. Amer. Soc. Hort. Sci.* 80:340–349.

Duhamel du Monceau. 1758. *La physique des arbres*, Vol. I.

Dybing, C. D., and Currier, H. B. 1959. A fluorescent dye method for foliar penetration studies. *Weeds* 7:214–222.

Eaks, I. L. 1964. The effect of harvesting and packing house procedures on rind staining of central California Washington Navel oranges. *Proc. Amer. Soc. Hort. Sci.* 85:245–256.

Eaton, F. M. 1957. Selective gametocide opens way to hybrid cotton. *Science* 126:1174–1175.

———. 1958–1959. Hybrid cotton . . . a scientific breakthrough. In *The cotton trade journal 26th international yearbook*, pp. 62, 73–74.

Eaves, C. A., and Forsyth, F. R. 1968. The influence of light, modified atmospheres and benzimidazole on brussels sprouts. *Jour. Hort. Sci.* 43: 317–322.

Ebert, V. A. 1955. Versuch über die quantitative Bestimmung der β-Indolylessigsäure (Heteroauxin) und ihre Lokalisierung im Gewebe der Pflanze. *Phytopathol. Zeitschr.* 24:216–242.

Eck, P. 1969. Effect of preharvest sprays of ethrel, alar, malathion on anthocyanin content of early black cranberry. *HortScience* 4:224–226.

———. 1970. Influence of ethrel upon highbush blueberry fruit ripening. *HortScience* 5:23–25.

Edgerton, L. J. 1966. Some effects of gibberellin and growth retardants on bud development and cold hardiness of peach. *Proc. Amer. Soc. Hort. Sci.* 88:197–203.

———. 1968. New compound aids regulation of fruit abscission. *New York's Food and Life Sci.* 1:19–20.

Edgerton, L. J., and Blanpied, G. D. 1968. Regulation of growth and fruit maturation with 2-chloroethanephosphonic acid. *Nature* 219:1064–1065.

———. 1970. Interaction of succinic acid 2,2-dimethyl hydrazide, 2-chloroethylphosphonic acid and auxins on maturity, quality, and abscission of apples. *Jour. Amer. Soc. Hort. Sci.* 95:664–666.

Edgerton, L. J., Forshey, C. G., and Blanpied, G. D. 1967. Effect of summer applications of Alar (B 995) in reducing drop and improving color and firmness of apples. *Proc. New York State Hort. Soc.* 112:204–206.

Edgerton, L. J., and Greenhalgh, W. J. 1969. Regulation of growth, flowering and fruit abscission with 2-chloroethanephosphonic acid. *Jour. Amer. Soc. Hort. Sci.* 94:11–13.

Edgerton, L. J., and Hatch, A. H. 1969. Promotion of abscission of cherries and apples for mechanical harvesting. *New York State Hort. Soc.* 114:109–113.

Edgerton, L. J., and Hoffman, M. B. 1965. Some physiological responses of apple to N-dimethyl amino succinamic acid and other growth regulators. *Proc. Amer. Soc. Hort. Sci.* 86:28–36.

———. 1966. Inhibition of fruit drop and colour stimulation with N-dimethylaminosuccinamic acid. *Nature* 209:314–315.

Edwards, J. A. 1961. Use of maleic hydrazide for cold protection of trees. *Citrus Grower* (South Africa) 327:11–15.

El-Antably, H. M. M., Wareing, P. F., and Hillman, J. 1967. Some physiological responses to D,L abscisin (dormin). *Planta* (Berlin) 73:74–90.

El Damaty, H., Kuhn, H., and Linser, H. 1964. A preliminary investigation on increasing salt tolerance of plants by application of (2-chloroethyl)-trimethyl ammonium chloride. *Agrochimica* 8:131–138.

El-Fouly, M. M., and Jung, J. 1966. Wirkung von Chlorcholinchlorid (CCC) auf das Wachstum des Pilzes *Rhizopus nigricans*. *Phytopathol. Zeitschr.* 57:192–194.

Ellison, J. H., and Smith, O. 1949. Retarding sprout growth of potato tubers by spraying the foliage with 2,4,5-trichlorophenoxyacetic acid. *Proc. Amer. Soc. Hort. Sci.* 54:447–451.

El-Zeftawi, B. M., and Weste, H. L. 1970a. Effects of some growth regulators on the fresh and dry yield of Zante currant (*Vitis vinifera* var.). *Vitis* 9:47–51.

———. 1970b. Effects of girdling and of parachlorophenoxyacetic acid and gibberellic acid sprays on yields and quality of Zante currant (*Vitis vinifera* L.) *Hort. Res.* 10:74–77.

Emden, van, H. F. 1964. Effect of (2-chloroethyl) trimethylammonium chloride on the rate of increase of the cabbage aphid (*Brevicorne brassicae*) L. *Nature* 201:946–948.

Erez, A., Samish, R. M., and Lavee, S. 1966. The role of light in leaf and flower bud break of the peach (*Prunus persica*). *Physiol. Plantarum* 19:650–659.

Esashi, Y., and Leopold, A. C. 1968. Regulation of tuber development in Begonia evansiava by cytokinin. In Wightman and Setterfield 1968, pp. 923–941.

Evans, H. 1951. Investigations on the propagation of cacao. *Trop. Agr. Trin.*, 28:147–203.

Evans, L. T., ed. 1969. *The induction of flowering*. Ithaca, N.Y.: Cornell Univ. Press.

Evans, L. T., 1971. Flower induction and the florigen concept. *Ann. Rev. Plant Physiol.* 22:365–394.

Fadl, M. S., and Hartmann, H. T. 1967a. Isolation, purification, and characterization of an endogenous root-promoting factor obtained from the basal sections of pear hardwood cuttings. *Plant Physiol.* 42:541–549.

Fadl, M. S., and Hartmann, H. T. 1967b. Relationship between seasonal changes in endogenous promoters and inhibitors in pear buds and cutting bases and the rooting of pear hardwood cuttings. *Proc. Amer. Soc. Hort. Sci.* 91:96–112.

Fawcett, C. H., Taylor, H. F., Wain, R. L., and Wightman, F. 1956. The degradation of certain phenoxy acids, amides, and nitriles within plant tissues. In Wain and Wightman 1956, pp. 187–194.

Fawcett, C. H., Wain, R. L., and Wightman, F. 1960. The metabolism of 3-indolylalkanecarboxylic acids, and their amides, nitriles, and methyl esters in plant tissues. *Proc. Royal Soc.* (series B) 152:231–254.

Filner, P., and Varner, J. E. 1967. A test for *de novo* synthesis of enzymes: Density labeling with H_2O^{18} of barley a-amylase induced by gibberellic acid. *Proc. Natl. Acad. Sci. U.S.* 58:1520–1526.

Filner, P., Wray, J. L., and Varner, J. E. 1969. Enzyme induction in higher plants. *Science* 165:358–367.

Fisher, D. V., and Looney, N. E. 1967. Growth, fruiting and storage response of five cultivars of bearing apple trees to N-dimethylaminosuccinamic acid (Alar). *Proc. Amer. Soc. Hort. Sci.* 90:9–19.

Fitting, H. 1907. Die Leitung tropistischer Reize in parallelotropen Pflanzenteilen. *Jahrb. Wiss. Bot.* 44:177–253.

———. 1909. Die Beeinflüssung der Orchideenblüten durch die Bestaubung und durch anderes Umstande. *Zeitschr. Bot.* 1:1–86.

———. 1910. Weitere entwicklungsphysiologische Untersuchungen an Orchideenblüten. *Zeitschr. Bot.* 2:225–267.

Fliegel, P., Parker, K. G., and Edgerton, L. J. 1966. Gibberellic-acid treatment of sour cherry infested by sour cherry yellows virus: Response to sprays applied throughout the growing season and the influence of environmental conditions. *Plant Disease Reporter* 50:240–242.

Fogle, H. W. 1958. Effects of duration of after-ripening, gibberellin and other pretreatments on sweet cherry germination and seedling growth. *Proc. Amer. Soc. Hort. Sci.* 72:129–133.

Fogle, H. W., and McCrory, C. S. 1960. Effects of cracking, after-ripening and gibberellin on germination of Lambert cherry seeds. *Proc. Amer. Soc. Hort. Sci.* 76:134–138.

Forshey, C. G. 1970. The use of Alar on vigorous 'McIntosh' apple trees. *Jour. Amer. Soc. Hort. Sci.* 95:64–67.

Foy, C. L. 1961. Absorption, distribution and metabolism of 2,2-dichloropropionic acid in relation to phytotoxicity, II: Distribution and metabolic fate of dalapon in plants. *Plant Physiol.* 36:698–709.

Frankland, B. 1961. Effect of gibberellic acid, kinetin and other substances on seed dormancy. *Nature* 192:678–679.

Frankland, B., and Wareing, P. F. 1960. Effect of gibberellic acid on hypocotyl growth of lettuce seedlings. *Nature* 185:255–256.

Freeman, J. A., and Carne, I. C. 1970. Use of succinic acid 2,2-dimethyl hydrazide (Alar) to reduce winter injury in strawberries. *Canadian Jour. Plant Sci.* 50:189–190.

Fuchs, Y., and Cohen, A. 1969. Degreening of citrus fruit with ethrel (Amchem 66–329). *Jour. Amer. Soc. Hort. Sci.* 94:617–618.

Furuta, T. 1961. Influence of gibberellin on germination of seeds. *Amer. camellia yearbook*, pp. 141–145.

———. 1965. Flowering azaleas the year around. *Florists' Rev.*, May 6.

———. 1967. *Chemical pinching agents for azaleas.* Univ. Calif. Ext. Serv. AXT-256.

Furuta, T., and Straiton, T. H., Jr. 1966. Synergism of kinetin and gibberellic acid in flowering of unchilled azalea, cultivar Red Wing. *Proc. Amer. Soc. Hort. Sci.* 88:591–594.

Galston, A. W., and Baker, R. S. 1953. Studies on the physiology of light action, V: Photoinductive alteration of auxin metabolism in etiolated peas. *Amer. Jour. Bot.* 40:512–516.

Galston, A. W., and Davies, P. J. 1969. Hormonal regulation in higher plants. *Science* 163:1288–1297.

———. 1970. *Control mechanisms in plant development.* Englewood Cliffs, N.J.: Prentice-Hall.

Galston, A. W., and Purves, W. K. 1960. The mechanism of action of auxin. *Ann. Rev. Plant Physiol.* 11:239–276.

Galun, E. 1959. Effects of gibberellic acid and naphthaleneacetic acid on sex expression and some morphological characters in the cucumber plant. *Phyton* 13:1–8.

Gardner, F. E., and Marth, P. C. 1937. Parthenocarpic fruits induced by spraying with growth-promoting chemicals. *Science* 86:246–247.

Gardner, F. E., Marth, P. C., and Batjer, L. P. 1939. Spraying with plant-growth substances for control of the pre-harvest drop of apples. *Proc. Amer. Soc. Hort. Sci.* 37:415–428.

Gates, C. T. 1957. The response of young tomato plant to brief period of water shortage, III: Drifts in nitrogen and phosphorus. *Australian Jour. Biol. Sci.* 10:125–146.

Gaur, B. K., and Leopold, A. C. 1955. The promotion of abscission by auxin. *Plant Physiol.* 30:487–490.

Gill, D. L. 1966. Camellia flowering response: Gibberellic acid bud treatments. *Amer. camellia yearbook*, pp. 187–191.

Gill, D. L., and Stuart, N. W. 1961. Stimulation of camellia flower-bud initiation by application of two growth retardants—a preliminary report. *Amer. camellia yearbook*, pp. 129–135.

Good, N. E., Andreae, W. A., and Van Ysselstein, M. W. H. 1956. Studies on 3-indoleacetic acid metabolism, II: Some products of the metabolism of exogenous indoleacetic acid in plant tissue. *Plant Physiol.* 31:231–235.

Goodin, J. R., McKell, C. M., and Webb, F. L. 1966. Influence of CCC on water use and growth characteristics of barley. *Agron. Jour.* 58:453–454.

Gordon, S. A. 1954. Occurrence, formation, and inactivation of auxins. *Ann. Rev. Plant Physiol.* 5:341–378.

Gordon, S. A., and Paleg, L. G. 1957. Observations on the quantitative determination of indoleacetic acid. *Physiol. Plantarum* 10:39–47.

Goren, R., and Monselise, S. P. 1969. Promotion of flower formation and fruit set in *Citrus* by antimetabolites of nucleic-acid and protein synthesis. *Planta* (Berlin) 88:364–368.

Gorter, C. J. 1958. Synergism of indole and indole-3-acetic acid in the root production of Phaseolus cuttings. *Physiol. Plantarum* 11:1–9.

———. 1961. Morphogenetic effects of synthetic auxins. In Ruhland 1961, pp. 1084–1109.

Gorter, C. J., and van der Zweep, W., 1964. Morphogenetic effects of herbicides. In Audus 1964a, pp. 235–275.

Gowing, D. P. 1956. An hypothesis of the role of naphthaleneacetic acid in flower induction in the pineapple. *Amer. Jour. Bot.* 43:411–418.

Gowing, D. P., and Leeper, R. W. 1955. Induction of flowering in pineapple by beta-hydroxyethylhydrazine. *Science* 122:1267.

———. 1961a. Studies on the relation of chemical structure to plant growth-regulator activity in the pineapple plant, III: Naphthalene derivatives and heterocyclic compounds. *Bot. Gaz.* 122:179–188.

———. 1961b. Studies on the relation of chemical structure to plant growth-regulator activity in the pineapple plant, IV: Hydrazine derivatives, compounds with an unsaturated aliphatic moiety and miscellaneous chemicals. *Bot. Gaz.* 123:34–43.

Granger, R. L., and Hogue, E. J. 1968. Effect of Alar on winter hardiness of raspberries. *Canadian Jour. Plant Sci.* 48:100–101.

Greenhalgh, W. J., and Edgerton, L. J. 1967. Interaction of Alar and gibberellin on growth and flowering of the apple. *Proc. Amer. Soc. Hort. Sci.* 91:9–17.

Greer, H. A. L. 1964. Effect of growth regulators on reproduction in soybeans. Ph.D. thesis, Iowa State Univ.

Griggs, W. H., Harris, R. W., and Iwakiri, B. T. 1956. The effectiveness of growth-regulators in reducing fruit loss of Bartlett pears caused by freezing temperatures. *Proc. Amer. Soc. Hort. Sci.* 67:95–101.

Griggs, W. H., and Iwakiri, B. T. 1961. Effects of gibberellin and 2,4,5-trichlorophenoxypropionic acid sprays on Bartlett pear trees. *Proc. Amer. Soc. Hort. Sci.* 77:73–89.

———. 1968. Effects of succinic acid 2,2-dimethyl hydrazide (Alar) sprays used to control growth in 'Bartlett' pear trees planted in hedgerows. *Proc. Amer. Soc. Hort. Sci.* 92:155–166.

Griggs, W. H., Iwakiri, B. T., and Bethell, R. S. 1965. B-nine fall sprays delay bloom and increase fruit set on Bartlett pears. *Calif. Agr.* 19:8–11.

Griggs, W. H., Iwakiri, B. T., and DeTar, J. E. 1951. The effect of 2,4,5-trichlorophenoxypropionic acid applied during the bloom period on the fruit set of several pear varieties and on the shape, size, stem length, seed content, and storage of Bartlett pears. *Proc. Amer. Soc. Hort. Sci.* 58:37–45.

Griggs, W. H., Iwakiri, B. T., Fridley, R. B., and Mehlschau, J. 1970. Effect of 2-chloroethylphosphonic acid and cycloheximide on abscission and ripening of 'Bartlett' pears. *HortScience* 5:264–266.

Griggs, W. H., Martin, G. C., and Iwakiri, B. T. 1970. The effect of seedless versus seeded fruit development on flower bud formation in pear. *Jour. Amer. Soc. Hort. Sci.* 95:243–248.

Grunwald, C., Mendez, J., and Stowe, B. B. 1968. Substrates for the optimum gas chromatographic separation of indolic methyl esters and the resolution of components of methyl 3-indolepyruvate solutions. In Wightman and Setterfield 1968, pp. 163–171.

Gupta, G. R. P., and Maheshwari, S. C. 1970. Cytokinins in seeds of pumpkin. *Plant Physiol.* 45:14–18.

Gustafson, F. G. 1936. Inducement of fruit development by growth-promoting chemicals. *Proc. Natl. Acad. Sci. U.S.* 22:628–636.

———. 1937. Parthenocarpy induced by pollen extracts. *Amer. Jour. Bot.* 24:102–107.

———. 1939a. The cause of natural parthenocarpy. *Amer. Jour. Bot.* 26:135–138.

———. 1939b. Auxin distribution in fruits and its significance in fruit development. *Amer. Jour. Bot.* 26:189–194.

———. 1941. The extraction of growth hormones from plants. *Amer. Jour. Bot.* 28:947–951.

———. 1961. Development of fruits. In Ruhland 1961, pp. 951–958.

Guthrie, J. D. 1938. Effect of ethylene thiocyanohydrin, ethyl carbylamine, and indoleacetic acid on the sprouting of potato tubers. *Contrib. Boyce Thompson Inst.* 9:265–272.

———. 1939. Inhibition of the growth of buds of potato tubers with vapor of the methyl ester of naphthaleneacetic acid. *Contrib. Boyce Thompson Inst.* 10:325–328.

Guttridge, C. G. 1962. Inhibition of fruit bud formation in apple with gibberellic acid. *Nature* 196:1008.

———. 1963. Inhibition of flowering in Poinsettia by gibberellic acid. *Nature* 197:920–921.

Gysin, H., and Knüsli, E. 1960. Chemistry and herbicidal properties of triazine derivatives. In R. L. Metcalf, ed., *Advances in pest control research*, III, 289–358. New York: Interscience.

Haagen-Smit, A. J., Dandliker, W. B., Wittwer, S. H., and Murneek, A. E. 1946. Isolation of 3-indoleacetic acid from immature corn kernels. *Amer. Jour. Bot.* 33:118–120.

Haagen-Smit, A. J., Leech, W. D., and Bergren, W. R. 1942. The estimation, isolation, and identification of auxins in plant materials. *Amer. Jour. Bot.* 29:500–506.

Haagen-Smit, A. J., and Went, F. W. 1935. A physiological analysis of the growth substance. *Proc. Kon. Ned. Akad. Wetensch. Amsterdam* 38:852–857.

Haberlandt, G. 1913. *Zur Physiologie der Zellteilung.* Sitz. Ber. K. Preuss. Akad. Wiss. 318–345.

———. 1921. Wundhormone als Erreger von Zellteilungen. *Beitr. Allg. Bot.* 2:1–53.

Hackett, W. P. 1970. The influence of auxin, catechol, and methanolic tissue extracts on root initiation in aseptically cultured shoot apices of the juvenile and adult forms of *Hedera helix. Jour. Amer. Soc. Hort. Sci.* 95:398–402.

Hackett, W. P., and Sachs, R. M. 1967. Chemical control of flowering in *Bougainvillea* 'San Diego Red.' *Proc. Amer. Soc. Hort. Sci.* 90:361–364.

Hale, C. R., Coombe, B. G., and Hawker, J. S., 1970. Effects of ethylene and 2-chloroethylphosphonic acid on the ripening of grapes. *Plant Physiol.* 45:620–623.

Hale, C. R., and Weaver, R. J. 1962. The effect of developmental stage on direction of translocation of photosynthate in *Vitis vinifera. Hilgardia* 33:89–131.

Halevy, A. H. 1962. Effect of hardening and chemical treatments on drought resistance of *gladiolus* plants. *Proc. 16th Int. Hort. Cong.* 4:252–258.

———. 1967. Effect of growth retardants on drought resistance and longevity of various plants. *Proc. 17th Int. Hort. Cong.* 3:277–283.

Halevy, A. H., and Kessler, B. 1963. Increased tolerance of bean plants to soil drought by means of growth-retarding substances. *Nature* 197:310–311.

Halevy, A. H., and Rudich, Y. 1967. Modification of sex expression in muskmelon by treatment with the growth retardant B-995. *Physiol. Plantarum* 20:1052–1058.

Halevy, A. H., and Shilo, R. 1970. Promotion of growth and flowering and increase in content of endogenous gibberellins in *Gladiolus* plants treated with the growth retardant CCC. *Physiol. Plantarum* 23:820–827.

Halevy, A. H., and Shoub, J. 1964. The effect of cold storage and treatment with gibberellic acid on flowering and bulb yields of Dutch iris. *Jour. Hort. Sci.* 39:120–129.

Halevy, A. H., and Wittwer, S. H. 1965a. Chemical regulation of leaf senescence. *Quar. Bull. Mich. Agr. Exptl. Sta.* 48:30–35.

————. 1965b. Growth promotion in the snapdragon by CCC, a growth retardant. *Naturwiss.* 52:310.

————. 1965c. Prolonging cut flower life by treatment with growth retardants B-Nine and CCC. *Florists' Rev.* 136:39–40.

————. 1966. Effect of growth retardants on longevity of vegetables, mushrooms, and cut flowers. *Proc. Amer. Soc. Hort. Sci.* 88:582–590.

Halfacre, R. G., and Barden, J. A. 1968. Anatomical responses of apple leaf and stem tissues to succinic acid 2,2-dimethylhydrazide (Alar). *Proc. Amer. Soc. Hort. Sci.* 93:25–32.

Halfacre, R. G., Barden, J. A., and Rollins, H. A., Jr. 1968. Effects of Alar on morphology, chlorophyll content, and net CO_2 assimilation rate of young apple trees. *Proc. Amer. Soc. Hort. Sci.* 93:40–52.

Hall, R. H. 1968. Cytokinins in the transfer-RNA: Their significance to the structure of t-RNA. In Wightman and Setterfield 1968, pp. 47–56.

Hamaker, J. W., Johnston, H., Martin, R. T., and Redemann, C. T. 1963. A picolinic acid derivative: a plant growth regulator. *Science* 141:363.

Hamner, C. L., and Rai, G. S. 1958. Growth inhibition studies with ornamentals and shade trees. *Hormolog* 2:11–13.

Hamner, C. L., and Tukey, H. B. 1944. The herbicidal action of 2,4-dichlorophenoxyacetic acid and 2,4,5-trichlorophenoxyacetic acid on bindweed. *Science* 100:154–155.

Hansen, C. J., and Hartmann, H. T. 1968. The use of indolebutyric acid and captan in the propagation of clonal peach and peach-almond hybrid rootstocks by hardwood cuttings. *Proc. Amer. Soc. Hort. Sci.* 92:135–140.

Hapitan, J. C., Jr., Shutak, V. G., and Kitchin, J. T. 1969. Vegetative and reproductive responses of highbush blueberry to succinic acid 2,2-dimethyl hydrazide (Alar). *Jour. Amer. Soc. Hort. Sci.* 94:26–28.

Harada, H., and Lang, A. 1965. Effect of some (2-chloroethyl) trimethyl-ammonium chloride analogs and other growth retardants on gibberellin biosynthesis in *Fusarium moniliforme*. *Plant Physiol.* 40:176–183.

Harada, H., and Nitsch, J. P. 1959. Flower induction in Japanese chrysanthemums with gibberellic acid. *Science* 129:777–778.

Hardenburg, R. E., Vaught, H. C., and Brown, G. A. 1970. Development and vase life of bud-cut Colorado and California carnations in preservative solutions following air shipment to Maryland. *Jour. Amer. Soc. Hort. Sci.* 95:18–22.

Harley, C. P., Moon, H. H., and Regeimbal, L. O. 1958. Evidence that postbloom apple thinning sprays of naphthaleneacetic acid increase blossom-bud formation. *Proc. Amer. Soc. Hort. Sci.* 72:52–56.

Harrington, J. F. 1960. The use of gibberellic acid to induce bolting and increase seed yield of tight-heading lettuce. *Proc. Amer. Soc. Hort. Sci.* 75:476–479.

Harrington, J. F., Rappaport, L., and Hood, K. J. 1957. Influence of gibberellins on stem elongation and flowering of endive. *Science* 125:601–602.

Hartley, G. S. 1964. Herbicide behavior in the soil, I: Physical factors and action through the soil. In Audus 1964a, pp. 111–161.

Hartmann, H. T. 1952. Spray thinning of olives with naphthaleneacetic acid. *Proc. Amer. Soc. Hort. Sci.* 59:187–195.

———. 1955. Induction of abscission of olive fruits by maleic hydrazide. *Bot. Gaz.* 117:24–28.

Hartmann, H. T., Fadl, M., and Whisler, J. 1967. Inducing abscission of olive fruits by spraying with ascorbic acid and iodoacetic acid. *Calif. Agr.* 21:5–7.

Hartmann, H. T., Heslop, A. J., and Whisler, J. 1968. Chemical induction of fruit abscission in olives. *Calif. Agr.* 22:14–16.

Hartmann, H. T., and Kester, D. E. 1968. *Plant propagation: Principles and practices,* 2d ed. Englewood Cliffs, N.J.: Prentice-Hall.

Hartmann, H. T., Tombesi, A., and Whisler, J. 1970. Promotion of ethylene evolution and fruit abscission in the olive by 2-chloroethanephosphonic acid and cycloheximide. *Jour. Amer. Soc. Hort. Sci.* 95:635–640.

Hayashi, F., Blumenthal-Goldschmidt, S., and Rappaport, L. 1962. Acid and neutral gibberellin-like substances in potato tubers. *Plant Physiol.* 37:774–780.

Hayashi, T. 1940. Biochemical studies on "Bakanae" fungus of rice, part 6: Effect of gibberellin on the activity of amylase in germinated cereal grains. *Jour. Agr. Chem. Soc. Japan.* 16:531–538.

Hayto, N., and Nobuhiro, K. 1963. *L'Élaircissage des Olives avec l'Amide thin.* Second Conference Oléicole, Nice, Oct. 1964.

Heide, O. M. 1969a. Environmental control of sex expression in *Begonia. Zeitschr. Pflanzenphysiol.* 61:279–285.

———. 1969b. Interaction of growth retardants and temperature in growth, flowering, regeneration, and auxin activity of *Begonia* × *cheimantha* Everett. *Physiol. Plantarum* 22:1001–1012.

Heide, O. M., and Øydvin, J. 1969. Effects of 6-benzylamino-purine on the keeping quality and respiration of glasshouse carnations. *Hort. Res.* 9:26–36.

Hejnowicz, A., and Tomaszewski, M. 1969. Growth regulators and wood formation in *Pinus silvestris. Physiol. Plantarum* 22:984–992.

Helgeson, J. P. 1968. The cytokinins. *Science* 161:974–981.

Hemberg, T. 1949a. Significance of growth-inhibitory substances and auxins for the rest period of the potato tuber. *Physiol. Plantarum* 2:24–36.

———. 1949b. Growth-inhibitory substances in terminal buds of *Fraxinus. Physiol. Plantarum* 2:37–44.

————. 1970. The action of some cytokinins on the rest period and the content of acid growth-inhibiting substances in potato. *Physiol. Plantarum* 23:850–858.

Hemphill, D. D. 1949. *The effects of plant growth regulating substances on flower-bud development and fruit set*. Univ. Missouri Agr. Exptl. Sta. Res. Bull. 434.

Hendershott, C. H. 1962. The influence of maleic hydrazide on citrus trees and fruits. *Proc. Amer. Soc. Hort. Sci.* 80:241–246.

Hendershott, C. H., and Matsolf, J. D. 1962. *Maleic hydrazide, a cold protection aid on Florida citrus*. Circular 241, Citrus Exptl. Sta., Lake Alfred, Florida.

Herman, D. E., and Hess, C. E. 1963. The effect of etiolation upon the rooting of cuttings. *Proc. Int. Plant. Prop. Soc.* 13:42–62.

Herrett, R. A., Hatfield, H. H., Jr., Crosby, D. G., and Vlitos, A. J. 1962. Leaf abscission induced by iodide ion. *Plant Physiol.* 37:358–363.

Heslop-Harrison, J. 1959. Growth substances and flower morphogenesis. *Jour. Linnean Soc. London Bot.* 56:269–281.

Hess, C. E. 1962. Characterization of the rooting co-factors extracted from *Hedera helix* L. and *Hibiscus rosa-sinensis*. *Proc. 16th Int. Hort. Cong.*, IV, 382–388.

————. 1964. Naturally occurring substances which stimulate root initiation. In Nitsch 1964, pp. 517–527.

Heyn, A. N. J. 1931. Der Mechanismus der Zellstreckung. *Rec. Trav. Bot. Néerland.* 28:113–244.

Hicks, J. R., and Crane, J. C. 1968. The effect of gibberellin on almond flower bud growth, time of bloom, and yield. *Proc. Amer. Soc. Hort. Sci.* 92:1–6.

Hield, H. Z., Burns, R. M., and Coggins, C. W., Jr. 1964. *Preharvest use of 2,4-D on citrus*.

Hield, H. Z., Coggins, C. W., Jr., and Garber, M. J. 1958. Gibberellin tested on citrus. *Calif. Agr.* 12:9–11.

————. 1965. Effect of gibberellin sprays on fruit set of Washington navel orange trees. *Hilgardia* 36:297–311.

Hield, H. Z., Lewis, L. N., and Palmer, R. L. 1969a. Orange fruit-stem separation . . . chemical influences. *Calif. Agr.* 23:12–13.

Hield, H. Z., Palmer, R. L., and Lewis, L. N. 1968. Progress report: Chemical aids to citrus fruit loosening. *Calif. Citrograph* 53:386–392.

————. 1969b. Ethrel effects on oranges. *Calif. Citrograph* 54:292, 324.

Higdon, R. J., and Westwood, M. N. 1963. Some factors affecting the rooting of hardwood pear cuttings. *Proc. Amer. Soc. Hort. Sci.* 83:193–198.

Hildebrand, E. M. 1944. The mode of action of the pollenicide, Elgetol. *Proc. Amer. Soc. Hort. Sci.* 45:53–58.

Hillman, W. S. 1962. *The physiology of flowering.* New York: Holt, Rinehart and Winston.

Hilton, J. L., Ard, J. S., Jansen, L. L., and Gentner, W. A. 1959. The pantothenate-synthezing enzyme, a metabolic site in the herbicidal action of chlorinated aliphatic acids. *Weeds* 7:381–396.

Hilton, J. L., Jansen, L. L., and Hull, H. M. 1963. Mechanisms of herbicide action. *Ann. Rev. Plant Physiol.* 14:353–384.

Hitchcock, A. E., and Zimmerman, P. W. 1940. Effects obtained with mixtures of root-inducing and other substances. *Contrib. Boyce Thompson Inst.* 11:143–160.

Hoffman, M. B., Edgerton, L. J., and Fisher, E. G. 1955. Comparisons of naphthaleneacetic acid and naphthaleneacetamide for thinning apples. *Proc. Amer. Soc. Hort. Sci.* 65:63–70.

Hoffman, O. L., and Smith, A. E. 1949. A new group of plant growth regulators. *Science* 109:588–590.

Holley, R. W., Apgar, J., Everett, G. A., Madison, J. T., Marquisee, M., Merrill, S. H., Penswick, J. R., and Zamir, A. 1965. Structure of a ribonucleic acid. *Science* 147:1462–1465.

Holmsen, T. W. 1966. *N,N*-dinitroethylenediamine, a plant growth stimulant. Address to Amer. Soc. Plant Physiol. AIBS Meeting, College Park, Maryland.

Holubowicz, T., and Boe, A. A. 1969. Development of cold hardiness in apple seedlings treated with gibberellic acid and abscisic acid. *Jour. Amer. Soc. Hort. Sci.* 94:661–664.

Holz, W. von. 1963. Neuere Ergebnisse bei der Duwock (*Equisetum palustre* L.). Bekämpfung mit dem Kortlandischen Untergrundschneider. *Proc. German Weed Conf. Stuttgart-Hohenheim* 7:191–194.

Hopp, R. J. 1962. Studies on the sex ratio in Butternut squash (*Cucurbita moschata* Poir). *Proc. Amer. Soc. Hort. Sci.* 80:473–480.

Hopp, R. J., and Rochester, H., Jr. 1967. Responses of butternut squash to *N*-dimethyl amino succinamic acid. *HortScience* 2:160–161.

Houghtaling, H. B. 1935. A developmental analysis of size and shape of tomato fruits. *Bull. Torrey Bot. Club* 62:243–252.

Howard, B. H. 1965. Increase during winter in capacity for root regeneration in detached shoots of fruit tree rootstocks. *Nature* 208:912–913.

Howell, M. J., and Wittwer, S. H. 1955. Further studies on the effects of 2,4-D on flowering in the sweetpotato. *Proc. Amer. Soc. Hort. Sci.* 66:279–283.

Howlett, F. S. 1949. Tomato fruit set and development with particular reference to premature softening following synthetic hormone treatment. *Proc. Amer. Soc. Hort. Sci.* 53:323–336.

Hruschka, H. W., Marth, P. C., and Heinze, P. H. 1965. External sprout inhibition and internal sprouts in potatoes. *Amer. Potato Jour.* 42:209–222.

Huang, H. 1961. A study on the application of gibberellin to celery production. *Natl. Taiwan Univ. College Agr. Res. Reports* 6:42–54.

———. 1963. The anatomical basis for the gibberellin responses in *Apium
graveolens*. *Memoirs College Agr. Natl. Taiwan Univ.* 7:36–48.

Huffaker, R. C., Miller, M. D., Baghott, K. G., Smith, F. L., and Schaller,
C. W. 1967. Effects of field application of 2,4-D and iron supplements
on yield and protein content of wheat and barley and yield of beans.
Crop Sci. 7:17–19.

Hughes, D. H. 1958. Discoloration findings. *Mushroom News* 4:9–13.

Hull, J., Jr., and Klos, E. J. 1958. Responses of healthy, ring, spot, and
yellows virus-infected Montmorency cherry trees to gibberellic acid.
Quar. Bull. Mich. Agr. Exptl. Sta. 41:19–23.

Hull, J., Jr., and Lewis, L. N. 1959. Response of one-year-old cherry and
mature bearing cherry, peach and apple trees to gibberellin. *Proc. Amer.
Soc. Hort. Sci.* 74:93–100.

Humphries, E. C., Welbank, P. J., and Witts, K. J. 1965. Effect of CCC
(chlorocholine chloride) on growth and yield of spring wheat in the
field. *Ann. Appl. Biol.* 56:351–361.

Ikekawa, N., Sumiki, Y., and Takahashi, N. 1963. Gas chromatographic
separation of gibberellins. *Chemistry and Industry*, no. 43, pp. 1728–
1729.

Imperial Chemicals Industries Ltd. 1955. *New organic compounds*. Accepted Patent Application, Commonwealth of Australia, no. 10190–1955.
Commonwealth of Australia.

Ingle, M., and Rogers, B. J. 1961. Some physiological effects of 2,2-di-
chloropropionic acid. *Weeds* 9:264–272.

Irvine, V. C. 1938. Studies in growth-promoting substances as related to
X-radiation and photoperiodism. *Univ. Colo. Stud.* 26:69–70.

Irving, R. M. 1969. Influence of growth retardants on development and
loss of hardiness of *Acer negundo*. *Jour. Amer. Soc. Hort. Sci.* 94:419–
422.

Irving, R. M., and Lanphear, F. O. 1968. Regulation of cold hardiness in
Acer Negundo. *Plant Physiol.* 43:9–13.

Ismail, M. A., Biggs, R. H., and Oberbacher, M. F. 1967. Effect of gibberellic acid on color changes in the rind of three sweet orange cultivars
(*Citrus sinensis* Blanco). *Proc. Amer. Soc. Hort. Sci.* 91:143–149.

Itakura, T., Kozaki, I., and Machida, Y. 1965. Studies with gibberellin
application in relation to response of certain grape varieties. *Bull. Hort.
Res. Sta., Ministry Agr. and Forest.*, Series A, no. 4 (Hiratsuka, Japan).

Ito, H., Motomura, Y., Konno, Y., and Hatayama, T. 1969. Exogenous
gibberellin as responsible for the seedless berry development of grapes,
I: Physiological studies on the development of seedless Delaware grapes.
Tohoku Jour. Agr. Res. 20:1–18.

Iwahori, S., and Lyons, J. M. 1970. Maturation and quality of tomatoes with preharvest treatments of 2-chloroethylphosphonic acid. *Jour. Amer. Soc. Hort. Sci.* 95:88–91.

Iwahori, S., Lyons, J. M., and Smith, O. E. 1970. Sex expression in cucumber plants as affected by 2-chloroethylphosphonic acid, ethylene, and growth regulators. *Plant Physiol.* 46:412–415.

Iwahori, S., Weaver, R. J., and Pool, R. M. 1968. Gibberellin-like activity in berries of seeded and seedless Tokay grapes. *Plant Physiol.* 43:333–337.

Jackson, D. I. 1968a. Gibberellin and growth in stone fruits: Induction of parthenocarpy in plum. *Australian Jour. Biol. Sci.* 21:1103–1106.

———. 1968b. Gibberellin and the growth of peach and apricot fruits. *Australian Jour. Biol. Sci.* 21:209–215.

Jackson, D. I., and Coombe, B. G. 1966. Gibberellin-like substances in the developing apricot fruit. *Science* 154:277–278.

Jackson, G. A. D., and Prosser, M. V. 1959. The induction of parthenocarpic development in *Rosa* by auxins and gibberellic acid. *Naturwiss.* 46:407–408.

Jacobs, W. P. 1968. Hormonal regulation of leaf abscission. *Plant Physiol.* 43:1480–1495.

Jacobsen, J. V., Scandalios, J. G., and Varner, J. E. 1970. Multiple forms of amylase induced by gibberellic acid in isolated barley aleurone layers. *Plant Physiol.* 45:367–371.

Jaffe, M. J., and Isenberg, F. M. 1965. Some effects of N-dimethylamino succinamic acid (B-nine) on the development of various plants, with special reference to the cucumber, *Cucumis sativus* L. *Proc. Amer. Soc. Hort. Sci.* 87:420–428.

Jensen, F. L. 1970. Effects of post-bloom gibberellin applications on berry shrivel and berry weight on seeded *Vitis vinifera* table grapes. Master's thesis, Univ. of Calif., Davis.

Jones, C. M. 1965. Effects of benzyladenine on fruit set in muskmelon. *Proc. Amer. Soc. Hort. Sci.* 87:335–340.

Jones, E. R. H., Henbest, H. B., Smith, G. F., and Bentley, J. A. 1952. 3-indolylacetonitrile: A naturally occurring plant hormone. *Nature* 169:485–487.

Jones, R. L. 1968a. Agar-diffusion techniques for estimating gibberellin production by plant organs; the discrepancy between extractable and diffusible gibberellins in pea. In Wightman and Setterfield 1968, pp. 73–84.

———. 1968b. Ethylene enhanced release of α-amylase from barley aleurone cells. *Plant Physiol.* 43:442–444.

Jones, R. L., and Phillips, I. D. J. 1964. Agar-diffusion technique for estimating gibberellin production by plant organs. *Nature* 204:497–499.

Jones, R. L., and Varner, J. E. 1967. The bioassay of gibberellins. *Planta* (Berlin) 72:155–161.

Judkins, W. P. 1945. The extraction of auxin from tomato fruit. *Amer. Jour. Bot.* 32:242–249.

Julliard, B., and Balthazard, J. 1965. Effets physiologiques de l'acide gibérellique sur quelques variétés de vigne (*Vitis vinifera* L.). *Ann. Amélior. Plantes* 15:61–78.

Kamal, A. L., and Marroush, M. 1971. Control of chocolate spot in potato tubers by foliar spray with 2-chloroethylphosphonic acid. *HortScience* 6:42.

Kasimatis, A. N., Weaver, R. J., and Pool, R. M. 1968. Effects of 2,4-D and 2,4-DB on the vegetative development of 'Tokay' grapevines. *Amer. Jour. Enol. Vitic.* 19:194–204.

Kasimatis, A. N., Weaver, R. J., Pool, R. M., and Halsey, D. D. 1971. Response of 'Perlette' grape berries to gibberellic acid applications applied during bloom and at fruit set. *Amer. Jour. Enol. Vitic.* 22:19–23.

Kasmire, R. F., Rappaport, L., and May, D. 1970. Effects of 2-chloroethylphosphonic acid on ripening of cantaloupes. *Jour. Amer. Soc. Hort. Sci.* 95:134–137.

Kato, J., Shiotani, Y., Tamura, S., and Sakurai, A. 1966. Physiological activities of helminthosporol in comparison with those of gibberellin and auxin. *Planta* (Berlin) 68:353–359.

Kato, T., and Ito, H. 1962. Physiological factors associated with the shoot growth of apple trees. *Tohoku Jour. Agric. Res.* 13:1–21.

Kato, Y., Fukuharu, N., and Kobayashi, R. 1958. Stimulation of flower bud differentiation of conifers by gibberellin. In *Abstr., 2nd meeting Japan Gibberellin Res. Assoc., Tokyo, Japan, Dec. 1*, pp. 67–68.

Kaufman, J., and Ringel, S. M. 1961. Tests of growth regulators to retard yellowing and abscission of cauliflower. *Proc. Amer. Soc. Hort. Sci.* 78:349–352.

Kawase, M. 1964. Centrifugation, rhizocaline, and rooting in *Salix alba* L. *Physiol. Plantarum* 17:855–865.

Kefeli, V. I., and Kadyrov, C. 1971. Natural growth inhibitors, their chemical and physiological properties. *Ann. Rev. Plant Physiol.* 22:185–196.

Kefford, N. P. 1955a. The growth substances separated from plant extracts by chromatography. *Jour. Exptl. Bot.* 6:129–151.

———. 1955b. The growth substances separated from plant extracts by chromatography, II: The coleoptile and root elongation properties of the growth substances in plant extracts. *Jour. Exptl. Bot.* 6:245–255.

Kelley, J. D., and Hamner, C. L. 1958. The effect of chelating agents and maleic hydrazide on the keeping qualities of snapdragon (*Antirrhinum majus*). *Quar. Bull. Mich. Agr. Exptl. Sta.* 41:332–343.

Kelley, J. D., and Schlamp, A. L. 1964. Keeping quality, flower size, and flowering response of three varieties of easter lilies to gibberellic acid. *Proc. Amer. Soc. Hort. Sci.* 85:631–634.

Kemmerer, H., and Butler, J. D. 1966. The effect of gibberellic acid in growth of woody ground cover plants. *Proc. Amer. Soc. Hort. Sci.* 88: 698–702.

Kende, H. 1964. Preservation of chlorophyll in leaf sections by substances obtained from root exudate. *Science* 145:1066–1067.

Kende, H., Ninnemann, H., and Lang, A. 1963. Inhibition of gibberellic acid biosynthesis in *Fusarium moniliforme* by Amo-1618 and CCC. *Naturwiss.* 50:599–600.

Kent, M., and Gortner, W. A. 1951. Effect of preillumination in the response of split-pea stems to growth substances. *Bot. Gaz.* 112:307–311.

Key, J. L. 1969. Hormones and nucleic acid metabolism. *Ann. Rev. Plant Physiol.* 20:449–474.

Khalifah, R. A., Lewis, L. N., and Coggins, C. W., Jr. 1965. Gradient elution column chromatography systems for the separation and identification of gibberellins. *Analyt. Biochem.* 12:113–118.

Khan, A. A. 1966. Morphactins destroy plant responses to gravity and light. *Farm Res.* 32(3):2–3.

———. 1967a. Dormancy, genes, and hormones. *Farm Res.* 32:10–11.

———. 1967b. Physiology of morphactins: Effect on gravi- and photoresponse. *Physiol. Plantarum* 20:306–313.

King, L. J. 1966. *Weeds of the world: Biology and control.* New York: Interscience.

Kishi, M., and Tasaki, M. 1958. The effect of gibberellin on grape varieties. In *Abstr., 2nd meeting Japan Gibberellin Res. Assoc., Tokyo, Japan, Dec. 1*, pp. 13–14.

Kitagawa, H., Sugiura, A., and Sugiyama, M. 1966. Effects of gibberellin spray on storage quality of kaki. *HortScience* 1:59–60.

Klein, R. M., and Klein, D. T. 1970. *Research methods in plant science.* Garden City, N.Y.: Natural History Press.

Klingman, D. L., and Shaw, W. C. 1967 (revised). *Using phenoxy herbicides effectively.* Farmers' Bull. No. 2183. U.S. Dept. Agr.

Klingman, G. C. 1961. *Weed control: As a science.* New York: Wiley.

Köckemann, A. 1934. Über eine keimungshemmende Substance in fleischigen Fruchten. *Ber. Deut. Bot. Ges.* 52:523–526.

Kofranek, A. M., and Leiser, A. T. 1958. Chemical defoliation of *Hydrangea macrophylla. Proc. Amer. Soc. Hort. Sci.* 71:555–562.

Kögl, F., and Elema, J. 1960. Wirkungsbeziehungen zwischen Indole-3-essigsäure und Gibberellinsäure. *Naturwiss.* 47:90.

Kögl, F., Haagen-Smit, A. J., and Erxleben, H. 1934a. Über die Isolierung der Auxine a und b aus pflanzlichen Materialien, IX Mitteilung. *Zeitschr. Physiol. Chem.* 225:215–229.

――――. 1934b. Über ein neues Auxin (Heteroauxin) aus Harn, XI Mitteilung. *Zeithschr. Physiol. Chem.* 228:90–103.

Kögl, F., and Kostermans, D. G. F. R. 1934. Heteroauxin als Stoffwechselprodukt niederer pflanzlicher Organismen, Isolierung aus Hefe. *Zeitschr. Physiol. Chem.* 228:113–121.

Köhler, D., and Lang, A. 1963. Evidence for substances in higher plants interfering with gibberellin response of dwarf peas to gibberellin. *Plant Physiol.* 38:555–560.

Koller, D. 1962. Preconditioning of germination in lettuce at time of fruit ripening. *Amer. Jour. Bot.* 49:841–844.

Koshimizu, K., Matsubara, S., Kusaki, T., and Mitsui, T. 1967. Isolation of a new cytokinin from immature yellow lupin seeds. *Agr. Biol. Chem.* 31:795–801.

Kovoor, R., and Klämbt, D. 1968. Cytokinins in transfer ribonucleic acids. In Wightman and Setterfield 1968, pp. 57–60.

Kramer, M., and Went, F. W. 1949. The nature of the auxin in tomato stem tips. *Plant Physiol.* 24:207–221.

Kraus, E. J., and Kraybill, H. R. 1918. *Vegetation and reproduction with special reference to the tomato.* Ore. Agr. Exptl. Sta. Bull. 149.

Kraus, E. J., and Mitchell, J. W., 1947. Growth-regulating substances as herbicides. *Bot. Gaz.* 108:301–350.

Krezdorn, A. H., and Cohen, M. 1962. The influence of chemical fruit-sprays on yield and quality of citrus. *Proc. Fla. State Hort. Soc.* 75:53–60.

Kriedemann, P. E. 1968. An effect of kinetin on the translocation of ^{14}C-labelled photosynthate in citrus. *Australian Jour. Biol. Sci.* 21:569–571.

Krishnamurthi, S., Randhawa, G. S., and Singh, J. P. 1959. Effect of gibberellic acid on fruit set, size, and quality in the Pusa Seedless variety of grapes (*Vitis vinifera* L.). *Indian Jour. Hort.* 16:1–4.

Kuiper, P. J. C. 1964. Inducing resistance to freezing and desiccation in plants by decenylsuccinic acid. *Science* 146:544–546.

Kuraishi, S., and Muir, R. M. 1963. Diffusible auxin increase in a rosette plant treated with gibberellin. *Naturwiss.* 50:337–338.

Kuraishi, S., and Okhumura, F. S. 1956. The effect of kinetin on leaf growth. *Bot. Mag.* (Tokyo) 69:300–306.

Kurosawa, E. 1926. Experimental studies on the secretion of *Fusarium heterosporum* on rice plants. *Trans. Nat. Hist. Soc. Formosa.* 16:213–227.

Kushman, L. J. 1969. Inhibition of sprouting in sweetpotatoes by treatment with CIPC. *HortScience* 4:61–63.

Kutacek, M., and Prochazka, Z. 1964. Méthodes de détermination et d'isolement des composés indolique chez les cruiciferés (Methods of determining and isolating indole compounds from crucifers). In Nitsch 1964, pp. 445–446.

Kuykendall, J. R., Sharples, G. C., Nelson, J. M., True, L. F., and Tate, H. F. 1970. Berry set response of 'Thompson Seedless' grapes to prebloom and bloom gibberellic-acid treatment. *Jour. Amer. Soc. Hort. Sci.* 95:697–699.

Laibach, F., and Kribben, F. J. 1950. Einfluss von Wuchstoff auf die Blutenbildung der Gurke. *Zeitschr. Naturforsch.* 56:160.

Lang, A. 1956. Induction of flower formation in biennial *Hyoscyamus* by treatment with gibberellin. *Naturwiss.* 43:284–285.

———. 1957. The effect of gibberellin upon flower formation. *Proc. Natl. Acad. Sci. U.S.* 43:709–717.

———. 1965a. Physiology of flower initiation. In Ruhland 1965a, pp. 1380–1536.

———. 1965b. Effects of some internal and external conditions on seed germination. In Ruhland 1965b, pp. 848–893.

———. 1970. Gibberellins: Structure and metabolism. *Ann. Rev. Plant Physiol.* 21:537–570.

Langer, C. A. 1952. Effect of maleic hydrazide on the thinning of peaches during three successive years. *Quar. Bull. Mich. Agr. Exptl. Sta.* 35:209–213.

Lanphear, F. O., and Soule, O. H. 1970. Injury to city plants from industrial emissions of herbicides. *HortScience* 5:215–217.

Larsen, F. E. 1966. Potassium iodide-induced leaf abscission of deciduous woody plants. *Proc. Amer. Soc. Hort. Sci.* 88:690–697.

———. 1967. Stimulation of leaf abscission of woody plants by potassium iodide and alanine. *HortScience* 2:19–20.

———. 1969a. The influence of N-dimethylamino succinamic acid pretreatment on chemically induced leaf abscission of certain deciduous woody plants. *HortScience* 4:208–209.

———. 1969b. Promotion of leaf abscission of deciduous tree fruit nursery stock with abscisic acid. *HortScience* 4:216–218.

———. 1970a. Prestorage promotion of leaf abscission of deciduous nursery stock with bromodine. *Jour. Amer. Soc. Hort. Sci.* 95:231–233.

———. 1970b. Promotion of leaf abscission of deciduous nursery stock with 2-chloroethylphosphonic acid. *Jour. Amer. Soc. Hort. Sci.* 95:662–663.

Larsen, F. E., and Cromarty, R. W. 1966. Effect of N-dimethyl amino succinamic acid (Alar) on microorganism growth in relation to cut-flower senescence. *Proc. Amer. Soc. Hort. Sci.* 89:723–726.

Larsen, F. E., and Frolich, M. 1969. The influence of 8-hydroxyquinoline citrate, N-dimethylamino succinamic acid, and sucrose on respiration and water flow in 'Red Sim' cut carnations in relation to flower senescence. *Jour. Amer. Soc. Hort. Sci.* 94:289–292.

Larsen, F. E., and Scholes, J. 1965. Effects of sucrose, 8-hydroxyquinoline citrate, and N-dimethyl amino succinamic acid on vase-life and quality of cut carnation. *Proc. Amer. Soc. Hort. Sci.* 87:458–463.

————. 1966. Effects of 8-hydroquinoline citrate, N-dimethyl amino succinamic acid, and sucrose on vase-life and spike characteristics of cut snapdragons. *Proc. Amer. Soc. Hort. Sci.* 89:694–701.

Larsen, P. 1944. 3-Indoleacetaldehyde as a growth hormone in higher plants. *Dansk. Bot. Arkiv.* 11:11–132.

————. 1955a. On the separation of acidic and non-acidic auxins. *Physiol. Plantarum* 8:343–357.

————. 1955b. Growth substances in higher plants. In K. Paech and M. V. Tracy, eds., *Modern methods of plant analysis*, III, 565–625. Berlin: Springer-Verlag.

————. 1961. Biological determination of natural auxins. In Ruhland 1961, pp. 521–582.

Larson, P. R. 1960. Gibberellic-acid induced growth of dormant hardwood cuttings. *Forest Sci.* 6:232–239.

Larter, E. N. 1967. The effect of (2-chloroethyl) trimethylammonium chloride (CCC) on certain agronomic traits of barley. *Canadian Jour. Plant Sci.* 47:413–421.

Larter, E. N., Samii, M., and Sosulski, F. W. 1965. The morphological and physiological effects of (2-chloroethyl) trimethylammonium chloride on barley. *Canadian Jour. Plant Sci.* 45:419–427.

La Rue, C. D. 1936. The effect of auxin on abscission of petioles. *Proc. Natl. Acad. Sci. U.S.* 22:254–259.

Laurie, A. 1936. Studies of the keeping quality of cut flowers. *Proc. Amer. Soc. Hort. Sci.* 34:595–597.

Lavee, S. 1958. The effect of some growth-regulating substances on the development of *Sclerotium rolfsii* (Sacc.). *Jour. Exptl. Bot.* 10:359–366.

————. 1959. Physiological aspects of postharvest berry drop in certain grape varieties. *Vitis* 2:34–39.

————. 1960. Effect of gibberellic acid on seeded grapes. *Nature* 185:395.

————. 1965. *Loosening of olive fruit on the trees to facilitate easy mechanical and hand harvesting.* Natl. Univ. Inst. Agr., Rehovoth, Israel. 1965 series, no. 910-E.

Lavee, S., and Spiegel, P. 1958. Spray thinning of olives with growth regulators. *Ktavim* 9:129–138.

Lavee, S., and Spiegel-Roy, P. 1967. Effect of time of application of two growth substances on the thinning of olive fruit. *Proc. Amer. Soc. Hort. Sci.* 91:180–186.

Leben, C., and Barton, L. V. 1957. Effects of gibberellic acid on growth of Kentucky bluegrass. *Science* 125:494–495.

Lee, C. I., McGuire, J. J., and Kitchin, J. T. 1969. The relationship between rooting cofactors of easy and difficult-to-root cuttings of three clones of *Rhododendron*. *Amer. Soc. Hort. Sci.* 94:45–48.

Leonard, O. A., and Crafts, A. S. 1956. Uptake and distribution of radioactive 2,4-D by brush species, III: Translocation of herbicides. *Hilgardia* 26:366–415.

Leonard, O. A., and Harvey, W. A. 1965. *Chemical control of woody plants*. Calif. Agr. Exptl. Sta. Bull. 812.

Leonard, O. A., Weaver, R. J., and Glenn, R. K. 1967. Effect of 2,4-D and picloram on translocation of ^{14}C-assimilates in *Vitis vinifera* L. *Weed Res.* 7:208–219.

Leopold, A. C. 1955. *Auxins and plant growth*. Berkeley and Los Angeles: Univ. of California Press.

———. 1958. Auxin uses in the control of flowering and fruiting. *Ann. Rev. Plant Physiol.* 9:281–310.

———. 1964a. *Plant growth and development*. New York: McGraw-Hill.

———. 1964b. Kinins and the regulation of leaf senescence. In Nitsch 1964, pp. 705–718.

———. 1967. Developmental aspects of plant senescence. *Proc. 17th Int. Hort. Cong.* 3:285–290.

Leopold, A. C., and Kawase, M. 1964. Benzyladenine effects on bean leaf growth and senescence. *Amer. Jour. Bot.* 51:294–298.

Letham, D. S. 1964. Isolation of a kinin from plum fruitlets and other tissues. In Nitsch 1964, pp. 104–119.

———. 1967. Chemistry and physiology of kinetin-like compounds. *Ann. Rev. Plant Physiol.* 18:349–364.

———. 1968. A new cytokinin bioassay and the naturally occurring cytokinin complex. In Wightman and Setterfield 1968, pp. 19–31.

———. 1969. Cytokinins and their relation to other phytohormones. *BioScience* 19:309–316.

Leuty, S. J., and Bukovac, M. J. 1968. The effect of naphthaleneacetic acid on abscission of peach fruits in relation to endosperm development. *Proc. Amer. Soc. Hort. Sci.* 92:124–134.

Levy, D., and Kedar, N. 1970. Effect of ethrel on growth and bulb initiation in onion. *HortScience* 5:80–82.

Lewis, L. N., and Coggins, C. W., Jr. 1964. The inhibition of carotenoid accumulation in navel oranges by gibberellin A_3, as measured by thin layer chromatography. *Plant Cell Physiol.* 5:457–463.

Lewis, L. N., Coggins, C. W., Jr., and Garber, M. J. 1964. Chlorophyll concentration in the navel orange rind as related to potassium gibberellate, light intensity, and time. *Proc. Amer. Soc. Hort. Sci.* 84:177–180.

Lewis, L. N., Coggins, C. W., Jr., Labanauskas, C. K., and Dugger, W. M., Jr. 1967. Biochemical changes associated with natural and gibberellin A_3 delayed senescence in the navel orange rind. *Plant Cell Physiol.* 8:151–160.

Lexander, K. 1953. Growth-regulating substances in roots of wheat. *Physiol. Plantarum* (Copenhagen) 6:406–443.

Libbert, E. 1954. Das Zusammenwirken von Wuchs-und Hemmstoffen bei der korrelativen Knosphehemmung. *Planta* (Berlin) 44:286–318.

———. 1955. Nachweis und chemische Trennung des Korrelationshemmstoffes und seiner Hemmstoffvorstufe. *Planta* (Berlin) 45:405–425.

———. 1957. Die hormonale und korrelative Steurrerung der Adventivwurzelbildung. *Wiss. Zeitschr. Humboldt. Univ. Berlin, Math.-Nat. Reihe* 6:315–347.

Lindstrom, R. S. 1967. Growth-regulating foliar sprays on poinsettias. *Quar. Bull. Mich. Agr. Exptl. Sta.* 49:412–415.

Lindstrom, R. S., and Asen, S. 1967. Chemical control of the flowering of *Chrysanthemum morifolium* Ram., I: Auxin and flowering. *Proc. Amer. Soc. Hort. Sci.* 90:403–408.

Lindstrom, R. S., and Wittwer, S. H. 1957. Gibberellin and higher plants, IX: Flowering in geranium (*Pelargonium Hortorum*). *Quar. Bull. Mich. Agr. Exptl. Sta.* 40:225–231.

Lindstrom, R. S., Wittwer, S. H., and Bukovac, M. J. 1957. Gibberellin and higher plants, IV: Flowering responses of some flower crops. *Quar. Bull. Mich. Agr. Exptl. Sta.* 39:673–681.

Linser, H. 1951. Versuche zur chromatographischen Trennung pflanzlicher Wuchstoffe. *Planta* (Berlin) 39:377–401.

Linser, H., and Kiermayer, O. 1957. *Methoden zur Bestimmung pflanzlicher Wuchstoffe.* Vienna: Springer-Verlag.

Linser, H., Mayr, H., and Maschek, F. 1954. Papierchromatographie von zellstreckend wirksamen Indolkörpern aus Brassica-Arten. *Planta* (Berlin) 44:103–120.

Linsmaier, E. M., and Skoog, F. 1965. Organic growth factor requirements of tobacco tissue cultures. *Physiol. Plantarum* 18:100–127.

Lipe, W. N., and Crane, J. C. 1966. Dormancy regulation in peach seeds. *Science* 153:541–542.

Lipton, W. J., and Ceponis, M. J. 1962. Retardation of senescence and stimulation of oxygen consumption in head lettuce treated with N^6 benzyladenine. *Proc. Amer. Soc. Hort. Sci.* 81:379–384.

Little, C. H. A., and Eidt, D. C. 1968. Effect of abscisic acid on budbreak and transpiration in woody species. *Nature* 220:448–499.

Liu, F. 1970. Storage of bananas in polyethylene bags with an ethylene absorbent. *HortScience* 5:25–27.

Lockhart, J. A. 1956. Reversal of the light inhibition of pea stem growth by the gibberellins. *Proc. Natl. Acad. Sci. U.S.* 42:841–848.

Lockhart, J. A., and Bonner, J. 1957. Effects of gibberellic acid on the photoperiod-controlled growth of woody plants. *Plant Physiol.* 32:492–494.

Loeffler, J. E., and van Overbeek, J. 1964. Kinin activity in coconut milk. In Nitsch 1964, pp. 77–82.

Loomis, W. E. and Evans, M. M. 1928. Experiments in breaking the rest period of corms and bulbs. *Proc. Amer. Soc. Hort. Sci.* 25:73–79.

Looney, N. E. 1967. Effect of N-dimethylaminosuccinamic acid on ripening and respiration of apple fruits. *Canadian Jour. Plant Sci.* 47:549–553.

———. 1968. Inhibition of apple ripening by succinic acid, 2,2-dimethyl hydrazide and its reversal by ethylene. *Plant Physiol.* 43:1133–1137.

———. 1969. Control of apple ripening by succinic acid 2,2-dimethyl hydrazide, 2-chloroethyltrimethylammonium chloride, and ethylene. *Plant Physiol.* 44:1127–1131.

Looney, N. E., Fisher, D.V., and Parsons, J. E. W. 1967. Some effects of annual applications of N-dimethylaminosuccinamic acid (Alar) to apples. *Proc. Amer. Soc. Hort. Sci.* 91:18–24.

Looney, N. E., and McIntosh, D. L. 1968. Stimulation of pear rooting by preplant treatment of nursery stock with indole-3-butyric acid. *Proc. Amer. Soc. Hort. Sci.* 92:150–154.

Lower, R. L., Miller, C. H., Baker, F. H., and McCombs, C. L. 1970. Effects of 2-chloroethylphosphonic acid treatment at various stages of cucumber development. *HortScience* 5:433–434.

Luckwill, L. C. 1948a. A method for the quantitative estimation of growth substances based on the response of tomato ovaries to known amounts of 2-naphthoxyacetic acid. *Jour. Hort. Sci.* 24:19–31.

———. 1948b. The hormone content of the seed in relation to endosperm development and fruit drop in the apple. *Jour. Hort. Sci.* 24:32–44.

———. 1953a. Studies of fruit development in relation to plant hormones, I: Hormone production by the developing apple seed in relation to fruit drop. *Jour. Hort. Sci.* 28:14–24.

———. 1953b. Studies of fruit development in relation to plant hormones, II: The effect of naphthaleneacetic acid on fruit set and fruit development in apples. *Jour. Hort. Sci.* 28:25–40.

———. 1956. Two methods for the bioassay of auxins in the presence of growth inhibitors. *Jour. Hort. Sci.* 31:89–98.

———. 1957. Hormonal aspects of fruit development in higher plants. In H. K. Porter, ed., *Symp. Soc. Expt. Biol.* no. 11, *The Biological Action of Growth Substances.* Cambridge: Cambridge Univ. Press, pp. 63–85.

———. 1959a. Fruit growth in relation to internal and external chemical stimuli. In D. Rudnick, ed., *Cell, organism, and milieu,* pp. 223–251. New York: Ronald Press.

———. 1959b. Factors controlling the growth and form of fruits. *Jour. Linnean Soc. London Bot.* 56:294–302.

————. 1959c. The effect of gibberellic acid on fruit set in apples and pears. In *Ann. Report Agr. Hort. Res. Sta.*, Univ. of Bristol (Long Ashton), pp. 59–64.

————. 1961. The effect of gibberellic acid on the cropping of pears following frost damage. In *Ann. Rept. Agr. Hort. Res. Sta.*, Long Ashton, Bristol, Univ. of Bristol, pp. 61–66.

Luckwill, L. C., and Lloyd-Jones, C. P. 1962. The absorption, translocation, and metabolism of 1-naphthaleneacetic acid applied to apple leaves. *Jour. Hort. Sci.* 37:190–206.

Luckwill, L. C., Weaver, P., and MacMillan, J. 1969. Gibberellins and other growth hormones in apple seeds. *Jour. Hort. Sci.* 44:413–424.

Lumis, G. P., and Davidson, H. 1967. Preventing fruit formation on landscape trees. *HortScience* 2:61–62.

MacLean, D. C., and Dedolph, R. R. 1962. Effects of N^6-benzylaminopurine on post-harvest respiration of *Chrysanthemum morifolium* and *Dianthus caryophyllus*. *Bot. Gaz.* 124:20–21.

MacLeod, A. M., and Millar, A. S. 1962. Effects of gibberellic acid on barley endosperm. *Jour. Inst. Brewing* 68:322–332.

MacMillan, J. 1968. Direct identification of gibberellins in plant extracts by gas chromatography-mass spectrometry. In Wightman and Setterfield 1968, pp. 101–107.

MacMillan, J., Seaton, J. C., and Suter, P. J. 1961. Isolation and structures of gibberellins from higher plants. *Advances Chem. Series* no. 28, pp. 18–25.

MacMillan, J., and Suter P. J. 1963. Thin layer chromatography of the gibberellins. *Nature* 197:790.

MacMillan, J., and Takahashi, N. 1968. Proposed procedure for the allocation of trivial names to the gibberellins. *Nature* 217:170–171.

Madison, J. H., Johnson, J. M., Davis, W. B., and Sachs, R. M. 1969. Testing fluorene compounds for chemical mowing of turf grass. *Calif. Agr.* 23:8–10.

Mainland, C. M., and Eck, P. 1968. Induced parthenocarpic fruit development in highbush blueberry. *Proc. Amer. Soc. Hort. Sci.* 92:284–289.

————. 1969a. Fruit and vegetative responses of the highbush blueberry to gibberellic acid under greenhouse conditions. *Jour. Amer. Soc. Hort. Sci.* 94:19–20.

————. 1969b. Fruiting response of the highbush blueberry to gibberellic acid under field conditions. *Jour. Amer. Soc. Hort. Sci.* 94:21–23.

Mann, J. D., Hield, H., Yung, K. H., and Johnson, D. 1966. Independence of morphactin and gibberellin effects upon higher plants. *Plant Physiol.* 41:1751–1752.

Mann, L. K., and Minges, P. A. 1949. Experiments on setting fruit with growth-regulating substances on field-grown tomatoes in California. *Hilgardia* 19:309–337.

Marcelle, R. 1966. Effects of (2-chloroethyl) trimethylammonium chloride and of maleic hydrazide on young apple trees of the 'Tydeman's Early Worcester' variety. *Hort. Res.* 6:100–112.

Marcelle, R., and Sironval, C. 1963. Effect of gibberellic acid on flowering of apple trees. *Nature* 197:405.

Marcus, A. 1971. Enzyme induction in plants. *Ann. Rev. Plant Physiol.* 22:313–336.

Marth, P. C. 1943. Retardation of shoot development on roses during common storage by treatment with growth regulating substance. *Proc. Amer. Soc. Hort. Sci.* 42:620–628.

———. 1952. Effect of growth regulators on the retention of color in green sprouting broccoli. *Proc. Amer. Soc. Hort. Sci.* 60:367–369.

———. 1963. Effect of growth retardants on flowering, fruiting, and vegetative growth of Holly (*Ilex*). *Proc. Amer. Soc. Hort. Sci.* 83:777–781.

———. 1965. Increased frost resistance by application of plant growth retardant chemicals. *Agr. Food Chem.* 13:331–333.

Marth, P. C., Audia, W. V., and Mitchell, J. W. 1956. Effect of gibberellic acid on growth and development of plants of various genera and species. *Bot. Gaz.* 118:106–111.

Marth, P. C., and Frank, J. R. 1961. Increasing tolerance of soybean plants to some soluble salts through application of plant growth-retardant chemicals. *Agr. Food Chem.* 9:359–361.

Marth, P. C., Havis, L., and Batjer, L. P. 1947. Further results with growth regulators in retarding flower opening of peaches. *Proc. Amer. Soc. Hort. Sci.* 49:49–54.

Martin, D., Lewis, T. L., and Cerny, J. 1968. The effect of Alar on fruit cell division and other characteristics in apples. *Proc. Amer. Soc. Hort. Sci.* 92:67–70.

Martin, G. C. 1971. 2-chloroethylphosphonic acid as an aid to mechanical harvesting of English walnuts. *Jour. Amer. Soc. Hort. Sci.* 96:434–436.

Martin, G. C., and Griggs, W. H. 1970. The effectiveness of succinic acid 2,2-dimethylhydrazide in preventing preharvest drop of 'Bartlett' pears. *HortScience* 5:258–259.

Martin, G. C., and Lopushinsky, W. 1966. Effect of N-dimethyl aminosuccinamic acid (B-995), a growth retardant, on drought tolerance. *Nature* 209:216–217.

Martin, G. C., Mason, M. I. R., and Forde, H. I. 1969. Changes in endogenous growth substances in the embryos of *Juglans regia* during stratification. *Jour. Amer. Soc. Hort. Sci.* 94:13–17.

Martin, G. C., and Nelson, M. 1969. The thinning effect of 3-chlorophenoxy-a-propionamide (3-CPA) in Paloro peach. *HortScience* 4:206–208.

Maryott, A. A. 1969. Gib tests on the national arboretum collection of *Camellia japonica*. *Amer. camellia yearbook,* pp. 35–39.

Mastalerz, J. W. 1965. The effect of gibberellic acid on the flowering shoots of Better Times roses. *Proc. Amer. Soc. Hort. Sci.* 87:525–530.

Masuda, Y. 1962. Effect of light on a growth inhibitor in wheat roots. *Physiol. Plantarum* 15:780–790.

Mavrodineanu, R., Stanford, W. W., and Hitchcock, A. E. 1955. Use of fluorescence for the estimation of substances separated on paper partition chromatography. *Contrib. Boyce Thompson Inst.* 18:167–172.

Maxie, E. C., and Crane, J. C. 1967. 2,4,5-Trichlorophenoxyacetic acid: Effect on ethylene production by fruits and leaves of fig tree. *Science* 155:1548–1550.

———. 1968. Effect of ethylene on growth and maturation of the fig, *Ficus carica* L., fruit. *Proc. Amer. Soc. Hort. Sci.* 92:255–267.

Mayr, H. H., and Bayzer, H. 1965. Untersuchungen über die Einlagerung von Gerbstsubstanzen in Weizenhalmen nach der Anwendung von Chlorcholinchlorid (CCC). *Zeitschr. Acker und Pflanzenbau* 121:295–299.

McComb, A. J., and Carr, D. J. 1958. Evidence from a dwarf pea bioassay for naturally occurring gibberellins in the growing plant. *Nature* 181: 1548–1549.

McConnell, D. B., and Struckmeyer, B. E. 1970. Effect of succinic acid 2,2-dimethylhydrazide on the growth of marigold in long and short photoperiods. *HortScience* 5:391–393.

McDonnell, P. F., and Edgerton, L. J. 1970. Some effects of CCC and Alar on fruit set and fruit quality of apple. *HortScience* 5:89–91.

McDowell, T. C., and Larson, R. A. 1966. Effects of (2-chloroethyl) trimethyl ammonium chloride (Cycocel), N-dimethyl succinamic acid (B-nine), and photoperiod on flower bud initiation and development in azaleas. *Proc. Amer. Soc. Hort. Sci.* 88:600–605.

McGuire, J. J., Albert, L. S., and Shutak, V. G. 1968. Effect of foliar applications of 3-indolebutyric acid on rooting of cuttings of ornamental plants. *Proc. Amer. Soc. Hort. Sci.* 93:699–704.

———. 1969. Uptake of IAA-2-^{14}C by cuttings of *Ilex crenata* Convexa. *Jour. Amer. Soc. Hort. Sci.* 94:44–45.

McIlrath, W. J., Ergle, D. R., and Dunlap, A. A. 1951. Persistence of 2,4-D stimulus in cotton plants with reference to its transmission to the seed. *Bot. Gaz.* 112:511–518.

McKee, M. W., and Forshey, C. G. 1966. Effects of chemical thinning on repeat bloom of McIntosh apple trees. *Proc. Amer. Soc. Hort. Sci.* 88:25–32.

McMurray, A. L., and Miller, C. H. 1969. The effect of 2-chloroethanephosphonic acid (ethrel) on the sex expression and yields of *Cucumis sativus*. *Jour. Amer. Soc. Hort. Sci.* 94:400–402.

McVey, G. R., and Wittwer, S. H., 1958. Gibberellin and higher plants, XI: Responses of certain woody ornamental plants. *Quar. Bull. Mich. Agr. Exptl. Sta.* 40:679–697.

Meredith, W. C., Joiner, J. N., and Biggs, R. H. 1970. Influences of indole-3-acetic acid and kinetin on rooting and indole metabolism of *Feijoa sellowiana. Jour. Amer. Soc. Hort. Sci.* 95:49–52.

Michniewicz, M., and Kentzer, T. 1965. The increase of frost resistance of tomato plants through application of 2-chloroethyl trimethylammonium chloride (CCC). *Experientia* 21:230–231.

Millborrow, B. V. 1967. The identification of (+)-abscisin II ((+)-dormin) in plants and measurement of its concentrations. *Planta* (Berlin) 76:93–113.

Miller, C. O. 1956. Similarity of some kinetin and red light effects. *Plant Physiol.* 31:318–319.

———. 1961. Kinetin and related compounds in plant growth. *Ann. Rev. Plant Physiol.* 12:395–408.

———. 1963. Kinetin and kinetin-like compounds. In H. F. Linkens and M. V. Tracey, eds., *Moderne Methoden der Pflanzenanalyse* VI:194–202. Berlin: Springer-Verlag.

———. 1968. Naturally occurring cytokinins. In Wightman and Setterfield 1968, pp. 33–45.

Miller, C. O., Skoog, F., Okhumura, F. S., Von Saltza, M. H., and Strong, F. M. 1955. Structure and synthesis of a kinin. *Jour. Amer. Chem. Soc.* 77:2662–2663.

———. 1956. Isolation, structure, and synthesis of kinetin, a substance promoting cell division. *Jour. Amer. Chem. Soc.* 78:1375–1380.

Mitchell, J. E., and Angel, C. R. 1950. Plant-growth-regulating substances obtained from cultures of *Fusarium moniliforme. Phytopathol.* 40:872.

Mitchell, J. W. 1966. Present status and future of plant regulating substances. *Agr. Sci. Rev., U.S. Dept. Agr.* 4:27–36.

Mitchell, J. W., Ezell, B. D., and Wilcox, M. 1949. Effect of p-chlorophenoxyacetic acid on the vitamin C content of snap beans following harvest. *Science* 109:202–203.

Mitchell, J. W., and Livingston, G. A. 1968. Methods of Studying Plant Hormones and Growth-Regulating Substances. *Agr. handbook* no. 336. Agr. Res. Serv., U.S. Dept. Agr.

Mitchell, J. W., Mandava, N., Plimmer, J. R., Worley, J. F., and Drowne, M. E. 1969. Plant growth-regulating properties of some acetone condensation products. *Nature* 223:1386–1387.

Mitchell, J. W., Mandava, N., Worley, J. F., and Plimmer, J. R. 1970. Brassins—a new family of plant hormones from rape pollen. *Nature* 225:1065–1066.

Mitchell, J. W., and Marth, P. C. 1944. Effects of 2,4-dichlorophenoxyacetic acid on the ripening of detached fruit. *Bot. Gaz.* 106:199–207.

————. 1947. *Growth regulators for garden, field, and orchard.* Chicago: Univ. of Chicago Press.

————. 1950. Effect of growth-regulating substances on the water-retaining capacity of bean plants. *Bot. Gaz.* 112:70–76.

————. 1962. Chemicals and plant growth. In *The yearbook of agriculture: After a hundred years,* pp. 137–146. U.S. Dept. Agr.

Mitchell, J. W., Wirwille, J. W., and Weil, L. 1949. Plant growth-regulating properties of some nicotinium compounds. *Science* 110:252–254.

Mitsuhashi, M., and Shibaoka, H. 1965. Isolation of an inhibitor of growth and root formation from *Portulaca grandiflora* leaves. *Plant Cell Physiol.* 6:87–99.

Miyamoto, T. 1962. Erhöhung der Widerstandsfähigkeit von Weizen gegenüber hohen Salzkonzentrationen durch Behandlung des Saatgutes mit (2-chloräthyl)-Trimethylammonium Chlorid. (Increase in the tolerance of wheat to high salt concentrations by seed treatment with 2-chloroethyltrimethyl ammonium chloride). *Naturwiss.* 49:213.

————. 1963. Increase of the resistance to herbicide by the seed treatment with 2-chloroethanol in spinach seedlings. *Naturwiss.* 50:24.

Modlibowska, I. 1963. Effect of gibberellic acid on fruit development of frost-damaged conference pears. *Ann. Rept. East Malling (England) Res. Sta. for 1962,* pp. 64–67.

————. 1965. Effects of (2-chloroethyl) trimethylammonium chloride and gibberellic acid on growth, fruit bud formation and frost resistance in one-year-old pear trees. *Nature* 208:503–504.

Modlibowska, I., and Ruxton, J. P. 1954. The effect of maleic hydrazide on the spring frost resistance of the Malling Exploit raspberry. *Jour. Hort. Sci.* 29:184–192.

Moewus, F. 1949. Der Kressewurzel Test, ein neuer quantitätiver Wuchsstofftest. *Biol. Zbl.* 68:118–140.

Molisch, H. 1938. *The longevity of plants.* Lancaster, Pa.: Science Press.

Monselise, S. P., Goren, R., and Goldschmidt, E. E. 1967. *The occurrence of natural growth promoters and inhibitors in citrus tree organs as influenced by season, age of leaves, nutritional disorders, pests and other factors.* Research conducted under grants authorized by Public Law 480 for the U.S. Dept. of Agr.

Monselise, S. P., Goren, R., and Halevy, A. H. 1966. Effects of B-Nine, Cycocel, and benzothiazole oxyacetate on flower bud induction of lemon trees. *Proc. Amer. Soc. Hort. Sci.* 89:195–200.

Monselise, S. P., and Halevy, A. H. 1962. Effects of gibberellin and Amo-1618 on growth, dry matter accumulation, chlorophyll content and peroxidase activity of citrus seedlings. *Amer. Jour. Bot.* 49:405–412.

————. 1964. Chemical inhibition and promotion of citrus flower-bud induction. *Proc. Amer. Soc. Hort. Sci.* 84:141–146.

Moore, J. N. 1970. Cytokinin-induced sex conversion in male clones of *Vitis* species. *Jour. Amer. Soc. Hort. Sci.* 95:387–393.

Moreland, D. E. 1967. Mechanism of action of herbicides. *Ann. Rev. Plant Physiol.* 18:365–386.

Morgan, D. G., and Mees, G. C. 1956. Gibberellic acid and the growth of crop plants. *Nature* 178:1356–1357.

Mothes, K. 1960. Über das altern der Blätter and die Möglichkeit ihrer Wiederverjüngung. *Naturwiss.* 47:337–350.

Mothes, K., Englebrecht, L., and Kulajewa, O. 1959. Über die Wirkung des Kinetins auf Stickstoffverteilung und Eiweissynthese in isolierte Blättern. *Flora (Jena)* 147:445–464.

Muir, R. M. 1947. The relationship of growth hormones and fruit development. *Proc. Natl. Acad. Sci. U.S.* 33:303–312.

Mukherjee, S. K. 1967. Recent advances in fruit tree propagation. *Indian Hort.*, July–September.

Mukherjee, S. K., and Bid, N. N. 1965. Propagation of mango (*Mangifera indica* L.), II: Effect of etiolation and growth-regulator treatments on the success of air-layering. *Indian Jour. Agr. Sci.* 35:309–34.

Mukherjee, S. K., and Majumder, P. K. 1963. Standardization of rootstocks of mango, I: Studies on the propagation of clonal rootstocks by stooling and layering. *Indian Jour. Hort.* 20:204–209.

Müller-Thurgau, H. 1898. Abhängigkeit der Ausbildung der Traubenbeeren und einiger anderer Früchte von der Entwicklung der Samen. *Landw. Jahrb. Schweiz.* 12:135–205.

Mullet, J. H. 1966. The effect of a growth regulator on seed production and retention in *Phalaris tuberosa* L. *Jour. Australian Inst. Agr. Sci.* 32: 218–219.

Murakami, Y. 1959. A paper chromatographic survey of gibberellins and auxins in immature seeds of leguminous plants. *Bot. Mag. Tokyo* 72:36–43.

———. 1961. Formation of gibberellin A_3 glucoside in plant tissues. *Bot. Mag. Tokyo* 74:424–425.

———. 1968. A new rice seedling test for gibberellins, 'micro-drop method,' and its use for testing extracts of rice and morning glory. *Bot. Mag. Tokyo* 81:33–43.

Nakagawa, S., Bukovac, M. J., Hirata, N., and Kurooka, H. 1968. Morphological studies of gibberellin-induced parthenocarpic and asymmetric growth in apple and Japanese pear fruits. *Jour. Jap. Soc. Hort. Sci.* 37:9–19.

Nakata, S. 1955. Floral initiation and fruit set in lychee with special reference to the effect of sodium naphthalene acetate. *Bot. Gaz.* 117:126–134.

Narasimham, P., Rao, M. M., Nagaraja, N., and Anandaswamy, B. 1967.

Effect of pre-harvest application of growth regulators on the control of berry drop in Bangalore Blue grapes (*Vitis labrusca* L.). *Jour. Food Sci. Technol.* 4:162–164.

Nash, A. S., and Mullaney, P. D. 1960. Commercial application of gibberellic acid to hops. *Nature* 185:25.

Naylor, A. W., and Davis, E. A. 1950. Maleic hydrazide as a plant growth inhibitor. *Bot. Gaz.* 112:112–126.

Neel, P. L. 1969. Growth factors in trunk development of young trees. *Proc. 45th Int. Shade Tree Conf.*, pp. 46–59.

Neely, P. M. 1959. The development and use of a bioassay for gibberellins. Ph.D. thesis, Univ. of Calif., Los Angeles.

Negi, S. S., and Olmo, H. P. 1966. Sex conversion of a male *Vitis vinifera* L. by a kinin. *Science* 152:1624–1625.

Nekrasova, T. V. 1960. The effect of gibberellic acid on the germination of seeds and the growth of seedlings of fruit plants. *Fiziol. Rastenii* 7:85–87.

Neljubow, D. 1901. Über die horizontal Nutation der Stengel von *Pisum sativum* und einiger anderen Pflanzen. *Bot. Zbl. Beih.* 10:128–139.

Nickell, L. G., and Tanimoto, T. T. 1967. Sugarcane ripening with chemicals. *Reports Hawaiian sugar technologists*. Jour. Series. Exptl. Sta. Paper no. 196, pp. 104–109.

Nielsen, N. 1928. Untersuchungen über Stoffe, die das Wachstum der *avenacoleoptile* Beschleunigen. *Planta* (Berlin) 6:376–378.

Nightingale, A. E. 1970. The influence of succinamic acid 2-2-dimethylhydrazide on the growth and flowering of pinched vs. unpinched plants of the *Kalanchoe* hybrid 'Mace.' *Jour. Amer. Soc. Hort. Sci.* 95:273–276.

Nigond, J. 1957. Le Retard au débourrément de le vigne par un traitement à l'acide—naphthalene acetique et la lutte contre les gelées. *C. R. Acad. Agr. France* 46:452–457.

Nijjar, G. S., and Bhatia, G. G. 1969. Effect of gibberellic acid and parachlorophenoxyacetic acid on the cropping and quality of Anab-e-Shahi grapes. *Jour. Hort. Sci.* 44:91–95.

Ninnemann, H., Zeevaart, J. A. D., Kende, H., and Lang, A. 1964. The plant growth retardant CCC as inhibitor of gibberellin biosynthesis in *Fusarium moniliforme*. *Planta* (Berlin) 61:229–235.

Nissl, D., and Zenk, M. H. 1969. Evidence against induction of protein synthesis during auxin-induced initial elongation of *Avena* coleoptiles *Planta* (Berlin) 89:323–349.

Nitsch, J. P. 1950. Growth and morphogenesis of the strawberry as related to auxin. *Amer. Jour. Bot.* 37:211–215.

——. 1952. Plant hormones in the development of fruits. *Quarterly Rev. Biol.* 27:33–57.

——. 1955. Free auxin and free tryptophane in the strawberry. *Plant Physiol.* 30:33–39.

Nitsch, J. P. 1956. Methods for the investigation of natural auxins and growth inhibitors. In Wain and Wightman 1956, pp. 3–31.

————. 1957. Growth responses of woody plants to photoperiodic stimuli. *Proc. Amer. Soc. Hort. Sci.* 70:512–525.

————. 1960. La chromatographie double à une dimension et son emploi à la séparation des substances de croissance. *Bull. de la Soc. France* 107:247–250.

————, ed. 1964. *Régulateurs naturels de la croissance végétale.* Paris: Centre National de la Recherche Scientifique.

————. 1965. Physiology of flower and fruit development. In Ruhland 1965a, pp. 1537–1647.

————. 1970. Hormonal factors in growth and development. In A. C. Hulme, ed., *The biochemistry of fruits and their products,* pp. 427–472. London and New York: Academic Press.

Nitsch, J. P., and Nitsch, C. 1956. Studies on the growth of coleoptile and first internode sections: A new, sensitive straight-growth test for auxins. *Plant Physiol.* 31:94–111.

Nixon, R. W., and Gardner, F. E. 1939. Effect of certain growth substances on inflorescences of dates. *Bot. Gaz.* 100:868–871.

Noodén, L. D. 1970. Metabolism and binding of ^{14}C-maleic hydrazide. *Plant Physiol.* 45:46–52.

Norman, A. G. 1946. Studies on growth-regulating substances. *Bot. Gaz.* 107:475.

Norris, R. F., and Bukovac, M. J. 1968. Structure of the pear leaf cuticle with special reference to cuticular penetration. *Amer. Jour. Bot.* 55:975–983.

Nussbaum, J. J. 1969. Chemical pinching for roots of container plants. *Calif. Agr.* 23:16–18.

Nutman, P. S., Thornton, H. G., and Quastel, J. H. 1945. Plant growth-substances as selective weed killers: Inhibition of plant growth by 2,4-D and other plant-growth substances. *Nature* 155:498–500.

Odebaro, O. A., and Smith, O. E. 1969. Effects of kinetin, salt concentration, and temperature on germination and early seedling growth of *Lactuca sativa* L. *Jour. Amer. Soc. Hort. Sci.* 94:167–170.

Odom, R. E., and Carpenter, W. J. 1965. The relationship between endogenous indole auxins and the rooting of herbaceous cuttings. *Proc. Amer. Soc. Hort. Sci.* 87:494–501.

Ogawa, Y. 1963. Studies on the conditions for gibberellin assay using rice seedling. *Plant Cell Physiol.* 4:227–237.

Ohkuma, K., Addicott, F. T., Smith, O. E., and Thiessen, W. E. 1965. The structure of abscisin II: *Tetrahedron letters* 29:2529–2535.

Osborne, D. J. 1955. Acceleration of abscission by a factor produced in senescent leaves. *Nature* 176:1161–1163.

———. 1962. Effect of kinetin on protein and nucleic acid metabolism in Xanthium leaves during senescence. *Plant Physiol.* 37:595–602.

———. 1965. Interactions of hormonal substances in the growth and development of plants. *Jour. Sci. Food Agr.* 16:1–13.

———. 1967. Hormonal regulation of leaf senescence. In Woolhouse 1967, pp. 305–321.

Osborne, D. J., and McCalla, D. R. 1961. Rapid bioassay for kinetin and kinins using senescing leaf tissue. *Plant Physiol.* 36:219–221.

Paál, A. 1918. Über phototropische Reizleitung. *Jahrb. Wiss. Bot.* 58:406–458.

Paleg, L. G. 1960. Physiological effects of gibberellic acid, I: On carbohydrate metabolism and amylase activity of barley endosperm. *Plant Physiol.* 35:293–299.

———. 1965. Physiological effects of gibberellins. *Ann. Rev. Plant Physiol.* 16:291–322.

Palevitch, D. 1970. Defoliation of snap beans with preharvest treatments of 2-chloroethylphosphonic acid. *HortScience* 5:224–226.

Parker, K. G., Edgerton, L. J., and Hickey, K. D. 1964. Gibberellin treatment for yellows-infected sour cherry trees. *Farm Res.* 29:8–9.

Penner, D., and Ashton, F. M. 1966. Biochemical and metabolic changes in plants induced by chlorophenoxy herbicides. *Residue Rev.* 14:39–113.

Pentzer, W. T. 1941. Studies on the shatter of grapes with special reference to the use of solutions of naphthaleneacetic acid to prevent it. *Proc. Amer. Soc. Hort. Sci.* 38:397–399.

Perley, J. E., and Stowe, B. B. 1966. An improvement in the sensitivity of the Salkowski reagent for tryptamine, tryptophan and indoleacetic acid. *Physiol. Plantarum* 19:683–690.

Peterson, C. E., and Anhder, L. D. 1960. Induction of staminate flowers on gynoecious cucumbers with gibberellin A_3. *Science* 131:1673–1674.

Pharis, R. P., and Morf, W. 1967. Experiments on the precocious flowering of western red cedar and four species of *Cupressus* with gibberellins A_3 and A_4/A_7 mixture. *Canadian Jour. Bot.* 45:1519–1524.

———. 1968. Physiology of gibberellin-induced flowering in conifers. In Wightman and Setterfield 1968, pp. 1341–1356.

———. 1969. Precocious flowering of coastal and giant redwood with gibberellins A_3, $A_{4/7}$, and A_{13}. *BioScience* 19:719–720.

Pharis, R. P., Ruddat, M. D. E., Phillips, C. C., and Heftmann, E. 1965. Precocious flowering of Arizona cypress with gibberellin. *Canadian Jour. Bot.* 43:923–927.

Phatak, S. C., Wittwer, S. H., Honma, S., and Bukovac, M. J. 1966. Gibberellin-induced anther and pollen development in a stamenless tomato mutant. *Nature* 209:635–636.

Phillips, I. D. J. 1962. Some interactions of gibberellic acid with naringenin (5,7,4′-trihydroxy flavone) in the control of dormancy and growth in plants. *Jour. Exptl. Bot.* 13:213–226.

———. 1971. *Introduction to the biochemistry and physiology of plant growth hormones.* New York: McGraw-Hill.

Phillips, I. D. J., and Wareing, P. F. 1958. Studies in dormancy of sycamore, I: Seasonal changes in the growth substance content of the shoot. *Jour. Exptl. Bot.* 9:350–364.

Phinney, B. O. 1956. Growth response of single-gene dwarf mutants in maize to gibberellic acid. *Proc. Natl. Acad. Sci. U.S.* 42:185–189.

Phinney, B. O., and West, C. A. 1960. Gibberellins as native plant growth regulators. *Ann. Rev. Plant Physiol.* 11:411–436.

———. 1961. Gibberellins and plant growth. In Ruhland 1961, pp. 1185–1227.

Pieringer, A. P., and Newhall, W. F. 1970. Growth retardation of citrus by quaternary ammonium derivatives of (+)-limonene. *Jour. Amer. Soc. Hort. Sci.* 95:53–55.

Pilet, P. E. 1961. *Les phytohormones de croissance.* Paris: Masson.

Pillay, D. T. N. 1966. Growth substances in developing Mazzard cherry seeds. *Canadian Jour. Bot.* 44:507–512.

Pillay, D. T. N., Brase, K. D., and Edgerton, L. J. 1965. Effects of pretreatments, temperature, and duration of after ripening on germination of Mazzard and Mahaleb cherry seeds. *Proc. Amer. Soc. Hort. Sci.* 86:102–107.

Pincus, G., and Thimann, K. V. 1948. *The hormones: Physiology, chemistry, and applications.* Vol. I, Chaps. 2 and 3. New York: Academic Press.

Pinthus, M. J., and Halevy, A. H. 1965. Prevention of lodging and increase in yield of wheat treated with CCC (2-chloroethyl trimethylammonium chloride). *Israel Jour. Agr. Res.* 15:159–161.

Platt, R. S., Jr., and Thimann, K. V. 1956. Interference in Salkowski assay of indoleacetic acid. *Science* 123:105–106.

Plaut, Z., and Halevy, A. H. 1966a. Increasing the regeneration of wheat plants following wilting by treatment with two growth-retarding chemicals. *Naturwiss.* 53:509.

———. 1966b. Regeneration after wilting, growth and yield of wheat plants, as affected by two growth-retarding compounds. *Physiol. Plantarum* 19:1064–1072.

Plaut, Z., Halevy, A. H., and Schmueli, E. 1964. The effect of growth-retarding chemicals on growth and transpiration of bean plants grown under various irrigation regimes. *Israel Jour. Agr. Res.* 14:153–158.

Pohl, R. 1951a. Das Hemmstoff-Wuchsstoffsystem des Maisskutellums. *Planta* (Berlin) 39:105–125.

———. 1951b. Die Wirkung von Wuchsstoff and Hemmstoff auf des Wachstum Pollenschläuche von Petunia. *Bio. Zbl.* 70:119–128.

Popoff, K. J., ed. 1969. Plant Stimulation. In *Proc. Intern. Symposium on Plant Stimulation,* 1966, held at M. Popoff Inst. Plant Physiol. Sofia: Bulgarian Academy Sci. Press.

Post, K. 1947. Hydrangea leaf removal—hydrangea fundamentals. *New York State Flower Growers Bull.* 25 (September): 1–3.

Powell, L. E. 1960. Separation of plant growth-regulating substances on silica gel columns. *Plant Physiol.* 35:256–261.

———. 1964. Preparation of indole extracts from plants for gas chromatography and spectrophotofluorometry. *Plant Physiol.* 39:836–842.

Pratt, H. K., and Goeschl, J. D. 1969. Physiological roles of ethylene in plants. *Ann. Rev. Plant Physiol.* 20:541–584.

Preston, W. H., Jr., and Link, C. B. 1958. Comparative effectiveness of several new quaternary ammonium compounds to induce dwarfing of plants. *Plant Physiol. Suppl.* 33, xlix.

Pridham, A. M. S. 1952. Preliminary report on defoliation of nursery stock by chemical means. *Proc. Amer. Soc. Hort. Sci.* 59:475–478.

Primost, E. 1965. Die Wirkung von Chlorcholinchlorid (CCC) auf das Wachstum von Weizenkeimpflanzen. *Zeitschr. Acker. Pflanzenbau* 121: 300–314.

———. 1967a. Der Einfluss von Chlorcholinchlorid (CCC) auf den Ertragsaufbau von Winter-und Sommerweisen. *Zeitschr. Acker. Pflanzenbau* 126:164–178.

———. 1967b. Der Einfluss von Chlorocholinchlorid (CCC) auf die Länge und Breite von Weissenblättern. *Die Bodenkultur* 18:127–140.

———. 1968. The effect of (chloroethyl)-trimethylammonium chloride (CCC) on wheat. (Results of long-term field experiments). *Euphytica Suppl.* no. 1, pp. 239–249.

———. 1970a. Der Einfluss des Standortes auf die Wirkung der CCC-Behandlung und Stickstoffdüngung bei Roggen. *Zeitschr. Acker. Pflanzenbau* 131:44–56.

———. 1970b. Die Reaktion von Winterweizen auf eine Behandlung mit Chlorcholinchlorid (CCC) in verschiedenen Wachstumsstadien. *Die Bodenkultur* 21:148–167.

Primost, E., and Rittmeyer, G. 1968a. Veränderungen im Aufbau der Weizenhalme bei CCC-Behandlung in Abhängigkeit von Sorte und Standort. *Zeitschr. Acker. Pflanzenbau* 128:117–138.

———. 1968b. Die Wirkung von Chlorcholinchlorid (CCC) auf Kornertrag und Qualität von Roggen in Feldversuchen. *Die Bodenkultur* 19: 112–126.

———. 1970. Veränderungen im Aufbau der Roggenhalme bei Stabilan-Behandlung. *Die Bodenkultur* 21:244–263.

Proebsting, E. L., Jr., and Mills, H. H. 1964. Gibberellin-induced hardiness responses in Elberta peach flower buds. *Proc. Amer. Soc. Hort. Sci.* 85: 134–140.

Proebsting, E. L., Jr., and Mills, H. H. 1966. Effect of gibberellic acid and other growth regulators on quality of Early Italian prunes (*Prunus domestica*, L.) *Proc. Amer. Soc. Hort. Sci.* 89:135–139.

————. 1969. Effects of 2-chloroethane phosphonic acid and its interaction with gibberellic acid on quality of 'Early Italian' prunes. *Jour. Amer. Soc. Hort. Sci.* 94:443–446.

Prosser, M. V., and Jackson, G. A. D. 1959. Induction of parthenocarpy in *Rosa arvensis* Huds. with gibberellic acid. *Nature* 184:108.

Quinlan, J. D., and Weaver, R. J. 1969. Influence of benzyladenine, leaf darkening, and ringing on movement of ^{14}C-labeled assimilates into expanded leaves of *Vitis vinifera* L. *Plant Physiol.* 44:1247–1252.

Rabinowitch, H. D., Rudich, J., and Kedar, N. 1970. The effect of ethrel on ripening of tomato and melon fruits. *Israel Jour. Agr. Research* 20: 47–54.

Randhawa, G. S., and Dhillon, B. S. 1965. Studies on fruit set and fruit drop in citrus, I: A review. *Indian Jour. Hort.* 22:33–45.

Randhawa, G. S., Dhuria, H. S., and Nair, P. K. R. 1964. A note on gibberellin-induced parthenocarpy in citrus. *Indian Jour. Hort.* 21:171–172.

Randhawa, G. S., and Negi, S. S. 1964. Preliminary studies on seed germination and subsequent seedling growth in grapes. *Indian Jour. Hort.* 21:186–196.

Randhawa, G. S., and Sharma, B. B. 1962. Effect of plant regulators on fruit set, drop quality of sweet oranges (*Citrus sinensis* Osbeck). *Indian Jour. Hort.* 19:83–91.

Randhawa, G. S., Singh, J. P., and Dhuria, H. S. 1959. Effect of gibberellic acid, 2,4-dichlorophenoxyacetic acid, and 2,4,5-trichlorophenoxyacetic acid on fruit set, drop, size and total yield in sweet lime (*Citrus limettioides* Tanaka). *Indian Jour. Hort.* 16:206–209.

Randhawa, G. S., Singh, J. P., and Khanna, S. S. 1959. Effect of gibberellic acid and some other plant regulators on fruit set, size, total yield and quality in phalsa (*Grewia asiatica* L.). *Indian Jour. Hort.* 16:202–205.

Randhawa, G. S., and Thompson, H. C. 1949. Effect of application of hormones on yield of tomatoes grown in the greenhouse. *Proc. Amer. Soc. Hort. Sci.* 53:337–344.

Rao, M. M. 1970. Effect of benzyladenine on postharvest berry drop in Anab-e-Shahi grapes (*Vitis vinifera* L.). *Vitis* 9:126–129.

Rao, S. N., and Wittwer, S. H. 1955. Further investigations on the use of maleic hydrazide as a sprout inhibitor for potatoes. *Amer. Potato Jour.* 32:51–59.

Rappaport, L. 1956. Growth regulating metabolites. Gibberellin compounds derived from rice disease-producing fungus exhibit powerful plant growth regulating properties. *Calif. Agr.* 10:4.

————. 1970. Chemical regulation of plant development. In D. G. Aldrich, ed., *Research for the World Food Crisis*, no. 92, pp. 147–180. Washington, D.C.: Amer. Assoc. for the Advancement of Science.

Rappaport, L., Lippert, L. F., and Timm, H. 1957. Sprouting, plant growth, and tuber production as affected by chemical treatment of white potato seed pieces, I: Breaking the rest period with gibberellic acid. *Amer. Potato Jour.* 34:254–260.

Rappaport, L., and Wolf, N. 1969. The problem of dormancy in potato tubers and related structures. In Woolhouse 1969, pp. 219–240.

Rasmussen, G. K., and Cooper, W. C. 1968. Abscission of citrus fruits induced by ethylene-producing chemicals. *Proc. Amer. Soc. Hort. Sci.* 93:191–198.

Read, P. E., and Fieldhouse, D. J. 1970. Use of growth retardants for increasing tomato yields and adaptation for mechanical harvest. *Jour. Amer. Soc. Hort. Sci.* 95:73–78.

Read, P. E., and Hoysler, V. 1969. Stimulation and retardation of adventitious root formation by application of B-Nine and Cycocel. *Jour. Amer. Soc. Hort. Sci.* 94:314–316.

————. 1971. Improving rooting of carnation and poinsettia cuttings with succinic acid-2,2-dimethylhydrazide. *HortScience* 6:350–351.

Rebeiz, C. A., and Crane, J. C. 1961. Growth regulator-induced parthenocarpy in the Bing cherry. *Proc. Amer. Soc. Hort. Sci.* 78:69–75.

Reed, D. J., Moore, T. C., and Anderson, J. D. 1965. Plant growth retardant B-995: A possible mode of action. *Science* 148:1469–1471.

Rennix, M. T., Dye, A. P., Hu, C. Y., and Neal, O. M. 1964. Shoot growth control and increased environmental tolerance of azaleas, obtained by spraying with retardants B-Nine and Cycocel. *Proc. West Va. Acad. Sci.* 36:63–68.

Rhodes, A., Sexton, W. A., Spencer, L. A., and Templeman, W. G. 1950. Use of isopropylphenylcarbamate to reduce sprouting of potato tubers during storage. *Research* (London) 3:189–190.

Richmond, A. E., and Lang, A. 1957. Effect of kinetin on protein content and survival of detached Xanthium leaves. *Science* 125:650–651.

Riddell, J. A., Hageman, H. A., J'Anthony, C. M., and Hubbard, W. L. 1962. Retardation of plant growth by a new group of chemicals. *Science* 136:391.

Riehl, L. A., Coggins, C. W., Jr., and Carman, G. E. 1966. Navel orange water spot protection by gibberellin. *Jour. Econ. Entomol.* 59:615–618.

Ries, S. K., Chmiel, H., Dilley, D. R., and Filner, P. 1967. The increase in nitrate reductase activity and protein content of plants treated with simazine. *Proc. Natl. Acad. Sci. U.S.* 58:526–532.

Ries, S. K., Schweizer, C. J., and Chmiel, H. 1968. The increase in protein content and yield of simazine-treated crops in Michigan and Costa Rica. *BioScience* 18:205–208.

Rinderknecht, H., Wilding, P., and Haverback, B. J. 1967. A new method for the determination of a-amalyse. *Experientia* 23:805.

Rives, M., and Pouget, R. 1959. Action de la gibberelline sur la compacité des grapes de deux variétés de vigne. *C. R. Séances Acad. Agr. France* 45:343–345.

Roberts, E. A., Southwick, M. D., and Palmiter, D. H. 1948. A micro-chemical examination of McIntosh apple leaves showing relationship of cell wall constituents to penetration of spray solutions. *Plant Physiol.* 23:557–559.

Roberts, E. H. 1964. A survey of the effects of chemical treatments on dormancy of rice seed. *Physiol. Plantarum* 17:30–43.

Roberts, J. B., and Stevens, R. 1962. Effect of gibberellic acid on the growth of hops. *Jour. Inst. Brew.* 68:247–250.

Robinson, R. W., Wilczynski, H., de la Guardia, M. D., and Shannon, S. 1970. Chemical regulation of fruit ripening and sex expression. *New York's Food and Life Sciences Quarterly* 3(1):10–11.

Robinson, R. W., Wilczynski, H., and Dennis, F. G., Jr. 1968. Chemical promotion of tomato fruit ripening. *Proc. Amer. Soc. Hort. Sci.* 93:823–830.

Rodriguez, A. B. 1932. Smoke and ethylene in fruiting of pineapple. *Jour. Dept. Agr. Puerto Rico* 25:5–18.

Rogers, B. L., and Thompson, A. H. 1968. Growth and fruiting response of young apple and pear trees to annual applications of succinic acid 2,2-dimethyl hydrazide. *Proc. Amer. Soc. Hort. Sci.* 93:16–24.

———. 1969. Chemical thinning of apple trees using concentrate sprays. *Jour. Amer. Soc. Hort. Sci.* 94:23–25.

Romberg, L. D., and Smith, C. L. 1938. Effects of indole-3-butyric acid in the rooting of transplanted pecan trees. *Proc. Amer. Soc. Hort. Sci.* 36:161–170.

Ross, J. D., and Bradbeer, J. W. 1968. Concentrations of gibberellin in chilled hazel seeds. *Nature* 220:85–86.

Rothert, W. 1894. Über Heliotropismus. *Beitr. Biol. Pflanzen* (Cohn) 7:1–212.

Rothwell, K., and Wain, R. L. 1964. Studies on a growth inhibitor in yellow lupin (*Lupinus lutens* L.). In Nitsch 1964, pp. 363–375.

Rowe, J. W. 1968 (with Addenda and Corrigenda, Feb., 1969). *The common and systematic nomenclature of cyclic diterpenes.* Madison, Wisc.: Forest Products Laboratory, U.S. Dept. Agr.

Rubinstein, B., and Leopold, A. C. 1962. Effects of amino acids on bean leaf abscission. *Plant Physiol.* 37:398–401.

———. 1964. The nature of leaf abscission. *Quar. Rev. Biol.* 39:356–372.

Ruddat, M., Lang, A., and Mosettig, E. 1963. Gibberellin activity of steviol, a plant terpenoid. *Naturwiss.* 50:23.

Rudich, J., Halevy, A. H., and Kedar, N. 1969. Increase in femaleness of three curcurbits by treatment with ethrel, an ethylene-releasing compound. *Planta* (Berlin) 86:69–76.

Rudnicki, R. 1969. Studies on abscisic acid in apple seeds. *Planta* (Berlin) 86:63–68.

Ruelke, O. C. 1961. Winter injury of Pangola grass reduced by use of maleic hydrazide. *Agron. Jour.* 53:405–406.

Ruhland, W., ed. 1961. *Handbuch der Pflanzenphysiologie* (*Encyclopedia of plant physiology*), vol. XIV. Berlin: Springer-Verlag.

———. 1965a. *Handbuch der Pflanzenphysiologie* (*Encyclopedia of plant physiology*), vol. XV/1. Berlin: Springer-Verlag.

———. 1965b. *Handbuch der Pflanzenphysiologie* (*Encyclopedia of plant physiology*), vol. XV/2. Berlin: Springer-Verlag.

Russo, L., Jr., Dostal, H. C., and Leopold, A. C. 1968. Chemical regulation of fruit ripening. *BioScience* 18:109.

Ryan, G. F., Frolich, E. F., and Kinsella, T. P. 1958. Some factors influencing rooting of grafted cuttings. *Proc. Amer. Soc. Hort. Sci.* 72:454–461.

Ryugo, K. 1966. Persistence and mobility of Alar (B-995) and its effect on anthocyanin metabolism in sweet cherries, *Prunus avium*. *Proc. Amer. Soc. Hort. Sci.* 88:160–166.

Ryugo, K., Kester, D. E., Rough, D., and Mikuckis, F. 1970. Effects of Alar on almonds . . . delayed flowering . . . shorter shoots. *Calif. Agr.* 24:14–15.

Ryugo, K., and Sachs, R. M. 1969. *In vitro* and *in vivo* studies of Alar (1,1-dimethylaminosuccinamic acid, B-995) and related substances. *Jour. Amer. Soc. Hort. Sci.* 94:529–533.

Saad, F. A., Crane, J. C., and Maxie, E. C. 1969. Timing of olive oil application and its probable role in hastening maturation of fig fruits. *Jour. Amer. Soc. Hort. Sci.* 94:335–337.

Sacher, J. A., and Salminen, S. O. 1969. Comparative studies of effect of auxin and ethylene on permeability and synthesis of RNA and protein. *Plant Physiol.* 44:1371–1377.

Sachs, J. 1880, 1882. Stoff und Form der Pflanzenorgane. I, II. *Arb. Bot. Inst. Würzburg* 2:452–488, 689–718.

Sachs, R. M. 1961. Gibberellin, auxin, and growth retardant effects upon cell division and shoot histogenesis. *Advances Chem. Series* 28:49–58.

Sachs, R. M., and Bretz, C. F. 1962. The effect of daylength, temperature and gibberellic acid upon flowering in *Fuchsia hybrida*. *Proc. Amer. Soc. Hort. Sci.* 80:581–588.

Sachs, R., Hackett, W., Maire, R., Baldwin, R., and Kretchun, T. 1967. Chemical control of vegetative growth in woody ornamental plants. In *Proc. Int. Symp. Subtropical and Tropical Horticulture*, pp. 426–433. Bangalore: Hort. Soc. India.

Sachs, R. M., Hackett, W. P., Maire, R. G., Kretchun, T. M., and de Bie, J. 1970. *Chemical control of plant growth in landscapes.* Calif. Agr. Exptl. Sta. Bull. 844.

Sachs, R. M., and Kofranek, A. M. 1963. Comparative cytohistological studies on inhibition and promotion of stem growth in *Chrysanthemum morifolium. Amer. Jour. Bot.* 50:772–779.

Sachs, R. M., Lang, A., Bretz, C. F., and Roach, J. 1960. Shoot histogenesis: Subapical meristematic activity in a caulescent plant and the action of gibberellic acid and AMO-1618. *Amer. Jour. Bot.* 47:260–266.

Sachs, R. M., and Maire, R. G. 1967. Chemical control of growth and flowering of woody ornamental plants in the landscape and nursery: Tests with maleic hydrazide and Alar. *Proc. Amer. Soc. Hort. Sci.* 91: 728–734.

Sakai, A. 1957. Effect of maleic hydrazide upon the frost resistance of mulberry twigs. *Nippon Sanshigaku Zasshi Hokkaido Univ.* 26:13–20.

Salkowski, E. 1885. Über das Verhalten der Skatolcarbonsäure im Organismus. *Zeitschr. Physiol. Chem.* 9:23–33.

Samish, R. M. 1954. Dormancy in woody plants. *Ann. Rev. Plant Physiol.* 5:183–204.

Samish, R. M., and Gur, A. 1962. Experiments with budding avocado. *Proc. Amer. Soc. Hort. Sci.* 81:194–201.

Samish, R. M., and Lavee, S. 1958. Spray thinning of grapes with growth regulators. *Ktavim* 8(3–4):273–285.

Sansavini, S., Martin, J., and Ryugo, K. 1970. The effect of succinic acid 2,2-dimethyl hydrazide on the uniform maturity of peaches and nectarines. *Jour. Amer. Soc. Hort. Sci.* 95:709–711.

Sastry, K. K. S., and Muir, R. M. 1963. Gibberellin: Effect on diffusible auxin in fruit development. *Science* 140:494–495.

Schneider, G. 1964. Eine neue Gruppe von synthetischen Pflanzen Wachstumsregulatoren. *Naturwiss.* 51:416–417.

———. 1970. Morphactins: Physiology and performance. *Ann. Rev. Plant Physiol.* 21:499–536.

Schneider, G., Erdman, D., Lust, S., Mohr, G., and Niethammer, K. 1965a. Morphactins, a novel group of plant growth regulators. *Nature* 208:1013.

Schneider, G., Sembner, G., and Schreiber, K. 1965b. Gibberelline, VI: Die Dünnschichtelektrophorese von Gibberellinen. *Jour. Chromatog.* 19: 358–363.

Schoene, D. L., and Hoffman, O. L. 1949. Maleic hydrazide, a unique growth regulant. *Science* 109:588–590.

Scott, D. H., and Draper, A. D. 1967. Light in relation to seed germination of blueberries, strawberries, and *Rubus. HortScience* 2:107–108.

Scott, R. A., Jr., and Liverman, J. L. 1956. Promotion of leaf expansion by kinetin and benzylaminopurine. *Plant Physiol.* 31:321–322.

Searle, N. E. 1965. Physiology of flowering. *Ann. Rev. Plant. Physiol.* 16: 97–118.

Seelby, S. D., and Powell, L. E. 1970. Electron capture gas chromatography for sensitive assay of abscisic acid. *Analyt. Biochem.* 35:530–533.

Seidman, G., and Ibrahim, J. H. 1962. Effects of some plant growth regulators on the sensitivity of plants exposed to irradiated auto exhaust. *Plant Physiol.* 37, Suppl. lxix–lxx.

Sembdner, G., and Schreiber, K. 1965. Gibberellins, IV: Über die Gibberelline von *Nicotiana tabacum* L. *Phytochem.* 4:49–56.

Sembdner, G., Schneider, G., Weiland, J., and Schreiber, K. 1964. Über ein gebundenes Gibberellin aus *Phaseolus coccinus* L. *Experientia* 20: 89–90.

Sen, S. P., and Leopold, A. C. 1954. Paper chromatography of plant growth regulators and allied compounds. *Physiol. Plantarum* (Copenhagen) 7: 98–108.

Seubert, E. 1925. Über Wachstumsregulatoren in der Koleoptile von *Avena*. *Zeitschr. Bot.* 17:49–88.

Shanks, J. B. 1969. Some effects and potential uses of ethrel on ornamental crops. *HortScience* 4:56–58.

Shantz, E. M. 1966. Chemistry of naturally occurring growth-regulating substances. *Ann. Rev. Plant Physiol.* 17:409–438.

Sharma, B. B., and Randhawa, G. S. 1967. Studies on fruit set and fruit drop in sweet orange. *Indian Jour. Hort.* 24:109–117.

Shaulis, N. 1959. Gibberellin trials for New York grapes. *New York Farm Res.* 25:11.

Sherwood, C. H., Weigle, J. L., and Denisen, E. L. 1970. 2,4-D as an air pollutant: Effects on growth of representative horticultural plants. *HortScience* 5:211–213.

Shibaoka, H., and Yamaki, T. 1959. A sensitized *Avena* curvature test and identification of the diffusible auxin in *Avena coleoptile*. *Bot. Mag. Tokyo* 72:152–158.

Shindy, W., and Weaver, R. J. 1967. Plant regulators alter translocation of photosynthetic products. *Nature* 214:1024–1025.

Shutak, V. G. 1968. Effect of succinic acid 2,2-dimethylhydrazide on flower bud formation of the 'Coville' highbush blueberry. *HortScience* 3:225.

Sill, L. Z., and Nelson, P. V. 1970. Relationship between azalea bud morphology and effectiveness of methyl decanoate, a chemical pinching agent. *Jour. Amer. Soc. Hort. Sci.* 95:270–273.

Sims, W. L. 1969. Effects of ethrel on fruit ripening of tomatoes . . . greenhouse, field and postharvest trials. *Calif. Agr.* 23:12–14.

Sims, W. L., Collins, H. B., and Gledhill, B. L. 1970. Ethrel effects on fruit ripening of peppers. *Calif. Agr.* 24:4–5.

Sims, W, L., and Gledhill, B. L. 1969. Ethrel effects on sex expression and growth development in pickling cucumbers. *Calif. Agr.* 23:4–6.

Singh, J. P., Randhawa, G. S., and Rajput, C. B. S. 1960. Effect of plant regulators on fruit development, ripening, drop, yield, size and quality of loquat (*Eriobotrya japonica* Lindl.) var. Improved Golden Yellow and Pale Yellow. *Indian Jour. Hort.* 17:156–164.

Sinha, A. K., and Wood, R. K. S. 1964. Control of *Verticillium* wilt of tomato plants with "Cyocel" ((2-chloroethyl)trimethylammonium chloride). *Nature* 202:824.

Sitton, B. G., Lewis, W. A., and Kilby, W. W. 1968. Chemical retardation of tung blossoming. *Proc. Amer. Soc. Hort. Sci.* 92:381–393.

Skene, K. G. M. 1970a. The gibberellins of developing bean seeds. *Jour. Exptl. Bot.* 21:236–246.

―――. 1970b. The relationship between the effects of CCC on root growth and cytokinin levels in the bleeding sap of *Vitis vinifera* L. *Jour. Exptl. Bot.* 21:418–431.

Skoog, F. 1937. A deseeded *Avena* test method for small amounts of auxin and auxin precursors. *Jour. Gen. Physiol.* 20:311–334.

―――. 1951. *Plant Growth Substances.* Madison: Univ. of Wisconsin Press.

Skoog, F., and Armstrong, D. J. 1970. Cytokinins. *Ann. Rev. Plant Physiol.* 21:359–384.

Skoog, F., Armstrong, D. J., Cherayil, J. D., Hampel, A. E., and Bock, R. M. 1966. Cytokinin activity: Localization in transfer RNA preparations. *Science* 154:1354–1356.

Skoog, F., Hamzi, H., Szweykowska, A. M., Leonard, N. J., Carraway, K. L., Fujii, T., Helgeson, J. P., and Loeppky, R. N. 1967. Cytokinins: Structure/activity relationships. *Phytochem.* 6:1169–1192.

Skoog, F., and Miller, C. O. 1957. Chemical regulation of growth and organ formation in plant tissue cultures *in vitro*. *Symp. Exptl. Biol.* 11:118–131.

Skoog, F., and Tsui, C. 1948. Chemical control of growth and bud formation in tobacco stem segments and callus cultured *in vitro*. *Amer. Jour. Bot.* 35:782–787.

―――. 1951. Growth substances and the formation of buds in plant tissues. In Skoog 1951, pp. 263–285.

Slade, R. E., Templeman, W. G., and Sexton, W. A. 1945. Plant-growth substances as selective weed killers: Differential effect of plant-growth substances on plant species. *Nature* 155:497–498.

Slife, F. W., Key, J. L., Yamaguchi, S., and Crafts, A. S. 1962. Penetration, translocation, and metabolism of 2,4-D and 2,4,5-T in wild and cultivated cucumber plants. *Weeds* 10:29–35.

Sloger, C., and Caldwell, B. E. 1970. Response of cultivars of soybean to synthetic abscisic acid. *Plant Physiol.* 45:634–635.

Smith, A. E., Zukel, J. W., Stone, G. M., and Ridell, J. E. 1950. Factors affecting the performance of maleic hydrazide. *Jour. Agr. Food Chem.* 7:341–344.

Smith, C. R. 1960. Effect of growth regulators on fruit drop of the Coville blueberry. In N. J. Moore and N. F. Childers, eds., *Blueberry research: Fifty years of progress,* p. 23. New Brunswick: New Jersey Agr. Exptl. Sta.

Smith, C. R., Soczek, Z., and Collins, W. B. 1961. Flowering and fruiting of strawberries in relation to gibberellins. *Advances Chem. Series* 28: 109–115.

Smith, O. E., Lyon, J. L., Addicott, F. T., and Johnson, R. E. 1968. Abscission physiology of abscisic acid. In Wightman and Setterfield 1968, pp. 1547–1560.

Smith, O. E., and Palmer, C. E. 1970. Cytokinin-induced tuber formation on stolons of *Solanum tuberosum. Physiol. Plantarum* 23:599–606.

Smith, O. E., Yen, W. W. L., and Lyons, J. M. 1968. The effects of kinetin in overcoming high-temperature dormancy of lettuce seed. *Proc. Amer. Soc. Hort. Sci.* 93:444–453.

Smock, R. M. 1969. Laboratory studies on the effects of chemicals on the coloration of apples. *Jour. Amer. Soc. Hort. Sci.* 94:49–51.

Snow, R. 1925. The correlative inhibition of the growth of axillary buds. *Ann. Bot.* 39:841–859.

————. 1935. Activation of cambial growth by pure hormones. *New Phytologist* 34:347–360.

Snyder, M. J., Welch, N. C., and Rubatzky, V. E. 1970. Influence of gibberellin upon time of bud development in Globe artichoke (*Cynara scolymus*). In *Abstracts Western Regional Meeting Amer. Soc. Hort. Sci., Berkeley, Calif., June 22–24.*

Söding, H. 1925. Zur Kenntnis der Wuchshormone in der Haferkoleoptile. *Jahrb. Wiss. Bot.* 64:587–603.

Soleimani, A., Kliewer, W. M., and Weaver, R. J. 1970. Influence of growth regulators on concentration of protein and nucleic acids in 'Black Corinth' grapes. *Jour. Amer. Soc. Hort. Sci.* 95:143–146.

Sondheimer, E., and Galson, E. C. 1966. Effects of abscisin II and other plant growth substances on germination of seeds with stratification requirements. *Plant Physiol.* 41:1397–1398.

Sondheimer, E., Tzou, D. S., and Galson, E. C. 1968. Abscisic acid levels and seed dormancy. *Plant Physiol.* 43:1443–1447.

Soost, R. K., and Burnett, R. H. 1961. Effects of gibberellin on yield and fruit characteristics of *Clementine mandarin. Proc. Amer. Soc. Hort. Sci.* 77:194–201.

Southwick, F. W., Demoranville, I. W., and Anderson, J. F. 1953. The influence of some growth regulating substances on preharvest drop, color and maturity of apples. *Proc. Amer. Soc. Hort. Sci.* 61:155–162.

Southwick, F. W., Lord, W. J., and Weeks, W. D. 1968. The influence of succinamic acid 2,2-dimethylhydrazide (Alar) on the growth, productivity, mineral nutrition, and quality of apples. *Proc. Amer. Soc. Hort. Sci.* 92:71–81.

Southwick, F. W., and Weeks, W. D. 1957. The influence of naphthaleneacetic acid and naphthaleneacetamide during a four-year period on thinning and subsequent flowering of apples. *Proc. Amer. Soc. Hort. Sci.* 69:28–40.

Southwick, F. W., Weeks, W. D., and Olanyk, G. W. 1964. The effect of naphthaleneacetic acid type materials and 1-naphthyl *N*-methylcarbamate (Sevin) on the fruiting, flowering and keeping quality of apples. *Proc. Amer. Soc. Hort. Sci.* 84:14–24.

Southwick, F. W., Weeks, W. D., Sawada, E., and Anderson, J. F. 1962. The influence of chemical thinners and seeds on the growth rate of apples. *Proc. Amer. Soc. Hort. Sci.* 80:33–42.

Sparks, D. 1967. Effect of potassium gibberellate on fruit characteristics and flowering of the pecan, *Carya illinoensis* Koch, cv. 'Stuart.' *Proc. Amer. Soc. Hort. Sci.* 90:61–66.

Sparks, D., and Chapman, J. W. 1970. Effect of indole-3-butyric acid on rooting and survival of air-layered branches of the pecan, *Carya illinoensis* Koch, cv. 'Stuart.' *HortScience* 5:445–446.

Spiegel, P. 1954. Auxins and inhibitors in canes of *Vitis*. *Bull. Res. Counc. Israel*, 4:176–183.

————. 1955. Some internal factors affecting rooting of cuttings. *Rpt. 14th Int. Hort. Cong.*, pp. 239–246.

Splittstoesser, W. E. 1970. Effects of 2-chloroethylphosphonic acid and gibberellic acid on sex expression and growth of pumpkins. *Physiol. Plantarum* 23:762–768.

Stahly, E. A., and Thompson, A. H. 1959. *Auxin levels of developing Halehaven peach ovules.* Univ. Md. Agr. Exptl. Sta. Bull. A-104.

Stahly, E. A., and Williams, M. W. 1967. Size control of nursery stock with *N*-dimethylaminosuccinamic acid. *Proc. Amer. Hort. Sci.* 91:792–794.

Stark, P. 1921. Studien über traumatotrope und haptotrope Reizleitungsvorgänge mit besonderer Berücksichtigung der Reizübertragung auf fremde Arten und Gattungen. *Jahrb. Wiss. Bot.* 60:67–134.

Steeves, T. A., Morel, G., and Wetmore, R. H. 1953. A technique for preventing inactivation at the cut surface in auxin-diffusion studies. *Amer. Jour. Bot.* 40:534–538.

Steffens, G. L., and Cathey, H. M. 1969. Selection of fatty acid derivatives surfactant formulations for the control of plant meristems. *Agr. Food Chem.* 17:312–317.

Steiger, R., Primost, E., and Rittmeyer, G. 1969. Die Möglichkeit der

Beurteilung des Einflusses pflanzenbaulicher Massnahmen auf die Back-qualität von Winterweizen mit Hilfe des Rapid-Mix-Testes. *Zeitschr. Acker. Pflanzenbau* 130:137–153.

Stembridge, C. E., and Gambrell, C. E. 1969. Thinning peaches with 3-chlorophenoxy-α-propionamide. *Jour. Amer. Soc. Hort. Sci.* 94:570–573.

————. 1970. Comparative effects of gibberellin and parthenocarpy on the shape and maturation of peaches. *HortScience* 5:156–158.

Stembridge, G. E., and La Rue, J. H. 1969. The effect of potassium gib-berellate on flower bud development in the Redskin peach. *Jour. Amer. Soc. Hort. Sci.* 94:492–495.

Stewart, I., and Leonard, C. D. 1960. Increased winter hardiness in citrus from maleic hydrazide sprays. *Proc. Amer. Soc. Hort. Sci.* 75:253–256.

Stewart, W. S., Halsey, D., and Ching, F. T. 1958. Effect of the potassium salt of gibberellic acid on fruit growth of Thompson Seedless grapes. *Proc. Amer. Soc. Hort. Sci.* 72:165–169.

Stewart, W. S., Hield, H. Z., and Brannaman, B. L. 1952. Effects of 2,4-D and related substances on fruit-drop, yield, size, and quality of Valencia oranges. *Hilgardia* 21:301–329.

Stewart, W. S., Klotz, L. J., and Hield, H. Z. 1951. Effects of 2,4-D and related substances on fruit-drop, yield, size, and quality of Washington navel oranges. *Hilgardia* 21:161–193.

Stewart, W. S., and Parker, E. R. 1954. Effects of 2,4-D and related sub-stances on fruit-drop, yield, size, and quality of grapefruit. *Hilgardia* 22:623–641.

Stodola, F. H. 1958. *Source book on gibberellin, 1828–1957.* Peoria, Illinois: U.S. Dept. Agr., Agr. Res. Serv.

Stolz, L. P., and Hess, C. E. 1966. The effect of girdling upon root initia-tion: Auxin and rooting cofactors. *Proc. Amer. Soc. Hort. Sci.* 89:744–751.

Stowe, B. B., and Schilke, J. F. 1964. Submicrogram identification and analysis of indole auxins by gas chromatography and spectrophotofluor-ometry. In Nitsch 1964, pp. 409–419.

Stowe, B. B., and Thimann, K. V. 1954. The paper chromatography of indole compounds and some indole-containing auxins of plant tissues. *Arch. Biochem.* 51:499–516.

Stowe, B. B., and Yamaki, T. 1959. Gibberellins: Stimulants of plant growth. *Science* 129:807–816.

Stuart, N. W. 1961. Initiation of flower buds in rhododendron after appli-cation of growth retardants. *Science* 134:50–52.

————. 1962. Stimulation of flowering in azaleas and camellias. *Proc. 16th Int. Hort. Cong.* 5:58–64.

————. 1965. Controlling the flowering of greenhouse azaleas. *Florist Nursery Exchange* 144 (September 11):22–23.

Stuart, N. W. 1967. Chemical pruning of greenhouse azaleas with fatty acid esters. *Florists' Rev.* 140:26–27, 68.

Stuart, N. W., and Cathey, H. M. 1961. The applied aspects of gibberellins. *Ann. Rev. Plant Physiol.* 12:369–394.

———. 1962. Control of growth and flowering of *Chrysanthemum morifolium* and *Hydrangea macrophylla* by gibberellin. In J. Garnaud, ed., *Advances in hort. sciences and their applications* II, 391–399. New York: Pergamon Press.

Sucoff, E. I. 1968. *N,N'*-Dinitroethylenediamine retards growth of red cedar and American arborvitae. *HortScience* 3:42–43.

Sullivan, D. T. 1968. The effect of *N*-dimethyl amino succinamic acid (Alar) on size, shape and maturity of Delicious apples. *HortScience* 3:18.

Sumiki, Y., and Kawarada, A. 1961. Occurrence of gibberellin A_1 in the water sprouts of citrus. In R. M. Klein, ed., *Plant growth regulation.* Ames: Iowa State Univ. Press, pp. 483–487.

Swanson, Carl P. 1946. A simple bio-assay method for the determination of low concentrations of 2,4-dichlorophenoxyacetic acid in aqueous solutions. *Bot. Gaz.* 107:507–509.

Symposia on Ethylene and Fruit Abscission. 1971. Miami Beach, Florida. *HortScience* 6(4):354–392.

Symposium on Leaf Abscission. 1968. Fort Detrick, Frederick, Maryland. *Plant Physiol.* 43(9), part B:1471–1586.

Synerholm, M. E., and Zimmerman, P. W. 1947. Preparation of a series of ω-(2,4-dichlorophenoxy)-aliphatic acids and some related compounds with a consideration of their biochemical role as plant growth regulators. *Contrib. Boyce Thompson Inst.* 14:369–382.

Tahori, A. S., Halevy, A. H., and Zeidler, G. 1965a. Effect of some plant growth retardants on the oleander aphid *Aphis nerii* (Boyer). *Jour. Sci. Food Agr.* 16:568–569.

———. 1965b. Effect of some plant growth retardants on the feeding of the cotton leaf worm. *Jour. Sci. Food Agr.* 16:570–572.

———. 1965c. Effect of some plant growth-retarding compounds on three fungal diseases and one viral disease. *Plant Disease Reporter* 49:775–777.

Takatori, F. H., Lorenz, O. A., and Zink, F. W. 1959. Gibberellin sprays on celery. *Calif. Agr.* 13:3–4.

Tanimoto, T., and Nickell, L. G. 1967. Effect of gibberellin on sugarcane growth and sucrose production. *Reports, Hawaiian sugar technologists.* Jour. Series Exptl. Sta. HSPA paper no. 199, pp. 137–146.

Templeman, W. G. 1955. The use of plant growth substances. *Ann. Appl. Biol.* 42:162–173.

Terpstra, W. 1953. Extraction and identification of growth substances, with special reference to the *Avena* coleoptile tip. Ph.D. thesis, Univ. of Utrecht, Holland.

Teubner, F. G., and Wittwer, S. H. 1957. Effect of N-arylphthalamic acids on tomato flower formation. *Proc. Amer. Soc. Hort. Sci.* 69:343–351.

Theriault, R. J., Friedland, W. C., Peterson, M. H., and Sylvester, J. C. 1961. Fluorometric assay for gibberellic acid. *Agr. Food Chem.* 9:21–23.

Thimann, K. V. 1935. On the plant growth hormone produced by Rhizopus suinus. *Jour. Biol. Chem.* 109:279–291.

———. 1948. In Pincus and Thimann 1948, pp. 6–119.

———. 1952. *The action of hormones on plants and invertebrates.* New York: Academic Press.

———. 1963. Plant growth substances: Past, present and future. *Ann. Rev. Plant Physiol.* 14:1–18.

———. 1969. The auxins. In Wilkins 1969, pp. 2–45.

Thimann, K. V., and Behnke-Rogers, J. 1950. *The use of auxins in the rooting of woody cuttings.* Harvard Forest, Petersham, Mass.: Maria Moors Cabot Foundation.

Thimann, K. V., and Delisle, A. L. 1939. The vegetative propagation of difficult plants. *Jour. Arnold Arb.* 20:116–136.

———. 1942. Notes on the rooting of some conifers from cuttings. *Jour. Arnold Arb.* 23:103–109.

Thimann, K. V., and Schneider, C. L. 1938. Differential growth in plant tissues. *Amer. Jour. Bot.* 25:627–641.

———. 1939. The relative activities of different auxins. *Amer. Jour. Bot.* 26:328–333.

Thimann, K. V., and Skoog, F. 1933. Studies on the growth hormones of plants, III: The inhibiting action of the growth substance on bud development. *Proc. Natl. Acad. Sci. U.S.* 19:714–716.

———. 1940. The extraction of auxins from plant tissues. *Amer. Jour. Bot.* 27:951–960.

Thimann, K. V., and Went, F. W. 1934. On the chemical nature of the root-forming hormone. *Proc. Kon. Ned. Akad. Wetensch. Amsterdam* 37:456–459.

Thomas, T. H. 1968. Studies on senescence-delaying substances extracted from *Brassica oleracea gemmifera* L. *Jour. Hort. Sci.* 43:59–68.

Thompson, A. H. 1957. *Six years' experiment on chemical thinning of apples.* Md. Agr. Exptl. Sta. Bull. A-88.

Thompson, P. A. 1962. Bioassay of gibberellins. In *Ann. Rept. Scottish Hort. Res. Inst. 1961–62,* pp. 43–46.

———. 1967. Promotion of strawberry fruit development by treatment with growth regulating substances. *Hort. Res.* 7:13–23.

Thompson, P. A. 1969. Comparative effects of gibberellins A_3 and A_4 on the germination of seeds of several different species. *Hort. Res.* 9:130–138.

Thompson, P. A., and Guttridge, C. G. 1959. Effect of gibberellic acid on the initiation of flowers and runners in the strawberry. *Nature* 184(B.A.): 72–73.

Thomson, W. W., Lewis, L. N., and Coggins, C. W., Jr. 1967. The reversion of chromoplasts to chloroplasts in Valencia oranges. *Cytologia* 32: 117–124.

Tija, B. O. S., Rogers, M. N., and Hartley, D. E. 1969. Effects of ethylene on morphology and flowering of *Chrysanthemum morifolium* Ramat. *Jour. Amer. Soc. Hort. Sci.* 94:35–39.

Timm, H., Rappaport, L., Bishop, J. C., and Hoyle, B. J. 1962. Sprouting, plant growth, and tuber production as affected by chemical treatment of white-potato-seed pieces, IV: Responses of dormant and sprouted seed potatoes to gibberellic acid. *Amer. Potato Jour.* 39:107–115.

Tognoni, F., DeHertogh, A. A., and Wittwer, S. H. 1967. The independent action of morphactins and gibberellic acid on higher plants. *Plant Cell Physiol.* 8:231–239.

Tolbert, N. E. 1960. (2-chloroethyl) trimethylammonium chloride and related compounds as plant growth substances. *Jour Biol. Chem.* 235:475–479.

Tomaszewski, M. 1964. The mechanism of synergistic effects between auxin and some natural phenolic substances. In Nitsch 1964, pp. 335–351.

Tompkins, D. R. 1965. Rhubarb rest period as influenced by chilling and gibberellin. *Proc. Amer. Soc. Hort. Sci.* 87:371–379.

———. 1966. Rhubarb petiole color and forced production as influenced by gibberellin, sucrose, and temperature. *Proc. Amer. Soc. Hort. Sci.* 89:472–477.

Trione, S. O., Pons, G. A., Tizo, R., and Trippi, V. S. 1963. Annual variation in rooting capacity and its relation to hormone treatments. *Phyton* 20:13–18.

Tso, T. E. 1964. Plant-growth inhibition by some fatty acids and their analogues. *Nature* 202:511–512.

Tsujita, M. J., and Andrew, W. T. 1966. The influence of N^6-benzyladenine on the storage, marketability and postharvest chlorophyll content of cabbage. *Canadian Jour. Plant Sci.* 47:193–195.

Tukey, H. B., ed. 1954. *Plant Regulators in Agriculture.* New York: Wiley.

Tukey, L. D., and Fleming, H. K. 1967a. Alar, a new fruit-setting chemical for grapes. *Pennsylvania Fruit News.* 46:12–31.

———. 1967b. Chemical to increase grape "set" may develop table-fruit industry. *Pennsylvania Agr. Exptl. Sta. Sci. for the Farmer* 14:11.

———. 1968. Fruiting and vegetative effects of N-dimethylaminosuccinamic acid on 'Concord' grapes, *Vitis labrusca* L. *Proc. Amer. Soc. Hort. Sci.* 93:300–310.

————. 1970. Post-year effects of *N*-dimethylaminosuccinamic acid on 'Concord' grapes, *Vitis labrusca* L. *HortScience* 5:161–163.

Tuli, V., Dedolph, R. R., and Wittwer, S. H. 1962. Effects of N^6-benzyladenine and dehydroacetic acid on the storage behavior of cherries and strawberries. *Quar. Bull. Mich. Agr. Exptl. Sta.* 45:223–226.

Unrath, C. R., and Kenworthy, A. L. 1968. The effects of Alar on Red Tart cherries (*Prunus cereasus*). *Abstr. 65th Ann. Meeting Amer. Soc. Hort. Sci., Davis, Calif.*, p. 93.

Uota, M., and Harris, C. M. 1964. Quality and respiration rates in stock flowers treated with N^6 benzylaminopurine. U.S. Dept. Agr., Agr. Marketing Serv., Marketing Quality Res. Div. Publication no. AMS-537.

Upper, C. D., Helgeson, J. P., Kemp, J. D., and Schmidt, C. J. 1970. Gas-liquid-chromatographic isolation of cytokinins from natural sources. *Plant Physiol.* 45:543–547.

Vaadia, Y., Raney, F. C., and Hagan, R. M. 1961. Plant water deficits and physiological processes. *Ann. Rev. Plant Physiol.* 12:265–292.

Valdovinos, J. G., Ernest, L. C., and Henry, E. W. 1967. Effect of ethylene and gibberellic acid on auxin synthesis in plant tissues. *Plant Physiol.* 42:1803–1806.

Van Doesburg, J. 1962. Use of fungicides with vegetative propagation. *Proc. 16th Int. Hort. Cong.* 1:365–372.

Van Eijden, J. 1965. Triomphe de Vienne acquires a new popularity: Earlier and better harvest with gibberellic acid. *RTC Roermond Groenten en fruit* 21:390 (translation).

Van Emden, H. F. 1964. Effect of (2-chloroethyl) trimethylammonium chloride on the rate of increase of the cabbage aphid (*Brevicoryne brassicae* L.). *Nature* 201:946–948.

Van Emden, H. F., and Cockshull, K. E. 1967. The effects of soil applications of (2-chloroethyl)-trimethylammonium chloride on leaf area and dry matter production by the brussels sprout plant. *Jour. Exptl. Bot.* 18:707–715.

Van Overbeek, J. 1947. Use of synthetic hormones as weed killers in tropical agriculture. *Econ. Bot.* 1:446–459.

————. 1956. Absorption and translocation of plant regulators. *Ann. Rev. Plant Physiol.* 7:355–372.

————. 1961. Applications of auxins in agriculture and their physiological bases. In Ruhland 1961, pp. 1137–1155.

————. 1964. Survey of mechanism of herbicide action. In Audus 1964a, pp. 387–400.

Van Overbeek, J. 1966. Plant hormones and regulators. *Science* 152:721–731.

Van Overbeek, J., *et al.* 1954. Nomenclature of chemical plant regulators. *Plant Physiol.* 29:307–308.

Van Overbeek, J., Conklin, M. E., and Blakeslee, A. F. 1942. Cultivation *in vitro* of small *Datura* embryos. *Amer. Jour. Bot.* 29:472–477.

Van Overbeek, J., Gordon, S. A., and Gregory, L. E. 1946. An analysis of the function of the leaf in the process of root formation in cuttings. *Amer. Jour. Bot.* 33:100–107.

Van Overbeek, J., and Loeffler, J. E. 1962. SD, 4901 N^6-benzyladenine delays postharvest deterioration of vegetables. Paper presented to 1st Int. Cong. Food Science Technol., London, September 18–21.

Van Overbeek, J., Olivo, D. G., and De Vasquez, E. M. S. 1945. A rapid extraction method for free auxin and its application in geotropic reactions of bean seedlings and sugar-cane nodes. *Bot. Gaz.* 106:440–451.

Van Staden, J., and Bornman, C. H. 1970. Cytokinin and gibberellin effects on abscisic acid induced inhibition of growth in *Spirodela*. *Jour. South Afr. Bot.* 36:207–214.

Van Steveninck, R. F. M. 1959. Factors affecting the abscission of reproductive organs in yellow lupins (*Lupinus luteus* L.), III: Endogenous growth substances in virus-infected and healthy plants and their effect on abscission. *Jour. Exptl. Bot.* 10:367–376.

Varner, J. E. 1961. Biochemistry of senescence. *Ann. Rev. Plant Physiol.* 12:245–264.

———. 1965. Seed development and germination. In J. Bonner and J. E. Varner, eds., *Plant biochemistry*, pp. 763–792. New York: Academic Press.

———. 1967. Hormonal control of enzyme production in barley endosperm. *Ann. New York Acad. Sci.* 144:219–222.

Varner, J. E., and Chandra, G. R. 1964. Hormonal control of enzyme synthesis in barley endosperm. *Proc. Natl. Acad. Sci. U.S.* 52:100–106.

Vegis, A. 1964. Dormancy in higher plants. *Ann. Rev. Plant Physiol.* 15:185–224.

Vergara, B. S., Miller, M., and Avelino, E. 1970. Effect of simazine on protein content of rice grain (*Oryza sativa* L.). *Agron. Jour.* 62:269–272.

Vliegenthart, J. A., and Vliegenthart, J. F. G. 1966. Reinvestigation of authentic samples of auxin a and b and related products by mass spectrometry. *Rucueil* 85:1266–1272.

Vlitos, A. J., and Meudt, W. 1953. The role of auxin in plant flowering, I: A quantitative method based on paper chromatography for the determination of indole compounds and of 3-indoleacetic acid in plant tissues. *Contrib. Boyce Thompson Inst.* 17:197–202.

Vogt, I. 1951. Über den Hemmstoff und Wuchsstoffgehalt in Getreidewurzelen. *Planta* (Berlin) 40:145–169.

Wain, R. L. 1958. Relation of chemical structure to activity for 2,4-D-type herbicides and plant growth regulator. In R. L. Metcalf, ed., *Advances in pest control research,* II:263–305. New York: Interscience.

————. 1964. The behaviour of herbicides in the plant in relation to selectivity. In Audus 1964a, pp. 465–481.

Wain, R. L., and Wightman, F. 1956. *The chemistry and mode of action of plant-growth substances.* London: Butterworths. New York: Academic Press.

Walhood, V. T., and Addicott, F. T. 1968. Harvest-aid programs: Principles and practices. In F. C. Elliot, M. Hoover, and W. K. Porter, Jr., eds., *Advances in production and utilization of quality cotton: Principles and practices,* pp. 407–431. Ames: Iowa State Univ. Press.

Wang, C. Y., and Hansen, E. 1970. Differential response to ethylene in respiration and ripening of immature 'Anjou' pears. *Jour. Amer. Soc. Hort. Sci.* 95:314–316.

Wang, S. Y., and Roberts, A. N. 1970. Physiology of dormancy in *Lilium longiflorum* 'Ace.' Thunb. *Jour. Amer. Soc. Hort. Sci.* 95:554–558.

Wangermann, E. 1965. Longevity and ageing in plants and plant organs. In Ruhland, 1965b, pp. 1026–1057.

Wareing, P. F. 1953. Growth studies in woody species, V: Photoperiodism in dormant buds of *Fagus sylvatica.* L. *Physiol. Plantarum* 6:692–706.

————. 1966. Natural inhibitors as growth hormones. In E. G. Cutter, ed., *Trends in morphogenesis.* London: Longmans, Green, pp. 235–252.

————. 1969. The control of bud dormancy in seed plants. In Woolhouse 1969, pp. 241–262.

Wareing, P. F., Eagles, C. F., and Robinson, P. M. 1964a. Natural inhibitors as dormancy agents. In Nitsch, 1964, pp. 377–386.

Wareing, P. F., Hanney, C. E. A., and Digby, J. 1964b. The role of endogenous hormones in cambial activity and xylem differentiation. In M. H. Zimmermann, ed., *The formation of wood in forest trees,* pp. 323–344. New York: Academic Press.

Wareing, P. F., and Phillips, I.D.J. 1970. *The control of growth and differentiation in plants.* Oxford: Pergamon Press.

Wareing, P. F., and Saunders, P. F. 1971. Hormones and dormancy. *Ann. Rev. Plant Physiol.* 22:261–288.

Wareing, P. F., and Seth, A. K. 1967. Ageing and senescence in the whole plant. In Woolhouse 1967, pp. 543–599.

Wear, S. 1959. Pruning with a chemical. *Western Fruit Grower* 13:30.

Weaver, R. J. 1954. Preliminary report on thinning grapes with chemical sprays. *Proc. Amer. Soc. Hort. Sci.* 63:194–200.

————. 1956. *Plant regulators in grape production.* Calif. Agr. Exptl. Sta. Bull. 752.

Weaver, R. J. 1957. Gibberellin on grapes. *The Blue Anchor* 34:10–11.

———. 1959. Prolonging dormancy in *Vitis vinifera* with gibberellin. *Nature* 183:1198–1199.

———. 1960. Toxicity of gibberellin to seedless and seeded varieties of *Vitis vinifera. Nature* 187:1135–1136.

———. 1962. The effect of benzothiozole-2-oxyacetic acid on maturation of seeded varieties of grapes. *Amer. Jour. Enol. Vitic.* 13:141–149.

———. 1963a. Experiments on thinning grapes with alpha-naphthalene-acetic acid and dinitro-sec-butylphenol. *Vitis* 4:1–10.

———. 1963b. Use of kinin in breaking rest of buds of *Vitis vinifera. Nature* 198:207–208.

Weaver, R. J., Kasimatis, A. N., and McCune, S. B. 1962. Studies with gibberellin on wine grapes to decrease bunch rot. *Amer. Jour. Enol. Vitic.* 13:78–82.

Weaver, R. J., and McCune, S. B. 1959. Response of certain varieties of *Vitis vinifera* to gibberellin. *Hilgardia* 28:297–350.

———. 1960. Further studies with gibberellin on *Vitis vinifera* grapes. *Bot. Gaz.* 121:155–162.

Weaver, R. J., McCune, S. B., and Coombe, B. G. 1961. Effects of various chemicals and treatments on rest periods of grape buds. *Amer. Jour. Enol. Vitic.* 12:131–142.

Weaver, R. J., and Pool, R. M. 1965a. Bloom spraying with gibberellin loosens clusters of Thompson Seedless grapes. *Calif. Agr.* 19:14–15.

———. 1965b. Relation of seededness and ringing to gibberellin-like activity in berries of *Vitis vinifera. Plant Physiol.* 40:770–776.

———. 1968. Induction of berry abscission in *Vitis vinifera* by morphactins. *Amer. Jour. Enol. Vitic.* 19:121–124.

———. 1969. Effect of ethrel, abscisic acid, and a morphactin on flower and berry abscission and shoot growth in *Vitis vinifera. Jour. Amer. Soc. Hort. Sci.* 94:474–478.

———. 1971. Effect of (2-chloroethyl) phosphonic acid on maturation of grapes (*Vitis vinifera* L.). *Jour. Amer. Soc. Hort. Sci.* 96:725–727.

Weaver, R. J., and Sachs, R. M. 1968. Hormonal-induced control of fruit-set and berry enlargement in *Vitis vinifera* L. In Wightman and Setterfield 1968, pp. 957–974.

Weaver, R. J., Shindy, W., and Kliewer, W. M. 1969. Growth regulator induced movement of photosynthetic products into fruits of 'Black Corinth' grapes. *Plant Physiol.* 44:183–188.

Weaver, R. J., and Van Overbeek, J. 1963. Kinins stimulate grape growth. *Calif. Agr.* 17:12.

Weaver, R. J., Van Overbeek, J., and Pool, R. M. 1966. Effect of kinins on fruit set and development in *Vitis vinifera. Hilgardia* 37:181–201.

Weaver, R. J., Yeou-Der, K., and Pool, R. M. 1968. Relation of plant regulators to bud rest in *Vitis vinifera* grapes. *Vitis* 7:206–212.

Webber, H. J. 1943. Plant characteristics and climatology. In H. J. Webber and L. D. Batchelor, eds., *The citrus industry*, I, 41–69. Berkeley: Univ. of Calif. Press, 1st ed.

Webster, B. D. 1968. Anatomical aspects of abscission. *Plant Physiol.* 43(9), Part B: 1512–1544.

Webster, D. H., and Crowe, A. D. 1969. Effects of gibberellic acid, N-dimethylaminosuccinamic acid (Alar), ringing and thinning on McIntosh apple shape. *Jour. Amer. Soc. Hort. Sci.* 94:308–310.

Weinberger, J. H. 1969. The stimulation of dormant peach buds by a cytokinin. *HortScience* 4:125–126.

Weiser, C. J., and Blaney, L. T. 1960. The effects of boron on the rooting of English holly cuttings. *Proc. Amer. Soc. Hort. Sci.* 75:704–710.

Wells, J. S. 1963. The use of Captan in rooting rhododendrons. *Proc. Int. Plant Prop. Soc.* 13:132–135.

Went, F. W. 1926. On growth-accelerating substances in the coleoptile of *Avena sativa*. *Proc. Kon. Ned. Akad. Wetensch. Amsterdam* 30:10–19.

———. 1928. Wuchsstoff und Wachstum. *Rec. Trav. Bot. Néerland.* 25:1–116.

———. 1934a. A test method for rhizocaline, the root-forming substance. *Proc. Kon. Ned. Akad. Wetensch. Amsterdam* 37:445–455.

———. 1934b. On the pea test method for auxin, the plant growth hormone. *Proc. Kon. Ned. Akad. Wetensch. Amsterdam* 37:547–555.

———. 1957. *The experimental control of plant growth.* Waltham, Mass.: Chronica Botanica.

Went, F. W., and Cosper, L. 1945. Plant growth under controlled conditions, VI: Comparison between field and air-conditioned greenhouse culture of tomatoes. *Amer. Jour. Bot.* 32:643–654.

Went, F. W., and Thimann, K. V. 1937. *Phytohormones.* New York: MacMillan.

West, C. A., Oster, M., Robinson, D., Lew, F., and Murphy, P. 1968. Biosynthesis of gibberellin precursors and related diterpenes. In Wightman and Setterfield 1968, pp. 313–332.

West, C. A., and Phinney, B. O. 1959. Gibberellins from flowering plants, I: Isolation and properties of a gibberellin from *Phaseolus vulgaris* L. *Jour. Amer. Chem. Soc.* 81:2424–2427.

West, C. A., and Reilly, T. 1961. Properties of gibberellins from flowering plants. *Advances Chem. Series* 28:37–41.

Wester, H. V., and Marth, P. C. 1950. Growth regulators prolong the bloom of oriental flowering cherries and dogwood. *Science* 111:611.

Westwood, M. N., and Bjornstad, H. O. 1968. Effects of gibberellin A_3 on fruit shape and subsequent seed dormancy. *HortScience* 3:19–20.

White, H. E. 1946. Fermate and its effect on rooting of geranium cuttings. *Proc. Amer. Soc. Hort. Sci.* 47:522–524.

White, J. W. 1970. Effects of cycocel, moisture stress and pinching on growth and flowering of F_1 hybrid geraniums (*Pelargonium* × *hortorum* Bailey). *Jour. Amer. Soc. Hort. Sci.* 95:546–550.

Whyte, P., and Luckwill, L. C. 1966. A sensitive bioassay for gibberellins based on retardation of leaf senescence in *Rumex obtusifolius* (L.). *Nature* 210:1360.

Wickson, M., and Thimann, K. V. 1958. The antagonism of auxin and kinetin in apical dominance. *Physiol. Plantarum* 11:62–74.

Wightman, F., and Setterfield, G., eds. 1968. *Biochemistry and physiology of plant growth substances.* Ottawa: Runge Press.

Wilde, M. H., and Edgerton, L. J. 1969. Some effects of a growth retardant on shoot meristems of apple. *Jour. Amer. Soc. Hort. Sci.* 94:118–122.

Wildman, S. G., and Bonner, J. 1948. Observations on the chemical nature and formation of auxin in the *Avena* coleoptile. *Amer. Jour. Bot.* 35:740–746.

Wildman, S. G., and Muir, R. M. 1949. Observations on the mechanism of auxin formation in plant tissues. *Plant Physiol.* 24:84–92.

Wilhelm, A. D. 1959. Über die Wirkung von Gibberellinsäure auf Reben. *Wein-Wiss. Beil. Fachz. Deut. Weinbau* 4:45–54.

Wilkins, M. B., ed. 1969. *The physiology of plant growth and development.* New York: McGraw-Hill.

Williams, M. W., Bartram, R. D., and Carpenter, W. S. 1970. Carryover effect of succinic acid 2,2-dimethylhydrazide on fruit shape of 'Delicious' apples. *HortScience* 5:257.

Williams, M. W., and Batjer, L. P. 1964. Site and mode of action of 1-naphthyl N-methyl carbamate (Sevin) in thinning apples. *Proc. Amer. Soc. Hort. Sci.* 85:1–10.

Williams, M. W., Batjer, L. P., and Martin, G. C. 1964. Effects of N-dimethyl amino succinamic acid (B-Nine) on apple quality. *Proc. Amer. Soc. Hort. Sci.* 85:17–19.

Williams, M. W., and Billingsley, H. D. 1970. Increasing the number and crotch angles of primary branches of apple trees with cytokinins and gibberellic acid. *Jour. Amer. Soc. Hort. Sci.* 95:649–651.

Williams, M. W., and Letham, D. S. 1969. Effect of gibberellins and cytokinins on development of parthenocarpic apples. *HortScience* 4:215–216.

Williams, M. W., and Stahly, E. A. 1968. Effect of cytokinins on apple shoot development from axillary buds. *HortScience* 3:68–69.

———. 1969. Effect of cytokinins and gibberellins on shape of 'Delicious' apple fruits. *Jour. Amer. Soc. Hort. Sci.* 94:17–19.

Wiltbank, W. J., and Krezdorn, A. H. 1969. Determination of gibberellins in ovaries and young fruits of navel oranges and their correlation with fruit growth. *Jour. Amer. Soc. Hort. Sci.* 94:195–201.

Winkler, A. J. 1931. *Pruning and thinning experiments with grapes.* Calif. Agr. Exptl. Sta. Bull. 519.

Wirwille, J. W., and Mitchell, J. W. 1950. Six new plant growth inhibiting compounds. *Bot. Gaz.* 111:491–494.

Withrow, R. B., and Price, L. 1957. A darkroom safelight for research in plant physiology. *Plant Physiol.* 32:244–248.

Wittwer, S. H. 1949. Effect of fruit-setting treatment, variety and solar radiation on yield and fruit size of greenhouse tomatoes. *Proc. Amer. Soc. Hort. Sci.* 53:349–354.

———. 1951. Growth substances in fruit setting. In Skoog 1951, pp. 365–377.

———. 1971. Growth regulants in agriculture. *Outlook on Agric.* 6:205–217. Imperial Chemicals, Ltd., Jealott's Hill Res. Sta., Bracknell, Berkshire, England.

Wittwer, S. H., and Bukovac, M. J. 1957a. Gibberellins—new chemicals for crop production. *Quar. Bull. Mich. Agr. Exptl. Sta.* 39:469–494.

———. 1957b. Gibberellin and higher plants, III: Induction of flowering in long-day annuals grown under short days. *Quar. Bull. Mich. Agr. Exptl. Sta.* 39:661–672.

———. 1957c. Gibberellin and higher plants, V: Promotion of growth in grass at low temperatures. *Quar. Bull. Mich. Agr. Exptl. Sta.* 39:682–686.

———. 1957d. Gibberellin and higher plants, VIII: Seed treatments for beans, peas, and sweet corn. *Quar. Bull. Mich. Agr. Exptl. Sta.* 40:215–224.

———. 1958. The effects of gibberellin on economic crops. *Econ. Bot.* 12:213–255.

Wittwer, S. H., Bukovac, M. J., and Grigsby, B. H. 1957. Gibberellin and higher plants, VI: Effects on the composition of Kentucky bluegrass (*Poa pratensis*) grown under field conditions in early spring. *Quar. Bull. Mich. Agr. Exptl. Sta.* 40:203–206.

Wittwer, S. H., Dedolph, R. R., Tuli, V., and Gilbart, D. 1962. Respiration and storage deterioration in celery (*Apium graveolens* L.) as affected by postharvest treatments with N^6-benzylaminopurine. *Proc. Amer. Soc. Hort. Sci.* 80:408–416.

———. 1962. Some effects of kinins on horticultural crops. *Proc. 16th Int. Hort. Cong.* 4:428–434.

Wittwer, S. H., and Hillyer, I. G. 1954. Chemical induction of male sterility in cucurbits. *Science* 120:893–894.

Wittwer, S. H., and Murneek, A. E. 1946. Further investigations on the value of "hormone" sprays and dusts for green bush snap beans. *Proc. Amer. Soc. Hort. Sci.* 47:285–293.

Wittwer, S. H., and Sharma, R. C. 1950. The control of storage sprouting in onions by preharvest foliage sprays of maleic hydrazide. *Science* 112: 597–598.

Wittwer, S. H., Stallworth, H., and Howell, M. J. 1948. The value of a "hormone" spray for overcoming delayed fruit set and increasing yields of outdoor tomatoes. *Proc. Amer. Soc. Hort. Sci.* 51:371–380.

Wittwer, S. H., and Teubner, F. G. 1956. Cold exposure of tomato seedlings and flower formation. *Proc. Amer. Soc. Hort. Sci.* 67:369–376.

Wittwer, S. H., and Tolbert, N. E. 1960. (2-chloroethyl) trimethylammonium chloride and related compounds as plant growth substances, III: Effect on growth and flowering of the tomato. *Amer. Jour. Bot.* 47:560–565.

Wood, H. N., Braun, A. C., Brandes, H., and Kende, H. 1969. Studies on the distribution and properties of a new class of cell division-promoting substances from higher plant species. *Proc. Natl. Acad. Sci. U.S.* 62:349–356.

Woolhouse, H. W., ed. 1967. *Aspects of the biology of ageing* (Symp. Soc. Exptl. Biol., no. 21). New York: Academic Press.

———. 1969. *Dormancy and survival* (Symp. Soc. Exptl. Biol., no. 23). New York: Cambridge Univ. Press. (Includes several interesting articles on dormancy.)

Wort, D. J. 1964a. Effects of herbicides on plant composition and metabolism. In Audus 1964a, pp. 291–334.

———. 1964b. Responses of plants to sublethal concentrations of 2,4-D, without and with added minerals. In Audus 1964a, pp. 335–342.

———. 1966. Growth, yield, composition, and metabolism of various crop plants as affected by the application of 2,4-D-nutrient dusts and sprays. *Proc. Int. Symp. Plant Stimulation, Sofia*, pp. 341–353.

Wort, D. J., and Patel, K. M. 1970a. Erhöhung des Buschbohnenertrages durch Naphthenate und die Auswirkungen der Anwendungsmethode. *Agnew. Bot.* 44:179–185.

———. 1970b. Response of plants to naphthenic and cycloalkanecarboxylic acids. *Agron. Jour.* 62:644–646.

Wright, S. T. C. 1956. Studies on fruit development in relation to plant hormones, III: Auxins in relation to fruit morphogenesis and fruit drop in the black currant *Ribes nigrum. Jour. Hort. Sci.* 31:196–211.

Wünsche, U. 1966. Influence of 2-chloroethyl trimethylammonium chloride and gibberellin A_3 on frost hardiness of winter wheat. *Naturwiss.* 53:386–387.

Yabuta, T. 1935. Biochemistry of the "bakanae" fungus of rice. *Agr. Hort.* (Tokyo) 10:17–22.

Yabuta, T., and Hayashi, T. 1939. Biochemical studies on "bakanae" fungus of rice, II: Isolation of gibberellin, the active principal which produces slender rice seedlings. *Jour. Agr. Chem. Soc. Japan* 15:257–266.

Yang, S. F. 1969. Ethylene evolution from 2-chloroethylphosphonic acid. *Plant Physiol.* 44:1202–1204.

Yasuda, S. 1934. The second report on the behavior of pollen tubes in the production of seedless fruits caused by interspecific pollination. *Jap. Jour. Genet.* 9:118–124.

Yeou-Der, K., Weaver, R. J., and Pool, R. M. 1968. Effect of low temperature and growth regulators on germination of seeds of 'Tokay' grapes. *Proc. Amer. Soc. Hort. Sci.* 92:323–330.

Young, R., and Cooper W. C. 1969. Effect of cycocel and abscisic acid on bud growth of Redblush grapefruit. *Jour. Amer. Soc. Hort. Sci.* 94:8–10.

Young, R., Jahn, O., Cooper, W. C., and Smoot, V. V. 1970. Preharvest sprays with 2-chloroethylphosphonic acid to degreen 'Robinson' and 'Lee' tangerine fruits. *HortScience* 5:268–269.

Zawawi, M. A., and Irving, R. M. 1968. Interaction of triiodobenzoic acid and indole-acetic acid in the development and flowering of *Kalanchoe*. *Proc. Amer. Soc. Hort. Sci.* 93:610–617.

Zelitch, I. 1964. Reduction of transpiration of leaves through stomatal closure induced by alkenylsuccinic acids. *Science* 143:692–693.

Zimmerman, C. E., Brooks, S. N., and Likens, S. T. 1964. Gibberellin A_3-induced growth responses of Fuggle hops (*Humulus lupulus* L.). *Crop Sci.* 4:310–313.

Zimmerman, P. W. 1942. Formative influences of growth substances on plants. *Cold Spring Harbor Symp. Quan. Biol.* 10:152–159.

Zimmerman, P. W., Crocker, W., and Hitchcock, A. E. 1933. Initiation and stimulation of roots from exposure of plants to carbon monoxide gas. *Contrib. Boyce Thompson Inst.* 5:1–17.

Zimmerman, P. W., and Hitchcock, A. E. 1933. Initiation and stimulation of adventitious roots caused by unsaturated hydrocarbon gases. *Contrib. Boyce Thompson Inst.* 5:351–369.

———. 1942. Substituted phenoxy and benzoic acid growth substances and the relation of structure to physiological activity. *Contrib. Boyce Thompson Inst.* 12:321–343.

Zimmerman, P. W., Hitchcock, A. E., and Wilcoxon, F. 1936. Several esters as plant hormones. *Contrib. Boyce Thompson Inst.* 8:105–112.

Zink, F. W. 1961. N^6-benzyladenine, a senescence inhibitor for green vegetables. *Jour. Agr. Food Chem.* 9:304–307.

Zucconi, F., Stösser, R., and Bukovac, M. J. 1969. Promotion of fruit abscission with abscisic acid. *BioScience* 19:815–817.

Zukel, J. W. 1957, 1963. *A literature summary on maleic hydrazide*, Vols. I and II. Naugatuck, Conn.: UniRoyal.

Zuluaga, E. M., Lumelli, J., and Christensen, J. H. 1968. Influence of growth regulators on the characteristics of berries of *Vitis vinifera* L. *Phyton* 25:35–48.

AUTHOR INDEX

SUBJECT INDEX

Abbreviations are generally used for chemical names. The student should refer to the list of Abbreviations and Trade Names on pp. xv to xviii.